Field Monitoring, GIS, Remote Sensing, Geophysical Techniques, and Hydrochemical Analysis in Groundwater Investigations

Field Monitoring, GIS, Remote Sensing, Geophysical Techniques, and Hydrochemical Analysis in Groundwater Investigations

Guest Editors

Ismael M. Ibraheem
Abdelazim Negm

MDPI
Basel • Beijing • Wuhan • Barcelona • Belgrade • Novi Sad • Cluj • Manchester

Guest Editors
Ismael M. Ibraheem
Institute of Geophysics
and Meteorology
University of Cologne
Cologne
Germany

Abdelazim Negm
Water and Water Structures
Engineering Department
Zagazig University
Zagazig
Egypt

Editorial Office
MDPI AG
Grosspeteranlage 5
4052 Basel, Switzerland

This is a reprint of the Special Issue, published open access by the journal *Water* (ISSN 2073-4441), freely accessible at: https://www.mdpi.com/journal/water/special_issues/K84CWDF852.

For citation purposes, cite each article independently as indicated on the article page online and as indicated below:

Lastname, A.A.; Lastname, B.B. Article Title. *Journal Name* **Year**, *Volume Number*, Page Range.

ISBN 978-3-7258-6810-0 (Hbk)
ISBN 978-3-7258-6811-7 (PDF)
https://doi.org/10.3390/books978-3-7258-6811-7

Contents

About the Editors

Ismael M. Ibraheem

Ismael M. Ibraheem is a Senior Researcher at the Institute of Geophysics and Meteorology, University of Cologne, Germany. His research interests focus on environmental studies, groundwater exploration, UXO detection, and archaeogeophysical prospecting, with specialized expertise in geophysical forward modeling, inverse problems, and multiphysics inversion of magnetic, gravity, geoelectric, and electromagnetic data. He has led and contributed to significant projects funded by the German Research Foundation (DFG) and the Volkswagen Foundation, focusing on subsurface imaging for landfill characterization and groundwater contamination. His achievements include authoring over 35 peer-reviewed publications and several book chapters, serving as a Guest Editor for the journal *Water*, serving as an Editorial Board Member of *Discover Geoscience* (Springer), and acting as an active reviewer for numerous leading international geoscience journals.

Abdelazim Negm

Abdelazim Negm is a Professor of Hydraulics (and Water Resources) at Zagazig University's Faculty of Engineering, Egypt, where he has been a full professor since 2004. He served as Vice Dean from 2008 to 2011 and later led the Environmental Engineering Department at the Egypt–Japan University of Science and Technology (E-JUST) from 2013 to 2016.

Since 2006, Prof. Negm has been involved in various funded research projects, both nationally and internationally, often taking the lead. He has published over 400 works in journals and conference proceedings, and contributed to more than 100 book chapters with Springer. His research has earned several best paper awards, and he has edited over 60 books for Springer, showcasing his extensive knowledge in academic publishing and peer review.

Prof. Negm has participated in more than 105 international conferences, delivered numerous keynote speeches, and served on scientific and organizing committees. He serves as a reviewer for over 50 journals indexed in Scopus and Web of Science. His editorial roles include serving as Editor-in-Chief of *the Egyptian International Journal of Engineering Sciences and Technology* (2017–2020), Editor of the *Euro-Mediterranean Journal for Environmental Integration* (Springer), and an Associate Editor for various Elsevier journals. He is also a member of several international editorial boards and has been an editorial board member of *the Handbook of Environmental Chemistry* (Springer) since 2019.

Additionally, he has chaired the Permanent Scientific Committee on Water Resources at the Supreme Council of Egyptian Universities (cycle 13) and has served as a member of cycles 12 and 15. Prof. Negm is recognized in notable biographical directories, including *Marquis Who's Who in the World* and the *International Biographical Center's Outstanding Intellectuals of the 21st Century*.

Preface

Groundwater systems constitute a vital component of the global water cycle and play a central role in sustaining human activities, agricultural production, and natural ecosystems. Increasing pressures from population growth, land use change, contamination, and climate variability have significantly altered groundwater availability and quality, exposing limitations in conventional investigation and management approaches. These challenges highlight the need for integrated frameworks that combine multiple data sources and complementary analytical techniques to better understand subsurface processes and support sustainable groundwater management.

This Reprint contains articles that advance integrated groundwater investigations by combining field monitoring, geophysical techniques, remote sensing, geographic information systems, and hydrochemical analysis. The collected studies address key groundwater issues, including aquifer characterization, salinity and contamination assessment, recharge potential zone mapping, groundwater storage variability, and groundwater-related impacts of climate extremes. By covering diverse hydrogeological and climatic settings, the Reprint illustrates how interdisciplinary approaches can be adapted to different environmental and socio-economic contexts.

This Reprint focuses on the interactions among subsurface properties, surface processes, and external drivers, including climate variability and human activities. Its scope ranges from methodological developments and data integration strategies to applied case studies that demonstrate practical solutions for groundwater assessment, monitoring, and planning. The aim is to strengthen scientific understanding of groundwater systems while providing actionable insights that support evidence-based water resources management and policy formulation.

The motivation for compiling this Reprint arises from the growing recognition that sustainable groundwater management depends on linking physical characterization with water quality assessment, spatial analysis, and data-driven modeling. This Reprint is addressed to researchers, practitioners, and decision-makers in hydrogeology, geophysics, water resources, and environmental sciences, as well as early-career scientists seeking a comprehensive perspective on modern, integrated groundwater investigation approaches.

Ismael M. Ibraheem and Abdelazim Negm
Guest Editors

water

MDPI

Editorial

Field Monitoring, GIS, Remote Sensing, Geophysical Techniques, and Hydrochemical Analysis in Groundwater Investigations

Ismael M. Ibraheem [1,*] and Abdelazim M. Negm [2]

1 Institute of Geophysics and Meteorology, University of Cologne, Pohligstrasse 3, 50969 Cologne, Germany
2 Water and Water structures Engineering Department, Faculty of Engineering, Zagazig University, Zagazig 44519, Egypt; amnegm85@gmail.com or amnegm@zu.edu.eg
* Correspondence: ismael.ibraheem@geo.uni-koeln.de or ismail.geo@gmail.com

1. Introduction

Groundwater represents a critical component of the Earth's freshwater system, sustaining human populations, agriculture, and ecosystems worldwide. However, increasing water demand; contamination from agricultural, industrial, and urban activities; and climate variability have increased the vulnerability of groundwater resources, creating an urgent need for comprehensive, data-driven assessment and management strategies. Effective groundwater management is, therefore, pivotal not only for local water security but also for achieving global sustainable development goals related to water availability, food security, and ecosystem resilience. In this context, combining field monitoring, remote sensing (RS), geophysical surveys, Geographic Information Systems (GIS), and hydrochemical analysis has emerged as a transformative strategy for understanding and managing groundwater resources at multiple scales.

RS provides synoptic and temporal perspectives on surface processes influencing groundwater, including land use changes, evapotranspiration, and recharge patterns. Geophysical methods, such as electrical resistivity tomography (ERT) and transient electromagnetic (TEM) surveys, enable non-invasive imaging of subsurface structures and groundwater dynamics, while hydrochemical and isotopic analyses reveal the quality and origin of groundwater. GIS serves as the integrative framework linking these datasets, enabling spatial modeling, visualization, and multi-criteria decision-making essential for sustainable groundwater management.

This Special Issue brings together thirteen papers that exemplify the convergence of these approaches across diverse hydrogeological and climatic settings, including arid and semi-arid basins, coastal aquifers, humid deltas, post-mining landscapes, and agricultural watersheds. The contributions illustrate how the integration of geophysical surveys, hydrochemical assessment, GIS-based mapping, remote sensing, and data-driven modeling can address key challenges in groundwater science, such as depletion, salinization, contamination, and climate-induced variability. Collectively, the studies highlight the global relevance of interdisciplinary approaches for sustainable groundwater management and provide methodological guidance for practitioners, researchers, and policymakers aiming to secure water resources in the face of increasing environmental and anthropogenic pressures.

2. Main Contributions of This Special Issue

The importance of this Special Issue lies in its content, which discusses four interconnected themes, namely, (i) geophysical and RS applications for subsurface characterization,

(ii) hydrochemical analysis and water quality assessment, (iii) groundwater recharge and potential zone mapping, and (iv) groundwater storage, climate variability, and flood dynamics.

2.1. Geophysical and Remote Sensing Applications for Subsurface Characterization

Several studies in this Special Issue showcase the power of geophysical methods for subsurface imaging and groundwater assessment, particularly in regions where borehole data are sparse or inaccessible.

Bou-Rabee et al. (Contribution 1) conducted a TEM survey in Kuwait's Subiya Peninsula, a strategic region for the Silk City megaproject. Using 63 TEM stations and applying both Occam and Levenberg–Marquardt inversions, they delineated saline intrusion pathways and fault-controlled fluid movement, revealing higher salinity toward the southern coast and fresher groundwater at depth in the north. These results underscore the growing relevance of TEM in coastal aquifer monitoring and management.

Extending this geophysical approach to broader contaminated landscapes, Siemon et al. (Contribution 2) applied helicopter-borne electromagnetic (HEM) surveys in the post-mining landscape of Lusatia, Germany, mapping low-resistivity zones associated with acid mine drainage. The correlation between conductivity and dissolved sulfate and iron concentrations demonstrated how airborne electromagnetic methods can effectively link physical parameters to hydrochemical processes in inaccessible terrains. Together, the Kuwaiti and German case studies illustrate the efficiency of electromagnetic techniques in detecting subsurface salinity and contamination at different scales.

Kelly and Hladik (Contribution 3) further demonstrated the capabilities of geophysical techniques in a coastal ecological context using multi-channel ERT to investigate salt marsh degradation in Georgia, USA. Their results revealed that tidal inundation rather than lithological variability governs vegetation loss and recovery, emphasizing hydrological controls on subsurface salinity dynamics. Also, Hu et al. (Contribution 4) introduced a multi-attribute channel wave tomography technique that integrates spectral and geometric attributes to detect water-bearing fractures and collapse columns in coal mines. Collectively, these studies show how advances in geophysical imaging provide vital insights into groundwater–surface water interactions, contamination, and hazard prevention.

2.2. Hydrochemical Analysis and Water Quality Assessment

While geophysical imaging defines the physical structure of aquifers, hydrochemical and statistical analyses reveal the processes shaping groundwater quality.

Arıman et al. (Contribution 5) assessed groundwater in Turkey's Kızılırmak Delta using Principal Component Analysis (PCA) and the Water Quality Index (WQI) in a GIS framework. Eleven physicochemical parameters from observation wells were analyzed to assess drinking suitability and spatial variability. Most samples were unsuitable for consumption due to elevated Ca^{2+}, Mg^{2+}, Na^{+}, and $SO_4{}^{2-}$ levels, with poor quality concentrated in the west and northeast. PCA revealed two main components that explain most variance, linking salinization and mineral dissolution to water deterioration. The study highlights the influence of geogenic and agricultural factors and demonstrates the effectiveness of integrated WQI–PCA for groundwater management in deltaic settings.

At a broader scale, Williams et al. (Contribution 6) connected hydrochemistry with socio-hydrological drivers by analyzing groundwater sustainability across 1881 drinking water systems in Arizona, USA. They found that rural and minority communities face higher depletion and contamination risks, linking aquifer stress to governance and social equity. The study bridges the physical–social gap by demonstrating that groundwater challenges are as much institutional as they are hydrogeological, an important consideration for future policy-oriented groundwater research.

Furthermore, Hussein et al. (Contribution 7) evaluated deep learning models for classifying the Irrigation Water Quality Index in Algeria's Naama region using data from 166 wells. Among 16 tested architectures, the CNN-BiLSTM achieved the highest accuracy (0.94) and AUC (0.994), effectively capturing spatial and temporal patterns. SHapley Additive exPlanations (SHAP) analysis identified sodium-related parameters as key predictors. Despite higher computational demands, the model proved robust and reliable, demonstrating the value of hybrid neural networks for data-driven irrigation management in semi-arid regions.

2.3. Groundwater Recharge and Potential Zone Mapping

Within this Special Issue, several studies employed GIS-based multi-criteria modeling to map groundwater recharge and potential zones, an essential step for sustainable water planning.

Hossain et al. (Contribution 8) mapped groundwater recharge zones in Bangladesh's Barind Tract using an integrated GIS–AHP approach with seven controlling factors. The results show that about 43% of the area has good-to-very-good recharge potential, driven mainly by geology, land use, and soil type. The study estimates annual recharge at 2554×10^6 m^3 (22.7% of rainfall) and highlights the value of multi-criteria modeling for groundwater conservation and climate-resilient water management. Similarly, El-Sorogy et al. (Contribution 9) identified groundwater potential zones in Al-Madinah Al-Munawarah, western Saudi Arabia, through weighted overlay analysis of seven thematic layers derived from RS, GIS, and meteorological data. The results highlight the dominance of drainage density, slope, and soil type in controlling recharge distribution, with favorable zones concentrated in alluvial deposits near Wadi Al-Hamd. Validation using well data confirmed the reliability of the delineation and demonstrated the applicability of GIS-based approaches for groundwater exploration in arid environments.

In addition, Rehman et al. (Contribution 10) applied Frequency Ratio (FR) and Weight of Evidence (WoE) models to delineate groundwater potential zones in Pakistan's Hangu District, integrating Sentinel-2, ALOS-DEM, and CHIRPS rainfall data. The FR model (AUC = 0.93) effectively identified high-potential zones near drainage networks, linking recharge potential to Sustainable Development Goals (SDGs) 6 (Clean Water and Sanitation) and 13 (Climate Action). In line with these efforts, Halder et al. (Contribution 11) introduced an integrated SWAT–VIKOR framework to prioritize agricultural land use in India's Upper Kansai Basin, explicitly coupling hydrological modeling with decision analysis. Their results highlight strong groundwater–surface water interactions, with baseflow contributing 64% of the total discharge, and identify sub-watersheds (SW4 and SW5) as optimal for sustainable agriculture and water management. Together, these four studies demonstrate how GIS, RS, and Multi-Criteria Decision Analysis (MCDA) frameworks, applied across varied climatic and geomorphological settings, can guide both recharge enhancement and agricultural planning.

2.4. Groundwater Storage, Climate Variability, and Flood Dynamics

While recharge mapping focuses on potential input, groundwater storage and its climatic drivers govern system resilience. Mohasseb et al. (Contribution 12) integrated GRACE satellite gravimetry and GPS-derived vertical displacement to estimate groundwater storage variations in the Main Karoo Aquifer, southern Africa, between 2013 and 2021. Their findings showed positive storage trends with fine-scale variability captured by GPS, validating the synergy between gravimetric and geodetic datasets for long-term groundwater monitoring in data-scarce regions.

Alobid et al. (Contribution 13) examined flood frequency trends in Germany (1990–2024) using ARIMA and ANN models to link climate variables with flood frequency. The results show strong correlations between rising temperatures, increased rainfall, and more frequent floods, with ANN outperforming ARIMA in prediction accuracy. The study underscores the need for adaptive, climate-informed flood management and nature-based solutions such as wetland restoration and groundwater recharge enhancement. This thematic continuity between climate resilience, recharge dynamics, and storage variability underscores the integrated hydrological vision that unites this Special Issue.

3. Conclusions

This Special Issue demonstrates that integrating geophysical, hydrochemical, and geospatial approaches is essential for advancing groundwater science. The combined use of field monitoring, RS, and modeling enhances spatial and temporal understanding of aquifer systems, improves predictive accuracy, and supports sustainable resource management. Data-driven and machine learning methods enable adaptive, scalable monitoring, while GIS- and RS-based analyses guide recharge mapping, groundwater zoning, and land use planning. Together, these studies highlight that groundwater sustainability depends on linking hydrological, environmental, and social dimensions. The collective outcome is a unified framework for groundwater assessment that directly supports global efforts toward SDGs 2, 6, 13, and 15.

Author Contributions: Conceptualization, I.M.I. and A.M.N.; writing—original draft preparation, I.M.I.; writing—review and editing, I.M.I. and A.M.N. All authors have read and agreed to the published version of the manuscript.

Funding: This research received no external funding.

Acknowledgments: We would like to express our sincere gratitude to all authors who submitted their manuscripts, contributing high-quality research that advances the integration of field monitoring, remote sensing, geophysical techniques, GIS, and hydrochemical analysis in groundwater investigations. Special thanks are due to the reviewers for their thorough and constructive evaluations, which significantly enhanced the clarity, rigor, and scientific value of the published papers.

Conflicts of Interest: The authors declare no conflicts of interest.

List of Contributions:

1. Bou-Rabee, F.; Yogeshwar, P.; Burberg, S.; Tezkan, B.; Duane, M.; Ibraheem, I.M. Imaging of Groundwater Salinity and Seawater Intrusion in Subiya Peninsula, Northern Kuwait, Using Transient Electromagnetics. *Water* **2025**, *17*, 652. https://doi.org/10.3390/w17050652
2. Siemon, B.; Cortés Arroyo, O.; Janetz, S.; Nixdorf, E. Benefits of an Airborne Electromagnetic Survey of Former Opencast Lignite Mining Areas in Lusatia, Germany. *Water* **2025**, *17*, 1000. https://doi.org/10.3390/w17071000
3. Kelly, J.L.; Hladik, C.M. Shallow Hydrostratigraphy Beneath Marsh Platforms: Insights from Electrical Resistivity Tomography. *Water* **2025**, *17*, 144. https://doi.org/10.3390/w17020144
4. Hu, Z.; Zhang, T.; Zhan, M. Multi-Attribute Analysis of Transmission Channel Waves: Applications in Mine Water Damage Prevention. *Water* **2025**, *17*, 1018. https://doi.org/10.3390/w17071018
5. Arıman, S.; Soydan-Oksal, N.G.; Beden, N.; Ahmadzai, H. Assessment of Groundwater Quality through Hydrochemistry Using Principal Components Analysis (PCA) and Water Quality Index (WQI) in Kızılırmak Delta, Turkey. *Water* **2024**, *16*, 1570. https://doi.org/10.3390/w16111570

6. Williams, S.A.; Zuniga-Teran, A.A.; Megdal, S.B.; Quanrud, D.M.; Christopherson, G. Assessing the Relationship Between Groundwater Availability, Access, and Contamination Risk in Arizona's Drinking Water Sources. *Water* **2025**, *17*, 1097. https://doi.org/10.3390/w17071097
7. Hussein, E.E.; Zerouali, B.; Bailek, N.; Derdour, A.; Ghoneim, S.S.M.; Santos, C.A.G.; Hashim, M.A. Harnessing Explainable AI for Sustainable Agriculture: SHAP-Based Feature Selection in Multi-Model Evaluation of Irrigation Water Quality Indices. *Water* **2025**, *17*, 59. https://doi.org/10.3390/w17010059
8. Hossain, M.Z.; Adhikary, S.K.; Nath, H.; Kafy, A.A.; Altuwaijri, H.A.; Rahman, M.T. Integrated Geospatial and Analytical Hierarchy Process Approach for Assessing Sustainable Management of Groundwater Recharge Potential in Barind Tract. *Water* **2024**, *16*, 2918. https://doi.org/10.3390/w16202918
9. El-Sorogy, A.S.; Alharbi, T.; Al-Kahtany, K.; Rikan, N.; Salem, Y. Identification and Validation of Groundwater Potential Zones in Al-Madinah Al-Munawarah, Western Saudi Arabia Using Remote Sensing and GIS Techniques. *Water* **2024**, *16*, 3421. https://doi.org/10.3390/w16233421
10. Rehman, A.; Xue, L.; Islam, F.; Ahmed, N.; Qaysi, S.; Liu, S.; Alarifi, N.; Youssef, Y.M.; Abd-Elmaboud, M.E. Unveiling Groundwater Potential in Hangu District, Pakistan: A GIS-Driven Bivariate Modeling and Remote Sensing Approach for Achieving SDGs. *Water* **2024**, *16*, 3317. https://doi.org/10.3390/w16223317
11. Halder, S.; Banerjee, S.; Youssef, Y.M.; Chandel, A.; Alarifi, N.; Bhandari, G.; Abd-Elmaboud, M.E. Ground–Surface Water Assessment for Agricultural Land Prioritization in the Upper Kansai Basin, India: An Integrated SWAT-VIKOR Framework Approach. *Water* **2025**, *17*, 880. https://doi.org/10.3390/w17060880
12. Mohasseb, H.A.; Shen, W.; Jiao, J.; Wu, Q. Groundwater Storage Variations in the Main Karoo Aquifer Estimated Using GRACE and GPS. *Water* **2023**, *15*, 3675. https://doi.org/10.3390/w15203675
13. Alobid, M.; Chellai, F.; Szűcs, I. Trends and Drivers of Flood Occurrence in Germany: A Time Series Analysis of Temperature, Precipitation, and River Discharge. *Water* **2024**, *16*, 2589. https://doi.org/10.3390/w16182589

water

MDPI

Article

Imaging of Groundwater Salinity and Seawater Intrusion in Subiya Peninsula, Northern Kuwait, Using Transient Electromagnetics

Firyal Bou-Rabee [1], Pritam Yogeshwar [2,*], Sven Burberg [2], Bülent Tezkan [2], Michael Duane [1] and Ismael M. Ibraheem [2,*]

[1] Department of Earth and Environmental Sciences, Kuwait University, P.O. Box 5969, Safat 13060, Kuwait; firyal.bourabee@ku.edu.kw (F.B.-R.); michael.duane@ku.edu.kw (M.D.)

[2] Institute of Geophysics and Meteorology, University of Cologne, Albertus-Magnus-Platz, 50923 Cologne, Germany; sburberg@smail.uni-koeln.de (S.B.); tezkan@geo.uni-koeln.de (B.T.)

* Correspondence: yogeshwar@geo.uni-koeln.de (P.Y.); ismael.ibraheem@geo.uni-koeln.de (I.M.I.)

Abstract: This study investigates the presence and spatial extent of saline water and seawater intrusion in the Subiya Peninsula, Kuwait, a region designated for the establishment of the new Silk City. We collected transient electromagnetic (TEM) data at 63 stations using a coincident loop setup on a regional, as well as local, scale. The data were analyzed through conventional 1D inversion techniques, including Occam and Levenberg–Marquardt methods, to create detailed resistivity models of the subsurface. Our findings indicate significant variations in groundwater salinity, with increased salinity towards the coast and partly decreasing resistivity with depth, suggesting a transition from brackish to saline water. In the northern region, close to the Abdali farms and Al-Raudhatain freshwater fields, groundwater remains fresher at greater depths, while in the south, saline conditions are encountered, occurring at shallower depths. Local scale analysis reveals potential saltwater intrusion pathways and highlighted geological features such as faults. A thorough understanding of the hydrogeological conditions is crucial, as saltwater injection for oil recovery is common in Kuwait, and may correlate with present-day seismic activity. These insights are critical for the sustainable planning and development of Silk City, emphasizing the necessity for further geophysical studies and borehole data to ensure construction safety and sustainable water supply management. This research provides a foundational understanding of the hydrogeological conditions essential for the successful implementation of the Silk City project and for groundwater management in northern Kuwait.

Keywords: coastal aquifer; saltwater–freshwater interface; aquifer characterization; electromagnetic surveying; groundwater evaluation; Silk City; Kuwait

1. Introduction

Seawater intrusion poses a significant threat to coastal aquifers, endangering freshwater resources essential for drinking water supply, agriculture, and ecological sustainability [1–3]. This phenomenon is primarily driven by over-extraction of groundwater, climate change-induced sea-level rises, and anthropogenic activities, leading to salinization of freshwater reserves [4,5]. In arid and semi-arid regions like Kuwait, where natural freshwater resources are scarce, the issue is particularly critical. The country heavily depends on desalination plants and groundwater extraction from shallow aquifers to meet its growing water demands. However, increasing groundwater exploitation, coupled with saltwater intrusion, poses significant risks to water security and sustainable development.

Kuwait's coastal aquifers are particularly vulnerable to two major sources of subsurface salinization. The first is deep subsurface saltwater injection, a process used in oil recovery that can potentially migrate upward and impact shallow groundwater systems. The second is direct brine infiltration into the shallow subsurface, which can lead to the contamination of water fields, such as those in the northern Al-Raudhatain and Abdali farms. However, saltwater intrusion along Kuwait's coastline has increased due to overexploitation of these aquifers and possible future hazards, such as rising sea levels and saltwater injection activities in Sabriya oilfields [6]. The intrusion not only threatens the availability of potable water, but also compromises agricultural productivity and ecosystem health in coastal areas.

Given the complexity of saltwater intrusion mechanisms, geophysical methods play a crucial role in delineating subsurface hydrogeological conditions. One such technique, loop source transient electromagnetics (TEM), has been widely used in groundwater exploration-related studies, due to its ability to detect resistivity variations in subsurface formations [7–12]. It offers a valuable tool for studying seawater intrusion in coastal aquifers [13–17]. TEM methods are particularly effective for identifying freshwater–seawater interfaces, mapping salinity distribution, and assessing the pathways and progression of saltwater intrusion [18,19]. This information is crucial for understanding the dynamics of saltwater intrusion, predicting its future impacts, and informing effective management and mitigation strategies.

While electric resistivity tomography (ERT) provides high-resolution 2D and 3D imaging of resistivity variations, making it effective for detecting lateral and vertical salinity gradients, it has lower depth penetration, and can be affected by electrode coupling issues in high-salinity environments. Nuclear magnetic resonance (NMR), on the other hand, directly measures water content and porosity, providing valuable hydrogeological information, but its depth penetration is limited, and it is highly sensitive to noise, making it less effective in urban and coastal environments. In contrast, TEM offers deeper penetration and is highly sensitive to conductive layers (ideal for saline groundwater mapping). It enables faster data acquisition compared to ERT, making it more efficient for regional-scale studies. However, TEM has lower near-surface resolution than ERT, and cannot directly measure water content like NMR.

The Subiya Peninsula (Figure 1), located in northern Kuwait, is undergoing rapid transformation due to the Silk City megaproject, one of the country's most ambitious urban development initiatives. Approved in 2010 and scheduled for completion between 2030 and 2035, the Silk City project aims to establish a major economic and residential hub featuring advanced infrastructure, including a 1001-meter-tall skyscraper, an international airport, and a major seaport on Bubiyan Island. This large-scale urban expansion, connected to Kuwait City via the 26-kilometer Jaber Causeway, is expected to drive economic growth and urbanization. However, the sustainability of such extensive development hinges on the availability and quality of groundwater resources, posing significant future socio-economic risks, particularly in light of potential seawater intrusion risks. By mapping affected areas, this study supports sustainable water use, safeguards freshwater reserves, aids urban planning, and informs climate resilience policies, ensuring long-term stability.

Approximately 75% of Kuwait's land is designated for grazing and tourism-related activities, while oil fields occupy 7% and military zones account for 4%. As one of the most water-stressed countries in the world, Kuwait relies heavily on transboundary aquifers to meet its water needs. Kuwait generally experiences low surface runoff due to sparse precipitation and the high permeability of sandy soils. During extreme weather events, such as winter storms, rainfall occurs, but is rapidly absorbed, forming small, saline ephemeral playa lakes. Historically, rainwater infiltration through the Dibdibba Formation of the

Kuwait Group was considered significant in models for the Al-Raudhatain and Umm Al-Aish agricultural zones [20]. However, a study by Al-Hurban et al. [21], which involved excavating six shallow pits in northern Kuwait's drainage basins and analyzing satellite imagery using ArcHydro GIS, revealed limited infiltration capacity and minimal interaction with freshwater lenses at depth. Rainwater that does not accumulate in playas eventually drains seaward.

The water balance in the Subiya Peninsula is highly constrained due to its arid climate. The region experiences low annual precipitation (100 to 120 mm) and high evaporation rates (~2500 mm/year), leading to minimal natural groundwater recharge [22]. Additionally, urban expansion and excessive groundwater extraction have intensified concerns over seawater intrusion into coastal aquifers, threatening water quality and availability [23]. As Subiya transforms from a desert landscape into an urban and industrial hub, large-scale projects such as Silk City and oil extraction activities in Bahra and Sabriya fields further increase water demand and groundwater stress [24]. The Jaber Al-Ahmad Causeway, facilitating regional connectivity, accelerates urbanization and economic growth, adding to water consumption pressures. Given that about 90% of Kuwait's water supply comes from desalination, with groundwater serving mainly agricultural and industrial purposes, addressing water sustainability challenges is critical [25]. Therefore, as Silk City advances, Kuwait must prioritize the expanding desalination capacity, improving water recycling, and protecting coastal aquifers to ensure long-term urban and industrial growth in Subiya.

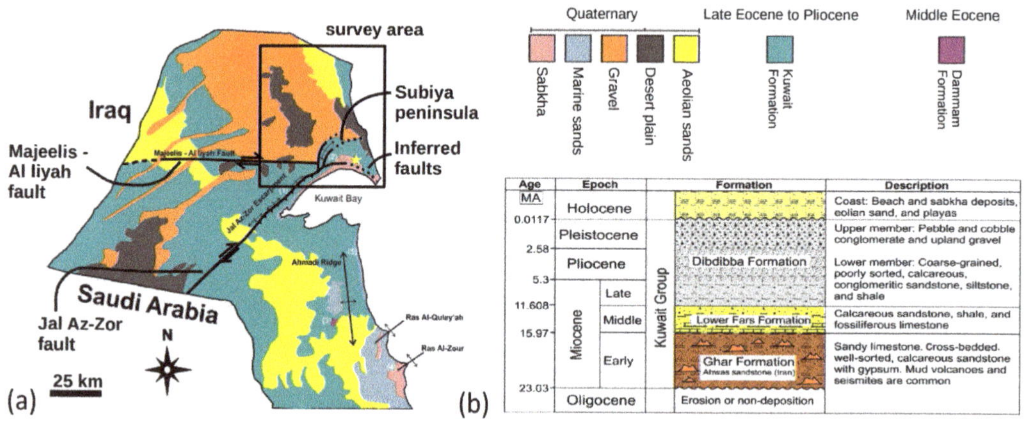

Figure 1. (**a**) A surface geological map of Kuwait illustrating its primary structural features, alongside the distribution and ages of outcrops and sedimentary cover [26]. The focus area of this study, Subiya, is outlined by a black square with coordinates at the lower-left corner at 29°30′50.22″ N, 47°35′43.32″ E, and at the upper-right corner at 30°02′44.28″ N, 48°09′37.26″ E. (**b**) A stratigraphic column of the major geological formations, from the Miocene to Holocene, in the Kuwait Group [27]. The white numbers on the geologic map indicate outcrops: (1) Jal Az-Zor Escarpment, (2) Subiyah, (3) Ahmadi Quarry, (4) Khairan Ridges, and (5) Enjefa Beach.

Despite the recognized threat of saltwater intrusion in Kuwait's coastal aquifers, there remains a notable research gap in assessing its implications for large-scale urban development in arid environments. While previous studies have focused on mapping groundwater salinity in northern Kuwait, comprehensive geophysical investigations specifically addressing the hydrogeological conditions of the Subiya Peninsula remain limited. The increasing reliance on groundwater, coupled with saltwater injection for oil recovery and the anticipated effects of future sea-level rises, underscores the urgent need for a detailed assessment of salinization risks in this region. This study seeks to bridge this critical knowledge gap by employing TEM surveys to achieve the following objectives: 1. delineate the distribution of saline water within the Subiya Peninsula's aquifers; 2. assess the extent and dynamics of seawater intrusion in the region; and 3. evaluate the potential impact on Silk City's infrastructure and water resource management strategies. By integrating geophysical data with hydrogeological analysis, this research aims to provide scientific insights necessary for developing sustainable groundwater management policies, ensuring long-term water security and infrastructure resilience for Kuwait's future urban expansion.

2. Geological and Hydrogeological Setting

Kuwait largely consists of desert areas. The topography of Kuwait is generally flat, with the surface elevation in the east of the country, near the coast, at 0 m, and in the southwest, at up to 280 m above sea level [28]. The surface is characterized by few elevations, with the Jal Az-Zor and Al-Subyiah elevations in the north and the Al-Ahmadi and Al-Kharian elevations in the south [26]. The most prominent elevation in the country is the Jal Az-Zor Escarpment [29]. The escarpment runs from the southwest to the northeast, up to the survey area in Subiya (cf. Figure 1a). There are various theories about the origin of the Jal Az-Zor elevation [30]. The near-surface structures in Kuwait mainly consist of sedimentary rocks from the early Miocene to the most recent Quaternary period (cf. Figure 1). The Quaternary sediments at the surface are sand deposits containing portions of salt carbonates and limestone. In the northeast of the country, along the coastline of Kuwait Bay, there are tidal flats with Sabkha deposits. The tidal flat sediments consist mainly of alluvial sandstones with a high salt content, and also consist partially of claystones [28]. Beneath the upper deposits lies the so-called Kuwait Group (cf. Figure 1b). This consists of clastic sediments, and can be divided into the following three sections, with increasing depth [31,32]: 1. the Dibdibba Formation (DF), 2. the Lower Fars Formation (LFF), and 3. the Ghar Formation (GF).

The three formations within the Kuwait Group differ in age and sediment composition. A stratigraphic representation of the geological formations within the Kuwait Group is shown in Figure 1b. The DF, from the Miocene to the Pleistocene, is the uppermost layer of the Kuwait Group, and can have a thickness ranging from a few meters to approximately 183 m in the north [33]. The formation consists of sand and gravel with portions of gypsum and clay, and can be divided into two sections [28]. The upper section of the formation primarily comprises coarse gravel, which differs from the deeper layers composed mainly of gravelly sandstone with calcareous carbonates [31,34]. Below the DF lies the LFF, which has a thickness ranging from 60 m in the west to 180 m in the north. This Miocene formation consists of shale, sandstone, and fossiliferous limestone [31,33]. The lowest layer of the Kuwait Group is the GF, from the Oligocene to the early Miocene. The sediments of the GF mainly comprise calcareous sandstone with gypsum components [31].

Kuwait's groundwater resources are limited and vulnerable to over-exploitation, as well as contamination, primarily due to the arid climate, rapid population growth, and extensive urbanization. Annual precipitation occurs predominantly from November to February, with a generally low amount of rainfall [22,35]. However, rainfall patterns vary across the different geographical locations. The groundwater level varies between 0 m along the coast of the Arabian Gulf and Kuwait Bay in the east, and up to 90 m below the surface towards the southwest [33]. Kuwait's groundwater is generally brackish to salty. The exceptions are the freshwater fields 'Al-Raudhatain' and 'Umm Al-Aish' in the north of the country, which represent the only freshwater resources in Kuwait [36]. The Kuwait Group and the underlying Dammam Formation act as groundwater aquifers.

In this work, we focus on understanding the aquifers, and relate the derived subsurface models to the value of 'Total Dissolved Solids' (TDS). TDS describes the number of solids dissolved in a fluid. In groundwater, mainly salts are dissolved, making the TDS value an indicator of whether the water is fresh or saline. Although the TDS value is usually given in milligrams per liter [mg/L], we use here the unit g/L, due to the generally very high values present in the area.

The LFF acts as an aquitard with lower permeability, separating the Dibdibba and Ghar formations [33,37]. The aquifers in the Kuwait Group show large variations, with TDS values ranging from below 4 g/L for brackish water in the southwest, to values over 100 g/L, indicating hypersaline water, in the northeast near the coast [38]. Below the Kuwait Group is the Dammam Formation, which mainly consists of limestone. It also serves as an aquifer, with TDS values ranging from 2.5 g/L to 200 g/L, increasing from southwest to northeast [33].

In the north lies Kuwait's largest freshwater field, Al-Raudhatain. The TDS values in this area vary from 0.2 g/L to 8 g/L. Water quality decreases with depth, and the freshwater field is surrounded by brackish and saline water, with values sometimes exceeding 14,000 mg/L at the periphery of the area [36]. Freshwater with salinity under 1 g/L can be found in the upper section of the DF, where the system can be distinguished in two aquifers. The upper aquifer has a thickness of 12 to 36 m, and is saturated with freshwater. Below follows an aquifer with a thickness of 11 to 18 m, containing brackish to saline water [36].

Fadlelmawla et al. [38] analyzed the groundwater TDS values in the Kuwait Group. The derived TDS map in Figure 2a is based on borehole data, and provides an overview for the entire country. The survey area presented in this study is marked by a black square. Along the coast of the Arabian Gulf, the TDS values range from 10 to 50 g/L, indicating hypersaline groundwater. Moving westward, the TDS concentration decreases, indicating fresher groundwater. This area includes the farming region of Abdali and the freshwater field of Al-Raudhatain. Al-Senafy et al. [39] analyzed and modeled the groundwater in northern Kuwait by integrating borehole information, as well as TEM measurements. Their study establishes a relationship between the electrical resistivity of groundwater and the TDS value. In our study, we use Archie's empirical law to transform the modeled bulk electrical resistivity into fluid resistivity (ρ_f), and subsequently use the relation proposed by [40] to further transform the ρ_f into TDS values.

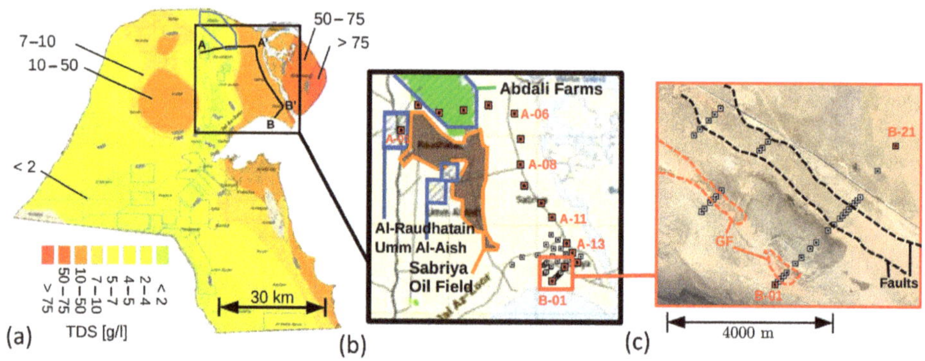

Figure 2. The TEM survey design in Subiya. (**a**) A TDS map, with values in g/L (after [38]). The location of the survey area is marked by a black square. The regional scale profile, with ~70 km length, is indicated by a black line, and is subdivided into three zones: A-A′ (Z1) in the north, A′-B′ (Z2) in the central part, and B′-B (Z3) in the south. (**b**) Soundings along the regional scale profile are marked as red dots. The background image shows the freshwater fields in blue and the Sabriya oil field in brown (after [38]). (**c**) Local scale soundings covering the tidal flats in the south of the Subiya Peninsula. The inferred faults are shown as black dashed lines, and the red GF as a red dashed outline.

3. TEM Survey Design and Data

The TEM method is widely used for imaging subsurface conductive structures [8–11,40,41]. It involves transmitting a current into square loops of insulated wire deployed on the Earth's surface. The transmitter loop's current alternates between time-on and time-off periods, generating a time-varying electromagnetic field, based on Faraday's Law. After the current's switch-off, the primary magnetic field decays. This decay induces eddy currents in the subsurface, which generate a secondary magnetic field recorded by a receiver positioned either at the center or at an off-set distance from the transmitter loop. TEM offers remarkable sensitivity to conductive formations, and enables high-resolution imaging of the subsurface without requiring direct galvanic contact with the ground. The efficiency and application of the TEM method are further detailed in [42,43]. With applications spanning various geological settings, including saline water mapping, TEM has demonstrated its efficiency and versatility in numerous studies worldwide. Despite its strengths, TEM may encounter limitations in highly resistive rock formations or in laterally highly heterogeneous media (e.g., [44]). Nevertheless, its relative simplicity, cost-effectiveness, and ability to probe depths ranging from the surface to depths of a few hundred meters make it a valuable tool for general geological and environmental studies, as well as mineral exploration.

TEM data were gathered utilizing the TEM-FAST portable device, using a coincident loop setup with a single 50 m × 50 m wire loop. The transmitting current was maintained at 4 Amperes, with a recording time ranging between 6 μs and 0.02 s. In total, we covered 63 TEM stations at both local and regional scales, as shown in Figure 2b,c. On the regional scale, 16 soundings, with a varying inter-station spacing between 5 and 10 km, cover a roughly 70 km long profile, from the tidal flats in the south of Subiya towards the freshwater fields around the Abdali farms in the north (cf. Figure 2b). A denser grid of soundings was acquired on a local scale, to image the saltwater intrusion at the coast and around the core area of the planned Silk City construction site (cf. Figure 2c). The station distance is 100 m along the local scale profiles, with a profile distance of roughly 2–3 km. The main profile (B-01–B-21) transects a possible fault zone, with an elevation change of approximately 20 m.

Four exemplary TEM transients are presented as induced voltage and transformed into late-time apparent resistivity ($\rho_{a,\ LT}$) in Figure 3a,b for soundings A-01, A-08, A-13, and B-01. The locations are marked on the map in Figure 2b. A clear change in data characteristics is observed, indicating low resistivity (below 1 Ωm) towards the tidal flats, and higher values towards the freshwater fields around the Abdali farms, with comparably low groundwater TDS values (cf. Figure 2a,b).

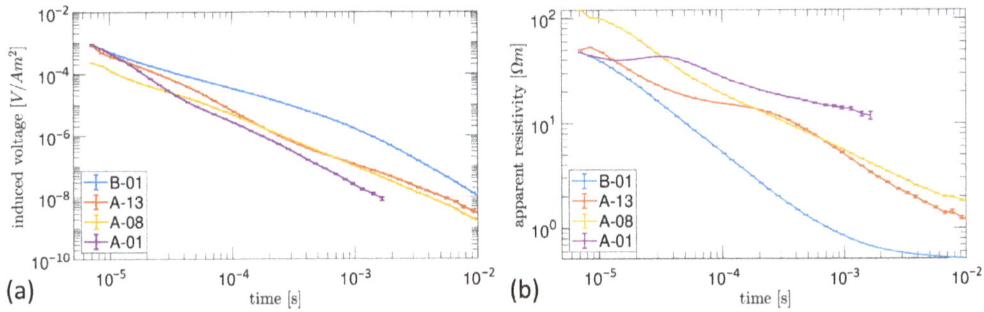

Figure 3. Exemplary TEM data for soundings A-01, A-08, A-13, and B-01. The locations are shown in Figure 2. (**a**) Observed induced voltage decay curves. (**b**) Late-time apparent resistivity curves.

In Figure 4, we present the color-coded late-time apparent resistivity for one transient time, t = 1 × 10^{-3} s. From the late-time apparent resistivity data, a very consistent behavior along the profile from south to north is observed. The apparent resistivity drops below 1 Ωm on the tidal flats, and surpasses ~10 Ωm towards the Abdali farms, indicating a decrease in subsurface salinity away from the coast. The general trend of the data is consistent with the TDS map in Figure 2a. Note that midway between A′ and B′, we observe a slight increase in $\rho_{a,\ LT}$.

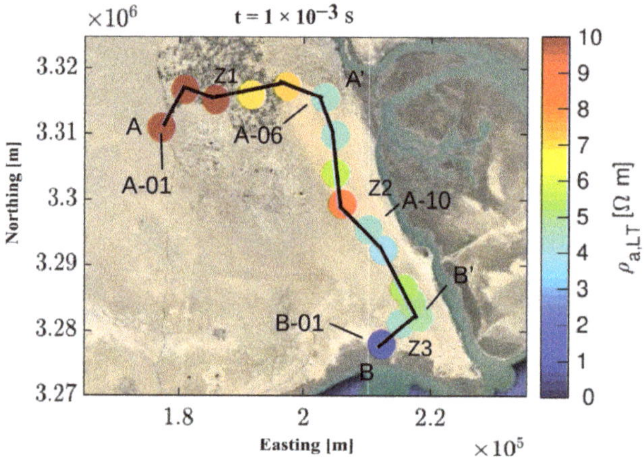

Figure 4. TEM data plotted for one time channel, t = 1 × 10^{-3} s, as late-time apparent resistivity for the regional scale transect. The transect is divided into AA′, A′B′, and B′B.

4. Results and Discussion

Subsequently, 1D inversion of TEM data was carried out using conventional Occam roughness 1 (R1) and roughness 2 (R2), as well as Levenberg–Marquardt (LM) inversion techniques, where Occam R1 applies a first-order smoothness constraint that minimizes changes between consecutive layers, providing a smooth model, while Occam R2 uses a second-order constraint, minimizing the second order difference (curvature) of the models [40]. Both models usually differ at depths where the model parameters are not supported by the data; hence, the divergence of R1 and R2 is often used to estimate the depth of investigation (DOI). Two exemplary 1D models are depicted, together with induced voltage data and model response, in Figure 5. The data are fitted well for both soundings, with chi (χ) values ~1 and reasonable equivalence ranges for the LM models. B-01 is located on the tidal flats, and indicates 2–3 subsurface layers with resistivities below 1 Ωm, already at a depth greater than 10 m below the surface. The lowest resistivity values are observed at 15 m depth, with values around 0.4 Ωm. The DOI estimate is roughly 80 m, using the approach suggested by Spies [45]. Sounding A-01 is located on the Al-Raudhatain freshwater field, and shows much larger resistivity values overall (~6 Ωm at 30 m depth), down to the DOI of roughly 115 m. Note that the divergence of R1/R2 suggests reduced DOI values, with 80 m for A-01 and roughly 50 m for B-01.

4.1. Spatial Distribution of Saline Groundwater

Figure 6 displays the resistivity extracted from the Occam R1 models for the 16 regional TEM stations, at three different depths, on a Google Earth satellite image (top row), on the TDS map from [38] (middle row), and on the topography map from [46] (bottom row). Here, we show the elevations −10 m (left), −30 m (middle), and −60 m (right). These graphics serve to assess the regional salinity in the groundwater, and discuss possible causes for the resistivity variations seen in the presented models, which are possibly related to TDS and altitude.

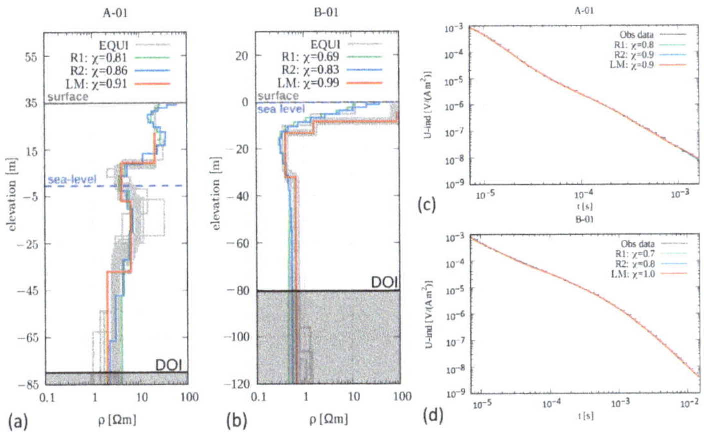

Figure 5. One-dimensional models, together with data and model response, for soundings A-01 (**a**,**c**) and B-01 (**b**,**d**). The data fit chi is displayed in the respective legend. The LM models are marked in red, and the equivalent models (EQUI) in gray. Occam R1/R2 models are marked in green and blue. The DOI, derived based on the approach suggested by Spies [45], is indicated by black lines.

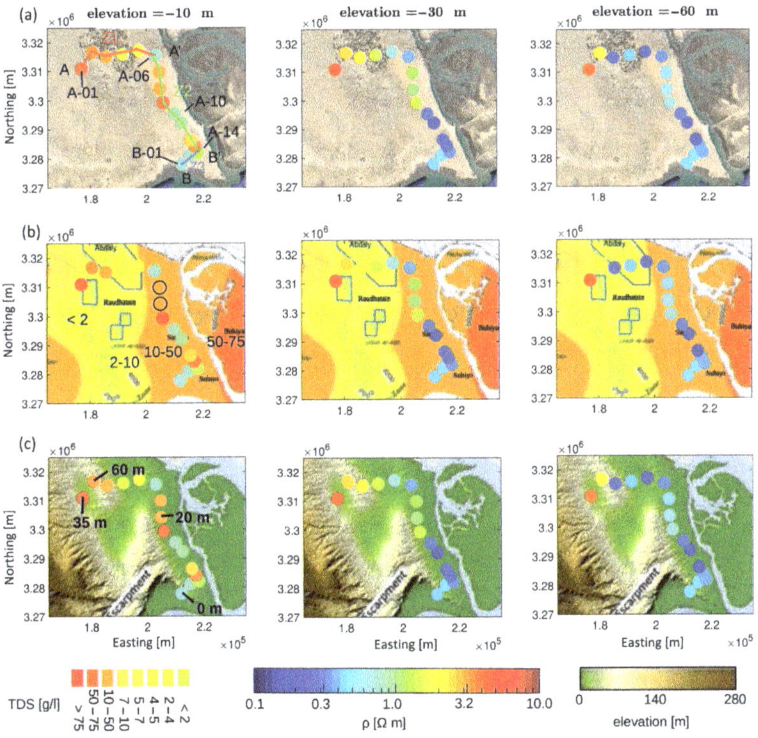

Figure 6. A scatter plot of the modeled subsurface resistivity, depicted on (**a**) the satellite image, together with station/zone labels (top row), (**b**) the TDS map from [38], with values in g/L (middle row), and (**c**) the topography map from [46], with labeled altitude (bottom row). The elevations for the model scatter plots are −10 m, −30 m, and −60 m. The maps are derived from the Occam R1 results.

According to [38], the groundwater level in the northeast of Kuwait ranges from an elevation of 0 m to 10 m. Thus, in the survey area, at all the depths shown in Figure 6, the soil is saturated with groundwater. Note the resistivity variation with increasing depths. At an elevation of −10 m, the resistivities for all soundings are above 1 Ωm, and decrease with depth. For the southern stations (A-10 to B-01), we observe a transition to a more conductive subsurface, with 0.2–0.5 Ωm at an elevation of −30 m. For the northern part of the profile (stations A-1 to A-10), the resistivity remains mainly above 1 Ωm down to an elevation of −30 m. Below this depth, there is a transition to resistivities below 0.8 Ωm. Since the subsurface resistivity decreases with depth, we suggest that the groundwater becomes more saline with increasing depth, indicating a transition from brackish to saline groundwater. The progression from initially fresh to brackish groundwater close to the surface, becoming saline at greater depths, is also reported in [47]. In the southern TEM stations, the transition from brackish to saline water is evident at shallower depths, whereas in the north, this transition occurs at greater depths (see also 1D models in Figure 4 for comparison).

Besides the general behavior of increasing resistivity with depth, a regional pattern is evident at all depths. In zone Z1 (A-A′), the resistivity decreases from west to east, and is higher overall than in the southern zones (Z2, Z3). This trend suggests that groundwater salinity decreases towards the inland areas and increases towards the coastal areas. In zone Z2 (A′-B′), along t'he eastern coast southward, three stations (A-07, A-08, and A-09) stand out with higher resistivity. At these stations, the salinity is lowered, possibly due to a freshwater recharge from terrain elevations in the west (Figure 6, bottom row). The resistivity increases towards Abdali, which aligns well with the decrease in TDS values of the groundwater. Noteworthy is station A-01, located in the Al-Raudhatain freshwater field. As visible in Figure 6 (bottom row), Al-Raudhatain is situated in a valley, and surrounded by 12 dry riverbeds that are flooded during heavy rains, ensuring natural groundwater recharge [33,36]. In Al-Raudhatain, the TDS values range from 0.2 to 8 g/L. Similarly, the TDS values are significantly lower in Abdali compared to zone Z2 along the profile. The fresher groundwater at Al-Raudhatain is confirmed by higher resistivities at all depths for soundings A-01 and A-02. For the Abdali soundings (A-03, A-04, and A-05), the groundwater remains fresher down to an elevation of −40 m. Below this, there is a transition to saline groundwater, with resistivities of roughly 0.3 Ωm.

4.2. Regional Scale Interpretation

The 1D Occam R1 models are stitched together as quasi-2D sections along the regional profile A-B (cf. Figure 7). We use a linear interpolation between the soundings. The profile is subdivided into three distinct sections (Z1 to Z3, cf. Figure 7). We observe three distinct layers that are interpreted in terms of geological and hydrogeological knowledge, namely Holocene deposits (HD) and groundwater aquifers (AQ1 and AQ2).

The northern DF, which is the uppermost geological formation of the Kuwait Group, has a thickness of up to 183 m [33]. Due to a DOI of maximum ~100 m, TEM resolves only information about the DF as the primary aquifer beneath the surface deposits.

- HD: The surface layer consists of Holocene sand and gravel deposits, with resistivities ranging from 20 Ωm to ~300 Ωm. Values vary due to locally different soil compositions and water saturation. This layer may already be partially saturated due to rain infiltration. The depth to the first aquifer ranges from a few meters, to 25 m in elevated areas.
- AQ1: This refers to the upper DF, serving as the shallow aquifer, and primarily consists of sand- and gravel-rich sediments. The transition between HD and AQ1 is as low as z = 0 m at location B-01, and reaches up to an elevation of 41 m at A-01 due to the

elevated topography. The resistivity ranges from 2 to 10 Ωm. Thus, the upper section of the DF is likely saturated with brackish water.

- AQ2: below AQ1, the resistivity decreases to between ~0.2 and 1.0 Ωm. The transition occurs at an elevation of 5 m to roughly −50 m (A-03), caused by increasing salinity. The lower DF differs in structure, being coarser and mainly composed of cemented sandstone. As seen in the 1D stitched models (Figure 7), there is a noticeable resistivity increase in zone Z2 between stations A-07 to A-09 down to roughly −40 m elevation, extending even further downwards. This possibly indicates freshwater recharge from the western elevated terrain (Figure 6). However, with values between 0.7 and 1.0 Ωm, the groundwater is still likely to be saline. In zone Z1, a lateral transition from saline to brackish groundwater occurs between A-03 and A-02, with resistivities varying from 0.2 Ωm to 4.5 Ωm at an elevation of −75 m. Conclusively, the groundwater is likely to be brackish near the freshwater field of Al-Raudhatain.

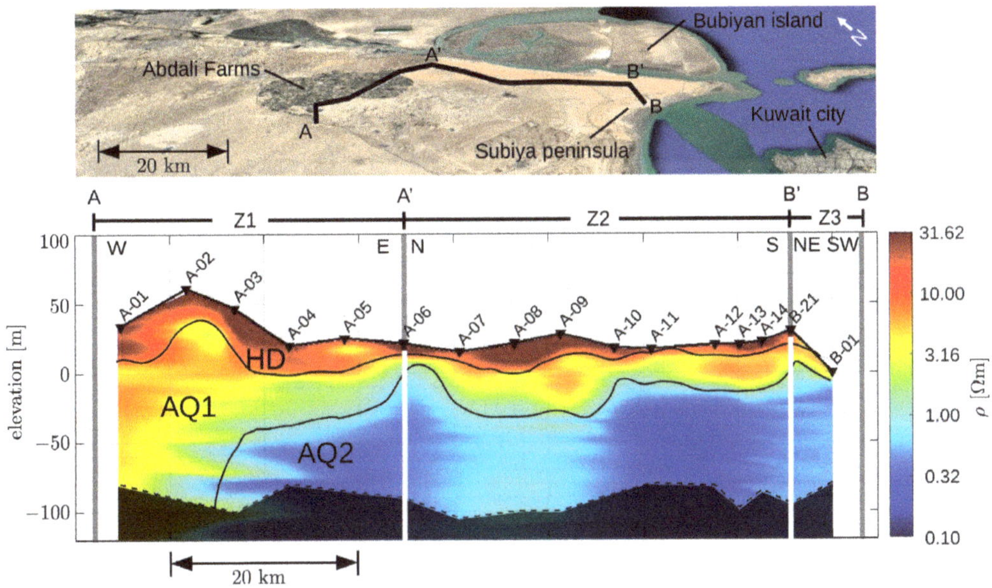

Figure 7. The stitched and interpolated 1D resistivity section derived from the Occam R1 results. For improved visualization, the models are linearly interpolated along the profile. The black lines indicate the transitions between the different layers in the DF. The formations are labeled. The vertical gray/white lines separate the zones Z1, Z2, and Z3 (cf. Figure 6). Note that Z3 is oriented NE–SW. The DOI is marked as the black shaded area.

4.3. Local Scale Interpretation

Figure 8a depicts the late-time apparent resistivity at $t = 1 \times 10^{-3}$ s as a scatter plot, and (c) depicts the stitched 1D models obtained in the southern zone Z3. The spatial data pattern indicates a significant resistivity decrease towards the coastline. We observe a few rather distinct zones in the model, marked as Res1, F1, IZ, C1 and C2. The surface layer Res1 shows resistivities larger than 5 Ωm, with a thickness ranging from a few meters in the south, to roughly 40 m towards location B-21. Note that the resistivity decreases between B-11 and B-18, which coincides with the visible known fault at the surface (F1). It is likely that due to the higher porosity around the fault zone, there is either an upwelling of saline water or higher freshwater recharge. Towards the north, we observe an interface zone (IZ) below the resistive zone Res1. From south to north (B-01–B-21), a more conductive layer (C1) appears, which may indicate a seawater intrusion pathway. The shape of the top of the seawater intrusion can be partly explained by the Ghyben–Herzberg relation, which describes the relationship between freshwater and saltwater in coastal aquifers, estimating that for every meter of freshwater above sea level, there is approximately 40 m of freshwater below sea level before reaching saline water [48]. Assuming the saltwater interface to be at an elevation of −40 m at sounding B-19, the freshwater level would be at an elevation of roughly 1 m. Below C1, the resistivity slightly increases to ~0.6 Ωm, indicating a more compact formation (C2), being shallowest right below the visible outcrops of the red GF (cf. Figure 8b,c). A unique interpretation is difficult, due to the absence of borehole information. However, a relation between the possible fluid pathways, the fault zones, and the known geological formations in this area is visually evident. This may have significant implications for the planning of the 'Silk City' regarding not only construction safety, but also water supply. Further geophysical studies could shed light on this matter.

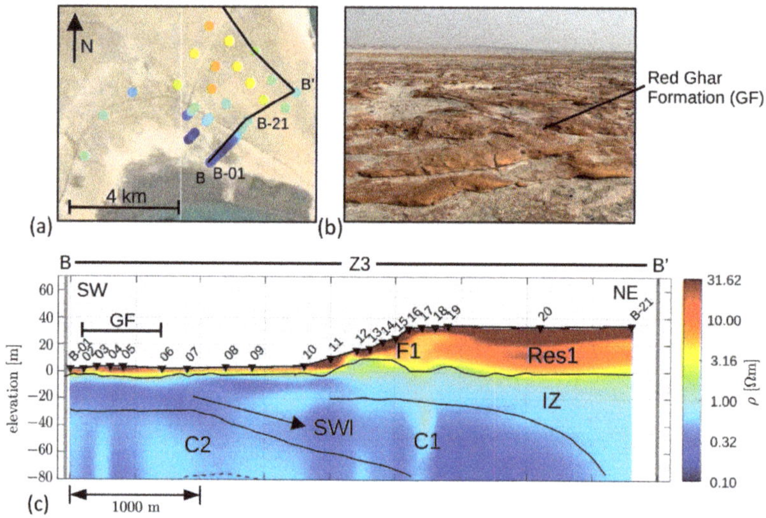

Figure 8. (**a**) A color-coded map of apparent resistivity late-time data for t = 1 ms. (**b**) An image of the outcrop of the red GF. (**c**) A stitched and interpolated 1D resistivity section derived from the Occam R1 results for the local scale transect (B-B′) at the Subiya coast. The inferred formations are labeled. Note that Z3 is oriented from SW to NE.

4.4. Transformation of Models to Fluid Resistivity and TDS

Based on Archie's formula, Al-Senafy et al. [39] derived a relationship between TDS values, water, and soil electrical resistivity for northern Kuwait, by the incorporation of TEM data, inferred borehole formation data, and well TDS:

$$\rho_f = 1/(1.3657 \cdot \text{TDS} + 146.82) \cdot 10^{-4} \quad (1)$$

Table 1 shows the various salinity categories and their corresponding TDS values [49]. Using Equation (1), the fluid resistivity values for each salinity category (fresh, brackish, and saline) were calculated.

Table 1. The different salinity categories with their corresponding TDS and fluid resistivity values.

Salinity Category	TDS (g/L)	Water ρ_f (Ωm)
Fresh	0–1	>6.61
Brackish	1–10	0.72–6.61
Saline	>10	<0.72

Figure 9a depicts the TEM sounding A-01, which is located nearby the Al-Raudhatain freshwater field. First, we calculated ρ_f from the model resistivities using Archie's equation ($\rho_f = \rho_{model}/F$), with the formation factors (F) proposed by Al-Senafy et al. [39], and plotted them versus the depth, as shown in Figure 9b. With Equation (1), proposed by Al-Senafy et al. [39], we subsequently solved for the TDS value of the fluid, and plotted it versus depth, as shown in Figure 9c. Note that for near-surface sands with higher porosity, a formation factor of F = 3.1–3.5 is specified, while for deeper areas, the formation factor for cemented sandstone can be up to F = 21. For the lower formation factors (e.g., F = 3.1, sand and gravel), the TDS values are in the brackish range, between 2 and 7 g/L. For higher formation factors (e.g., F = 21, sandstone), the TDS can also be saline, with a TDS value over 10 g/L. According to [36], there are two aquifers in the DF at Al-Raudhatain. The upper aquifer (AQ1) contains fresh groundwater, while the lower aquifer (AQ2) is saturated with brackish to saline water, and with increasing depth, there is cemented sandstone, with a higher formation factor, that is saturated with saltwater. Note that the formation factors are based on estimates provided by Al-Senafy et al. [39], which are derived from distinct borehole analysis. For a more accurate interpretation, data from boreholes along the survey profile line A-B are required, and could allow for more accurate information about the formation parameters necessary for using Archie's empirical equation.

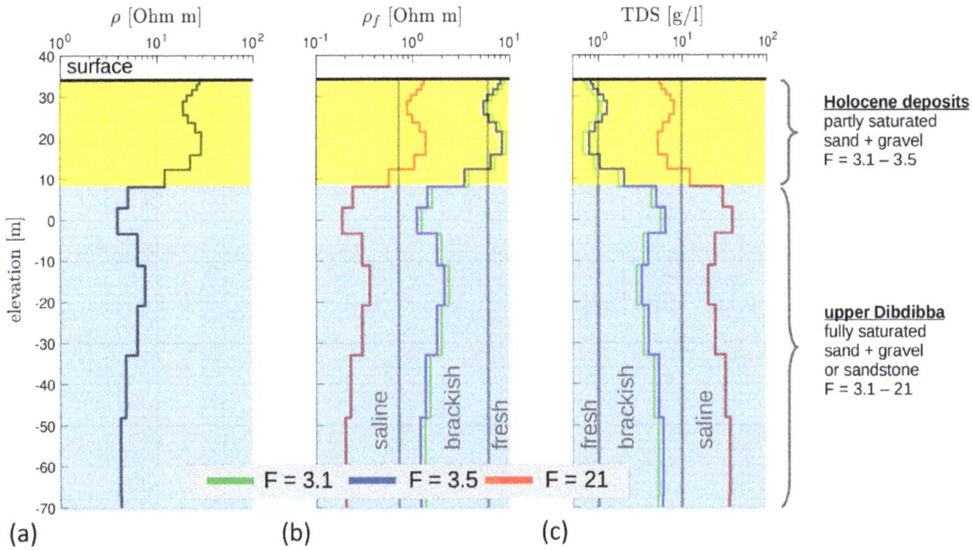

Figure 9. (**a**) One-dimensional models for sounding A-01, (**b**) the derived fluid resistivity ρ_f [Ωm], and (**c**) the TDS value [g/L] of the fluid. The values of ρ_f and TDS are shown for formation factors F = 3.1, 3.5, and 21. The yellow area indicates partially saturated soil; the blue area indicates fully saturated soil. Note that the DOI is at roughly −80 m elevation, and, therefore, is not plotted (see the DOI in Figure 5a).

4.5. Geophysical Insights for Sustainable Engineering in Silk City

The study's findings regarding groundwater salinity and seawater intrusion in the Subiya Peninsula provide critical insights for engineering construction and urban development in the Silk City project area. Variations in groundwater conditions, such as high salinity in coastal zones and fresher water in northern regions like the Al-Raudhatain and Abdali farms, directly impact construction stability and material durability. Saline water poses a corrosion risk to infrastructure, necessitating corrosion-resistant designs or avoidance of high-salinity areas. The identified fault features further highlight construction safety concerns, guiding site selection and mitigation strategies. Additionally, the correlation between terrain elevation and freshwater recharge zones informs sustainable water resource management, ensuring reliable freshwater supply for urban and agricultural needs. Data on water content and bulk resistivity also offer valuable insights into subsurface porosity, which are essential for assessing soil properties and infrastructure design.

However, this study is only a preliminary step; large-scale projects like Silk City require comprehensive geophysical surveys and deep borehole data to refine interpretations and ensure long-term safety. Government agencies should prioritize providing baseline geophysical data, and support further investigations to address risks such as saltwater intrusion, groundwater depletion, and potential seismic impacts from saltwater injection practices. By integrating these findings into urban planning frameworks, the research will enhance construction resilience, support sustainable resource allocation, and mitigate risks, ultimately ensuring the success of multi-billion-dollar developments in the region.

5. Conclusions

This study concludes that the Subiya Peninsula exhibits significant variations in groundwater salinity, which are critical for the proposed Silk City development in terms of sustainable freshwater supply, but also with respect to agricultural progress at the Abdali farms. Regional- and local-scale surveys, well-validated with TDS salinity maps, reveal a clear spatial trend: coastal areas exhibit higher salinity, while northern regions towards the Al-Raudhatain and Abdali farms maintain fresher groundwater at greater depths. The spatial trend correlates with the terrain elevation, possibly indicating zones of increased freshwater recharge. Although Al-Raudhatain shows the least subsurface conductivities, fresh-to brackish water occurs predominantly only in the Holocene deposits, whereas water in the upper DF is likely to be brackish, or even saline. In contrast, southern regions experience saline conditions at shallower depths, likely due to saltwater intrusion pathways influenced possibly by faults. These also indicate increased freshwater recharge or upwelling of fluids. The saltwater intrusion pathway aligns roughly with the Ghyben–Herzberg relation. Towards the coast, a subsurface resistivity pattern below the GF is slightly visible. However, no unique interpretation can be made without deep borehole data to better understand formation factors.

Our findings confirm the need for more spatial geophysical data on a regional scale and detailed hydrogeological assessments, in order to improve regional interpretation and ensure construction safety and sustainable water resource management in the Silk City project area. This deeper understanding of saltwater intrusion pathways provides actionable insights into mitigating salinization risks for planned developments such as Silk City. Specifically, integrating geophysical findings into urban planning frameworks ensures construction resilience and sustainable resource allocation. Implementing a monitoring concept at the Abdali farms and Al-Raudhatain freshwater fields is highly advised to track changes over time, especially to monitor whether the saltwater interface moves towards the freshwater fields due to extensive groundwater abstraction. Further groundwater depletion poses a significant threat to the agricultural production at Abdali and the food supply for the Kuwait population. Saltwater injection for enhanced oil recovery is a common practice in northern Kuwait. Although the correlation between injection and recent shallow seismic movements (less than 2 km depth) is not yet quantified, further investigation is needed.

By integrating these insights into engineering construction practices and complementing them with further detailed studies, the research enhances its application value, offering actionable data to mitigate risks, optimize resource use, and ensure the long-term resilience and sustainability of the Silk City project and surrounding agricultural areas. While TEM alone provides a strong basis for salinity assessment, future studies could integrate ERT for improved near-surface resolution, and NMR for improved hydrogeological characterization. This combined approach would enhance the accuracy of salinity detection and provide a more comprehensive understanding of groundwater conditions.

Author Contributions: Conceptualization, F.B.-R., P.Y., B.T., M.D. and I.M.I.; data curation, P.Y. and I.M.I.; formal analysis, P.Y., S.B. and I.M.I.; funding acquisition, F.B.-R. and B.T.; investigation, P.Y., S.B. and M.D.; methodology, F.B.-R., P.Y., S.B., B.T., M.D. and I.M.I.; project administration, F.B.-R. and B.T.; resources, F.B.-R., P.Y., B.T. and M.D.; software, P.Y. and S.B.; supervision, P.Y., B.T. and I.M.I.; validation, P.Y., B.T. and I.M.I.; visualization, P.Y., S.B. and I.M.I.; writing—original draft, P.Y., S.B. and I.M.I.; writing—review and editing, P.Y. and I.M.I. All authors have read and agreed to the published version of the manuscript.

Funding: This research received no external funding.

Data Availability Statement: The raw data supporting the conclusions of this article will be made available by the authors on request.

Acknowledgments: The authors express their gratitude to the editor and the five anonymous reviewers for their valuable and constructive feedback.

Conflicts of Interest: The authors declare no conflicts of interest.

References

1. Akhtar, J.; Sana, A.; Tauseef, S.M. Review: Assessment and modeling of seawater intrusion in coastal aquifers of the Arabian Peninsula. *Hydrogeol. J.* **2023**, *31*, 1121–1145. [CrossRef]
2. BinMakhashen, G.M.; Benaafi, M. Predicting seawater intrusion in coastal areas using machine learning: A case study of arid coastal aquifers, Saudi Arabia. *Groundw. Sustain. Dev.* **2024**, *26*, 101300. [CrossRef]
3. Hossain, I.; Reza, S.; Shafiuzzaman, S.M.; Sultan-Ul-Islam, M. Effect of seawater intrusion on water quality of coastal aquifer of Bagerhat district, Bangladesh. *Int. J. Energy Water Res.* **2024**, *8*, 123–140. [CrossRef]
4. Lyra, A.; Loukas, A.; Sidiropoulos, P.; Mylopoulos, N. Climatic Modeling of Seawater Intrusion in Coastal Aquifers: Understanding the Climate Change Impacts. *Hydrology* **2024**, *11*, 49. [CrossRef]
5. Moorthy, P.; Sundaramoorthy, S.; Roy, P.D.; Usha, T.; Dash, S.K.; Gowrappan, M.; Chokklingam, L. Evaluation of spatial and temporal dynamics of seawater intrusion in coastal aquifers of southeast India: Insights from hydrochemical facies analysis. *Environ. Monit. Assess.* **2024**, *196*, 179. [CrossRef]
6. Kuwait Oil Company. Lessons Learned from Two Decades of Water Injection in North Kuwait, Report. 2018. Available online: https://www.kockw.com/sites/EN/EMagazine/Pages/Technology/Lessons-Learned-from-Two-Decades-of-Water-Injection-in-North-Kuwait-.aspx (accessed on 10 April 2024).
7. Ibraheem, I.M. Geophysical Potential Field Studies for Developmental Purposes at El-Nubariya—Wadi El-Natrun Area, West Nile Delta, Egypt. Ph.D. Thesis, Mansoura University, Mansoura, Egypt, 2009; pp. 155–171.
8. Danielsen, J.E.; Auken, E.; Jørgensen, F.; Søndergaard, V.; Sørensen, K.I. The application of the transient electromagnetic method in hydrogeophysical surveys. *J. Appl. Geophys.* **2003**, *53*, 181–198. [CrossRef]
9. Auken, E.; Jørgensen, F.; Sørensen, K.I. Large-scale TEM investigation for groundwater. *Explor. Geophys.* **2003**, *34*, 188–194. [CrossRef]
10. Lévesque, Y.; Walter, J.; Chesnaux, R. Transient Electromagnetic (TEM) Surveys as a First Approach for Characterizing a Regional Aquifer: The Case of the Saint-Narcisse Moraine, Quebec, Canada. *Geosciences* **2021**, *11*, 415. [CrossRef]

11. Ibraheem, I.M.; Othman, A.; Ghazala, H. Pliocene Aquifer Characterization Using TEM and VES Geophysical Techniques: Case Study at the Area to the East of Wadi El-Natrun City, West Nile Delta, Egypt. In *Sustainability of Groundwater in the Nile Valley, Egypt*; Negm, A.M., El-Rawy, M., Eds.; Earth and Environmental Sciences Library; Springer: Cham, Switzerland, 2022; pp. 235–266. [CrossRef]
12. Zamora-Luria, J.C.; McLachlan, P.; Maurya, P.K.; Liu, L.; Grombacher, D.; Christiansen, A.V. A feasibility study on time-lapse transient electromagnetics for monitoring groundwater dynamics. *Geophysics* **2023**, *88*, E135–E146. [CrossRef]
13. El-Kaliouby, H. Mapping Sea water intrusion in coastal area using time-domain electromagnetic method with different loop dimensions. *J. Appl. Geophys.* **2020**, *175*, 103963. [CrossRef]
14. Kalisperi, D.; Kouli, M.; Vallianatos, F.; Soupios, P.; Kershaw, S.; Lydakis-Simantiris, N. A Transient ElectroMagnetic (TEM) Method Survey in North-Central Coast of Crete, Greece: Evidence of Seawater Intrusion. *Geosciences* **2018**, *8*, 107. [CrossRef]
15. Ez-zaouy, Y.; Bouchaou, L.; Schreiber, H.; Montcoudiol, N.; Kalberkamp, U.; Danni, S.O.; Tauab, A.; Abourrig, F.; Hssaisoune, M. Combined geophysical methods to investigate seawater intrusion in the Souss-Massa coastal area, Morocco. *Groundw. Sustain. Dev.* **2023**, *21*, 100915. [CrossRef]
16. Wang, L.; Dai, Y.; Liu, W.; Zhou, S.; Long, X.; Xi, Z.; Xue, J.; Wang, W. Deep Learning Transient Electromagnetic Inversion for Seawater Intrusion. *J. Geophys. Eng.* **2024**, *21*, 1810–1821. [CrossRef]
17. Zhu, Z.; Shan, Z.; Pang, Y.; Wang, W.; Chen, M.; Li, G.; Sun, H.; Revil, A. The transient electromagnetic (TEM) method reveals the role of tectonic faults in seawater intrusion at Zhoushan islands (Hangzhou Bay, China). *Eng. Geol.* **2024**, *330*, 107425. [CrossRef]
18. AbuRajab, J.; El-Kaliouby, H.; Al-Tarazi, E.; AlAmoush, H. Multiscale geoelectrical characteristics of seawater intrusion along the eastern coast of the Gulf of Aqaba, Jordan. *J. Appl. Geophys.* **2023**, *208*, 104868. [CrossRef]
19. El-Kaliouby, H.; Abdallah, O. Application of time-domain electromagnetic method in mapping saltwater intrusion of a coastal alluvial aquifer, North Oman. *J. Appl. Geophys.* **2015**, *115*, 59–64. [CrossRef]
20. Al-Ruwaih, F.; Hadi, K. Water quality trends and management of fresh groundwater at Al-Rawdhatain, Kuwait. *Eur. J. Sci. Res.* **2005**, *9*, 40–64.
21. Al-Hurban, A.; Al-Ruwaih, F.; Al-Dughairi, A. Quantitative Geomorphological and Hydromorphometric Analysis of Drainage Basins of As Sabriyah (Kuwait) Using GIS Techniques. *J. Geogr. Inf. Syst.* **2021**, *13*, 166–193. [CrossRef]
22. Sabarathinam, C.; Bhandary, H.; Ali, A. Strategies to characterize the geochemical interrelationship between coastal saline groundwater and seawater. *Environ. Earth Sci.* **2021**, *80*, 642. [CrossRef]
23. Bhandary, H.; Sabarathinam, C.; Al-Khalid, A. Occurrence of hypersaline groundwater along the coastal aquifers of Kuwait. *Desalination* **2018**, *436*, 15–27. [CrossRef]
24. Kamal, H.J.; Thomas, M.; Abdulhadi, A. Perspectives of Geography, Environment, and Physiography of Kuwait. In *Terrestrial Environment and Ecosystems of Kuwait*; Suleiman, M.K., Shahid, S.A., Eds.; Springer: Cham, Switzerland, 2024; pp. 21–55. [CrossRef]
25. Hadi, K. Current Status, Challenges, and Future Management Strategies for Water Resources of Kuwait. In *Terrestrial Environment and Ecosystems of Kuwait*; Suleiman, M.K., Shahid, S.A., Eds.; Springer: Cham, Switzerland, 2024; pp. 141–169. [CrossRef]
26. Al-Refaei, Y.; Najem, A.; Amer, A.; Al-Qattan, F. Surface Geology of Kuwait. In *The Geology of Kuwait. Regional Geology Reviews*; Abdel-aal, A.K., Al-Awadhi, J.M., Al-Dousari, A., Eds.; Springer: Cham, Switzerland, 2023; pp. 1–25. [CrossRef]
27. Al-Helal, A.; AlRefai, Y.; AlKandari, A.; Abdullah, M. Subsurface Stratigraphy of Kuwait. In *The Geology of Kuwait. Regional Geology Reviews*; Abdel-aal, A.K., Al-Awadhi, J.M., Al-Dousari, A., Eds.; Springer: Cham, Switzerland, 2023. [CrossRef]
28. Milton, D.I. *Geology of the Arabian Peninsula; Kuwait*; Professional Paper; US Geological Survey: Reston, VA, USA, 1967; pp. F1–F7.
29. Amer, A.; Al-Hajeri, M.; Najem, A.; Al-Qattan, F. Facies architecture of Lower Fars Formation at Jal Az-Zor escarpment, Kuwait. *Arab. J. Geosci.* **2019**, *12*, 502. [CrossRef]
30. Al-Sarawi, M. The origin of the Jal-Az Zor escarpment. *J. Univ. Kuwait* **1982**, *9*, 151–162.
31. Duane, M.J.; Reinink-Smith, L.; Eastoe, C.; Al-Mishwat, A.T. Mud volcanoes and evaporite seismites in a tidal flat of northern Kuwait—Implications for fluid flow in sabkhas of the Persian (Arabian) Gulf. *Geo-Mar. Lett.* **2015**, *35*, 237–246. [CrossRef]
32. Owen, R.M.; Naser, S.N. Stratigraphy of the Kuwait-Basrah area. Habitat of Oil. *Am. Assoc. Pet. Geol.* **1958**, *42*, 1252–1278.
33. Misak, R.; Hussain, W. Groundwater in Kuwait. In *The Geology of Kuwait. Regional Geology Reviews*; Abdel-aal, A.K., Al-Awadhi, J.M., Al-Dousari, A., Eds.; Springer: Cham, Switzerland, 2023; pp. 199–214.
34. Al-Sulaimi, J.; Mukhopadhyay, A. An overview of the surface and near-surface geology, geomorphology and natural re-sources of Kuwait. *Earth-Sci. Rev.* **2000**, *50*, 227–267. [CrossRef]
35. Alkandari, A.J. Climate and Climate Change Aspects of Kuwait. In *Terrestrial Environment and Ecosystems of Kuwait*; Suleiman, M.K., Shahid, S.A., Eds.; Springer: Cham, Switzerland, 2023; pp. 57–91. [CrossRef]
36. Hadi, K.M.; Al-Ruwaih, F.M. Geochemical evolution of the fresh groundwater in Kuwait desert. *Emir. J. Eng. Res.* **2008**, *13*, 37–45.
37. Fadlelmawla, A.; Al-Otaibi, M. Analysis of the water resources status in Kuwait. *Water Resour. Manag.* **2005**, *19*, 555–570. [CrossRef]

38. Fadlelmawla, A.A.; Fayad, M.; El-Gamily, H.; Rashid, T.; Mukhopadhyay, A.; Kotwicki, V. A Land Surface Zoning Approach Based on Three-Component Risk Criteria for Groundwater Quality Protection. *Water Resour. Manag.* **2011**, *25*, 1677–1697. [CrossRef]
39. Al-Senafy, M.; Fadlelmawla, A.; Bhandary, H.; Al-Khalid, A.; Rashid, T.; Al-Fahad, K.; Al-Salma, B. Assessment of Usable Groundwater Reserve in Northern Kuwait. *Int. J. Sci. Eng. Res.* **2013**, *4*, 2427–2436.
40. Yogeshwar, P.; Küpper, M.; Tezkan, B.; Rath, V.; Kiyan, D.; Byrdina, S.; Cruz, J.; Andrade, C.; Viveiros, F. Innovative boat-towed transient electromagnetics—Investigation of the Furnas volcanic lake hydrothermal system, Azores. *Geophysics* **2020**, *85*, E41–E56. [CrossRef]
41. Ibraheem, I.M.; Yogeshwar, P.; Bergers, R.; Tezkan, B. Joint interpretation of magnetic, transient electromagnetic, and electric resistivity tomography data for landfill characterization and contamination detection. *Sci. Rep.* **2024**, *14*, 30616. [CrossRef] [PubMed]
42. Kaufman, A.A.; Keller, G.V. *Frequency and Transient Soundings*; Elsevier: Amsterdam, The Netherlands, 1983.
43. Nabighian, M.N.; Macnae, J.C. Time Domain Electromagnetic Prospecting Methods. In *Investigations in Geophysics No 3*; Nabighian, M.N., Ed.; Electromagnetic Methods in Applied Geophysics; Society of Exploration Geophysicists: Tulsa, OK, USA, 1991; pp. 427–514. [CrossRef]
44. Yogeshwar, P.; Tezkan, B. Analysing two-dimensional effects in central loop transient electromagnetic sounding data using a semi-synthetic tipper approach. *Geophys. Prospect.* **2018**, *66*, 444–456. [CrossRef]
45. Spies, B.R. Depth of Investigation in Electromagnetic Sounding Methods. *Geophysics* **1989**, *54*, 872–888. [CrossRef]
46. Naqi, M.; Amer, A. Abdel-aal, A.K., Al-Awadhi, J.M., Al-Dousari, A., Eds.; Structures and Tectonics of Kuwait. In *The Geology of Kuwait. Regional Geology Reviews*; Springer: Cham, Switzerland, 2023; pp. 99–115. [CrossRef]
47. Hadi, K.; Saravana Kumar, U.; Al-Senafy, M.; Bhandary, H. Environmental isotope systematics of the groundwater system of southern Kuwait. *Environ. Earth Sci.* **2016**, *75*, 1096. [CrossRef]
48. Bear, J. *Hydraulics of Groundwater*; McGraw-Hill Series in Water Resources and Environmental Engineering; McGraw-Hill: New York, NY, USA, 1979; pp. 384–385.
49. Aliewi, A.; Al-Kandari, J.; Al-Khalid, A.; Bhandary, H.; Al-Qallaf, H. Modelling the effect of high level of total dissolved solids (TDS) for the sustainable utilization of brackish groundwater from saline aquifers in Kuwait. *Environ. Dev. Sustain.* **2021**, *23*, 2204–2223. [CrossRef]

water

MDPI

Article

Benefits of an Airborne Electromagnetic Survey of Former Opencast Lignite Mining Areas in Lusatia, Germany

Bernhard Siemon [1,*], Olaf Cortés Arroyo [2], Silvio Janetz [2] and Erik Nixdorf [2]

1 Department of Groundwater and Soil Science, Federal Institute for Geosciences and Natural Resources (BGR), 30655 Hannover, Germany
2 Department of Research and Development Centre for Post-Mining Areas, Federal Institute for Geosciences and Natural Resources (BGR), 03048 Cottbus, Germany; olaf.cortesarroyo@bgr.de (O.C.A.); silvio.janetz@bgr.de (S.J.); erik.nixdorf@bgr.de (E.N.)
* Correspondence: bernhard.siemon@bgr.de; Tel.: +49-511-6433488

Abstract: A large contiguous former opencast lignite-mining district is located in Lusatia, Germany, about 100 km south of Berlin. Since this region has been heavily polluted by acid mine drainage from opencast lignite mines, time- and cost-efficient investigation methods are required to obtain comprehensive information on aquifers. In summer 2021, BGR carried out a helicopter-borne electromagnetic (HEM) survey with its helicopter and a RESOLVE system covering an area of about 250 km^2. While current surface water levels are derived from in-flight altitude measurements, the spatial resistivity distribution shows indications for the water table in the entire survey area. Apparent resistivities agree very well with water resistivities measured in larger lakes, where the distance to the lakeshores outreaches the system's footprint. While groundwater sampling requires observation wells, we use HEM results to close the gaps between the sparsely distributed wells. The analysis of groundwater samples shows that groundwater EC correlates with dissolved iron (Fe^{2+}) and sulfate ($SO_4{}^{2-}$) content. HEM resistivities help to approximate groundwater EC, and further Fe^{2+} and $SO_4{}^{2-}$ contents at the observation wells, and to estimate these within the mining area where oxidative pyrite weathering processes dominate other contamination processes.

Keywords: airborne electromagnetics; helicopter-borne groundwater survey; resistivity; water EC; opencast lignite mining; acid mine drainage; pyrite oxidation; water table

1. Introduction

Hydrochemical contaminations such as acid mine drainage (AMD) from opencast mines as well as increased susceptibility to soil instabilities pose major challenges for the sustainable rehabilitation and reclamation of post-mining landscapes. In this paper, we focus on groundwater related problems using airborne electromagnetic (AEM) data for a large-scale survey within the Lusatian mining district in Germany. This area (Figure 1) forms the largest contiguous opencast lignite-mining district (814 km^2) in the EU and has been mined for around two centuries. Most of the former lignite mines have been closed during the last three decades and a post-mining recultivation is ongoing [1–4].

In Lusatia, loose Cenozoic sediments (sands, silts, clays, and glacial tills) form several tens of meters of heterogeneous lithological successions in extensive opencast dumps. However, the effects of newly formed depositional structures, spatial heterogeneities, the rising of the water table, and the pore water mineralization of the opencast dumps on groundwater flow and mass transport remain largely unknown in the post-mining areas of Lusatia. In particular, acid mine drainage is a severe problem in the lignite mining area for both post-mining lakes and groundwater [4]. The oxidation of sulfides such as pyrite (e.g., [5,6]), being present in the lignite coal itself as well as in the hosting material, is responsible for causing AMD [7]. Therefore, comprehensive, meaningful, time- and cost-efficient geoscientific methods for the investigation of former opencast lignite mines are required.

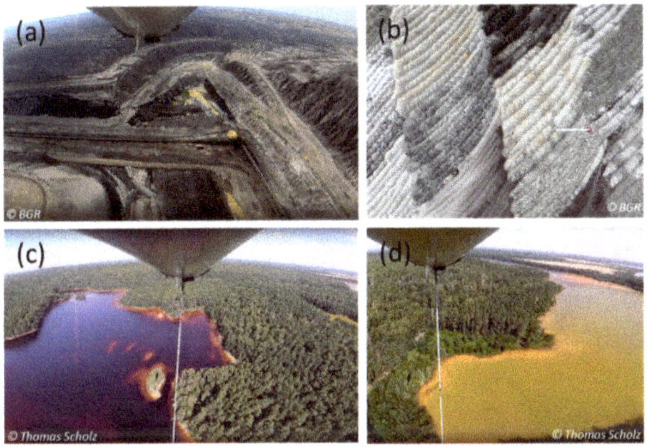

Figure 1. Examples of landscapes in the opencast lignite-mining district in Lusatia: (**a**) active opencast lignite mine, (**b**) nearby dump with towed helicopter-borne system, (**c**,**d**) post-mining lakes with contaminated water (photos taken in July 2021 by cameras attached to the bottom of the BGR helicopter).

The effect of former mining areas on the environment is usually investigated at selected measuring points using boreholes and in situ samples. This means that the most accurate data can be collected at certain points, but can only be conducted there, resulting in spatial data gaps that must be interpolated to obtain a continuous spatial representation [8]. Geophysics can help to close this gap. For example, Gondolfo [9] discusses the determination of water table depth and several papers describe the mapping of ACM distribution using geophysical methods [10–12]. Among the geophysical methods applicable on the ground [13], only those that do not require direct ground contact are suitable for airborne surveys [14].

Airborne electromagnetic surveys served and still serve to acquire large baseline datasets in groundwater and soil exploration (e.g., [15–17]). AEM results combined with data and models from external sources (e.g., boreholes, groundwater samples, ground-based geophysics, geological models) enable the derivation of spatial estimates, which then may serve as the baseline data for advanced (hydro-)geological modeling (e.g., [18–21]). While AEM has already been frequently used in groundwater surveys (e.g., [22,23]), only little is known about airborne geophysical investigations of (former) opencast lignite mining areas [24,25]. Most of the airborne (and spaceborne) surveys reported in the literature applied remote sensing technologies to investigate abandoned opencast mines and their impact on the environment (e.g., [7,26,27]).

Due to there only being local (in situ data) or non-existent (soil geophysical data) information on groundwater contamination caused by mining, the Federal Institute for Geosciences and Natural Resources (BGR) conducted an airborne geophysical survey in a former opencast lignite mining region near the towns of Cottbus and Senftenberg in Lusatia, Germany, in summer 2021 [28]. The main objective of BGR's airborne survey was to outline the benefits of an airborne geophysical investigation for the characterization of opencast dumps in terms of their (hydro-)geological setting. The investigation focuses on revealing information on aquifers and, in particular, on their depth and mineralization by AMD down to a depth of about 100 m below ground level.

2. Materials and Methods

2.1. Geology and Hydrogeology of the Survey Area

The survey area is located in the western part of the Lusatian lignite mining area between the Elster River lowland in the south and the terminal moraine and outwash plains of the Lower Lusatian border wall in the northeast (Figure 2).

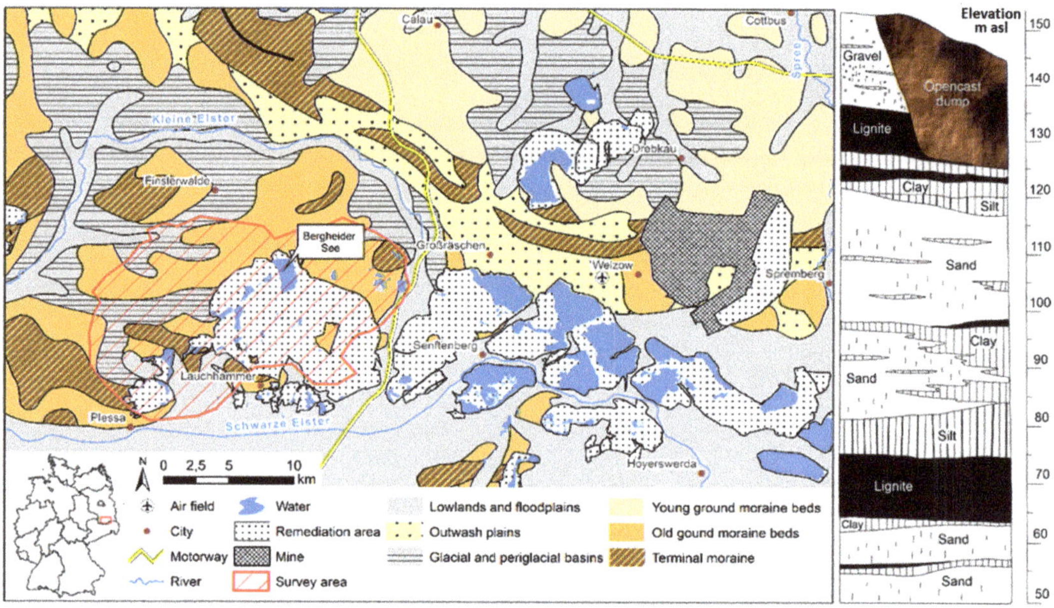

Figure 2. Geological overview map (**left**) and typical lithological units (**right**) of the survey area (after: [29–33]). The airborne survey area is shown in red.

During the Palaeogene and Neogene, Lusatia lay on the southern edge of the Palaeo-Northern Sea. As a result, the area was particularly exposed to the influences of rising and falling sea levels. The frequent spatial and temporal changes in the marine, brackish, and terrestrial depositional areas can be attributed to global and climatically induced sea level fluctuations [34]. This led to the formation of five lignite seam horizons in the area of the large coastal bogs [35]. Locally, the uplift and subsidence of the Mesozoic–Paleozoic basement occurred in the geological subsurface, which is associated with tectonic activity during the Neogene [36]. The Pleistocene deposits are up to 150 m thick and consist mainly of glacial, interglacial, fluvial, and lacustrine sediments (sand, silt, clay). The repeated advance of the inland ice from northern and northeastern Europe into the Lusatia region led to complicated bedding conditions and subglacial valleys that cut into the Cenozoic and Mesozoic layers [36].

As a result of the commercial open pit mining of the second Miocene lignite seam since the 1920s, the original geological layer sequence above the lignite seam was removed to depths of about 80 m above sea level (asl) and relocated on overburden dumps. The surveyed post-mining landscape is part of the LMBV (Lusatian and Central German Mining Administration Company) remediation area. The extensive remediation areas of lignite mining in Lusatia consist of former opencast dumps with anthropogenically altered deposition conditions and the adjacent post-mining lakes.

The shallow aquifers of the investigation area drain their water to the tributaries of the Schwarze Elster, which is a second-order stream of the Elbe Basin with a total length of 179 km. In the central region of the investigation area, the surface water system consists of 18 post-mining lakes, which are connected by 9 km of constructed water channels forming the so-called "Kleine Restlochkette" (small chain of post-mining lakes). The surface water quality in the central region is heavily impacted by the AMD leachate. The groundwater declines towards the west, except in the northeast of the survey area at higher elevations [37], where no lignite mining has taken place (Figure 2). There, higher lake water levels occur and perched aquifers are to be expected.

2.2. Airborne Geophysical Survey

In July 2021, BGR conducted a one-week airborne survey (1736 line km, nominal separation of the NW–SE flight lines: 250 m) over an area of 250 km^2 between Finsterwalde and Lauchhammer about 50 km southwest of the city of Cottbus in eastern Germany ([28], Figure 2). The airborne geophysical system used consists of a Sikorsky S-76B helicopter, three geophysical (electromagnetic, magnetic, and gamma-ray spectrometric), navigation, and positioning (incl. inertial measurement unit) systems, a data acquisition system, and a base station recording diurnal variations. The helicopter-borne electromagnetic (HEM) system (RESOLVE) is operated at six frequencies ranging from 0.4 kHz to 128 kHz with onboard calibration facilities. The transmitter and receiver coils are oriented in a horizontal–coplanar (5 $\times$ HCP) or vertical–coaxial (1 $\times$ VCX) position. The maximum investigation depth generally depends on the electrical conductivity of the surface water, the groundwater, and the sediments/rocks in the subsurface. Both the penetration depth [38] and the footprint [39] increase with decreasing system frequency and conductivity of the earth. On average, a depth of investigation of 50–150 m and a footprint of 100–300 m (belonging to the highest to lowest frequency of the RESOLVE system and a mean sensor altitude of about 60 m) can be expected [40].

2.3. Airborne Data

The general airborne data processing is described in detail in [28,40]. Here, we focus on position and electromagnetic data. All raw and processed datasets as well as various products (maps, vertical sections) are publicly available via BGR Geoportal. The flow chart in Figure 3 shows the various processing steps, their interaction, and the comparison with external reference data described below.

2.3.1. Position Data Processing

Position data processing includes checking and, if necessary, correcting the position and distance data measured (GPS data of the helicopter and the flying probe (bird) as well as the distance-to-ground measurements with a laser altimeter). For the post-processing of raw data delivered by the GNSS (Global Navigation Satellite System), we use the Precise Point Positioning (PPP) web service of the Geological Survey of Canada (Natural Resources Canada). This post-processing provides a significant improvement for the position data, especially for the critical vertical component. The bird attitude is corrected with the help of an inertial measurement unit (IMU) integrated into the GNSS. We derive the terrain elevation (Topo) from the difference between the GPS elevation and the laser altitude of the bird. As the tree canopy can affect the laser data, a digital elevation model (DEM) is used to check and correct the laser data [28].

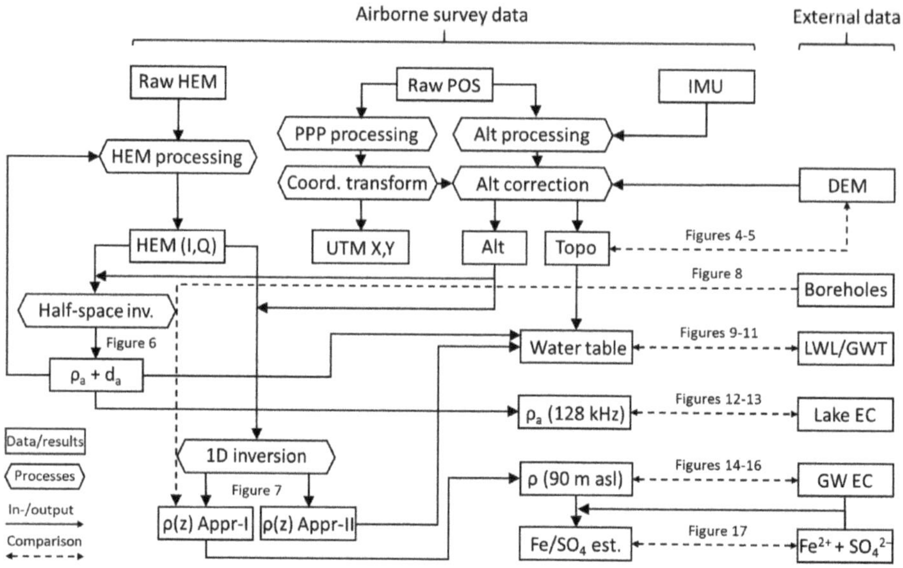

Figure 3. Flow chart showing processing steps for helicopter-borne electromagnetic (HEM), position (POS), and attitude (IMU) data. After the correction of the position data, coordinates (UTM) and sensor altitudes (Alt) together with the processed HEM data are used for inversion (homogeneous half-space and multi-layer 1D inversion). The product (terrain elevation (Topo), water table, apparent resistivity ρ_a at the highest HEM frequency, resistivity ρ at 90 m asl, and estimated Fe^{2+}/SO_4^{2-} content) are compared (see Figures) with the external reference data (borehole data, digital elevation model (DEM), lake water level (LWL), groundwater table (GWT), the electrical conductivity (EC) of lake and groundwater, and Fe^{2+}/SO_4^{2-} content in wells.

2.3.2. HEM Data Processing

The processing of the HEM data, in-phase (I) and quadrature (Q), aims to determine the electrical conductivities (or their reciprocal values, the electrical resistivities) of the earth's subsurface. The data processing also includes the identification and correction of man-made effects (cultural noise), which are not caused by the (hydro-)geological subsurface, e.g., lightning, interference with power lines or metallic installations (cables, fences, rails). System-related causes, such as incorrect calibration, thermal electronic drift, data noise, and motion-induced effects, can also affect the HEM data [41]. Weak local effects can usually be reduced by suitable filters (non-linear and low-pass filters), but larger effects have to be marked, and the corresponding data sections are removed. Smaller data gaps can often be filled by suitable interpolation. In order to reduce the system-related effects, however, plausibility checks, drift correction, and line leveling are necessary. Transformed HEM data derived from homogeneous half-space models [42] are used for this purpose. In contrast to the HEM data (I and Q), the half-space parameters, apparent resistivity (ρ_a), and apparent depth (d_a), are only slightly dependent on variations of the sensor altitude (above ground) and can therefore be used much better for data quality control and correction [22].

2.3.3. HEM Data Inversion

Homogeneous (single-frequency inversion) and layered (multi-frequency inversion) half-space models are used to derive spatial resistivity distributions. At each frequency of the HEM system, the homogeneous half-space inversion provides an apparent resistivity (ρ_a) for the half-space and an apparent distance from the bird to the top of the half-space. The difference in the apparent distance to the measured bird altitude, the apparent depth

(d_a), indicates an increase or decrease in resistivity with depth (within frequency-dependent depth ranges). This value is negative for a more conductive cover and positive for a more resistive cover with respect to the substratum. In the latter case, d_a tends to approximate the thickness of the resistive cover layer, provided the corresponding penetration depths are much larger. A measure for the mean penetration depth, the centroid depth (z^*), is derived from ρ_a and d_a [42].

The multi-layer (1D) inversion applies a Levenberg–Marquardt procedure in combination with general matrix inversion that is able to use both few and many (resistivity, thickness) model layers [43]. Layer interfaces and resistivities are best determined for a few-layer inversion (all model parameters are free) if the subsurface also consists of a few dominant layers with clear resistivity contrasts. Otherwise, the inversion result is often strongly dependent on the starting model used and layer resistivities and thicknesses may jump from model to model along a survey line [43,44]. A smooth inversion (only the resistivities and the cover layer thickness are free) can better represent gradual transitions in resistivities and complicated structures are easier to recognize.

The starting models for the 1D inversion are derived from apparent resistivity vs. centroid depth sounding curves. The layer thicknesses increase continuously in a range determined by the centroid depths of the highest (z_h) and lowest (z_l) frequency. The standard approach I subdivides the area above z_h using the apparent depth of the highest frequency, whereby the minimum layer thickness is 0.5 m. The resistivity of the cover layer is either set to a high value for a positive apparent depth or extrapolated from the sounding curve [40]. Alternatively, the boundaries z_h and z_l can be shifted upwards or downwards depending on local conditions, e.g., to reveal near-surface resistivity structures. In approach II, the depth range of the starting model layers with increasing thickness begins close to the surface, i.e., the shallow model region, including the cover layer contains more layers than before.

2.4. External Material

A multitude of datasets from public and closed sources has been used to evaluate and interpret the HEM results (Table 1). LMBV has provided most of the sample data. An excerpt used for the comparison with HEM results is listed in Appendix A.

Table 1. Characteristics of primary data sources used within this study. The column "Time" describes the timestamp of the acquired information.

Dataset	Reference	Type	Extent	Resolution	Time
DEM	BKG [45]	Grid	Germany	25 m	2012
Topography	BKG [30]	Polylines	Germany	-	
Water table/level	LMBV [46]	Points	Local	-	2021-07
Water quality	LMBV [46]	Points	Local	-	2021-07
Stream topography	LfU [31]	Polylines	Brandenburg	-	
Lake topography	LMBV [32]	Polygons	Brandenburg	-	
Mining area	LMBV [33]	Polylines	Brandenburg	-	
Borehole/well data	LMBV [46]	Points	Brandenburg	-	

2.5. Interpretation of HEM Data

2.5.1. Water Table Estimation

Assuming that the resistivity in the unsaturated (vadose) zone is usually higher compared to the saturated zone [47], a water table should be indicated by a resistivity decrease with depths. This expected resistivity decrease should be observable in both homogeneous half-space and multi-layer inversion models. However, it has to be noted that low resistivities can also occur above the water table, e.g., caused by increased soil moisture or clayey layers. We focus on the elevation of the water table, because this surface (in m asl) is presumably smoother than the corresponding depth of the water table (in m below ground level) and therefore less prone to artifacts caused by gridding or smoothing filters.

First, we try to estimate the water table directly from the resistivity models. A simple and fast way to do this is to use the apparent depth d_a, provided that the corresponding penetration depth lies significantly below the water table. For shallow water tables, d_a at the highest frequency is suitable. Deeper water tables, however, also require d_a values at the other frequencies (e.g., d_{am} = maximum of d_a). Small or negative apparent depths often indicate surface water. We therefore set negative d_a values to zero, i.e., the terrain elevation derived from the flight data (Figure 4) is used to approximate the water level there. Another option for the water table estimation is to use the cover layer thickness of the 1D inversion models, which also use d_a at the highest frequency as a starting parameter.

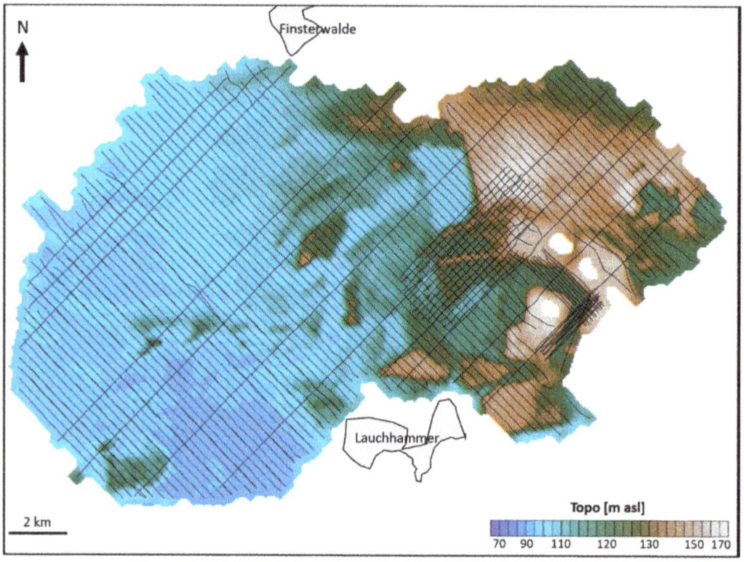

Figure 4. Terrain elevations derived from flight data (Topo), corrected with the help of a digital elevation model (DEM, [45]) in areas without recent elevation changes. The grids (50 m cell size) are based on flight line data.

For water tables occurring below the cover layer, approaches that are more sophisticated are necessary to reveal the appropriate resistivity increase, which should be strong in a few-layer inversion and weaker in a smooth inversion. However, the layer boundaries of an inversion with few layers often do not sufficiently coincide with resistivity interfaces, especially if the number of model layers differs from the real (geological/hydrogeological) layering. A smooth inversion, on the other hand, is less prone to this problem, but sharp resistivity boundaries tend to appear smooth. In order to reveal the freshwater–saltwater interface from smooth HEM inversion models in Zeeland (NL), Siemon et al. [40] developed a steepest gradient approach, which we apply here to derive the water table in the entire survey area. This approach consists of three steps: (1) A search of the steepest gradients in all 1D HEM inversion models. (2) An evaluation (selection, smoothing) of the corresponding depth (elevation) of these gradients within the entire survey area. (3) A comparison of the airborne results with water level/table data measured at lakes and in boreholes. Since this approach approximates a spline function to resistivity–depth values of the smooth inversion models below a (sometimes thick) cover layer, shallow boundaries affecting the subsurface resistivity (e.g., water table) may be hard to resolve using the standard starting model (approach I). Therefore, we also apply approach II, which provides several thin model layers close to the surface and thus extends the depth range of the spline function upwards.

A water table should be indicated by the steepest negative resistivity–depth gradient. The corresponding depths (SG) and elevations (SGE) are first selected with respect to an appropriate range of the water table within the current survey area (e.g., known from measurements or literature); in this instance, this is an SG less than 60 m depth and an SGE above 80 m asl. The uppermost SGE focuses on the shallowest water table. However, if there are several aquifers, the automatic evaluation is often unable to distinguish between the low-lying regional and higher-lying local (perched) aquifers. For this reason, the automatic evaluation is not only limited downwards (to 80 m asl) but also upwards (to 135 m asl). A coarse grid (CG, 1000 m cell size) is created to check the SGE values with respect to outliers. Only those values where the deviation of SGE and CG is less than 10 m are used. SGE values with higher deviations are erased and—if existent—replaced by a deeper SG (second steepest negative gradients) if the corresponding SGE lie within the 10 m distance to the CG. Finally, coarse gridding is applied again, followed by resampling along the flight lines or at any other location (e.g., water wells).

The SG approach is not applicable to lake areas because the water levels equal the lake surfaces. Therefore, we replace the SGE values beforehand by the terrain elevation (Topo) derived from the airborne data for the lake areas. The areas of the lakes are estimated from thresholds for the half-space parameters at the highest frequency (apparent resistivities $\rho_a < 30\ \Omega m$ and apparent depths $d_a \leq 0$ m), but only for elevations below the highest lake level (Topo < 130 m asl).

2.5.2. Water Quality Estimation

The groundwater in the mining area is heavily polluted by pyrite weathering. This should also be seen in the HEM results due to an increase in the electrical conductivity (EC). Therefore, it is first examined whether the EC values in the lakes correlate with near-surface resistivities derived from HEM. Apart from the transformation of EC values into resistivity values, no further conversions are required. However, the influence of the lake size and the location of the sampling points on the HEM results must be examined.

While surface water contamination can easily be sampled in situ at the surface, observation wells are required for groundwater sampling. The first task is to estimate the formation factor F (Equation (1)), which is carried out through using a simplified Archie Equation [48] to discover the ratio of bulk resistivity ρ_b to water resistivity ρ_w

$$F = \rho_b/\rho_w \approx \rho \cdot EC/10{,}000. \quad (1)$$

The water resistivity can be derived from EC values as before, but the bulk resistivity must be estimated using HEM resistivities ρ, which are influenced by both groundwater mineralization and lithology. Since the filter sections in the wells are generally located in sandy sediments, a formation factor typical for sand can be expected there. However, nearby layers of clay, silt or lignite can influence the HEM data. This effect must be checked. If the HEM resistivities correlate with the resistivities of the groundwater at the observation wells, this result could be transferred to the entire survey area, but only for the sand-dominated areas. The further estimation of certain groundwater mineralization is only possible if the EC values in the groundwater are characterized by a specific element or element pair, such as the salt content in seawater. Here, we focus on iron (Fe^{2+}) and sulfate ($SO_4{}^{2-}$) content.

3. Results

3.1. Terrain Elevation

The current terrain elevation (Topo) derived from flight data was checked and corrected in places using a DEM (DGM25, [45]), particularly outside the mining areas, where trees and other obstacles on the ground can affect the measured data. Inside, however, the DEM used was often unable to correctly map the actual elevation of the changing landscape (Figure 4). The differences between the resulting Topo and DEM values are shown in Figure 5. They are generally very small and only prominent in areas with recent surface elevation changes (e.g., lake water levels, dumps, pits, and slopes).

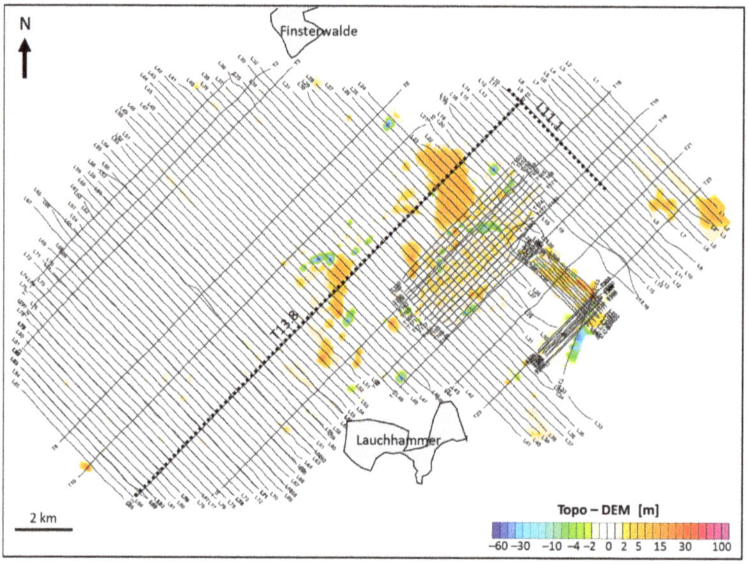

Figure 5. Differences between Topo and DEM. The grids (50 m cell size) are based on flight line data. Flight lines are drawn in black.

3.2. General HEM Results

The main aim of the electromagnetic survey is to obtain information about the regional groundwater distribution. The focus is set on the current water table and the mineralization of the groundwater as well as the identification of aquifers and aquitards. The HEM results are displayed as colored maps (50 m grid cell size) showing the apparent resistivities at each frequency and the resistivities derived from the 1D inversion models with 20 layers at selected elevations. The inversion models are also displayed as cross-sections consisting of colored resistivity columns stitched together along flight lines with respect to Topo.

Figure 6 provides a first overview of the conductivity distribution in the survey area showing maps of apparent resistivities (half-space resistivities) at all frequencies. As the corresponding centroid depths increase with decreasing frequency (f = 128–0.4 kHz) and increasing resistivity, the maps display the prominent apparent resistivity structures at increasing depths. Higher apparent resistivities occur in predominantly sandy areas with weakly mineralized groundwater (blue colors, $\rho_a \geq 60\ \Omega m$) and lower values are typical for clay sediments or higher mineralized groundwater (yellow to green colors $10\ \Omega m < \rho_a < 60\ \Omega m$) [15]. The lowest values (orange colors, $\rho_a \leq 10\ \Omega m$) are mostly observed in post-mining lakes with usually significantly increased mineralization.

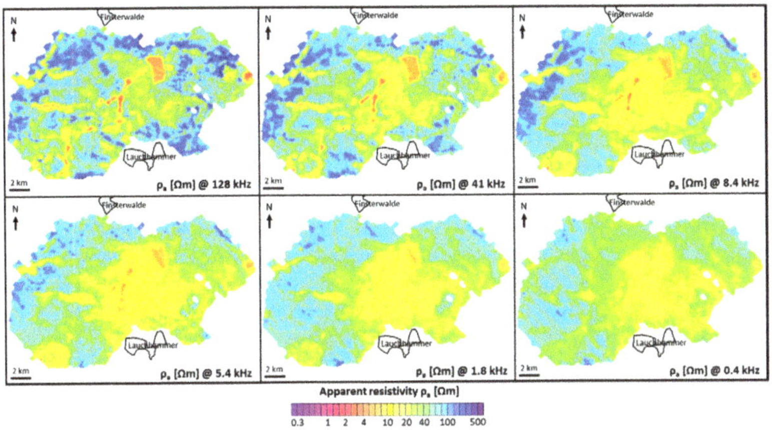

Figure 6. Apparent resistivities at each frequency of the HEM system. The corresponding mean depth values (±standard deviations) are approximately 4–11 m (f = 128 kHz), 6–15 m (f = 41 kHz), 14–28 m (f = 8.4 kHz), 16–32 m (f = 5.4 kHz), 28–53 m (f = 1.8 kHz), and 60–100 m (f = 0.4 kHz).

Generally, we perform both an inversion with few layers and a smooth inversion of the HEM data with starting models consisting of 6 and 20 layers, respectively. As an example, Figure 7 shows the 1D inversion results along an approximately 18 km long SW–NE tie line (T13.8, see Figure 5 for location) that crosses the largest post-mining lake in the survey area (Bergheider See) at 14–16 km on the profile. The color-coded models are shown in relation to the terrain elevation (Figure 4). Both inversion settings lead to similar results. The lowest resistivities occur at about 70–110 m asl in the south-western and central parts of the profile, i.e., in the abandoned mining area. The vertical shapes of the lakes, especially those of Bergheider See, are recognizable best in the smooth version.

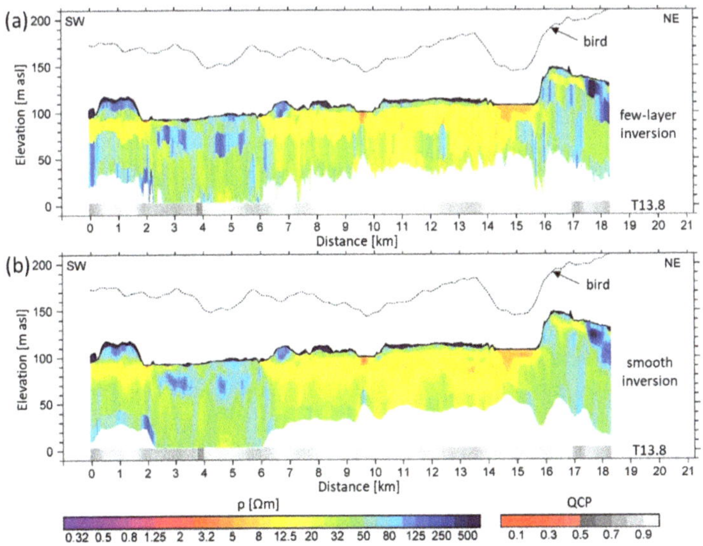

Figure 7. Resistivity cross-sections along tie line T13.8 (location see Figure 5) for (**a**) the inversion with few layers and (**b**) the smooth inversion (approach I). The path (elevation) of the HEM system (bird), the topographic elevation (black line above the colored models), and the quality control parameter (QCP, white: high quality, red: low quality) of the HEM data are also shown.

Outside the mining area, especially at higher altitudes (e.g., at 16–18 km on T13.8), we do not expect groundwater contamination from mining. Therefore, a section (about 4 km) of a flight line (L11.1) over a small hill is considered here (Figure 8) to show the influence of the lithology on the calculated resistivity. The drilled lithology derived from nearby boreholes (≤50 m distance) is drawn in the resistivity cross-section. In most cases, the conductive areas correlate with silt, clay, and lignite, and the resistive areas with sand and gravel. While layers with increased resistivities bound the shallow conductors, the deeper conductors do not appear to be bounded by depth due to the limited depth of the investigation of the HEM system. On the other hand, the drilled silt, clay, and lignite layers do not appear as continuous layers near the survey line. This indicates—at least in places—a complex lithology in the survey area.

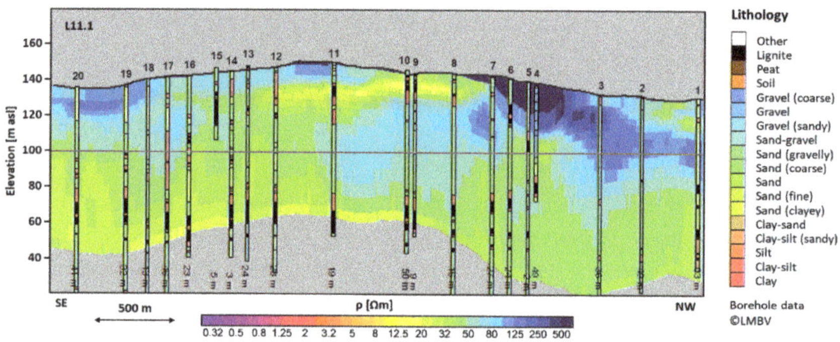

Figure 8. A resistivity cross-section of line L11.1 (location see Figure 5) for the smooth inversion with color-coded drilled lithology (data provided by LMBV [46]) on top (distance to L11.1 ≤ 50 m).

3.3. Special HEM Results

3.3.1. Water Table

The water table in sandy sediments normally varies little and follows the ground surface slightly. In hilly terrain, however, locally varying water tables can occur and a large-scale recording of the water table would require numerous sampling points. Since these sampling points are sparsely distributed within the survey area, particularly outside the mining area, we try to derive the water table from HEM.

First, we estimate the water table directly from the resistivity models. Table 2 shows the mean values (Δ) and the standard deviations (σ) of the elevation differences in the estimated water tables derived from apparent depths d_a (homogeneous half-space inversion) and cover layer thicknesses d_1 (multi-layer inversion) with respect to the measured water levels of the lakes and water tables in the wells (and total). Elevations are derived from d_a (at 128 kHz), d_{am} (the maximum d_a at all frequencies), and d_1 (cover layer thickness of 6 and 20 layer resistivity models). Most of the water tables recorded at the location of the boreholes belong to the main regional aquifer, but some originate from upper aquifers, especially in the eastern survey area where the terrain elevation is high (Figure 4) and confining layers (clay/silt/lignite) are present (Figure 8). Table 2 also contains the statistical parameters, excluding those 19 samples belonging to perched aquifers (_sel) or lake levels ≥125 m asl. All estimated water tables are too high on average and show a wide variation. The best values can be derived from the elevations of the maximum of the apparent depth values (Figure 9a, total Δd_{am}_sel = 1.8 ± 3.9 m).

Table 2. Mean values (Δ) and standard deviations (σ) of differences in estimated elevations of water tables derived from d_a (f = 128 kHz), d_{am} (maximum), $d_{1(6)}$ (6 layers), $d_{1(20)}$ (20 layers), and measured values [46] in lakes, wells, and both are listed. Statistical parameters excluding 19 samples belonging to perched aquifers or high (≥125 m asl) lake levels are indicated by "_sel". For locations, see Tables A1 and A2.

Elevation Parameter	Number	Lakes Δ [m]	σ [m]	Number	Wells Δ [m]	σ [m]	Number	Total Δ [m]	σ [m]
d_a	37	0.80	1.71	71	7.36	8.42	108	5.11	7.56
d_a_sel	34	0.67	1.54	55	5.66	6.44	89	3.75	5.68
d_{am}	37	0.32	1.97	71	3.87	6.69	108	2.65	5.78
d_{am}_sel	34	0.25	1.89	55	2.69	4.56	89	1.76	3.94
$d_{1(6)}$	37	1.26	2.03	71	7.78	8.06	108	5.54	7.32
$d_{1(6)}$_sel	34	1.10	1.81	55	6.14	6.45	89	4.22	5.73
$d_{1(20)}$	37	1.19	1.74	71	6.74	7.61	108	4.83	6.77
$d_{1(20)}$_sel	34	0.79	1.91	55	6.05	6.49	89	4.04	5.81

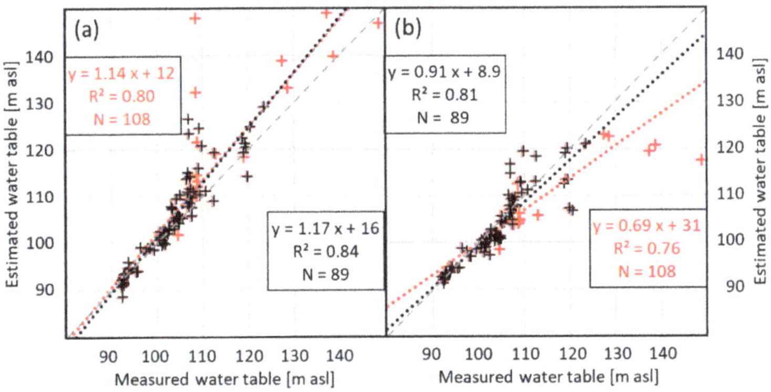

Figure 9. A comparison of the estimated and measured [46] water tables using all (red) or selected (black) sample data. (**a**) Derived from the homogeneous half-space inversion (maximum apparent depth d_{am}). (**b**) Derived from the modified smooth 1D inversion (SGE-II). The dotted lines indicate the best-fit lines; the dashed gray line indicates the 1:1 line.

The results listed in Table 2 show that the methods relating directly to the cover layer in the resistivity models do not provide satisfactory results, with the exception of d_{am}. We now apply the steepest gradient method to derive the water table from all of the 1D inversion models. Compared to the measured water table (Tables A1 and A2), approach I (with partially thick overburden) provides overestimated thicknesses of the vadose zone, i.e., the water tables are below the measured values (Table 3, total ΔSGE-I = −5.5 ± 5.2 m), even with a focus on the selected samples. The statistical results of approach II are shown in Figure 9b. The mean deviation is now significantly lower (ΔSGE-II = −1.8 ± 5.0 m), but the slope of the linear regression is not steep enough. The shallow slopes are caused by measured water tables belonging to higher lying aquifers and lakes, which both approaches try to ignore. Focusing on the previously selected samples further reduces the deviation (ΔSGE-II_sel = −0.7 ± 2.2 m) and increases both the slope and the R^2 value in Figure 9b (black).

Table 3. Mean values (Δ) and standard deviations (σ) of differences in estimated elevations of water tables (SGE-I, SGE-II) derived from steepest gradient analyses of 1D inversion models and measured values [46] in lakes, wells, and both are listed for two approaches: Appr. I (SGE-I) with thick cover layer, Appr. II (SGE-II) with thin cover layer. Differences between terrain (Topo) and surface elevations (SE) are also shown. Statistical parameters, excluding 19 samples belonging to perched aquifers or high (≥125 m asl) lake levels, are indicated by "_sel". For locations and water levels/tables, see Tables A1 and A2.

Elevation Parameter	Number	Lakes Δ [m]	σ [m]	Number	Wells Δ [m]	σ [m]	Number	Total Δ [m]	σ [m]
SGE-I	37	−5.71	7.22	71	−5.39	3.88	108	−5.50	5.24
SGE-I_sel	34	−4.20	3.59	55	−5.19	3.12	89	−4.81	3.32
SGE-II	37	−2.44	6.56	71	−1.39	3.92	108	−1.75	4.98
SGE-II_sel	34	−1.02	3.43	55	−0.47	3.21	89	−0.68	3.29
Topo-SE	37	1.11	1.82	71	−0.04	3.75	108	0.35	3.26
Topo-SE_sel	34	1.01	1.69	55	0.48	2.51	89	0.68	2.24

Table 3 also shows the differences between the terrain elevations derived from the airborne data (Figure 4) and the surface elevations of the boreholes. Note that some of the high water table deviations listed in Tables 2 and 3 could also be due to the differences between the terrain and surface elevations (total ΔTopo-SE_sel = 0.7 ± 2.2 m).

Figure 10 shows an example of a vertical section (T13.8), in which the estimated water table (purple line) derived from the HEM models is displayed in comparison with measured values (blue section in colored columns). While the calculated water table based on the cover layer depth of the standard smooth inversion ($d_{1(20)}$, bottom of dark blue cover layer) is generally too high, the values derived from the maximum apparent depth (d_{am}) are often close to the measured values, except for some resistive areas (Figure 10a). The best results are achieved with approach II (SGE-II, Figure 10b).

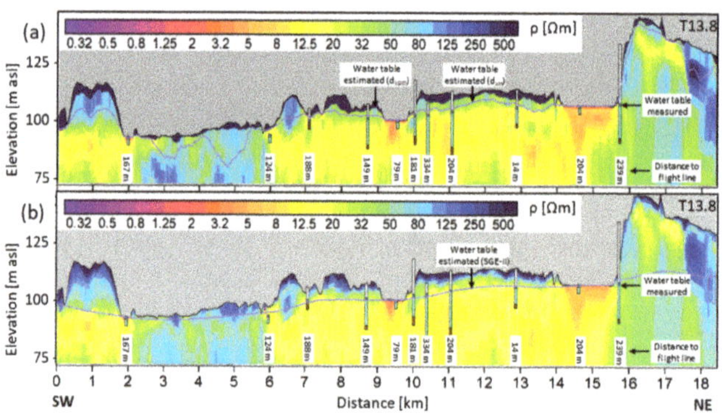

Figure 10. Resistivity cross-sections belonging to (**a**) approach I and (**b**) approach II along tie line T13.8 (for location see Figure 11) together with the estimated elevation of the water table (purple line) and samples (colored columns, after LMBV [46]) indicating the vadose zone (white), the water table (blue), and the saturated zone below. The maximum distance of the samples to the flight line is less than 350 m.

Figure 11 displays an elevation map of the estimated water table belonging to approach II (Figure 10b) together with the locations of the water samples (wells and lakes), aquifer code, and elevation of the measured water table/level. Black crosses indicate the deselected samples. Similar to [37], the water table mostly falls in a westerly direction, except in the northeast of the survey area. This map also demonstrates that there are only a few measuring points outside the mining area (in particular, there are almost no samples available in the north-west), while there are many inside.

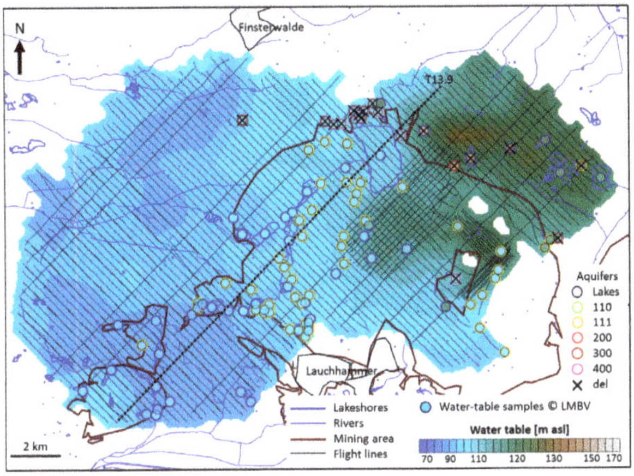

Figure 11. Elevations of the estimated water table derived from smooth HEM models using approach II (grid) in comparison with water tables provided by LMBV [46]. These values (face color of the dots) belong to different aquifers (edge color of the dots). Rivers [31], lakeshores [32], mining area [33], and flight lines are marked. The grid (50 m cell size) is based on flight line data.

3.3.2. Lake Water EC

Figure 1c,d shows two examples of post-mining lakes in the study area that appear to be contaminated due to their color and presumably also due to their water EC values. The question is now whether the HEM results can reflect the water EC values. These lakes can also be used to check the results of the HEM analysis in comparison with water samples.

In order to check the calibration of the HEM system at the highest frequency, we compare the apparent resistivities in Figure 12 with the EC values of lake water samples provided by LMBV [46] and check the apparent depth values, which should be close to zero (or negative for shallow lakes) in such a case.

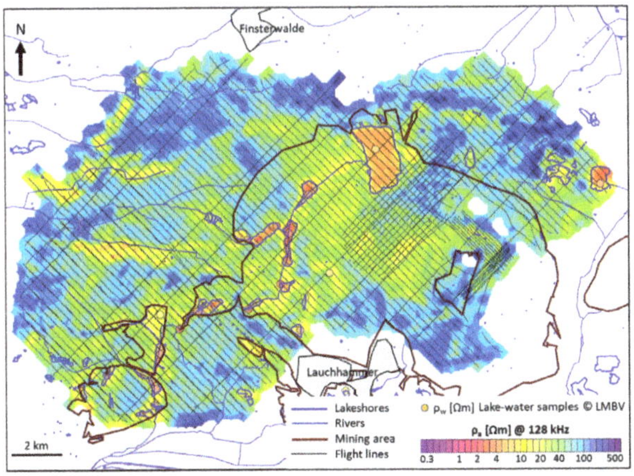

Figure 12. Apparent resistivities (ρ_a) at the highest frequency of f = 128 kHz (map) compared to measured electrical conductivities (EC) of water (provided by LMBV [46]) after conversion to water resistivities (ρ_w) at sample points in lakes (dots). Rivers [31], lakeshores [32], mining areas [33], and flight lines are marked.

The apparent resistivities at 128 kHz (ρ_a) vary from lake to lake. While most of the lakes within the mining area have lower ρ_a-values (3–5 Ωm, orange to red colors on the map), the ρ_a-values at most of the other lakes are quite normal (>10 Ωm, yellow to green colors)—except the lake at the eastern corner of the survey area. These apparent resistivities fit very well with the water resistivities (ρ_w) in the lakes (converted from measured water EC values: ρ_w [Ωm] = 10,000/EC [µS/cm], Table A3) as long as the widths of the lakes exceed 200 m so that the distance to the lakeshores outreaches the footprint of the HEM system at the highest frequency (Figure 13). The corresponding apparent depths (d_a) at the lakes are small, on average being $\Delta d_a = -0.1 \pm 1.1$ m, i.e., close to the expected value ($\leq$0 m).

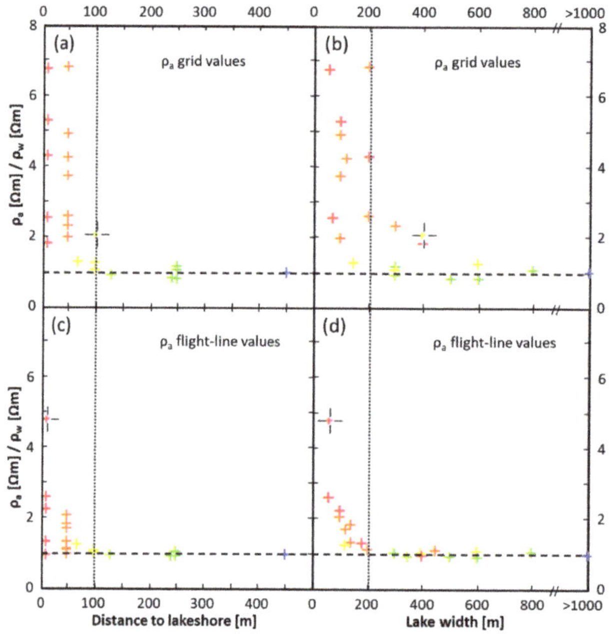

Figure 13. Ratios of apparent resistivities at 128 kHz and resistivities of lake water [46]. (**a**,**b**): values picked from grids, (**c**,**d**): values picked on closest flight lines. Colors refer to the distance to the lakeshore. Dashed lines indicate the 1:1 line; dotted lines indicate 100 m and 200 m distance to lakeshore and lake width, respectively.

3.3.3. Groundwater EC

As the observation wells are sparsely distributed in the survey area, the question arises whether HEM results can help to close the gaps. Figure 14 shows the resistivity distribution at 90 m asl, derived from 1D inversion models with 20 layers including a cover layer (cf. Figure 7b). The majority of the well screens (in most cases 2 m thick) in the observation wells are located in a sandy aquifer near this elevation. The resistivities ρ at 90 m asl are plotted against the resistivity ρ_w of the groundwater samples (Table A4) in Figure 15, where a linear relationship can be observed if the two extreme values are ignored. The slope, which should roughly correspond to the formation factor F, is below the expected value of F = 4 for sand.

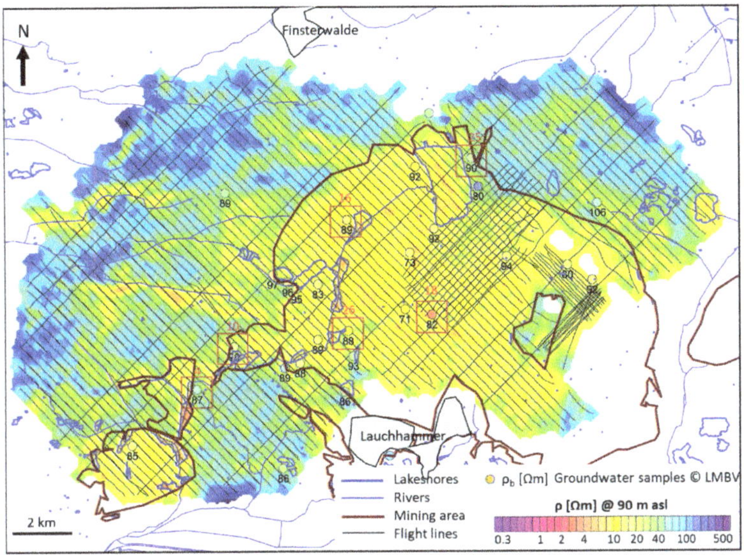

Figure 14. Resistivity ρ at 90 m asl (grid) vs. bulk resistivity ρ_b derived from groundwater samples (dots [46]). The mean well screen depth (m asl) is shown below the borehole dots. Red squares indicate the location of the observation wells (red numbers) shown in Figure 16. Rivers [31], lakeshores [32], mining areas [33], and flight lines are marked.

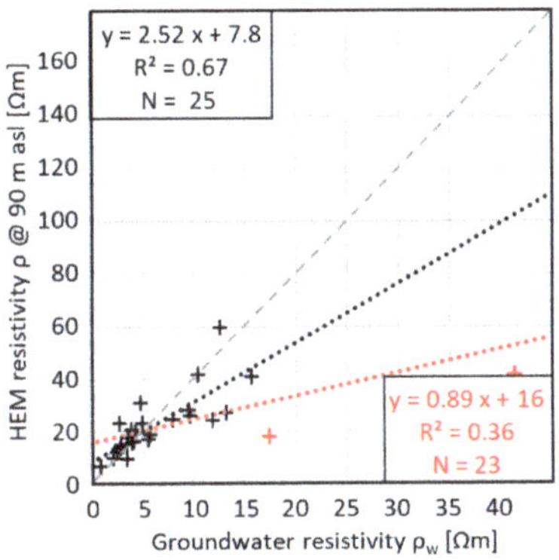

Figure 15. Resistivities ρ_b at 90 m asl vs. groundwater resistivities ρ_w [46]. Dotted lines indicate the best-fit lines for datasets with (red) and without (black) extreme values; the dashed line indicates the 4:1 line.

However, when looking directly at the ratios of ρ and ρ_w derived from Equation (1), we see a formation factor of F = 4.3 ± 1.9 (N = 25). Therefore, the use of F = 4 seems appropriate here and the colored dots in Figure 14, which represent the assumed bulk resistivity at the observation wells (ρ_b [Ωm] = 40,000/EC [μS/cm]), generally agree with the resistivities derived from HEM data. For example, Figure 16 shows six resistivity cross-sections together with borehole results (for the location, see Figure 14). Deviations occur where the well screens are deeper/shallower or the lithology (clay or silt layers near the well screen sections) affect the HEM results.

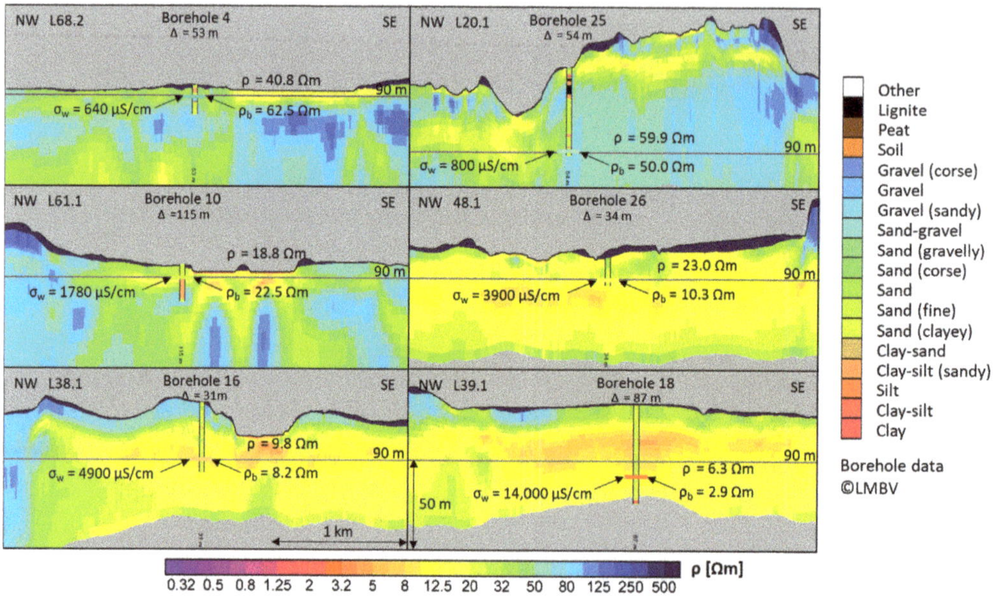

Figure 16. Resistivity cross-sections (exaggeration ≈ 13) near selected groundwater observation wells (LMBV [46]). The electrical conductivity of the sampled groundwater (EC) and the derived bulk resistivity (ρ_b) using a formation factor of F = 4 are indicated at the well screen position (arrowheads) on the borehole column (lithological profile). Above ρ_b, the corresponding HEM resistivity value (ρ) is shown. Δ indicates the distance of the well to the next flight line.

3.3.4. Groundwater Fe^{2+} and SO_4^{2-} Content

The analysis of 25 groundwater samples (Table A4) shows that groundwater EC [µS/cm] appears to correlate on the log–log scale with dissolved iron (Fe^{2+}) and sulfate (SO_4^{2-}) content [mg/L], which is displayed in Figure 17a. The ratios between the EC values calculated from (Equation (2))

$$EC = 131.67 \cdot (Fe^{2+})^{0.5131},\ EC = 10.229 \cdot (SO_4^{2+})^{0.7441} \tag{2}$$

and the EC values measured in situ are, on average, $EC(Fe^{2+})/EC(Fe^{2+})_{meas} = 1.05 \pm 0.29$ and $EC(SO_4^{2-})/EC(SO_4^{2-})_{meas} = 1.00 \pm 0.09$. Applying somewhat simplified relationships (Equation (3))

$$EC \approx 100 \cdot (2 \cdot Fe^{2+})^{1/2},\ EC \approx 10 \cdot (SO_4^{2-})^{3/4} \tag{3}$$

provides similar results: $EC(Fe^{2+})/EC(Fe^{2+})_{meas} = 1.05 \pm 0.30$ and $EC(SO_4^{2-})/EC(SO_4^{2-})_{meas} = 1.02 \pm 0.09$. Since the differences are small, we use Equation (3) for further analyses. In order to estimate the content of Fe^{2+} and SO_4^{2-} at the observation wells (N = 25), we rearrange Equation (3) and use the transformation EC [µS/cm] = F·10,000/ρ [mg/L]:

$$Fe^{2+}{}_{est} = 10{,}000/2 \cdot (F/\rho)^2,\ SO_4^{2-}{}_{est} = (1000 \cdot F/\rho)^{4/3}. \tag{4}$$

Figure 17b displays the results using a formation factor of F = 4 and a resistivity at 90 m asl (ρ). The corresponding best-fit lines for the estimated vs. measured values deviate from the 1:1 line and the ratios are quite high on average: $Fe^{2+}{}_{est}/Fe^{2+}{}_{meas} = 1.63 \pm 1.89$ and $SO_4^{2-}{}_{est}/SO_4^{2-}{}_{meas} = 1.35 \pm 1.32$. Using Equation (4) to estimate the distribution of Fe^{2+} and SO_4^{2-} content in groundwater at 90 m asl throughout the survey area results in maps similar to that shown in Figure 14 with $Fe^{2+}{}_{est}$ and $SO_4^{2-}{}_{est}$ (instead of ρ) color scales.

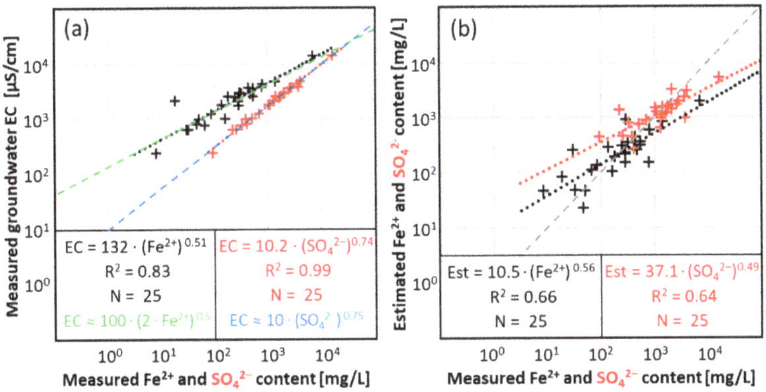

Figure 17. (**a**) Groundwater EC vs. Fe^{2+} (black) and $SO_4{}^{2-}$ (red) content with calculated (doted) and estimated (dashed) trend lines. (**b**) Estimated Fe^{2+} (black) and $SO_4{}^{2-}$ (red) content derived from resistivities ρ at 90 m asl vs. Fe^{2+} (black) and $SO_4{}^{2-}$ (red) content in groundwater [46]. Dotted lines indicate the best-fit lines; dashed line indicates the 1:1 line.

4. Discussion

4.1. Methodological Validation

4.1.1. Positioning and Elevations

In an airborne geophysical survey, the spatial allocation of the measured data to the position of the geophysical sensors is of great importance. The BGR helicopter system enables this by using GPS receivers in the helicopter and in the bird as well as using an altimeter and an inclinometer in the bird to record the GPS position and the distance to the ground, respectively. This is important for the Finsterwalde survey, as the terrain surface in the mining area as well as the water levels in the post-mining lakes have changed over time and the digital elevation models did not correctly reflect the surface elevation everywhere at the time of the survey. Therefore, we used a hybrid terrain elevation, which mainly refers to a DEM outside the mining area, especially where forests occur, which was derived from the flight data inside the mining area. All resistivity cross-sections and depth slices refer to this elevation. In turn, the corrected terrain elevation helped to check and correct the measured bird altitude. Without these corrections, 1D HEM inversion models and their elevation would be strongly affected by false bird altitudes.

4.1.2. Resistivity Distribution

The resistivity pattern in the survey area can basically be divided into two areas (Figure 14). Within the mining area, generally, lower resistivities dominate, which are apparently caused by the deposition and rearrangement of fine-grained sediments and more highly mineralized lake and groundwater. Outside the mining area, the resistivities are significantly higher, indicating rather sandy subsoil and only slightly mineralized groundwater. The areas with lower resistivities there, which mostly run from east to west and correlate with river courses, are most likely due to deposited clayey sediments; however, outflow from the mining area cannot be excluded.

4.1.3. Water Table Estimation

The knowledge of the water levels of the lakes is important for the estimation of the water table in the entire survey area, as both are connected to each other. The HEM inversion models best reveal the water table if it is not too deep and is covered by dry and sandy sediments. In this way, there is a clear contrast between higher (vadose zone) and lower resistivities (water-saturated zone). In the Finsterwalde survey area, however, lower resistivities can occur at shallow depths, for example, in river lowlands and bogs, but also in the mining areas. In addition, the resistivities at the surface of the lakes, where increased water temperatures are to be expected, tend to be lower than below. A first attempt using the thickness of the cover layer derived from homogeneous half-space and 1D inversion did not lead to satisfactory results (Table 2), since the apparent depth at the highest frequency, which also serves as the initial value for the cover layer thickness of the 1D inversion, did not prove to be good enough. As explained for a two-layer case [42], the apparent depth tends to the cover layer thickness only if (i) the resistivity of the upper layer is many times greater than that of the lower layer and (ii) the centroid depth is much greater than the apparent depth. These assumptions often do not hold true in the survey area and the water table was generally estimated at shallower depths. Therefore, the maximum apparent depth d_{am} (at all frequencies) was used instead, which often provided better results, but this was not everywhere due to deeper layers of fine-grained sediments (Figure 10a). The second attempt using standard (approach I) and modified (approach II) smooth inversion models in combination with a steepest gradient search led to significantly better results, but only for approach II (Table 3, Figure 10b). The water table estimated from airborne measurements can help to predict the possible flow directions of contaminated groundwater from the mining area. According to Figure 11, these are to be expected in a westerly direction, in line with the river courses.

4.1.4. Contamination of Lakes and Groundwater

The apparent resistivities at the highest frequency were used to estimate the contamination of the post-mining lakes (Figure 12). For this, an accurate calibration of the HEM system is essential, which was checked using the highest frequency data in comparison with water samples. The results showed that the calibration of the HEM system and the transformation based on a homogeneous half-space model were satisfactory—at least for the highest frequency HEM data relevant here. Therefore, the results presented in Figure 12 can be used to estimate the resistivity in the lakes as an approximation for the water quality, at least for the larger lakes. For smaller or shallow lakes, however, a 3D inversion is required to obtain accurate water quality parameters.

The comparison of the estimated groundwater resistivities derived from the HEM results with borehole data (Figure 16) shows that there is a fundamental uncertainty: The HEM resistivities depend not only on the mineralization of the groundwater, but also on the lithology, especially if clayey sediments occur in the vicinity of the well screens.

The uncertainty described above must of course also be taken into account in the further interpretation of the HEM results with regard to the estimation of Fe^{2+} and $SO_4{}^{2-}$ content in groundwater. We assume that a clear correlation between EC and dissolved Fe^{2+} or $SO_4{}^{2-}$ in groundwater can only be observed in post-mining landscapes such as Lusatia, where pyrite weathering dominates other contamination processes. However, these approximations cannot be applied to the lakes, as the lake waters are affected by inflowing water from different external sources, and no clear correlations between Fe^{2+} or $SO_4{}^{2-}$ content and the EC values were found. Outside the mining area, where the contamination of groundwater by Fe^{2+} or $SO_4{}^{2-}$ is generally lower, the significance of other, particularly non-mining, contaminants and the effect of lithology on resistivity values must also be considered.

4.2. Comparison with Ground Data

Despite careful data processing, there are still some uncertainties in determining the location and estimating groundwater-related parameters. The greatest differences in elevation to the DEM (up to several meters, Figure 5) occur in the post-mining lakes, as these have been flooded for several years. In other places, spoil heaps have been removed and leveled. While the terrain elevations derived from the flight data appear more reliable in these cases, the interpolation between the flight lines (250 m line spacing, 50 m grid cell size) causes uncertainties there that cannot be further reduced. This is also true for the resistivity grids, but due to the footprint of the HEM system, the lateral resolution is limited and the survey parameters are appropriate. This is acceptable for large-scale groundwater investigations.

The inversion of AEM data always provides resistivity models, which can be ambiguous and uncertain. However, the biggest source of error is the quality of the processed data [49,50]. Therefore, we have focused on reducing the data errors and using smoothed models in which data errors have less influence. Of course, this also reduces the vertical resolution, which affects the determination of interfaces (e.g., water table) and their comparison with borehole data. Thus, the HEM results shown here (smoothed spatial model data) are compared with reference data (point data), which have a much higher accuracy locally, but are only available at a few locations. Nevertheless, quite high correlations have been achieved, as the groundwater parameters considered (water table, EC values, and Fe^{2+}/SO_4^{2-} content) are also rather smoothly distributed.

Despite the high correlations already observed, it is important to take a critical look at the reference data used for comparison. This is illustrated, for example, by the assessment of the derived water table. The mean deviations from the water table belonging to the regional aquifer (Table 2, Δd_{am}_sel = 1.8 ± 3.9 m, Table 3, ΔSGE-II_sel = −0.7 ± 3.3 m) are of the same order as the differences between calculated and measured surface elevations at the selected sites (Table 3, ΔTopo-SE_sel = 0.7 ± 2.2 m). A further improvement in the statistic values (e.g., for ΔSGE-II_sel = −0.4 ± 2.6 m) would be possible with the exclusion of some further samples. These are five sample points at the edge of the survey area in the southeast, where the grid shown in Figure 11 was extrapolated, and two sample points at small lakes, where the SGE results were not automatically replaced due to the terrain elevation.

A further uncertainty can occur if the distance between the groundwater sample locations and the flight lines or the depth sections under consideration is too large. This is less critical if the resistivities change only slightly within these distances. This is usually the case for the horizontal distances, as the locations of the reference data are mostly within the HEM footprint, which is ensured by the half profile distance (approx. 125 m). For the vertical distances, it can be more critical, e.g., if a lithology change takes place near the filter section, which influences the HEM results but not the in situ data. We have checked this for the estimation of the formation factor (Section 3.3.3). Since the well screens are not always located close to the considered elevation of 90 m asl (Figure 14), a change in lithology cannot be excluded. Using only the water samples for well screen depths of 90 ± 10 m asl, the resulting formation factor is only slightly changed (F = 4.3 ± 1.7, N = 20) compared to the previous result (F = 4.3 ± 1.9, N = 25). This shows that it is not the vertical distance, but rather an inhomogeneous lithology that leads to the observed scatter.

In general, only one chemical indicator can be derived from one HEM parameter (here, this is EC derived from resistivity (at 90 m asl) assuming F = 4 and neglecting lithological differences). Since iron and sulfate are often used to characterize water quality in post-mining landscapes and show a high correlation with EC values in groundwater (Figure 17), we use both parameters. The correlation between Fe^{2+} and SO_4^{2-} is also high (R^2 = 0.85). All other chemical indicators show significantly lower correlations. The use of the approximate values in Equations (3) and (4) is appropriate, as the differences to the use of the calculated regression coefficients are small. The largest uncertainty arises from the smoothed resistivity distribution (ρ at 90 m asl) and the formation factor F. We have assumed that sandy sediments dominate in the areas of the well screens and used F = 4. Although this is mostly true, clayey sediments exist above and below the well screens at some borehole locations, which affect the HEM results but not the in situ measurements. The application of a lower but still constant formation factor of F = 3 (typical for clayey sand) leads to a downshift of the estimated values, as shown in Figure 17b, and significantly lower ratios, which are, on average, as follows: $Fe^{2+}{}_{est}/Fe^{2+}{}_{meas} = 0.91 \pm 1.06$ (instead of 1.63 ± 1.89) and $SO_4{}^{2-}{}_{est}/SO_4{}^{2-}{}_{meas} = 0.92 \pm 0.90$ (instead of 1.35 ± 1.32). The standard deviation of the ratios is still high and the slopes of the best-fit lines remain unchanged. Therefore, we also checked the residuals. Two extreme values occur for wells located at the edge of the mining area or close to the Bergheider See (Well No 20 and 23 in Table A4). Very low Fe^{2+} and $SO_4{}^2$ values were measured there. Repeating the statistical analyses without these two wells results in $Fe^{2+}{}_{est}/Fe^{2+}{}_{meas} = 1.15 \pm 0.96$ and $SO_4{}^{2-}{}_{est}/SO_4{}^{2-}{}_{meas} = 0.99 \pm 0.49$ and the R^2 values are slightly higher (R^2 (Fe^{2+}) = 0.72 and R^2 ($SO_4{}^{2-}$) = 0.73). We have to assume that neither changing the constant formation factor nor omitting some local reference data is sufficient to achieve high correlations between estimated (from HEM) and measured well data, as there are principally differences between large-scale HEM resistivities and local in situ measurements. Particularly, in the case of a smooth HEM inversion, it should be noted that thin (lithological or hydrogeological) layers with strongly varying resistivities appear flattened. Consequently, a comparison with discrete borehole data leads to a flattening of the fit line of a linear regression.

5. Conclusions

The focus of the study described here was the investigation of a post-mining landscape in the Lusatian mining district, which used helicopter-borne electromagnetics to obtain information about the aquifers, their depth, and pore water mineralization. With the airborne survey, the post-mining landscape could be mapped over a large area and in a relatively short time without having to enter the partially inaccessible terrain. This is a clear advantage of airborne over ground-based geophysical surveys, which often have to be limited to smaller survey areas and specific questions.

The first airborne electromagnetic survey carried out in this former opencast lignite mining region provided the spatial resistivity distribution with the following benefits:

- The regional water table was derived from the HEM resistivity models, which could be verified by measurements in lakes and boreholes.
- Significant variations in the water conductivity of the post-mining lakes were mapped using the apparent resistivities at the highest system frequency. For the larger lakes, the results are in good agreement with the conductivity measurements in the lakes. For smaller lakes, however, more detailed measurements (reduced flight line separation) and complex evaluations (modeling) are required.
- Water analyses in the study area showed that the electrical conductivity in the groundwater correlates with the content of dissolved Fe^{2+} and $SO_4{}^{2-}$. This content could be estimated from airborne resistivity data at 90 m asl within the entire area of the former mining site. The results using a formation factor of F = 4 can be improved with the help of—if available—more borehole data (lithology) and improved techniques (e.g., machine learning), in order to better take into account the contribution of lithology to electrical conductivity and the distribution of the formation factor. However, here we have focused on using the borehole data only to verify the airborne results.
- The airborne results can also be used to close data gaps between boreholes, where information on the (hydro-)geological subsurface is often limited.

- The results of non-invasive airborne surveys are well suited for the (hydro-)geological description of post-mining landscapes. Once the airborne geophysical results have been evaluated, a uniform and extensive dataset is available that can be used to obtain valuable information on the (hydro-)geological subsurface conditions.
- The data resulting from the geophysical investigations can also be used as baseline data for further evaluation, interpretation, and modeling.

Supplementary Materials: The following supporting information can be downloaded at https://www.mdpi.com/article/10.3390/w17071000/s1: Lithology data S1: Lithology.dat.

Author Contributions: Conceptualization, B.S.; methodology, B.S.; software, B.S.; validation, B.S., O.C.A., S.J. and E.N.; formal analysis, B.S.; investigation, B.S.; resources, B.S., O.C.A., S.J. and E.N.; data curation, B.S., O.C.A., S.J. and E.N.; writing—original draft preparation, B.S., S.J. and E.N.; writing—review and editing, B.S., O.C.A., S.J. and E.N.; visualization, B.S. and S.J.; supervision, B.S.; project administration, B.S. and S.J.; funding acquisition, B.S. and S.J. All authors have read and agreed to the published version of the manuscript.

Funding: This research received no external funding.

Data Availability Statement: The data are available in a publicly accessible repository: Raw Data on Helicopter-borne Electromagnetics (HEM) Area 191 Finsterwalde comprises the following zip file: 191FinsterwaldeHEM-RD.zip, accessible from following the link: https://doi.org/10.25928/zb8x-ke65. Helicopter-borne Electromagnetics (HEM) Area 191 Finsterwalde comprises the following zip file: 191FinsterwaldeHEM.zip, accessible from following the link: https://doi.org/10.25928/t7pz-1369. The following dataset is available on request from the authors: Inversion Models derived from Helicopter-borne Electromagnetics (HEM) Area 191 Finsterwalde. In terms of the third party data, all in situ measurements have been provided by LMBV [46]. LMBV has agreed to the publication of the data presented in the figures, in Appendix A, and provided in Supplementary Materials (27 January 2025). Information on the terms of use for the following download products: [31]: "dl-de/by-2-0" referring to the licence text available at www.govdata.de/dl-de/by-2-0); [32,33]: Adapted with permission from LMBV (27 January 2025).

Acknowledgments: We would like to thank LMBV for providing geological and hydrogeological data for comparisons with the results of the airborne geophysical survey. We also thank the academic editor and two anonymous reviewers for their helpful suggestions to improve the paper.

Conflicts of Interest: The authors declare no conflicts of interest.

Abbreviations

The following abbreviations are used in this manuscript:

AEM	Airborne Electromagnetic;
AMD	Acid Mine Drainage;
BGR	Bundesanstalt für Geowissenschaften und Rohstoffe;
BKG	Bundesamt für Kartographie und Geodäsie;
DEM	Digital Elevation Model;
EC	Electrical Conductivity;
GNSS	Global Navigation Satellite System;
GPS	Global Positioning System;
HEM	Helicopter-borne Electromagnetic;
IMU	Inertial Measurement Unit;
I, Q	In-phase, Quadrature of HEM data;
LBGR	Landesamt für Bergbau, Geologie und Rohstoffe Brandenburg;

LfU	Landesamt für Umwelt Brandenburg;
LMBV	Lausitzer und Mitteldeutsche Bergbau-Verwaltungsgesellschaft mbH;
PPP	Precise Point Positioning;
SG(E)	Depth (Elevation) of Steepest Gradient;
Topo	Terrain elevation.

Appendix A

Appendix A.1

Table A1. Measured lake water levels (LMBV [46]) in comparison with estimated water tables derived from airborne data (approaches I + II, deselected values in italics).

Lake No	Northing [UTM33]	Easting [UTM33]	Water Level [m asl]	Est. Water Table Appr. I [m asl]	Est. Water Table Appr. II [m asl]
1	416080	5714196	107.58	107.27	107.60
2	424875	5713288	119.02	113.61	119.15
3	409268	5708310	92.94	90.76	93.39
4	407034	5705210	92.52	88.76	92.41
5	408457	5708023	92.61	88.63	93.36
6	411421	5711155	101.29	97.97	98.89
7	410508	5711137	101.56	94.59	99.07
8	413098	5712787	102.66	102.72	102.81
9	423244	5713485	119.34	119.53	120.15
10	408881	5708436	92.67	91.25	93.53
11	412380	5710877	100.56	100.61	100.54
12	412320	5709099	99.33	93.81	98.66
13	411036	5708434	97.44	93.48	98.10
14	407034	5707796	92.52	91.83	92.72
15	405223	5705871	94.20	91.66	94.88
16	404084	5705046	96.32	93.43	98.76
17	419691	5714071	138.94	*113.21*	*120.83*
18	418632	5708286	119.80	107.02	107.28
19	409946	5708173	93.70	90.80	94.88
20	407158	5706486	92.52	89.54	92.09
21	412738	5707365	104.84	95.52	100.85
22	407088	5704322	92.21	86.46	91.98
23	405510	5707631	92.52	90.23	92.80
24	422670	5714770	123.43	122.12	121.07
25	410822	5704602	92.62	89.73	93.22
26	410414	5704994	92.65	87.82	92.59
27	413004	5708472	103.81	93.12	101.80
28	406637	5704145	91.99	88.82	92.70
29	421273	5714449	128.81	*123.17*	*122.73*
30	415396	5711040	109.43	107.57	110.52
31	417806	5715149	148.84	*111.91*	*117.54*
32	416107	5716223	120.51	104.89	106.81
33	410156	5711775	102.43	94.71	97.34
34	414676	5713710	105.02	101.66	105.59
35	414254	5709584	105.59	101.00	104.21
36	417106	5710541	110.74	107.71	111.20
37	415594	5711462	112.43	107.27	113.10

Appendix A.2

Table A2. Water tables measured in boreholes (LMBV [46]) in comparison with estimated water tables derived from airborne data (approaches I + II, deselected values in italics).

Well No	Northing [UTM33]	Easting [UTM33]	Water Table [m asl]	Est. Water Table Appr. I [m asl]	Est. Water Table Appr. II [m asl]
1	407023	5704111	92.22	86.50	91.98
2	419641	5708364	106.83	100.66	106.21
3	424099	5713756	127.58	*120.44*	*123.11*
4	414530	5715447	107.94	*104.46*	*104.06*
5	410553	5715591	104.76	*94.80*	*98.46*
6	415273	5713126	107.62	102.52	108.03
7	415112	5712697	107.62	102.73	108.95
8	415018	5712387	107.94	103.53	109.61
9	414585	5711045	107.20	106.06	109.33
10	419025	5713775	137.56	*112.89*	*119.45*
11	411577	5708382	101.33	94.84	99.76
12	414748	5714660	107.32	103.37	104.80
13	413295	5712061	105.08	102.99	104.12
14	412903	5711042	103.17	102.08	101.47
15	419053	5709379	108.73	*109.60*	*112.57*
16	419143	5711416	112.57	111.13	118.39
17	414495	5710654	106.78	105.21	107.88
18	420117	5708736	106.53	99.50	110.15
19	413001	5707117	104.56	95.40	100.84
20	411980	5710188	101.67	99.24	100.64
21	415455	5711643	109.08	106.84	112.89
22	421008	5706491	101.12	97.71	97.92
23	412607	5708680	101.82	93.18	101.16
24	412471	5709368	101.19	95.25	100.80
25	420643	5709384	106.99	103.85	114.87
26	420764	5709708	106.97	106.10	116.85
27	420878	5710433	109.81	111.34	119.50
28	413024	5707640	104.00	95.28	101.24
29	413166	5707696	103.65	95.37	101.29
30	413156	5708749	103.78	92.45	102.02
31	412858	5708433	102.58	93.32	101.64
32	416956	5712960	109.29	110.77	113.02
33	420006	5707440	104.28	96.67	100.63
34	414595	5709969	106.87	104.34	106.98
35	413139	5707506	104.24	95.52	101.11
36	411046	5708197	100.46	93.50	98.87
37	412786	5711663	103.37	102.20	102.45
38	413659	5713096	104.85	103.71	104.49
39	412188	5709852	101.28	98.54	100.87
40	422617	5710566	118.71	108.86	112.60
41	415409	5715924	109.15	*104.58*	*104.84*
42	410853	5708987	100.60	93.58	99.67
43	411371	5708068	101.25	94.61	99.53
44	411588	5708093	101.64	95.32	99.98
45	411386	5708190	101.19	94.55	99.58
46	413141	5707709	103.70	95.31	101.31
47	413204	5707818	104.73	95.28	101.43
48	413145	5707679	103.68	95.36	101.27
49	413132	5707357	104.62	95.51	101.01
50	412394	5707406	103.48	96.22	100.37
51	423133	5710928	119.15	*107.73*	*113.56*

Table A2. *Cont.*

Well No	Northing [UTM33]	Easting [UTM33]	Water Table [m asl]	Est. Water Table Appr. I [m asl]	Est. Water Table Appr. II [m asl]
52	422836	5710873	119.51	108.67	113.48
53	407584	5704631	92.36	85.82	91.37
54	410542	5708154	98.27	92.17	97.05
55	413862	5715540	107.88	*104.37*	*104.01*
56	415780	5716217	113.08	*104.50*	*105.76*
57	415998	5715885	109.08	*105.48*	*106.19*
58	413181	5714410	106.65	104.08	105.41
59	416897	5714958	108.69	*108.69*	*110.60*
60	414212	5715417	107.90	*104.57*	*103.80*
61	413837	5714737	107.36	103.85	103.90
62	409822	5709248	97.95	94.19	97.00
63	415025	5715942	108.89	*104.17*	*104.00*
64	415244	5715783	109.10	*104.53*	*104.63*
65	415224	5715781	109.06	*104.51*	*104.60*
66	415310	5715647	108.90	*104.74*	*104.88*
67	406466	5706827	93.75	90.24	93.64
68	414708	5713135	107.13	101.88	107.04
69	408711	5708486	95.73	91.70	94.67
70	408724	5708346	95.34	90.79	94.59
71	408688	5708253	93.63	90.20	94.02

Appendix A.3

Table A3. Water EC measured in lakes (LMBV [46]) converted to resistivity ρ_w in comparison with apparent resistivities ρ_a (f = 128 kHz) derived from airborne data (picked from grid or next flight line). Additional parameters: Approximate distance of the sampling location to the lakeshore and maximum width of the lakes.

Lake No	Northing [UTM33]	Easting [UTM33]	EC [µS/cm]	ρ_w [Ωm]	ρ_a (Grid) [Ωm]	ρ_a (Line) [Ωm]	Distance [m]	Width [m]
1	415912	5714464	2010	4.98	5.01	4.90	450	1200
2	413203	5712690	3110	3.22	4.11	3.50	100	600
3	412405	5711200	3240	3.09	2.59	2.90	250	500
4	411918	5711480	3240	3.09	6.41	3.20	100	400
5	410759	5710952	2730	3.66	24.99	4.20	50	200
6	412058	5709196	3330	3.00	7.82	3.00	50	400
7	414015	5709624	1350	7.41	27.71	13.60	50	140
8	410774	5708228	3370	2.97	14.63	4.00	50	140
9	413041	5708671	945	10.58	21.13	21.90	50	100
10	407217	5708036	788	12.69	13.86	13.50	250	800
11	405861	5707732	671	14.90	19.43	18.90	70	120
12	408857	5708066	2600	3.85	16.57	5.10	10	180
13	408717	5708140	3120	3.21	16.95	7.20	10	100
14	408007	5707465	3200	3.12	21.12	15.00	10	60
15	407218	5706502	2980	3.36	3.13	3.20	130	350
16	407043	5705132	2530	3.95	4.73	4.10	250	300
17	405060	5705669	2730	3.66	15.60	6.30	50	120
18	406960	5704493	2370	4.22	10.75	11.00	10	60
19	419735	5714201	379	26.39	29.07	27.00	100	400
20	424912	5713076	2070	4.83	4.14	4.50	240	600
21	423374	5713202	908	11.01	25.60	12.40	50	450
22	412847	5707389	2380	4.20	7.70	4.10	10	400

Appendix A.4

Table A4. Groundwater EC (converted to resistivity ρ_W) and Fe^{2+} and SO_4^{2-} content in observation wells (LMBV [46]) in comparison with resistivities ρ derived from airborne data (picked from resistivity grid at 90 m asl) and estimated Fe^{2+} and SO_4^{2-} content (deselected values in italics).

Well No	Northing [UTM33]	Easting [UTM33]	EC [µS/cm]	ρ_w [Ωm]	ρ (90 m asl) [Ωm]	Fe^{2+} [mg/L]	SO_4^{2-} [mg/L]	Est. Fe^{2+} [mg/L]	Est. SO_4^{2-} [mg/L]
1	412857	5708433	1029	9.43	27.82	162	511	103	753
2	420131	5711427	1853	5.46	16.74	138	1030	285	1483
3	420965	5710926	3000	3.37	15.05	318	2000	353	1709
4	421150	5713476	764	13.16	26.89	67	329	111	788
5	407545	5707330	638	15.62	40.77	32	257	48	453
6	408541	5713823	993	10.31	41.58	51	381	46	441
7	411677	5710796	2920	3.44	17.56	470	2040	259	1391
8	414595	5709969	3650	2.78	14.87	535	2550	362	1736
9	411046	5708197	2400	4.20	19.21	282	1610	217	1234
10	408817	5708747	1790	5.62	18.80	287	1100	226	1270
11	418061	5711706	2620	3.77	20.51	181	1380	190	1131
12	410132	5711115	2120	4.88	22.97	285	1330	152	972
13	410638	5710867	2200	4.69	30.97	19	1280	83	653
14	410935	5710643	1260	7.87	24.47	87	690	134	894
15	412659	5712930	4810	2.04	9.75	1390	4050	842	3048
16	414757	5711834	3020	3.33	9.33	296	2040	919	3233
17	415511	5709795	14150	0.71	6.34	7160	16600	1990	5411
18	405339	5705494	2540	4.00	16.11	245	1410	308	1561
19	411662	5708975	2600	3.85	15.96	534	1760	314	1580
20	410526	5708046	650	17.54	*17.79*	30	221	253	1367
21	414960	5714673	3740	2.70	13.51	435	2790	438	1973
22	415597	5712629	4360	2.33	11.69	803	3590	585	2393
23	417129	5714026	247	41.49	*41.57*	8	93	46	441
24	416896	5714959	802	12.50	59.92	47	405	22	271
25	412690	5709263	3870	2.56	23.05	770	3730	151	968

References

1. von Bismarck, F. Land in motion–opencast restoration and recultivation in Lusatia. In *Post-Mining Landscape, Conference Documentation*; Internationale Bauausstellung (IBA) Fürst-Pückler-Land, Ed.; Jovis: Berlin, Germany, 2010; pp. 37–41. ISBN 978-3-86859-043-2.
2. Krümmelbein, J.; Bens, O.; Raab, T.; Anne Naeth, M. A history of lignite coal mining and reclamation practices in Lusatia, eastern Germany. *Can. J. Soil Sci.* **2012**, *92*, 53–66. [CrossRef]
3. Drebenstedt, C.; Kuyumcu, M. (Eds.) *Braunkohlesanierung: Grundlagen, Geotechnik, Wasserwirtschaft, Brachflächen, Rekultivierung, Vermarktung*; Springer: Berlin/Heidelberg, Germany, 2014. [CrossRef]
4. Benthaus, F.-C.; Totsche, O. The groundwater raise in lignite mining induced areas in Lusatia–actions taken to reduce the follows ups. *Min. Rep.* **2015**, *151*, 522–530. Available online: https://mining-report.de/wp-content/uploads/2015/11/MiRe_1506_Grundwasserwiederanstieg_151124.pdf (accessed on 12 February 2025).
5. Frei, S.; Klostermann, L.; Gilfedder, B.S. Iron retention in Lusatia: A spatially differentiated analysis of the Spree. *Grundwasser* **2024**, *29*, 213–228. [CrossRef]
6. Schultze, M.; Pokrandt, K.-H.; Hille, W. Pit lakes of the Central German lignite mining district: Creation, morphometry and water quality aspects. *Limnologica* **2010**, *40*, 148–155. [CrossRef]
7. Gläßer, C.; Groth, D.; Frauendorf, J. Monitoring of hydrochemical parameters of lignite mining lakes in Central Germany using airborne hyperspectral casi-scanner data. *Int. J. Coal Geol.* **2011**, *86*, 40–53. [CrossRef]
8. Bamisaiye, O.A. Subsurface mapping: Selection of best interpolation method for borehole data analysis. *Spat. Inf. Res.* **2018**, *26*, 261–269. [CrossRef]
9. Gondolfo, O.C.B. Determination of water table depth using geophysical methods. *Braz. J. Geophys.* **2022**, *40*, 1–15. [CrossRef]

10. King, A.R.; Hynes, T. Applications of geophysical methods for monitoring acid mine drainage. In Proceedings of the International Land Reclamation and Mine Drainage Conference and the Third International Conference on the Abatement of Acidic Drainage, Pittsburgh, PA, USA, 24 April 1994; pp. 317–326. [CrossRef]
11. Shokri, B.J.; Ardejani, F.D.; Ramazi, H.; Moradzadeh, A. Predicting pyrite oxidation and multi-component reactive transport processes from an abandoned coal waste pile by comparing 2D numerical modeling and 3D geo-electrical inversion. *Int. J. Coal Geol.* **2016**, *164*, 13–24. [CrossRef]
12. Moreira, C.A.; Casagrande, M.F.S.; Büchi, F.M.S.; Targa, D.A. Hydrogelogical characterization of a waste rock pile and bedrock affected by acid mine drainage from geophysical survey. *SN Appl. Sci.* **2020**, *2*, 1236. [CrossRef]
13. Kirsch, R. (Ed.) *Groundwater Geophysics—A Tool for Hydrogeology*; Springer: Berlin, Heidelberg, Germany, 2006. [CrossRef]
14. Paine, J.G.; Minty, B.R.S. Airborne hydrogeophysics. In *Hydrogeophysics*; Rubin, Y., Hubbard, S.S., Eds.; Springer: Dordrecht, The Netherlands, 2005; pp. 333–357. [CrossRef]
15. Siemon, B.; Ibs-von Seht, M.; Steuer, A.; Deus, N.; Wiederhold, H. Airborne electromagnetic, magnetic, and radiometric surveys at the German North Sea coast applied to groundwater and soil investigations. *Remote Sens.* **2020**, *12*, 1629. [CrossRef]
16. Harrison, E.J.; Baranwal, V.C.; Pfaffhuber, A.A.; Christensen, C.W.; Skurdal, G.H.; Rønning, J.S.; Anschütz, H.; Brönner, M. AEM in Norway: A Review of the Coverage, Applications and the State of Technology. *Remote Sens.* **2021**, *13*, 4687. [CrossRef]
17. Knight, R.; Smith, R.; Asch, T.; Abraham, J.; Cannia, J.; Viezzoli, A.; Fogg, G. Mapping aquifer systems with airborne electromagnetics in the Central Valley of California. *Groundwater* **2018**, *56*, 893–908. [CrossRef]
18. Jørgensen, F.; Møller, M.M.; Nebel, L.; Jensen, N.-P.; Christiansen, A.V.; Sandersen, P.B.E. A method for cognitive 3D geological voxel modelling of AEM data. *Bull. Eng. Geol. Environ.* **2023**, *72*, 421–432. [CrossRef]
19. Høyer, A.-S.; Jørgensen, F.; Foged, N.; He, X.; Christiansen, A.V. Three-dimensional geological modelling of AEM resistivity data—A comparison of three methods. *J. Appl. Geophys.* **2015**, *115*, 65–78. [CrossRef]
20. Marker, P.A.; Foged, N.; He, X.; Christiansen, A.V.; Refsgaard, J.C.; Auken, E.; Bauer-Gottwein, P. Performance evaluation of groundwater model hydrostratigraphy from airborne electromagnetic data and lithological borehole logs. *Hydrol. Earth Syst. Sci.* **2015**, *19*, 3875–3890. [CrossRef]
21. King, J.; Mulder, T.; Essink, G.O.; Bierkens, M.F.P. Joint estimation of groundwater salinity and hydrogeological parameters using variable-density groundwater flow, salt transport modelling and airborne electromagnetic surveys. *Adv. Water Res.* **2022**, *160*, 104118. [CrossRef]
22. Siemon, B.; Christiansen, A.V.; Auken, E. A review of helicopter-borne electromagnetic methods for groundwater exploration. *Near Surf. Geophys.* **2009**, *7*, 629–646. [CrossRef]
23. Auken, E.; Boesen, T.; Christiansen, A.V. Chapter Two—A review of airborne electromagnetic methods with focus on geotechnical and hydrological applications from 2007 to 2017. *Adv. Geophys.* **2017**, *58*, 47–93. [CrossRef]
24. Hammack, R.; Love, E.; Veloski, G.; Ackman, T.; Harbert, W. Using helicopter electromagnetic surveys to identify environmental problems at coal mines. *Mine Water Environ.* **2003**, *22*, 80–84. [CrossRef]
25. Love, E.; Hammack, R.; Harbert, W.; Sams, J.; Veloski, G.; Ackman, T. Using airborne thermal infrared imagery and helicopter EM conductivity to locate mine pools and discharges in the Kettle Creek watershed, north-central Pennsylvania. *Geophysics* **2005**, *70*, B73–B91. [CrossRef]
26. Demirel, N.; Emil, M.K.; Duzgun, H.S. Surface coal mine area monitoring using multi-temporal high-resolution satellite imagery. *Int. J. Coal Geol.* **2011**, *86*, 3–11. [CrossRef]
27. Schroeter, L.; Gläβer, C. Analyses and monitoring of lignite mining lakes in Eastern Germany with spectral signatures of Landsat TM satellite data. *Int. J. Coal Geol.* **2011**, *86*, 27–39. [CrossRef]
28. Siemon, B.; Ibs-von Seht, M.; Pielawa, J. *Gebietsbericht D-AERO-Finsterwalde HEM/HMG/HRD-Befliegung 191 Finsterwalde Mai/Juli 2021*; Bundesanstalt für Geowissenschaften und Rohstoffe: Hannover, Germany, 2022. [CrossRef]
29. LBGR. *Geologische Übersichtskarte 1:100000 (GÜK100)*; Landesamt für Bergbau, Geologie und Rohstoffe: Cottbus, Germany, 2015; Available online: https://geo.brandenburg.de/?page=Geologische-Karten (accessed on 11 November 2024).
30. BKG. *Digitale Topographische Karte 1:1000000 (DTK1000)*; Bundesamt für Kartographie und Geodäsie: Frankfurt, Germany, 2021; Topographic Map Generated From: Spatial Data Base: ©GeoBasis-DE/BKG (2021); Available online: https://gdz.bkg.bund.de/index.php/default/catalog/product/view/id/1454/s/digitale-topographische-karte-1-100-000-dtk100 (accessed on 11 November 2024).
31. LfU. *Gewässernetz*; Landesamt für Umwelt Brandenburg: Potsdam, Germany, 2024; Available online: https://data.geobasis-bb.de/geofachdaten/Wasser/Hydrologie/gewnet25.zip (accessed on 11 November 2024).
32. LMBV. *Bergbaufolgeseen*; Lausitzer und Mitteldeutsche Bergbau-Verwaltungsgesellschaft mbH: Senftenberg, Germany, 2024; Available online: https://lmbv.maps.arcgis.com/apps/webappviewer/index.html?id=64068d71103d40a9a0a07f6b0682db1c&extent=1425029.5094,6647119.0691,1718547.698,6795254.0299,102100&showLayers=Wasser_View_5083;Ueberleiter_View_2598 (accessed on 11 November 2024).

33. LMBV. *Landinanspruchnahme*; Lausitzer und Mitteldeutsche Bergbau-Verwaltungsgesellschaft mbH: Senftenberg, Germany, 2024; Available online: https://geodatenportal-lmbv.hub.arcgis.com/maps/62bfad18b48244eebc81bca44f337cb8 (accessed on 11 November 2024).
34. Standke, G.; Rascher, J.; Volkmann, N. Lowstand cycles and coal formation in paralic environments: New aspects in sequence statigraphy. In *Northern European Cenozoic Stratigraphy—Proceedings of the 8th Biannual Meeting of the RCNNS/RCNPS*; Guers, K., Ed.; Landesamt für Natur und Umwelt des Landes Schleswig-Holstein: Flintbek, Germany, 2002; pp. 153–161.
35. Barth, W.; Bönisch, R.; Krentz, O.; Kühner, R.; Lapp, M.; Petzold, H.; Schneider, W.; Schulze, H.; Seibel, B.; Standke, G. *Die geologische Entwicklung der Lausitz*; Vattenfall Europe Mining AG: Cottbus, Germany, 2010.
36. Stackebrandt, W.; Franke, D. *Geologie von Brandenburg*; Schweizerbart Science Publishers: Stuttgart, Germany, 2015; ISBN 978-3-510-65295-2.
37. LfU. *Grundwassermessnetz*; Landesamt für Umwelt Brandenburg: Potsdam, Germany; Available online: https://maps.brandenburg.de/WebOffice/synserver?project=GWM_www_CORE (accessed on 14 February 2025).
38. Christiansen, A.V.; Auken, E. A global measure for depth of investigation. *Geophysics* **2012**, *77*, WB171–WB177. [CrossRef]
39. Tølbøll, R.J.; Christensen, N.B. Sensitivity functions of frequency-domain magnetic dipole-dipole systems. *Geophysics* **2007**, *72*, F45–F56. [CrossRef]
40. Siemon, B.; van Baaren, E.; Dabekaussen, W.; Delsman, J.; Dubelaar, W.; Karaoulis, M.; Steuer, A. Automatic identification of fresh-saline groundwater interfaces from airborne electromagnetic data in Zeeland, the Netherlands. *Near Surf. Geophys.* **2019**, *17*, 3–25. [CrossRef]
41. Siemon, B.; Steuer, A.; Ullmann, A.; Vasterling, M.; Voß, W. Application of frequency-domain helicopter-borne electromagnetics for groundwater exploration in urban areas. *Phys. Chem. Earth* **2011**, *36*, 1373–1385. [CrossRef]
42. Siemon, B. Improved and new resistivity-depth profiles for helicopter electromagnetic data. *J. Appl. Geophys.* **2001**, *46*, 65–76. [CrossRef]
43. Sengpiel, K.-P.; Siemon, B. Advanced inversion methods for airborne electromagnetic exploration. *Geophysics* **2000**, *65*, 1983–1992. [CrossRef]
44. Annetts, A.; George, R.; Munday, T.; Ibrahimi, T.; Cahill, K.; Paul, R.; Davis, A. The application of AEM to mapping sea-water intrusion at La Grange, WA. In Proceedings of the ASEG-PESA 2015, Perth, Australia, 15–18 February 2015. [CrossRef]
45. BKG. *Digitales Geländemodel (DGM25)*; Bundesamt für Kartographie und Geodäsie: Frankfurt, Germany, 2021; Spatial Data Base: ©GeoBasis-DE/BKG (2021); Available online: https://gdz.bkg.bund.de (accessed on 1 February 2012).
46. LMBV. *Lithology Data, Water Table Data, Hydrochemical Data*; Personal communication; adapted with permission from LMBV (27 January 2025); Lausitzer und Mitteldeutsche Bergbau-Verwaltungsgesellschaft mbH: Senftenberg, Germany, 2022.
47. Kang, S.; Knight, R.; Greene, T.; Buck, C.; Fogg, G. Exploring the model space of airborne electromagnetic data to delineate largescale structure and heterogeneity within an aquifer system. *Water Resour. Res.* **2021**, *57*, e2021WR029699. [CrossRef]
48. Archie, G.E. The electrical resistivity log as an aid in determining some reservoir characteristics. *Trans. AIME* **1942**, *146*, 54–62. [CrossRef]
49. Yin, C.-C.; Ren, X.-Y.; Liu, Y.-H.; Qi, Y.-F.; Qiu, C.-K.; Cai, J. Review on airborne electromagnetic inverse theory and applications. *Geophysics* **2015**, *80*, W17–W31. [CrossRef]
50. Brodie, R.; Sambridge, M. Holistic inversion of frequency-domain airborne electromagnetic data with minimal prior information. *Explor. Geophys.* **2009**, *40*, 8–16. [CrossRef]

water

MDPI

Article

Shallow Hydrostratigraphy Beneath Marsh Platforms: Insights from Electrical Resistivity Tomography

Jacque L. Kelly * and Christine M. Hladik

School of Earth, Environment and Sustainability, Georgia Southern University, Statesboro, GA 30458, USA; chladik@georgiasouthern.edu
* Correspondence: jkelly@georgiasouthern.edu

Abstract: Salt marshes are ecologically and economically valuable ecosystems, yet are vulnerable to marsh dieback, the rapid death of marsh vegetation, which has affected coastal areas along the southeastern and Gulf coasts of the United States in recent decades. This study used multichannel electrical resistivity tomography (ERT) surveys to investigate the shallow hydrostratigraphy (up to 39.2 m depth) of three dieback-affected salt marshes along the Georgia coast to evaluate the influence of site location, vegetation status (dieback versus healthy), and tidal conditions on ERT profiles. ERT profiles revealed consistent subsurface resistivity patterns across the marsh platforms, with low resistivity (0.2 ohm-m) at shallow depths indicating saltwater saturation and a transition to higher resistivity (up to 8.1 ohm-m) at greater depths, potentially signifying a shift to brackish conditions and/or sandy strata. The ERT data indicated that the hydrostratigraphy is similar across all study sites. Furthermore, the ERT data remained consistent regardless of vegetation status, tidal variations, and seasonal changes, suggesting that the processes driving the recovery of marsh dieback are independent of the shallow marsh stratigraphy. These findings enhance our understanding of marsh subsurface conditions, supporting efforts to better understand marsh resilience and guide future research on salt marshes.

Keywords: groundwater; marsh dieback; electrical resistivity tomography; salt marsh; *Spartina alterniflora*

1. Introduction

Salt marshes are valuable ecological and economic systems. They provide critical breeding and foraging habitats for a diverse array of aquatic and terrestrial organisms. They also improve water quality through nutrient and metal removal, sequester carbon, and provide crucial protection to developed coastal areas from shoreline erosion and storms [1,2]. Despite their ecological significance, salt marshes are vulnerable ecosystems that have experienced marsh dieback throughout the United States (US) Atlantic and Gulf Coasts [3–7], including coastal Georgia, since at least 2001 [4,8]. Marsh dieback is characterized by a rapid senescence of salt marsh vegetation. It particularly impacts *Spartina alterniflora* (marsh cordgrass) and *Juncus roemerianus* (black needlerush). This phenomenon has been extensively documented, affecting over 8 km^2 in coastal Georgia [8,9]. Marsh dieback leads to the complete loss of vegetation structures, resulting in large mudflats that are vulnerable to erosion and loss of elevation, which further hinders marsh recovery [3,10–12]. Observations have shown dieback across various elevations, from low-lying creek banks to higher-elevation upland borders [9].

The causes of marsh dieback in coastal Georgia have been attributed to abiotic and biotic variables, including soil biogeochemistry, fungal pathogens, and consumer food webs [8]. While a single definitive cause is rarely identified, a multi-stressor hypothesis associated with drought conditions is most widely accepted for dieback in Georgia [8]. Groundwater dynamics influence salt marsh ecosystems, impacting soil salinity, nutrient cycling, marsh productivity, and plant species distribution [4,13,14]. In marshes, dissolved oxygen in shallow groundwater is impacted by vegetation roots, which typically grow 12 to 20 cm deep for *S. alterniflora* [15] and 30 to 60 cm deep for *J. roemerianus* [16]. This interaction also contributes nutrients to the ecosystem [17]. Despite its importance, the subsurface environment of salt marshes remains underexplored. In particular, little is known about how the shallow marsh hydrostratigraphy interacts with marsh dieback and recovery processes. Previous studies have primarily focused on surface conditions, overlooking the potential influence of subsurface hydrology. During droughts, for instance, shallow aquifers become highly susceptible to salinization due to reduced recharge from local precipitation [4]. Additionally, drought conditions can dry the marsh surface and increase porewater salinity when evapotranspiration proceeds in the absence of precipitation [4]. Moreover, coastal aquifers provide hydrologic connections between marshes and estuaries [17]. The aquifers influence estuarine water chemistry, which increases dynamic mixing in the reaction zones, where the water chemistry changes drastically with tidal fluctuations [17]. Given the association between drought, changes in groundwater dynamics, and marsh dieback, a more comprehensive investigation of groundwater below the salt marsh platform is needed. This study, therefore, aims to examine the shallow hydrostratigraphy of salt marsh platforms and their relationship to marsh dieback and recovery.

Multichannel electrical resistivity tomography (ERT) is a powerful geophysical tool for noninvasive exploration of shallow subsurface environments [18]. ERT surveys deliver detailed subsurface images, making them highly effective in detecting clay-dominated strata and differentiating between freshwater- and saltwater-saturated sediments [18]. Such surveys have characterized subsurface geology and hydrogeology, revealed stratigraphic heterogeneities, and defined freshwater-seawater boundaries in coastal aquifers [18–20]. This technique works well in coastal environments because freshwater and seawater differ in their electrical resistivity properties [18,19]. Additionally, ERT surveys provide insights into the geometry of the groundwater seepage conduits and the movement and mixing of groundwater and seawater in coastal aquifers [18,19]. Resistivity profiles are a function of the underlying lithology, porosity, water salinity, and temperature [20] and, therefore, reflect bulk resistivity (resistivity is the inverse of conductivity) of the subsurface materials. ERT's capability to measure bulk resistivity makes it a valuable tool for examining the hydrostratigraphy of aquifers and aquitards in coastal environments. In this study, we present the shallow hydrostratigraphy (<40 m) as revealed by ERT surveys from three salt marshes located along a latitudinal gradient of the Georgia coast, US. This study uses bulk resistivity to identify subsurface resistivity patterns beneath salt marshes and evaluates whether the resistivity patterns vary based on site location, marsh vegetation type, marsh vegetation health and zonation, seasonal dynamics, and tidal flooding. Our results, therefore, offer insights into the geological and hydrological characteristics of these vulnerable salt marsh ecosystems.

2. Materials and Methods

2.1. Study Sites

The Georgia coast is characterized by a subtropical climate with long, hot summers and short, mild winters [21] and spans a 160 km section of southeastern North America along the Western North Atlantic. The region's lower coastal plain gives way to an extensive network of coastal wetlands, salt marshes, river deltas, and meandering tidal rivers, leading to an outer chain of barrier islands facing the Atlantic Ocean [22]. Within this setting, we collected field data at three marsh dieback locations (Figure 1A) that were dominated by the foundation species *S. alterniflora*. Dieback at our field sites was restricted to the interior marsh at distances exceeding 75 m from the tidal creeks, consistent with findings from Schepers et al. [7]. In the subtropical climate, *S. alterniflora* production is responsive to abiotic drivers year-round. Thus, we sampled quarterly (October–November, March, and July) from November 2014 to July 2016.

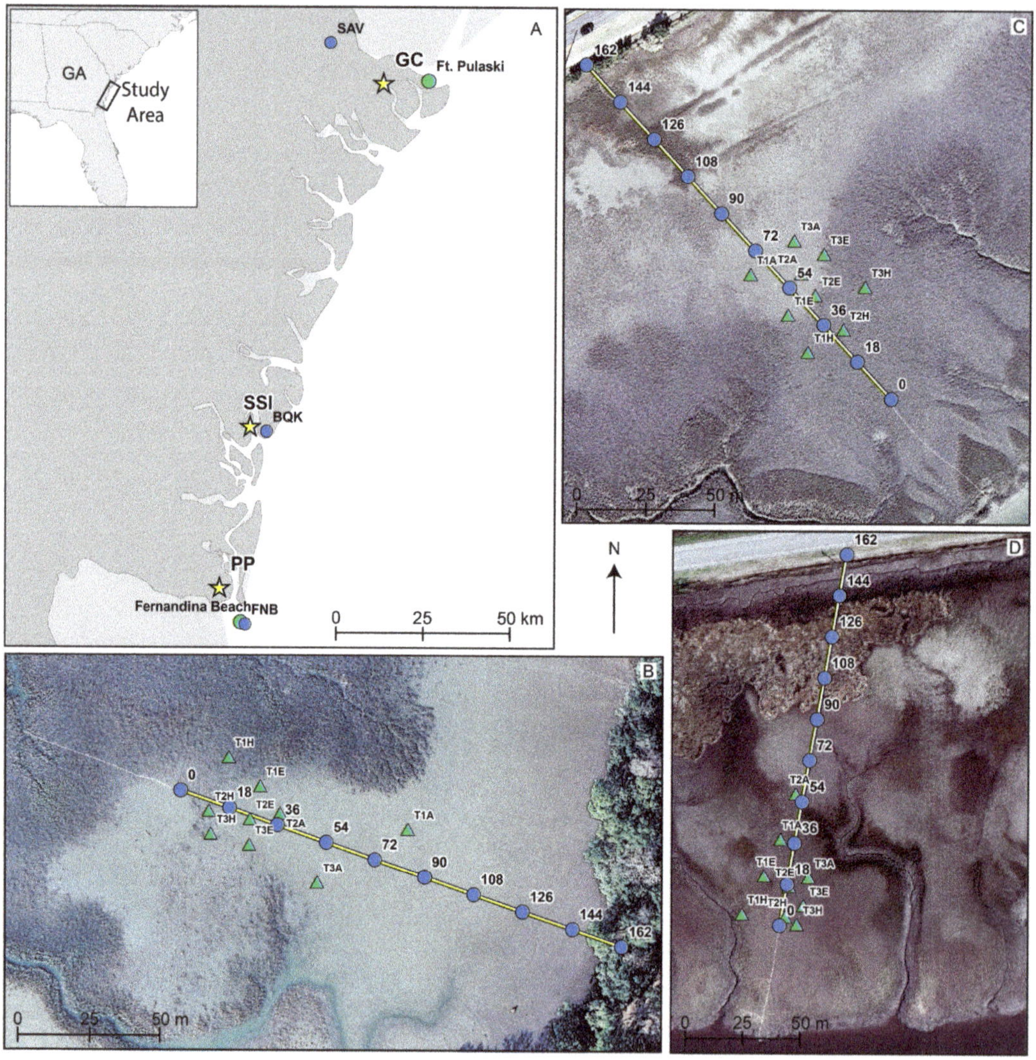

Figure 1. (**A**) Site location map showing Grays Creek (GC) located at 32.030289° N, 81.037223° W, St. Simons Island (SSI) located at 31.165996° N, 81.441708° W, and Point Peter (PP) located at 30.759840° N, 81.531281° W field sites (yellow stars). The Fort Pulaski and Fernandina Beach tide stations (green circles) and the Savannah (SAV), Brunswick (BQK), and Fernandina Beach (FNB) weather stations (blue circles) are also shown. The weather stations are located within the boundaries of the cities of Savannah, Brunswick, and Fernandina Beach, respectively. Field site maps showing the plot locations (green triangles and plot labels), electrical resistivity cable placement (yellow line), and cable distances (blue dots and associated numbers) for (**B**) GC, (**C**) SSI, and (**D**) PP.

Site names were assigned based on location (Figure 1A), and abbreviations are as follows from north to south: Gray's Creek (GC), St. Simons Island (SSI), and Point Peter (PP). For geospatial analyses presented below, all data used a common horizontal (North American Datum (NAD) 83(2011)/Universal Transverse Mercator (UTM) Zone 17N, European Petroleum Survey Group (EPSG): 6346) and North American Vertical Datum (NAVD) 88, GEOID 12A, EPSG: 5703) in unit meters. The GC site is located at 496,468.97mE 3,543,772.09mN (32.030289° N, 81.037223° W) near the city of Savannah. It is bounded by an urban housing development on the east and a tidal river called Grays Creek on the west (Figure 1B). *S. alterniflora* is the dominant vegetation at this site. The SSI site is located at 457,978.70mE 3,448,034.98mN (31.165996° N, 81.441708° W) off the F.J. Torras Causeway between St. Simons Island and the city of Brunswick (Figure 1C). This site is bordered by the F.J. Torras Causeway on the north–northwest and the tidal Mackay River on the south–southeast. Data from SSI were collected exclusively from *S. alterniflora*-dominated areas, avoiding areas with salt pans. *Borrichia frutescens* (sea ox-eye) and *J. roemerianus* are found along the upland border of the SSI site. The PP site is located at 449,154.00mE 3,403,002.81mN (30.759840° N, 81.531281° W) near the city of St. Marys and is bounded by the North River Causeway to the north and the tidal St. Marys River to the south (Figure 1D). In addition to *S. alterniflora*, *J. roemerianus* is present at this site near the causeway (Figure 1D), and both species are included in the ERT data at this site.

We established three shore-perpendicular transects (212 to 221 m in length; Table 1) at each site. The transects (T1, T2, and T3) were spaced ~15 m lateral distance from each other and extended from the upland border to the creek bank. Each transect bisected healthy (H), transitional edge (E), and affected dieback (A) condition classes of the marsh. For clarity, only T2 is shown as yellow lines in Figure 1B–D, as this is the transect from which the ERT data were collected. A permanent monitoring plot was established for each condition class (e.g., GC T1A, GC T1E, GC T1H) along each transect at each field site (green triangles in Figure 1B–D). Field data consisted of (1) real-time kinematic (RTK) Global Positioning System (GPS) measurements of elevation along each transect for all sampling dates, (2) standard water quality measurements (temperature, salinity, specific conductivity) of water from a target depth of 90 cm beneath the marsh from T1 plots for all sampling dates, and (3) ERT profiles along T2 for bulk resistivity measurements in October 2015 and March 2016.

Table 1. Real-time kinematic summary statistics for transect two at each study site. Transects were measured from the upland border to the nearest creek, with elevations relative to the North American Vertical Datum 88. Refer to Figure 1B (Grays Creek), Figure 1C (St. Simons Island), and Figure 1D (Point Peter). Electrode 0 distance represents the closest point of the resistivity cable to the creek.

Parameter	Grays Creek (GC)	St. Simons Island (SSI)	Point Peter (PP)
Minimum Elevation (m)	0.03	−0.50	−0.48
Maximum Elevation (m)	0.92	1.79	0.98
Elevation Change (m)	0.88	2.28	1.45
Median Elevation (m)	0.72	0.70	0.58
Mean Elevation (m)	0.70	0.74	0.55
Standard Deviation (m)	0.18	0.36	0.28
Transect Length (m)	221	216	212
Electrode 0 Distance to Creek (m)	56	51	47

The nearest weather stations to each field site were Savannah International Airport (station #USW00003822), located ~19.24 km NW of GC; the Brunswick Malcolm McKinnon Airport (station #USW000013878), located ~4.67 km ESE of SSI; and Fernandina Beach (station #USC00082944), located ~12.22 km SE of PP (Figure 1A). As summarized in Table 2, the weather stations indicated between 0.4 and 3.0 cm of precipitation in the seven days prior to ERT profile collection.

Table 2. Total precipitation and average tidal amplitude in the seven days prior to each field sampling date.

Sample Date	Precipitation (cm)	Tidal Amplitude (m)
	Gray's Creek (GC)	
10/09/2015	1.2	2.3
03/04/2016	1.7	1.8
	St. Simons Island (SSI)	
10/11/2015	0.4	1.9
03/06/2016	1.0	1.7
	Point Peter (PP)	
10/10/2015	3.0	1.7
03/05/2016	0.5	1.5

The National Atmospheric and Oceanic Administration maintains two harmonic tide stations that bound the northern and southern ends of our field sites. The Fort Pulaski, Savannah River Entrance (gage #8670870) is located ~12.60 km E of GC, and the Fernandina Beach, Amelia River (gage #8720030) is located ~11.56 km SE of PP (Figure 1A). Fort Pulaski experiences a great diurnal range of 2.3 m [23], while Fernandina Beach experiences a 2.0 m great diurnal range [24]. Based on monthly mean sea level data from 1897 to 2023, the Fort Pulaski station is registering higher relative sea-level rise (3.61 $\pm$ 0.27 mm/yr) [25] compared to the Fernandina Beach station (2.28 $\pm$ 0.17 mm/yr) [26]. Georgia marshes experience mesotidal semidiurnal tides with an average tidal period of 12.4 h and an average tidal amplitude of ~1.3 m, while spring tides exceed 3.5 m. The average tidal amplitudes in the week leading up to each ERT field sampling date were typical for the region, varying between 1.5 and 2.3 m, with GC experiencing larger tidal amplitudes than PP (Table 2). All field sites experienced flooding to the upland border during high tides.

Investigation of historic high-resolution multispectral orthoimages from 1999 onward revealed marsh dieback at the GC, SSI, and PP sites starting in 2005, 2003, and 2006, respectively. However, poor image quality and infrequent image availability precluded thorough analysis of PP before 2006, except for one image from 2003, which showed no dieback [27]. Based on the orthoimagery analysis, the field sites have been recovering from dieback (dieback area shrinking) since 2012 for GC and SSI and since 2006 for PP [27]. Thus, field sampling at all sites occurred during a phase of dieback recovery.

2.2. Data Collection and Processing

Figure 2 shows the data collection, data processing, and data overlay flowchart. During each field site visit, ground elevations along all transects were surveyed every 5 to 10 m from the marsh upland to the creek bank. A Trimble R8 RTK GPS receiver was used to collect these data. The RTK data had vertical and horizontal accuracies of 0.020 and 0.0010 m, respectively, and are reported as the root mean square (RMS) error (68% confidence level) [28]. RTK data are in the NAD83 UTM Zone 17 horizontal coordinate system (EPSG 26917), and elevations are NAVD 88 orthometric heights (in meters) computed using the National Geodetic Survey GEOID 12A. All RTK data were collected within two hours of low tide. RTK data are real-time and are not post-processed.

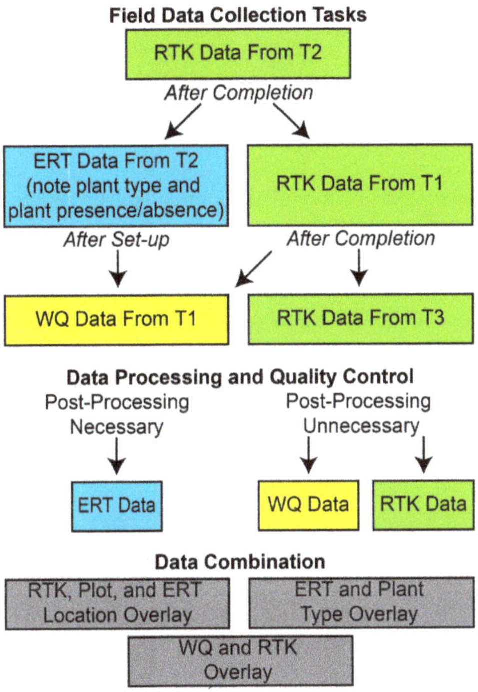

Figure 2. Flowchart showing field data collection tasks, data processing, and quality control, as well as data combination tasks to create data overlays (gray boxes) for the real-time kinematic (RTK) elevation and location information (green boxes), the electrical resistivity tomography (ERT) bulk resistivity data (blue boxes), and the water quality (WQ) data (yellow boxes).

After RTK measurements were complete (Figure 2), water quality data were collected from T1 plots (green triangles in Figure 1B–D) within two hours of low tide using a 90 cm long push-point piezometer (MHE Products) coupled to a peristaltic pump (Geotech Environmental Equipment, Inc., Denver, CO, USA) and measured with a YSI-Xylem EXO1 water quality meter (Table 3). The water quality meter was calibrated to factory-supplied solutions following manufacturer specifications before each field sampling session. The metal push-point piezometer has a 5 cm screen length and a 0.5 cm inner diameter. Water was typically drawn from 90 cm depth but was occasionally drawn from shallower depths (Table 3) if 90 cm depth yielded inadequate water for measurement. All water quality data were collected after >500 mL of water (>3 times the piezometer volume) was purged from the piezometer, and water quality parameters were stabilized.

After RTK data were collected (Figure 2), bulk resistivity was measured concurrently with water quality data. To avoid interference between metal piezometers and resistivity measurements, ERT data were collected from T2 (yellow line in Figure 1B–D). To image bulk resistivity, two-dimensional electrical resistivity profiles were collected during daylight hours. A stationary eight-channel, 168 m long SuperSting R8/IP Marine Electrical Resistivity Meter equipped with 56 electrodes spaced 3 m apart (Advanced Geosciences, Inc., Austin, TX, USA) was used to collect the bulk resistivity data. Before collecting data, we tested the receiver, relay, and cable to verify optimal equipment functioning. After equipment checks, the cable was positioned shore perpendicular in a straight line to image as much of the dieback area as possible, with any remainder of the cable extending toward the nearest creek. At all sites, the length of the marsh platform exceeded the cable length. A sandbag was placed atop each electrode to optimize contact resistance. Contact resistance checks ensured values were typically below 100 ohm-m. All data were collected using a dipole-dipole array with a strong gradient to maximize the signal-to-noise ratio. The dipole-dipole array was selected because of its horizontal and vertical exploration capabilities, ability to resolve near-surface features, and potential to detect lateral resistivity variations [29,30]. This ERT configuration has a 3.0 m horizontal resolution and a 1.0 m vertical resolution near the Earth's surface, increasing to 4.5 m horizontal and 1.5 m vertical resolution with depth. When time allowed, time-series resistivity profiles were collected on the same day but at tide conditions ranging from low to high tide. These surveys provide information about temporal and spatial changes in the marsh platform [18]. We also recorded the presence or absence of marsh grasses at each electrode. Depression storage of surface water on the marsh platform was always minimal, as were elevation changes along the profile.

Table 3. Water quality data for the study sites were organized by plot and sample date. Depth is in centimeters, and specific conductivity (SPC) is in microsiemens per centimeter. Some plots did not always yield water and are indicated with "-".

Sample Date	Depth	Salinity	SPC	Depth	Salinity	SPC	Depth	Salinity	SPC
Grays Creek		GC T1A			GC T1E			GC T1H	
11/14/2014	40	37.5	56,329	30	36.3	54,050	30	35.8	54,024
03/20/2015	90	29.0	44,801	90	26.8	41,650	90	30.0	46,206
07/31/2015	90	24.3	38,580	90	23.8	37,590	90	22.2	35,390
10/09/2015	-	-	-	90	26.9	33,854	90	23.7	32,063
03/04/2016	90	25.2	39,379	90	23.1	36,438	90	22.8	36,118
07/22/2016	90	19.8	31,766	90	19.8	31,820	90	18.5	29,880
St. Simons		SSI T1A			SSI T1E			SSI T1H	
11/01/2014	70	35.6	53,867	50	33.7	51,160	30	32.6	49,761
03/21/2015	90	38.3	55,286	60	34.4	51,829	90	36.6	52,648
07/25/2015	90	35.2	53,796	90	27.5	42,830	90	31.3	48,200
10/11/2015	90	36.0	48,002	-	-	-	-	-	-
03/06/2016	90	35.9	54,289	-	-	-	-	-	-
07/24/2016	90	34.1	51,882	60	30.9	47,596	90	28.6	44,410
Point Peter		PP T1A			PP T1E			PP T1H	
11/02/2014	40	20.3	32,409	50	15.5	25,282	50	17.0	27,629
03/28/2015	90	26.9	42,665	90	32.2	54,160	90	22.5	35,562
07/26/2015	90	25.0	39,412	90	22.8	36,084	90	18.4	24,773
10/10/2015	-	-	-	90	22.8	38,115	90	23.0	-
03/05/2016	-	-	-	90	24.6	38,710	90	22.6	35,721
07/23/2016	-	-	-	90	21.6	34,300	90	17.3	28,308

All measured apparent resistivity profiles were processed identically using Earth Imager 2D, which uses Gauss–Newton and quasi-Newton methods (Advanced Geosciences, Inc.). Inversion parameters were set to default values for smooth inversion and included data spike removal [31]. The inversion parameters were set to allow no more than seven model iterations; however, all profiles converged in three iterations. The inversion process was considered converged when RMS was <6% and L2-norm was less than 1 [31,32]. RMS represents the difference between the calculated and measured apparent resistivity values [33] in the profile, while L2-norm is a smoothness constraint that measures data misfit between observed and predicted data [31,32].

3. Results

At the beginning of the study, the median intertidal elevation of the marsh platform was 0.72, 0.70, and 0.58 m relative to the NAVD 88 for GC, SSI, and PP, respectively (Table 1). Elevation changes across the platform from the upland border to the creek bank for each site ranged from 0.88, 2.28, and 1.45 m for GC, SSI, and PP, respectively (Table 1). The lowest elevations were near creek banks, where tall *S. alterniflora* was found, and the highest elevations were along the upland border. Similar average water salinity existed across the marsh platform (A, E, and H condition classes) within each site (Figure 3). PP had the least saline water (range 15.5–32.2, average ± standard deviation 22.2 ± 4.3, *n* = 15), followed by GC (range 18.5–37.5, 26.2 ± 5.8, n = 17), and then SSI (range 27.5 to 38.3, 33.6 ± 3.1, n = 14), as shown in Table 3. Approximate specific conductivities and resistivities for these waters are also shown in Figure 3.

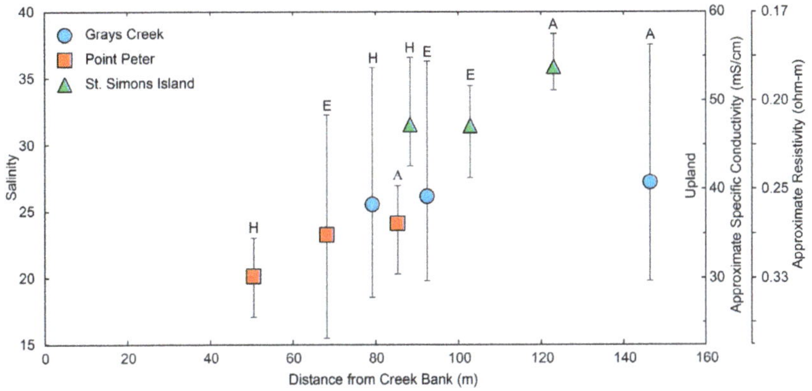

Figure 3. Salinity, approximate specific conductivity, and approximate resistivity of the transect one (T1) plot at each field site relative to the distance from the creek. Error bars show the range of salinity values measured quarterly at each plot from November 2014 through July 2016. "H" indicates healthy plots, "E" indicates transitional edge plots, and "A" indicates affected (dieback) plots.

The ERT cable was placed in the same location along T2 for both collection dates, and the data collection and inversion parameters were held constant over time. Thus, changes in resistivity across different sampling times resulted from changes in water chemistry [18]. The cable straddled the dieback areas at the SSI and PP sites (Figure 4C–F). The dieback extended beyond the cable length (Figure 4A,B) at GC. The ERT profiles from all sites showed consistent subsurface characteristics for October and March (Figure 4A–F) and through ebb-to-flood tide conditions (not shown). During the modeling process, the electrodes closest to the upland border (horizontal distance = 162 m) at all sites typically had lower contact resistance than electrodes closer to the creek. All data from all sites were processed with RMS varying between 3.75 and 5.64% and L2-norm varying between 0.44 and 0.74 (Figure 4A–F). These statistics are well within model acceptance tolerances [19]. All models had low overall resistivities, with PP having the highest maximum resistivity of 8.1 ohm-m (Figure 4E,F), followed by GC at 4.8 ohm-m (Figure 4A,B) and SSI at 2.0 ohm-m (Figure 4C,D).

Generally speaking, all sites showed consistent resistivity at shallow depths across each transect regardless of plant species and dieback status (Figure 4A,F). Although the resistivity scales vary between sites, several generalizations regarding the ERT data can be made. Bulk resistivity within the top ~0.8 m of the platform (green hues) at all sites was similar to the bulk resistivity at depths ranging from ~5 to ~6.5 m for GC and PP and from ~5 to ~13 m for SSI (Figure 4A–F). Between ~0.8 and ~5 m depth, bulk resistivity decreased to its lowest values, indicated by blue hues in Figure 4A–F. Bulk resistivity increased to its highest values (yellow, orange, and red hues) below ~6.5 m at the GC and PP sites and below ~13 m at the SSI site (Figure 4A–F).

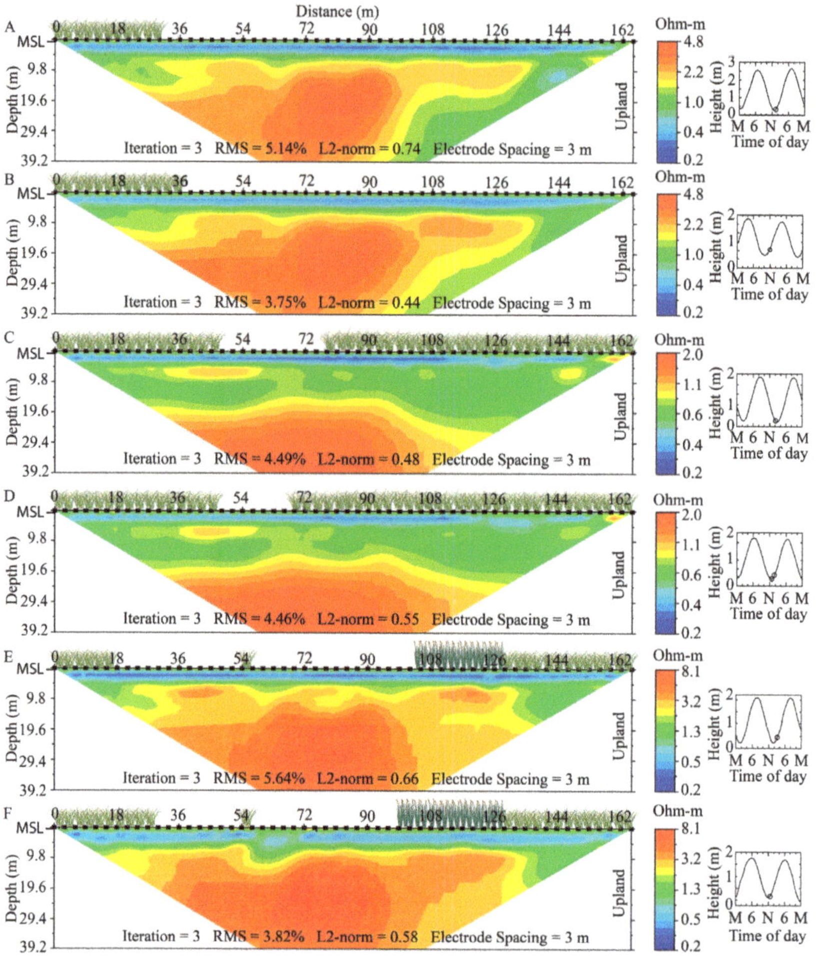

Figure 4. Bulk electrical resistivity results for Gray's Creek (GC) on (**A**) 9 October 2015 and (**B**) 4 March 2016; for St. Simons Island (SSI) on (**C**) 11 October 2015 and (**D**) 6 March 2016; and for Point Peter (PP) on (**E**) 10 October 2015 and (**F**) 5 March 2016. For all panels, the x-axis shows the distance across the marsh platform, with a distance of 0 m closest to the creek bank (but not at the creek bank) and a distance of 162 m at the upland border. The y-axis shows depth. All panels are displayed with a 0.85 vertical exaggeration with mean sea level (MSL) indicated on the x-axis of each panel. Model convergence parameters, including iterations, root mean square (RMS) error, and L2-norm, are shown in each panel. The coloration is a nonlinear electrical resistivity scale (ohm-m), which varies by site and was selected to maximize the resistivity scale for each location. Red hues represent higher bulk resistivity and lower conductivity compared to blue hues, which represent lower bulk resistivity and higher conductivity. Plant coverage is also shown along the x-axis, with the shorter light green plants representing *Spartina alterniflora*, the taller dark green plants representing *Juncus roemerianus*, and marsh dieback areas shown with no plant coverage. All panels correspond to low or nearly low tide conditions, as shown on the tidal stage insets. The tidal stage inset shows the time of day (M = Midnight and N = Noon) and tidal height (m) relative to mean lower low water (note, the y-axis scale changes from inset to inset). Tide data for GC were compiled from the Fort Pulaski station (#8670870), while tide data from SSI and PP were compiled from the Fernandina Beach station (#8720030).

4. Discussion

ERT models produce non-unique solutions from inherent limitations in the inversion process, which relies on a finite number of electrodes [30]. This characteristic allows for multiple resistivity models that fit the observed data while adhering to data accuracy constraints [34]. Additionally, as data density decreases with increasing depth, the potential for alternative solutions increases, further reducing model constraints, particularly at depth [32,35]. While the horizontal and vertical resolution of the ERT method smooths and averages the resistivity signal, the data nevertheless suggest a coherent interpretation of the shallow hydrostratigraphy beneath all three marsh platforms.

For all sites, the ERT data suggest a hydrostratigraphic sequence starting with a layer of low permeability unsaturated mud and peat from the surface down to ~0.8 m depth, shown as green hues in Figure 4A–F. Biological processes such as root growth, organic matter decomposition, and fiddler crab burrows occur within this unsaturated zone [14,36]. At shallow depths, Hemond and Fifield [37] hypothesized that evapotranspiration could lead to a cascade of events, including air entry into shallow marsh muds through crab burrows and old root channels. Dense mats of *S. alterniflora* roots, rhizomes, and aerenchyma modify the physical structure of the shallow subsurface and contribute to an environment conducive to air trapping and enhanced aeration [38]. Li et al. [39] also demonstrated that during inundation, air can become trapped in finer marsh soils, such as silt and clay. Xin et al. [40] further suggested that crab burrows can increase water flow and enhance aeration. Enhanced aeration, trapped air, and root networks act as insulators, increasing bulk resistivity compared to saturated materials [41]. The bulk resistivities at <0.8 m depth observed across each platform at all three field sites (green hues in Figure 4A–F) are consistent with these hypotheses.

Below the unsaturated zone, the marsh hydrostratigraphic sequence at all sites indicated a resistivity change at ~0.8 m depth, transitioning to the lowest measured resistivity values (0.2 to 0.4 ohm-m), represented by blue hues in Figure 4A–F. We interpret this change as the top of the water table, which is located near the Earth's surface, where the ERT resolution is higher due to the greater density of current paths. Although this depth is shallower than the vertical resolution of the ERT data, this interpretation is supported by field measurements from piezometers, which consistently yielded water from depths of 0.9 m and occasionally shallower (Table 3). These measurements also align with the expected location of the water table, which fluctuates around the mean sea level (MSL; Figure 4A–F). Both the piezometer data and the alignment with MSL provide ground-truth validation for the resistivity interpretation.

Since water is conductive, its presence reduces bulk resistivity. Consequently, the lowest resistivity values indicate marsh muds or mud–sand mixtures saturated with brackish water or saltwater [42] at ~0.8 m depth. The lower salinities observed at PP are consistent with its higher resistivity values compared to the other sites (Table 1). These findings align with previous studies, where resistivity values below 2.5 ohm-m typically indicate marine sediments saturated with seawater, while higher resistivity values suggest marine sediments saturated with brackish or freshwater [43]. The ERT data suggest that this saturated zone forms a continuous, unconfined aquifer in marsh muds at all sites (Figure 4A–F). During site visits, water extracted from the piezometer was often mixed with air. We do now know the definitive source of the air; however, possibilities include air trapped in the marsh during flooding [37–39], biogenic gas [44], or air resulting from reduced inflow rates during sampling. As expected, given the distance from the creek, no slope of the water table toward the creek bank was observed, as each profile was collected at least 47 m from the nearest creek (Table 1).

Below ~5 m depth, each site exhibited a gradual increase in resistivity represented by green, then yellow, orange, and finally red hues. This pattern suggests that, with increasing depth, salty water may gradually change to brackish water, that sandy materials associated with a confined aquifer may become more prevalent, and that greater sediment compaction decreased porosity [42]. At greater depths, Harvey et al. [45] hypothesize that fine decomposed organic matter could fill pore spaces within sandy strata, reducing porosity. As resistivity is inversely related to porosity and the conductivity of the fluids [42], this reduction in porosity would result in increased resistivity. The observed pattern of increasing resistivity with depth (Figure 4A–F) aligns with this hypothesis. Additionally, the consistently low bulk resistivity values (max of 8.1 ohm-m) across all sites suggest that brackish water is likely present at depth at each site [42], with PP being the least saline site (Figure 3).

The ERT data provide valuable insights, but overlapping resistivity ranges and the absence of sediment core data reduce the interpretability of bulk resistivity in the deeper sections of the salt marshes. Nevertheless, the typical stratigraphy of marshes in the southeastern United States often consists of 1 to 4 m of fine-grained marsh mud overlying sandy layers, of which all materials are Quaternary in age [4,14,46–50]. These sandy layers are frequently relict beach ridge deposits or sandy tidal channel deposits [47]. They function as confined aquifers and serve as conduits for groundwater flow from upland marsh areas to tidal creeks [45,49].

At the SSI site (Figure 4C,D), a notable resistivity feature is characterized by light green hues. This feature extends 9 m horizontally, between 60 and 69 m horizontal distance, and vertically from a depth of 4.9 to 14.7 m (approximately 10 m tall). This feature is positioned beneath the dieback-affected area of the marsh but does not connect to any distinguishing characteristic on the marsh surface (Figure 1C). The exact nature of this feature remains inconclusive without additional data. This feature also connects to a shallower, higher-resistivity zone, indicated by yellow and orange hues, that is located at horizontal distances of 15 to 60 m and depths of 6 to 10 m. This shallower feature may represent a higher permeability zone that facilitates water movement beneath the marsh toward the creek. Such a zone could consist of a sandy layer interspersed within the presumably predominantly fine-grained muddy substrate. Alternatively, this feature may be associated with higher resistivity less-saline water movement, a plausible interpretation if the stratigraphy consists predominantly of homogeneous fine-grained materials. Additionally, the ERT data from the SSI site suggest the presence of a brackish lens at the upland border between horizontal distances of 156 to 162 m and depths of 0 and 5 m (Figure 4C,D). This lens is depicted by yellow and orange hues (Figure 4C,D). A recently installed 90 cm deep well at the upland border yielded water with an average salinity of 3.7, supporting the growth of *B. frutescens*, a species that prefers fresher water environments [51]. However, we cannot rule out that this resistivity feature could be related to fill material associated with the adjacent F.J. Torras Causeway. The causeway directly abuts the upland border plots, which are outside of the scope of this study and are not shown in Figure 1C. Notably, both the vertical conduit and upland border resistivity features are absent from the GC and PP sites.

At the PP site, a ~1 m wide, shallow creek is located between horizontal distances of 147 and 150 m (Figures 1D and 4E,F). Although this creek held water at low tide during data collection, it is not visible in the ERT profiles. Another small creek was forming between horizontal distances of 78 and 84 m (Figure 1D), but this creek also does not appear in the ERT data (Figure 4E,F). Additionally, *J. roemerianus* was present in the ERT profile between horizontal distances of 99 and 129 m (Figure 1D); however, there was no detectable difference in the hydrostratigraphy beneath *J. roemerianus* compared to *S. alterniflora* (Figure 4E,F). The ERT profiles reveal the highest overall resistivity (8.1 ohm-m) at the PP field site, suggesting the presence of less saline groundwater compared to GC and SSI. This result aligns with the presence of a large stand of *J. roemerianus*, which typically indicates systems with fresher water conditions [52].

Our sampling protocol involved collecting water quality data along T1 and ERT data along T2. This approach created a limitation for data interpretation because the shallow ground truth measurements (Table 3) were not collected directly underneath the cable. However, average salinities for all condition classes at each field site were not significantly different (repeated measures ANOVA, F2,40 = 0.611, $p = 0.574$; Figure 3), suggesting similar water quality parameters across the platform of the marsh through time. Thibodeau [53] also found little variation in the salinity spatial pattern over time at a marsh site in South Carolina. Several studies [53–55] found lower salinities in their respective salt marsh field sites during the winter months when rainfall was higher and evapotranspiration rates were lower compared to summer months. Except for PP (Table 3), our field data do not suggest a similar relationship (repeated measures ANOVA, F1,34 = 0.044, $p = 0.845$), though there was a significant interaction between site and season ($p = 0.00108$). However, note that our sampling protocol was limited to one data point seasonally for only two years, making statistical inferences difficult.

Flooding of the marsh platform during tidal inundation is the dominant driver of the saturation state [56], and therefore, groundwater resistivity changes. Thus, it is unsurprising that the precipitation (Table 2) prior to collecting ERT profiles did not noticeably impact the bulk resistivity measurements. The bulk resistivity at depth also remained similar at each field site over time. This is likely due to the predictable tidal flooding regime at the field sites, which creates predictable groundwater flow regimes [57]. The ERT data also show little variation across seasons, suggesting that seasonal changes do not greatly impact the groundwater conductivity at any of the study sites. Collectively, the frequent and dominant tidal flushing signature of the platform might obscure any such relationships if they exist. Alternatively, while the ERT profiles image relatively deep beneath the marsh platform, the observed similarities may suggest that, at the horizontal and vertical resolution of the ERT data, the data are insufficient to detect changes in soil aeration conditions. This limitation includes macroporosity from crab burrows and roots, as well as crab burrow density. Furthermore, the ERT data are insufficient to detect changes in water conductivity resulting from evapotranspiration in the root zone, as observed by field researchers [37,58]. Alternatively, such signatures may be masked by the tidal signal, as suggested by Thibodeau [53].

The ERT data across the marsh platform were similar beneath all three vegetation condition classes (dieback, affected, and healthy; Figure 4A–F). Furthermore, we detected no changes to the shallow hydrostratigraphy as the marsh vegetation began recolonizing the dieback areas during the period of dieback recovery between the October 2015 and March 2016 ERT collection dates (Figure 4A–F). Additionally, changes in vegetation type or marsh dieback and recovery status do not appear to be affected by depth to groundwater. However, any potential relationships between initial dieback and resistivity would have been missed as our sampling occurred well after the onset of dieback.

We found that vegetation species followed a general zonation pattern across the topographical gradient from low to high marsh. Observed plant distributions corresponded to those previously described [59] for Southeastern salt marshes with tall *S. alterniflora* at lower elevations and *J. roemerianus* and *B. frutescens* at higher elevations. McKee and Patrick [60] found that species elevations can vary widely among salt marshes, making elevation comparisons difficult between marsh sites. The mean RTK elevation and range for each site correspond with previous observations in Georgia salt marshes [61,62]. However, despite the elevation difference among the studied marsh sites, we did not detect corresponding changes in bulk resistivity, suggesting that elevation may not influence resistivity patterns.

ERT profiles show similar bulk resistivities across different tidal heights at each field site, suggesting that the profiles were collected too far from the creek to detect tidally driven groundwater flow or depth to any mud/sand contact at the creek bank. The cable was positioned 56, 51, and 47 m from the low tide line of the creek at GC, SSI, and PP, respectively (Table 1). Several studies have modeled [48,56,57,63] or measured [64,65] groundwater flow and tidal signal propagation from the creek for various marsh geometries in the southeastern United States. Except for storm-driven groundwater flow [48], these studies generally agree that groundwater flow typically occurs within 2 to 15 m of the creek, and tidal signal propagation into the marsh platform rarely exceeds 35 m from the creek. Thus, the ERT results align with previously published models and field data.

Finally, the deployment of the ERT cable on the marsh had ecological consequences. Despite our efforts to minimize disturbance, the process left a visible path through the marsh due to the large crew required to deploy the equipment. This resulted in unintended damage to vegetation and soil structure. Based on our experience and the consistent ERT results obtained at each field site, we recommend conducting only one ERT survey at any given marsh site with the smallest crew possible to minimize ecological impact.

5. Conclusions

This study used multichannel ERT surveys to identify subsurface resistivity patterns beneath three salt marshes. The results for all sites suggest typical marsh hydrostratigraphy consisting of unsaturated mud and peat overlying an unconfined muddy aquifer saturated with saltwater. There is minimal evidence of freshwater influence. We established that the ERT data reveal consistent subsurface resistivity patterns across seasons and tidal flooding. Bulk resistivity patterns were similar under *S. alterniflora* and *J. roemerianus*. Further, all three marsh vegetation condition classes (healthy, affected, dieback) showed similar bulk resistivity patterns. Vegetation status, seasonal changes, and tidal variation had minimal influence on water depth or chemistry, suggesting that tidal flooding predominantly governs shallow subsurface dynamics in these marshes. Our findings suggest that the processes driving the recovery of marsh dieback are independent of the shallow marsh stratigraphy. These findings contribute to a better understanding of salt marsh subsurface conditions and the role they play in marsh resilience and dieback recovery, informing future studies on the hydrological factors influencing marsh health.

Author Contributions: Conceptualization, J.L.K. and C.M.H.; methodology, J.L.K. and C.M.H.; investigation, J.L.K. and C.M.H.; resources, J.L.K. and C.M.H.; data curation, J.L.K. and C.M.H.; writing—original draft preparation, J.L.K. and C.M.H.; writing—review and editing, J.L.K. and C.M.H.; project administration, J.L.K. and C.M.H.; funding acquisition, J.L.K. and C.M.H. All authors have read and agreed to the published version of the manuscript.

Funding: This research was funded under grant award #NA15NOS4190160 to the Department of Natural Resources from the Office of Ocean and Coastal Resource Management, National Oceanic and Atmospheric Administration. The statements, findings, conclusions, and recommendations are those of the authors and do not necessarily reflect the views of DNR, OCRM, or NOAA.

Data Availability Statement: Some data for the St. Simons Island field site are openly available in interactive form via a Story Map titled "A Case Study on Salt Marsh Dieback", which is available online at: https://storymaps.arcgis.com/stories/aabb57ed8a524e9f8a0d56b129d41460 (accessed on 3 January 2025). The raw data supporting the conclusions of this article will be made available by the authors on request.

Acknowledgments: We thank the many students from Georgia Southern University who helped us collect the data presented herein. We thank the Coastal Resources Division and Jan Mackinnon for their help with selecting field sites. Comments by two anonymous reviewers greatly improved this manuscript.

Conflicts of Interest: The authors declare no conflicts of interest.

References

1. Costanza, R.; d'Arge, R.; de Groot, R.; Farber, S.; Grasso, M.; Hannon, B.; Limburg, K.; Naeem, S.; O'Neill, R.V.; Paruelo, J.; et al. The value of the world's ecosystem services and natural capital. *Nature* **1997**, *387*, 253–260. [CrossRef]
2. Mitsch, W.J.; Gosselink, J.G. *Wetlands*, 3rd ed.; John Wiley and Sons: Hoboken, NJ, USA, 2000.
3. Elsey-Quirk, T.; Lynn, A.; Jacobs, M.D.; Diaz, R.; Cronin, J.; Wang, L.; Huang, H.; Justic, D. Vegetation dieback in the Mississippi River Delta triggered by acute drought and chronic relative sea-level rise. *Nat. Commun.* **2024**, *15*, 3518. [CrossRef] [PubMed]
4. Hughes, A.L.; Wilson, A.M.; Morris, J.T. Hydrologic variability in a salt marsh: Assessing the links between drought and acute marsh dieback. *Estuar. Coast. Shelf Sci.* **2012**, *111*, 95–106. [CrossRef]
5. Marsh, A.; Blum, L.K.; Christian, R.R.; Ramsey III, E.; Rangoonwala, A. Response and resilience of *Spartina alterniflora* to sudden dieback. *J. Coast. Conserv.* **2016**, *20*, 335–350. [CrossRef]
6. Morton, R.A.; Tiling, G.; Ferina, N.F. Causes of hot-spot wetland loss in the Mississippi delta plain. *Environ. Geosci.* **2003**, *10*, 71–80. [CrossRef]
7. Schepers, L.; Kirwan, M.; Guntenspergen, G.; Temmerman, S. Spatio-temporal development of vegetation die-off in a submerging coastal marsh. *Limnol. Oceanogr.* **2017**, *62*, 137–150. [CrossRef]
8. Alber, M.; Swenson, E.M.; Adamowicz, S.C.; Mendelssohn, I.A. Salt marsh dieback: An overview of recent events in the US. *Estuar. Coast. Shelf Sci.* **2008**, *80*, 1–11. [CrossRef]
9. Ogburn, M.B.; Alber, M. An investigation of salt marsh dieback in Georgia using field transplants. *Estuaries Coasts* **2006**, *29*, 54–62. [CrossRef]
10. Day, J.W.; Christian, R.R.; Boesch, D.M.; Yáñez-Arancibia, A.; Morris, J.; Twilley, R.R.; Naylor, L.; Schaffner, L.; Stevenson, C. Consequences of Climate Change on the Ecogeomorphology of Coastal Wetlands. *Estuaries Coasts* **2008**, *31*, 477–491. [CrossRef]
11. Baustian, J.J.; Mendelssohn, I.A.; Hester, M.W. Vegetation's importance in regulating surface elevation in a coastal salt marsh facing elevated rates of sea level rise. *Glob. Change Biol.* **2012**, *18*, 3377–3382. [CrossRef]
12. Coleman, D.J.; Kirwan, M.L. The effect of a small vegetation dieback event on salt marsh sediment transport. *Earth Surf. Process. Landf.* **2019**, *44*, 944–952. [CrossRef]
13. Krest, J.M.; Moore, W.S.; Gardner, L.R.; Morris, J.T. Marsh nutrient export supplied by groundwater discharge: Evidence from radium measurements. *Glob. Biogeochem. Cycles* **2000**, *14*, 167–176. [CrossRef]
14. Morris, J.T.; Sundareshwar, P.V.; Nietch, C.T.; Kjerfve, B.; Cahoon, D.R. Responses of coastal wetlands to rising sea level. *Ecology* **2002**, *83*, 2869–2877. [CrossRef]
15. Materne, M.; Bush, T.; Houck, M. Plant Guide for Smooth Cordgrass (*Spartina alterniflora*). USDA-Natural Resources Conservation Service. 2022. Available online: https://www.nrcs.usda.gov/plantmaterials/njpmcpg13933.pdf (accessed on 3 January 2025).
16. Eleuterius, L.N. Vegetative Morphology and Anatomy of the Salt Marsh Rush, *Juncus roemerianus*. *Gulf Res. Rep.* **1976**, *5*, 1–10. [CrossRef]
17. Moore, W.S.; Blanton, J.O.; Joye, J. Estimates of flushing times, submarine groundwater discharge, and nutrient fluxes to Okatee Estuary, South Carolina. *J. Geophys. Res.* **2006**, *111*, C09006. [CrossRef]
18. Swarzenski, P.W.; Izbicki, J.A. Coastal groundwater dynamics off Santa Barbara, California: Combining geochemical tracers, electromagnetic seepmeters, and electrical resistivity. *Estuar. Coast. Shelf Sci.* **2009**, *83*, 77–89. [CrossRef]
19. Dimova, N.T.; Swarzenski, P.W.; Dulaiova, H.; Glenn, C.R. Utilizing multichannel electrical resistivity methods to examine the dynamics of the fresh water-seawater interface in two Hawaiian groundwater systems. *J. Geophys. Res.* **2012**, *117*, C02012. [CrossRef]
20. Swarzenski, P.W.; Burnett, W.C.; Greenwood, W.J.; Herut, B.; Peterson, R.; Dimova, N.; Shalem, Y.; Yechieli, Y.; Weinstein, Y. Combined time-series resistivity and geochemical tracer techniques to examine submarine groundwater discharge at Dor Beach, Israel. *Geophys. Res. Lett.* **2006**, *33*, L24405. [CrossRef]
21. U.S. Climate Data. Available online: https://www.usclimatedata.com/ (accessed on 13 November 2024).
22. Wiegert, R.G.; Pomeroy, L.R.; Wiebe, W.J. Ecology of salt marshes: An introduction. In *The Ecology of a Salt Marsh*; Pomeroy, L.R., Wiegert, R.G., Eds.; Springer: New York, NY, USA, 1981; pp. 3–19.
23. Tides and Currents (Datums for 8670870, Fort Pulaski GA). Available online: https://tidesandcurrents.noaa.gov/datums.html?datum=MLLW&units=1&epoch=0&id=8670870&name=Fort+Pulaski&state=GA (accessed on 13 November 2024).

24. Tides and Currents (Datums for 8720030, Fernandina Beach FL). Available online: https://tidesandcurrents.noaa.gov/datums.html?datum=MLLW&units=1&epoch=0&id=8720030&name=Fernandina+Beach&state=FL (accessed on 13 November 2024).
25. Tides and Currents (Relative Sea Level Trend Fort Pulaski). Available online: https://tidesandcurrents.noaa.gov/sltrends/sltrends_station.shtml?id=8670870 (accessed on 13 November 2024).
26. Tides and Currents (Relative Sea Level Trend Fernandina Beach). Available online: https://tidesandcurrents.noaa.gov/sltrends/sltrends_station.shtml?id=8720030 (accessed on 13 November 2024).
27. Dibble, M. Examination of Salt Marsh Dieback Development and Recovery Using Historical Aerial Photography and Climatic Conditions. Bachelor's Thesis, Georgia Southern University, Statesboro, GA, USA, 2017.
28. Trimble R8 GNSS Receiver User Guide. Available online: https://trl.trimble.com/docushare/dsweb/Get/Document-666215/R8-R6-5800_v400A_UserGuide.pdf (accessed on 23 July 2024).
29. Rinaldi, V.; Guichon, M.; Ferrero, V.; Serrano, C.; Ponti, N. Resistivity Survey of the Subsurface Conditions in the Estuary of the Rio de la Plata. *J. Geotech. Geoenviron. Eng.* **2006**, *132*, 72–79. [CrossRef]
30. Loke, M.H.; Chambers, J.E.; Rucker, D.F.; Kuras, O.; Wilkinson, P.B. Recent developments in the direct-current geoelectrical imaging method. *J. Appl. Geophys.* **2013**, *95*, 135–156. [CrossRef]
31. Advanced Geosciences Incorporated Instruction Manual for EarthImager 2D Resistivity and IP Inversion Software. Available online: http://www.agiusa.com/ (accessed on 13 November 2024).
32. Nguyen, F.; Garambois, S.; Jongmans, D.; Pirard, E.; Loke, M.H. Image processing of 2D resistivity data for imaging faults. *J. Appl. Geophys.* **2005**, *57*, 260–277. [CrossRef]
33. Caputo, R.; Piscitelli, S.; Oliveto, A.; Rizzo, E.; Lapenna, V. The use of electrical resistivity tomographies in active tectonics: Examples from the Tyrnavos Basin, Greece. *J. Geodyn.* **2003**, *36*, 19–35. [CrossRef]
34. Hoffmann, R.; Dietrich, P. An approach to determine equivalent solutions to the geoelectrical 2D inversion problem. *J. Appl. Geophys.* **2004**, *56*, 79–91. [CrossRef]
35. Oldenburg, D.W.; Li, Y. Estimating depth of investigation in dc resistivity and IP surveys. *Geophysics* **1999**, *64*, 403–416. [CrossRef]
36. Michaels, R.E.; Zieman, J.C. Fiddler crab (*Uca* spp.) burrows have little effect on surrounding sediment oxygen concentrations. *J. Exp. Mar. Biol. Ecol.* **2013**, *448*, 104–113. [CrossRef]
37. Hemond, H.F.; Fifield, J.L. Subsurface flow in salt marsh peat: A model and field study. *Limnol. Oceanogr.* **1982**, *27*, 126–136. [CrossRef]
38. Granse, D.; Titschack, J.; Ainouche, M.; Jensen, K.; Koop-Jackobsen, K. Subsurface aeration of tidal wetland soils: Root-system structure and aerenchyma connectivity in *Spartina* (Poaceae). *Sci. Total Environ.* **2022**, *802*, 149771. [CrossRef]
39. Li, H.; Li, L.; Lockington, D. Aeration for plant root respiration in a tidal marsh. *Water Resour. Res.* **2005**, *41*, W06023. [CrossRef]
40. Xin, P.; Jin, G.; Li, L.; Barry, D.A. Effects of crab burrows on pore water flows in salt marshes. *Adv. Water Resour.* **2009**, *32*, 439–449. [CrossRef]
41. Silva Filho, A.M.; Silva, J.R.S.; Fernandes, G.M.; Morais, L.D.; Coimbra, A.P.; Calixto, W.P. Root system analysis and influence of moisture on soil electrical properties. *Energies* **2021**, *14*, 6951. [CrossRef]
42. Archie, G.E. The Electrical Resistivity Log as an Aid in Determining Some Reservoir Characteristics. *Trans. AIME* **1942**, *146*, 54–62. [CrossRef]
43. Papen, M.; Hanssens, D.; De Smedt, P.; Walraevens, K.; Hermans, T. Combining resistivity and frequency domain electromagnetic methods to investigate submarine groundwater discharge in the littoral zone. *Hydrol. Earth Syst. Sci.* **2020**, *24*, 3539–3555. [CrossRef]
44. Koontz, E.L.; Parker, S.M.; Stearns, A.E.; Roberts, B.J.; Young, C.M.; Windham-Myers, L.; Oikawa, P.Y.; Megonigal, J.P.; Noyce, G.L.; Buskey, E.J.; et al. Controls on spatial variation in porewater methane concentrations across United States tidal wetlands. *Sci. Total Environ.* **2024**, *957*, 177290. [CrossRef] [PubMed]
45. Harvey, J.W.; Germann, P.F.; Odum, W.E. Geomorphological control of subsurface hydrology in the creekbank zone of tidal marshes. *Estuar. Coast. Shelf Sci.* **1987**, *25*, 677–691. [CrossRef]
46. Gardner, L.R.; Porter, D.E. Stratigraphy and geologic history of a southeastern salt marsh basin, North Inlet, South Carolina, USA. *Wetl. Ecol. Manag.* **2001**, *9*, 371–385. [CrossRef]
47. Gardner, L.R. Role of stratigraphy in governing pore water seepage from salt marsh sediments. *Water Resour. Res.* **2007**, *43*, W07502. [CrossRef]
48. Wilson, A.M.; Moore, W.S.; Joye, S.B.; Anderson, J.L.; Schutte, C.A. Storm-driven groundwater flow in a salt marsh. *Water Resour. Res.* **2011**, *47*, W02535. [CrossRef]
49. Wilson, A.M.; Evans, T.; Moore, W.; Schutte, C.A.; Joye, S.B.; Hughes, A.H.; Anderson, J.L. Groundwater controls ecological zonation of salt marsh macrophytes. *Ecology* **2015**, *96*, 840–849. [CrossRef]
50. Alexander, C.R.; Hodgson, J.Y.S.; Brandes, J.A. Sedimentary processes and products in a mesotidal salt marsh environment: Insights from Groves Creek, Georgia. *Geo-Mar. Lett.* **2017**, *37*, 345–359. [CrossRef]

51. Lonard, R.I.; Judd, F.W.; Stalter, R. Biological flora of coastal dunes and wetlands: *Borrichia frutescens* (L.) DC. *J. Coast. Res.* **2015**, *31*, 749–757. [CrossRef]
52. Pennings, S.C.; Grant, M.-B.; Bertness, M.D. Plant zonation in low-latitude salt marshes: Disentangling the roles of flooding, salinity and competition. *J. Ecol.* **2005**, *93*, 159–167. [CrossRef]
53. Thibodeau, P.M.; Gardner, L.R.; Reeves, H.W. The role of groundwater flow in controlling the spatial distribution of soil salinity and rooted macrophytes in a southeastern salt marsh, USA. *Mangroves Salt Marshes* **1998**, *2*, 1–13. [CrossRef]
54. Gardner, L.R.; Reeves, H.W. Seasonal patterns in the soil water balance of a *Spartina* marsh site at North Inlet, South Carolina, USA. *Wetlands* **2002**, *22*, 467–477. [CrossRef]
55. Montalvo, M.S.; Grande, E.; Graswell, A.E.; Visser, A.; Arora, B.; Seybold, E.C.; Tatariw, C.; Haskiins, J.C.; Endris, C.A.; Gerbl, F.; et al. A Fresh Take: Seasonal Changes in Terrestrial Freshwater Inputs Impact Salt Marsh Hydrology and Vegetation Dynamics. *Estuaries Coasts* **2024**, *47*, 2389–2405. [CrossRef]
56. Wilson, A.M.; Morris, J.T. The influence of tidal forcing on groundwater flow and nutrient exchange in a salt marsh-dominated estuary. *Biogeochemistry* **2012**, *108*, 27–38. [CrossRef]
57. Wilson, A.M.; Gardner, L.R. Tidally driven groundwater flow and solute exchange in a marsh: Numerical simulations. *Water Resour. Res.* **2006**, *42*, W01405. [CrossRef]
58. Xin, P.; Gibbes, B.; Ling, L.; Song, Z.; Lockington, D. Soil saturation index of salt marshes subjected to spring-neap tides: A new variable for describing marsh soil aeration condition. *Hydrol. Process.* **2010**, *24*, 2564–2577. [CrossRef]
59. Wiegert, R.G.; Freeman, B.J. *Tidal Salt Marshes of the Southeast Atlantic Coast: A Community Profile*; U.S. Department of the Interior, Fish and Wildlife Service: Washington, DC, USA, 1990. [CrossRef]
60. McKee, K.L.; Patrick, W.H., Jr. The relationship of smooth cordgrass (*Spartina alterniflora*) to tidal datums: A review. *Estuaries* **1988**, *11*, 143–151. [CrossRef]
61. Hladik, C.M.; Alber, M.; Schalles, J.F. Salt marsh elevation and habitat mapping using hyperspectral and LIDAR data. *Remote Sens. Environ.* **2013**, *139*, 318–330. [CrossRef]
62. Hladik, C.; Alber, M. Accuracy assessment and correction of a LIDAR-derived salt marsh digital elevation model. *Remote Sens. Environ.* **2012**, *121*, 224–235. [CrossRef]
63. Gardner, L.R. Role of geomorphic and hydraulic parameters in governing pore water seepage from salt marsh sediments. *Water Resour. Res.* **2005**, *41*, W07010. [CrossRef]
64. Schultz, G.; Ruppel, C. Constraints on hydraulic parameters and implications for groundwater flux across the upland-estuary interface. *J. Hydrol.* **2002**, *260*, 255–269. [CrossRef]
65. Moffett, K.B.; Gorelick, S.M. Relating salt marsh pore water geochemistry patterns to vegetation zones and hydrologic influences. *Water Resour. Res.* **2016**, *52*, 1729–1745. [CrossRef]

water

MDPI

Article

Multi-Attribute Analysis of Transmission Channel Waves: Applications in Mine Water Damage Prevention

Zean Hu [1,2], Tianhao Zhang [1,*] and Mengjie Zhan [1]

[1] School of Earth and Environment, Anhui University of Science and Technology, Huainan 232001, China; zahu@aust.edu.cn (Z.H.); zhanmj123@163.com (M.Z.)
[2] State Key Laboratory for Safe Mining of Deep Coal Resources and Environmental Protection, Huainan 232000, China
* Correspondence: ztianhao123@163.com; Tel.: +86-13912367789

Abstract: In-seam seismics is one of the main methods for detecting small geostructures in a coal seam working face, using the velocity and energy attenuation characteristics. When the receiver coupling is poor or geological anomaly is strong, the stability and accuracy of the inversion results are affected greatly. In order to improve the stability and enrich the parameters of the inversion, this paper proposes a channel wave multi-attribute tomography method. Based on the theoretical analysis, the formulas for calculating the spectral ratio, arc length, and bandwidth parameters of the channel wave were obtained. A two-dimensional numerical simulation was used to analyze the attenuation characteristic and exploration potential of three attribute parameters of the channel wave. Through the field measured experiment, the exploration effect of the three attributes was realized and compared. In conclusion, the effectiveness of the channel wave multi-attribute tomography method in characterizing geological structures within coal seams was successfully demonstrated and verified. This approach offers a novel and robust solution for channel wave data processing and exploration, significantly enhancing the prevention of mine water damage.

Keywords: in-seam seismics; attribute analysis; spectral ratio; arc length; bandwidth

1. Introduction

The primary challenge in achieving green and intelligent coal mining lies in constructing detailed and transparent geological models that account for factors such as faults, goaf areas, and in situ stress and mine water [1]. The occurrence and migration of mine water in coal seams and the surrounding rock masses are complex. It will interact with other concealed disaster-causing bodies, further intensifying safety risks [2,3]. Accidents caused by these hidden disaster-inducing factors are the primary barriers to the intelligent mining and safe production of coal mines [4]. Minor faults within coal seams are a significant factor in coal mine safety incidents, making an accurate prediction of these faults an urgent necessity [5–7]. While 3D seismic exploration is effective for detecting faults with offsets greater than 3 m, new methods are needed to detect faults with offsets of less than 3 m [8–10]. Channel wave exploration, with its high resolution and extended detection range, has proven effective for identifying smaller faults [11–14].

Channel wave signals, however, often suffer from noise interference due to factors like geophone coupling during data acquisition, resulting in poor amplitude preservation [15]. Simultaneously, the presence of mine water can alter the electrical and acoustic properties of coal seams and surrounding rocks, further disrupting the propagation and reception of channel wave signals. Large geological anomalies can cause severe channel wave energy attenuation, creating excessive energy discrepancies between adjacent signal channels. This, in turn, leads to imaging results that disproportionately highlight certain anomalies while neglecting smaller ones. Geological anomalies in a coal mining face will cause differences in channel wave velocity and energy attenuation. At present, channel wave exploration faults are mainly velocity and attenuation coefficient tomography. This method is widely used in the exploration of geological anomalies in working faces such as faults, collapse columns, and stress concentration areas. By analyzing and comparing the exploration of small faults by transmission and reflection channel wave methods, the response characteristics of in-seam wave to small and medium faults in thick coal seams and extremely thick coal seams are obtained [16,17]. Through the joint exploration of seismic in-seam wave transmission and reflection, the internal structure of coal seam can be better obtained [18]. But they lack the precision needed for accurate time synchronization in channel wave transmission. These issues reduce the accuracy of both energy attenuation and velocity imaging. Channel wave single-attribute imaging suffers from ambiguity, failing to provide definitive judgments on coal seam structures. Moreover, the interference and influences from factors related to mine water further complicate the process of making such judgments [19].

To address these challenges, this paper proposes a multi-attribute analysis and imaging approach for channel waves. Building upon previous work, we selected three attributes of channel waves for research: namely, the spectral ratio, arc length, and bandwidth [20,21]. The multi-attribute characteristics of channel waves aid in identifying geological structures within coal seams, thereby enhancing the potential of channel wave exploration and offering new technologies for preventing mine water damage.

2. Materials and Methods

Figure 1 illustrates the technical route and successive stages of the study.

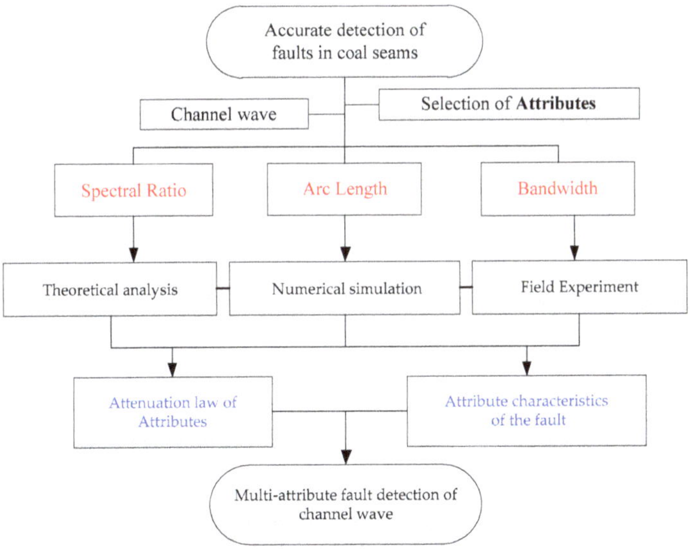

Figure 1. Technical route of the study.

2.1. Theoretical Approach

2.1.1. Spectral Ratio

The spectral ratio is one of the commonly used attributes in reflected-wave seismic exploration and is often applied to reservoir interpretation. The spectral ratio refers to the ratio relationship between different frequency components in the spectrum analysis of seismic signals. Anomalous bodies are geological formations that exhibit significant differences in physical properties, such as density and elastic modulus, when compared to the surrounding medium. Anomalous bodies produce frequency-selective absorption and scattering of seismic waves. High-frequency seismic waves are typically more susceptible to absorption or scattering, whereas low-frequency seismic waves tend to pass through more readily. This results in a change in the seismic wave spectrum following the abnormal body, with a relative reduction in the high-frequency component and a relative increase in the low-frequency component, thereby altering the spectral ratio [22]. The channel wave signal undergoes FFT (Fast Fourier Transform) to obtain the frequency-amplitude spectrum $R(f_j)$, $f_j = (j-1)/T$, Here, j represents the number of sampling points, where $j = 1, \ldots, N$, T is the total sampling time [23]. The frequency corresponding to the maximum energy in the amplitude spectrum is the dominant frequency of the seismic wave f_p. The spectral ratio represents the ratio of low-frequency energy to high-frequency energy. The calculation formula for a single-trace signal is as follows [24]:

$$H = \sum_{f=1}^{f_p} R^2(f_j) / \sum_{f=f_p}^{(N-1)/T} R^2(f_j) \tag{1}$$

2.1.2. Arc Length

Arc length is one of the commonly used seismic waveform attributes and is often used in the interpretation of sandstone and shale interfaces in reflected seismic waves [25]. In geological structures, such as faults and collapse columns, the propagation path of seismic waves becomes complexly affected. The arc length attribute can highlight these structural areas because seismic waves near the structure will exhibit phenomena such as diffraction and scattering, causing an abnormal change in arc length [26]. The arc length is defined as the waveform length of a seismic trace, which is a proportional measurement of the variation range of the seismic trace within a time window. Imagine using a rope to draw the curve of the seismic trace, and the arc length of the seismic trace is the total length when the rope is stretched out. Practice has shown that this attribute parameter is highly applicable to high-frequency signals. Channel wave signals exhibit dispersion, with abundant amplitude information in different frequency bands, and their characteristics fall within the applicable range of the arc length attribute.

The calculation formula for arc length is as follows [27]:

$$S = \frac{1}{N\Delta T} \sum_{i=1}^{N} \sqrt{[A(i+1) - A(i)]^2 + \Delta T^2} \tag{2}$$

ΔT is the sampling interval time, N is the total number of sampling points, and $A(i)$ is the amplitude value of the i-th sampling point.

2.1.3. Bandwidth

Bandwidth is a commonly used seismic attribute in reflection seismic exploration, particularly for interpreting oil and gas reservoirs [28]. Due to the development of internal fissures and loose fillings within the fault or fracture zone, there is strong scattering and absorption of seismic waves during their propagation. This results in significant attenuation of the high-frequency components, leading to a narrower bandwidth (with a reduced main frequency and a downward shift in the high-frequency cut-off frequency) [29]. The bandwidth parameter reflects the frequency range in which the vast majority of the seismic wave energy is concentrated on the frequency spectrum curve. The channel wave energy disperses as the propagation distance increases, and the bandwidth parameter can well describe this process.

Bandwidth is defined as the absolute value of the change in the logarithm of the amplitude. Its mathematical calculation formula is as follows [30]:

$$\sigma = \frac{1}{2\pi}\sum_{i=1}^{N}\left|\frac{d}{di}\ln A(i)\right| \tag{3}$$

$A(i)$ represents the amplitude value of the i-th sampling point.

2.2. Experimental Approach

2.2.1. Numerical Simulation Experiment

To investigate the theoretical characteristics of the transmission channel wave attributes, this study employs a high-order staggered-grid finite-difference method to simulate the 3D elastic wave equation. We designed two 3D models, each measuring 1000 m × 200 m × 200 m (Figure 2), and used numerical simulation data to calculate and analyze the three attributes of channel wave described in the previous section. This analysis aims to explore the relationship between channel wave attribute parameters and propagation distance. To minimize calculation errors in the numerical simulation, the coal seam thickness in each model is set to 10 m.

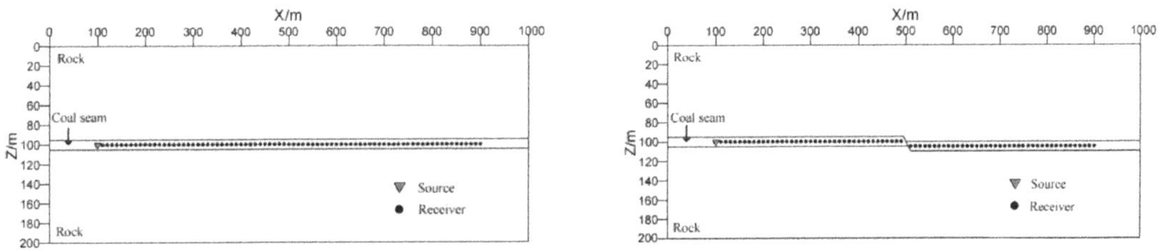

(a) Faultless (b) Fault-bearing

Figure 2. Cross-section of the 3D numerical model (at y = 100 m).

Figure 2a shows a model with a uniform horizontal coal seam, while Figure 2b includes a fault structure located at X = 500 m, with a fault throw of 5 m. The specific physical parameters of the models are detailed in Table 1.

Table 1. Physical parameters of the numerical model.

Medium	P-Wave Velocity	S-Wave Velocity	Densities	Quality Factor
Coal seam	1700 m/s	1000 m/s	1300 Kg/m^3	50
Rock	3500 m/s	2000 m/s	2600 Kg/m^3	150

2.2.2. Field Experiment

To verify the effectiveness of the attenuation feature imaging based on the transmission channel wave attribute parameters, an in-field experiment was conducted at a coal mine working face where channel waves are developed.

The experiment was conducted at the No. 4108 working face, located in the southwest part of the mining area. The working face is in the first mining zone, with a mining level at the first horizon. The strike length is 1444 m, the dip length is 274 m, and the available mining length is 1315 m. The primary sources of direct water inrush at the 4108 working face are attributed to fissure water in the sandstone roof and floor of the No. 4 coal seam. The indirect sources of water inrush include the floor limestone water and the deep Ordovician limestone water. Therefore, identifying minor faults in the working face that may lead to water inrush is crucial for preventing mine water damage in that area.

The experimental setup involved placing seismic sources and geophones along the tunnel in the working face. Both the geophone and shot point spacing were set to 10 m. Considering the tunnel equipment and production conditions, a total of 133 receivers and 76 effective sources were arranged (Figure 3). Data collection was carried out using the "one shot, multiple recordings" method, and a total of 76 Common Shot Point gathers were collected.

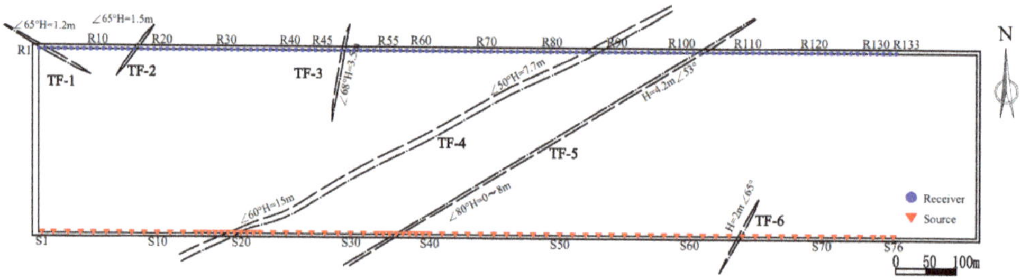

Figure 3. The transmission channel wave experimental observation system.

3. Results

3.1. Numerical Simulation Results and Analysis

The channel wave records of three-dimensional models of two types of coal-rock media were obtained through simulation, as shown in Figure 4. By comparing the channel wave records of the two models, it was found that in the channel wave signal of the model with a fault, there was a sudden decrease in energy starting from the 40th trace.

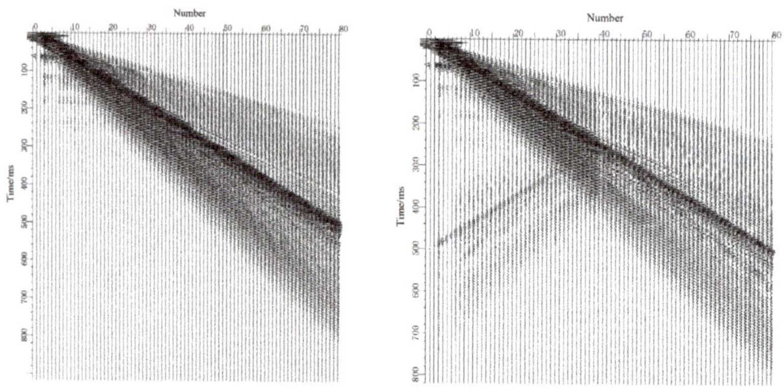

(**a**) Faultless (**b**) Fault-bearing

Figure 4. Channel wave recording from the numerical simulation.

To validate the reliability of the numerical simulation data, we analyzed the total energy characteristics of the simulated channel wave signals. Figure 5 shows the total energy calculation results for the channel wave. Since the source energy in the numerical simulation was not assigned units, the total energy data is dimensionless. The total energy of the channel wave exhibits a significant linear decay with propagation distance, confirming the reliability of the numerical simulation data. Further analysis can be conducted on other attribute characteristics.

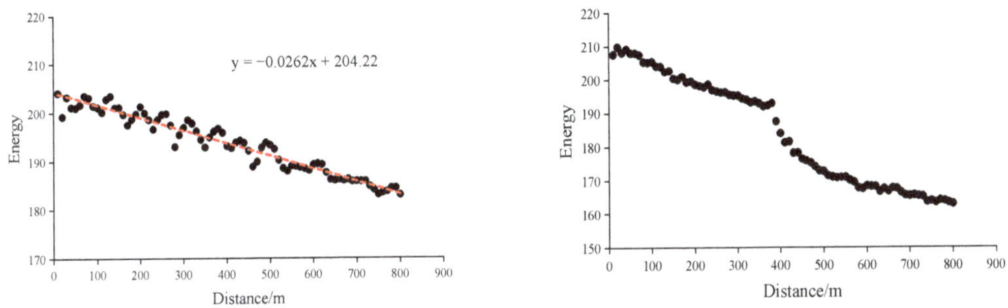

(a) Faultless (b) Fault-bearing

Figure 5. Calculation results of total energy from numerical simulation signals.

Based on Formula (1), the spectral ratio of the numerical simulation signals is depicted in Figure 6. Figure 6a presents the calculation results of the numerical simulation signals from the faultless model depicted in Figure 2a, whereas Figure 6b showcases the results from the fault-bearing model as illustrated in Figure 2b. The black dots represent the spectral ratio of the channel wave signal, and the red dashed line indicates its trend. In Figure 6a, the spectral ratio exhibits a linear growth pattern, increasing with propagation distance. Figure 6b shows a significant increase in the spectral ratio at X = 400. These phenomena indicate that the growth characteristics of the channel wave spectral ratio exhibit a linear relationship with propagation distance and are sensitive to faults, making them useful parameters for channel wave tomography.

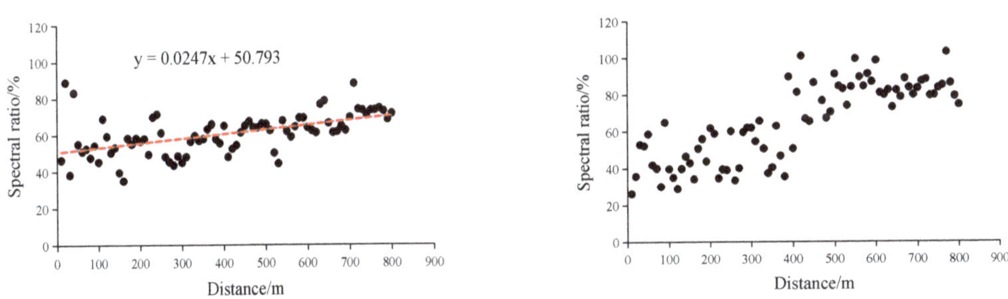

(a) Faultless (b) Fault-bearing

Figure 6. Calculation results of the spectral ratio from numerical simulation signals.

According on Formula (2), The arc length calculation results are shown in Figure 7. Since the arc length is a measure of the channel wave waveform length without a defined unit, the arc length data is dimensionless. Figure 7a presents the calculation results for the model without a fault (as in Figure 2a), while Figure 7b shows the results for the model with a fault (as in Figure 2b). In both figures, black dots represent the arc lengths of the channel wave signals, and the red dashed line indicates the trend line. As seen in Figure 7a, the arc length of the channel wave signals shows a linear decay characteristic, decreasing with increasing propagation distance. In Figure 7b, a clear arc length attenuation anomaly appears at X = 400 m. These phenomena suggest that the arc length decay characteristics of channel wave signals are also sensitive to geological anomalies and maintain a linear relationship with propagation distance, making it another valuable parameter for channel wave tomography.

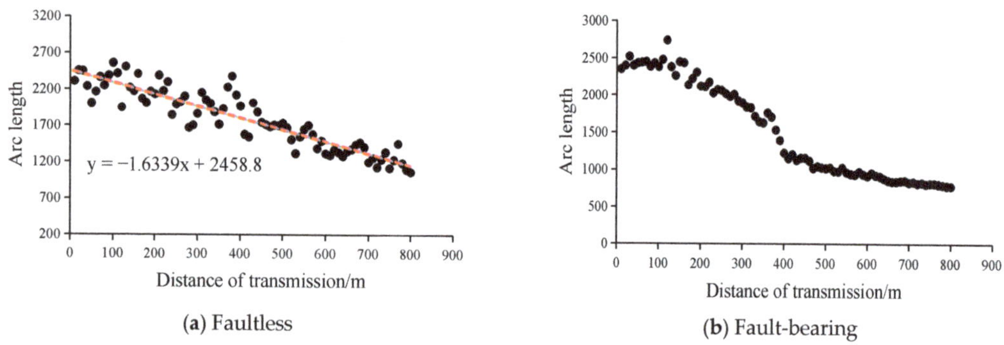

Figure 7. Calculation results of arc length from numerical simulation signals.

The bandwidth calculation results are shown in Figure 8. Figure 8a presents the calculation results for the model without a fault (as in Figure 2a), while Figure 8b shows the results for the model with a fault (as in Figure 2b). In both figures, black dots represent the bandwidth of the channel wave signals, and the red dashed line indicates its trend. As seen in Figure 8a, the bandwidth of the channel wave signals exhibits a linear increase with increasing propagation distance. In Figure 8b, a clear bandwidth increase anomaly appears at X = 400 m. This indicates that the bandwidth is also a useful parameter for channel wave tomography.

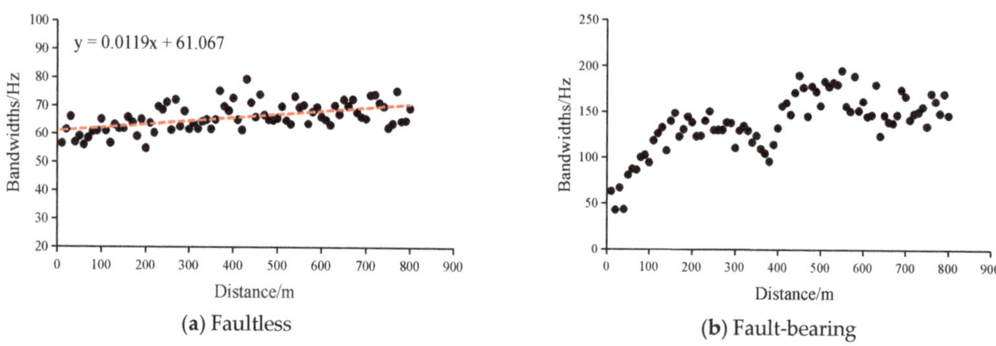

Figure 8. Calculation results of Bandwidth from numerical simulation signals.

3.2. Experimental Data and Tomographic Imaging Results

A group of common source point data was selected for preliminary analysis. The measured common shot point gathers and attribute calculation results of the 37th shot are shown in Figure 9.

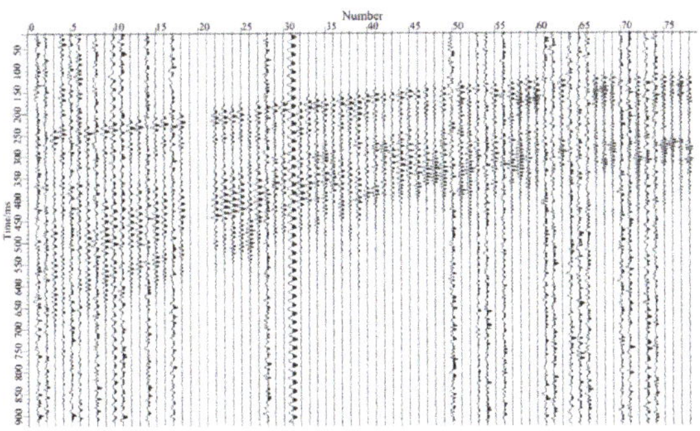

(**a**) The 37th Common Shot Point gather

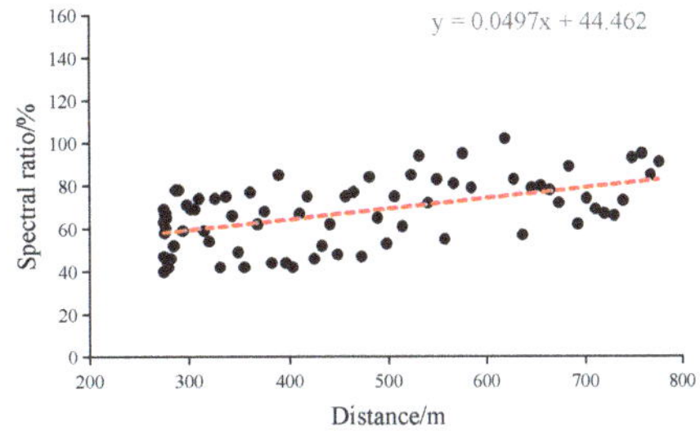

(**b**) Spectral Ratio

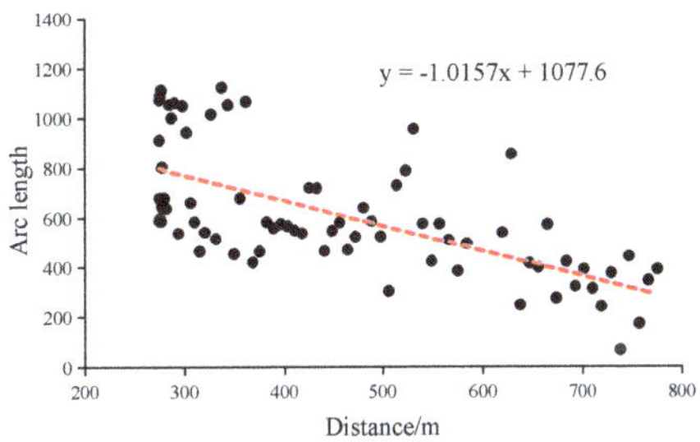

(**c**) Arc length

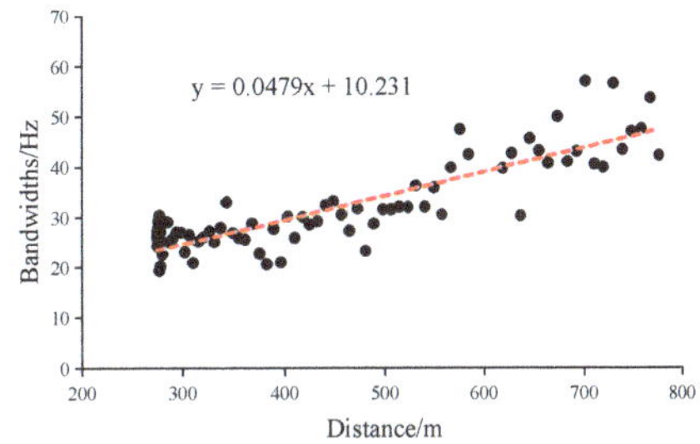

(**d**) Bandwidths

Figure 9. Three attribute characteristics of field channel wave data.

Figure 9a shows a typical shot gathered from the coal seam transmission seismic data, normalized for display. The signal has a high overall signal-to-noise ratio, and the transmission channel wave is well-developed, meeting the data quality requirements for transmission channel wave imaging. Figure 9b, Figure 9c, and Figure 9d show the spectral ratio, arc length, and bandwidth of the measured channel wave data from Shot 37, respectively. Their variations with propagation distance exhibit a clear linear relationship, making them suitable for direct use in linear tomography imaging. Since seismic source signals are typically not collected in transmission channel wave exploration, the spectral ratio, arc length, and bandwidth of the source cannot be accurately estimated and are usually assigned empirical values. From Figure 9b, Figure 9c, and Figure 9d, The empirical values for the seismic source's spectral ratio, arc length, and bandwidth are found to be 40, 1000, and 10, respectively. We then performed imaging processing on the three parameters and total energy of the measured transmission channel wave data, resulting in tomography images for each parameter, as shown in Figure 10.

Figure 10a–c illustrate the tomography results for the three-channel wave attribute parameters, whereas Figure 10d depicts the tomography result for the conventional total energy attenuation coefficient. In Figure 10d, three anomalies are predicted. However, TF-1 and TF-3 were not revealed during the actual mining, and the results in Figure 10d indicate false positives. When compared with the Total energy attenuation coefficient tomography, the spectral ratio tomography result (Figure 10a) accurately predicts the faults TF-1, TF-2, TF-4, TF-5, but fails to predict TF-3 and TF-6. The tomography results for the arc length (Figure 10b) successfully predict faults TF-3, TF-4, and TF-5. However, the bandwidth attenuation tomography results (Figure 10c) also successfully predict faults TF-3, TF-4, and TF-5 but fail to predict other faults.

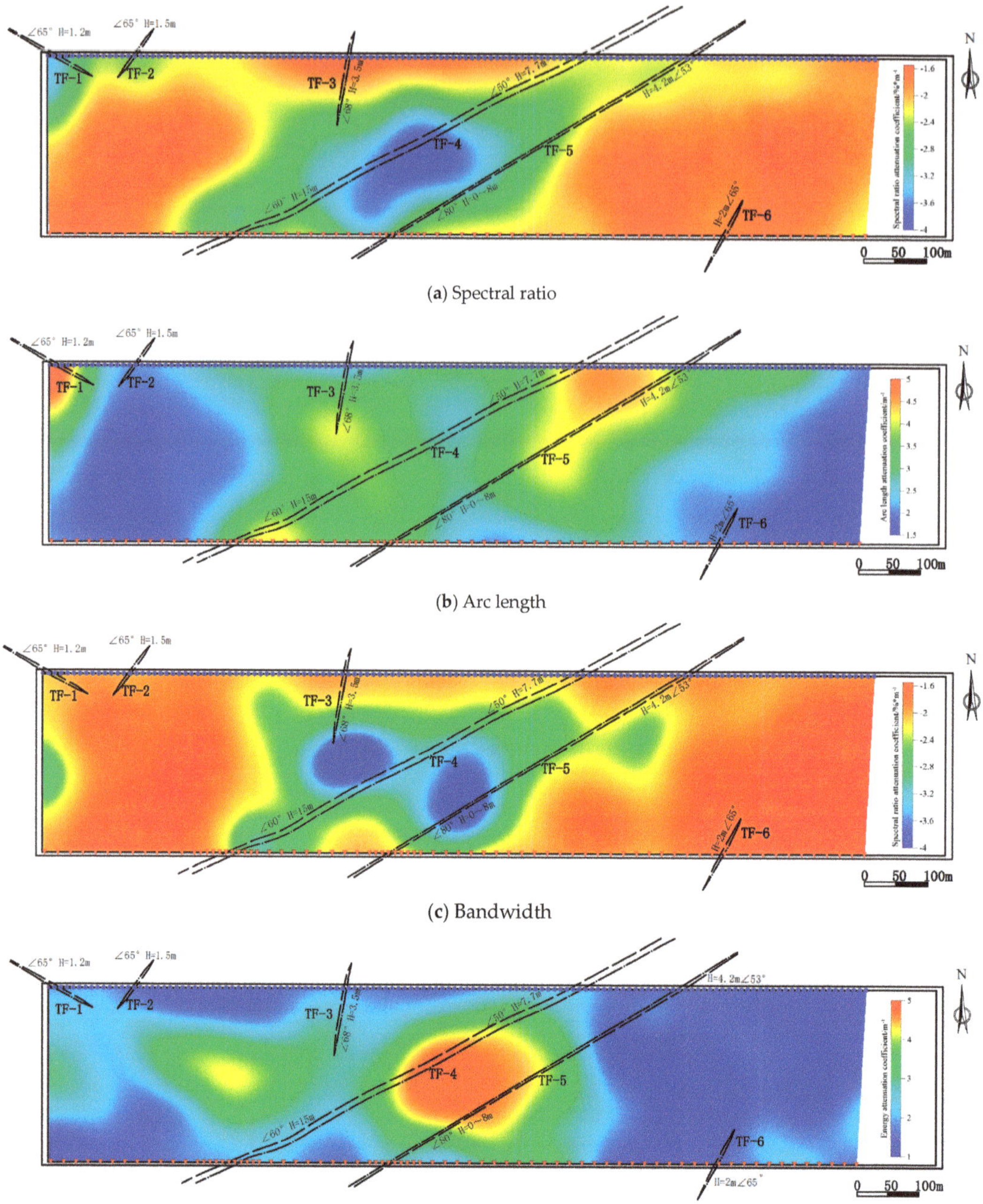

(a) Spectral ratio

(b) Arc length

(c) Bandwidth

(d) Total energy

Figure 10. Comparison of channel wave attributes' tomography result.

4. Discussion

Through theoretical analysis and numerical simulation experiments, it was discovered that the spectral ratio, arc length, and bandwidth of channel waves exhibit significant linear correlations with propagation distance and are quite sensitive to the fault of coal seams.

When the channel wave passed through a fault, all three attributes exhibited a positive response to it. In field experiments, the spectral ratio and arc length have proven effective in predicting large-scale faults. However, bandwidth and total energy are more sensitive to minor faults in coal seam working faces. It is important to emphasize that this study has only verified the response characteristics of three attributes of channel waves to faults, while the sensitivity and differences to other geological anomalous bodies, such as collapse columns and coal seam thinning areas, have not yet been supported by experimental data. Utilizing the multi-attributes of channel waves for the qualitative identification of geological anomalies in coal seams still requires further effort.

This study has, for the first time, achieved the calculation of various attributes of channel waves, obtained the response characteristics of three attribute parameters to coal seam faults, and conducted a preliminary evaluation of their effectiveness on faults. No further research has been conducted on the various attribute combinations, multi-attribute fusion, and intelligent detection models for fault identification. However, the authors acknowledge that with the deep integration of artificial intelligence, the multi-attribute technology of channel waves holds significant potential for intelligent qualitative identification and quantitative prediction of coal seam geological anomalies. It will play a more important role in coal seam geological structure and floor water prevention and control.

5. Conclusions

(1) Based on the analysis of attribute calculations from numerical simulation and measured data, the spectral ratio, arc length, and bandwidth of channel waves exhibited an obvious linear variation pattern with the propagation distance. The variation characteristics of the spectral ratio, arc length, and bandwidth of channel waves were highly sensitive to geological anomalies, making them effective parameters for tomographic imaging in channel wave exploration.

(2) This paper proposes a method for multi-attribute calculation and tomography of transmitted channel waves. The feasibility of this method has been verified through numerical simulation experiments and field measurements. This method addresses issues related to traditional single-attribute channel wave exploration, enhancing the stability and reliability of geological interpretation, and provides new technologies for mine water damage prevention.

Author Contributions: Conceptualization, Z.H. and T.Z.; methodology, Z.H.; software, T.Z.; validation, Z.H., T.Z. and M.Z.; formal analysis, Z.T; investigation, M.Z.; resources, Z.H.; data curation, T.Z.; writing—original draft preparation, T.Z.; writing—review and editing, Z.H.; visualization, T.Z.; supervision, Z.H.; project administration, Z.H.; funding acquisition, Z.H. All authors have read and agreed to the published version of the manuscript.

Funding: This work was supported by the National Natural Science Foundation (No. 42074148), University Synergy Innovation Program of Anhui Province (GXXT-2021-016), and Anhui Provincial Natural Science Foundation (No. 2208085Y14).

Data Availability Statement: The original contributions presented in this study are included in the article. Further inquiries can be directed to the corresponding authors.

Acknowledgments: Special thanks are given to the anonymous reviewers for their assistance, comments and suggestions.

Conflicts of Interest: The authors declare no conflicts of interest.

References

1. Yuan, L. Scientific conception of precision coal mining. *J. China Coal Soc.* **2017**, *42*, 1–7.
2. Xu, M.Z.; Sun, Y.J. Mine Water Safety and Environment: Chinese Experience. *Water* **2024**, *16*, 2833. [CrossRef]
3. Gu, H.L.; Shang, X.Y.; Zhao, H.T. Theory and Technology for the Prevention of Mine Water Disasters. *Water* **2024**, *16*, 2952. [CrossRef]
4. Wang, G.F.; Ren, H.W.; Zhao, G.R.; Zhang, D.; Wen, Z.; Meng, L.; Gong, S. Research and practice of intelligent coal mine technology systems in China. *Int. J. Coal Sci. Technol.* **2022**, *9*, 24.
5. Yang, T.; Xue, Y.; Juan, Y.; Mao, K.; Yao, Y.; Liang, H. Evolution dynamic of intelligent construction strategy of coal mine enterprises in China. *Heliyon* **2022**, *8*, e10933.
6. Wang, G.F.; Zhang, D.S. Innovation practice and development prospect of intelligent fully mechanized technology for coal mining. *J. China Univ. Min. Technol.* **2018**, *47*, 4–12.
7. Yang, X.H.; Cao, S.Y.; Li, D.C.; Yu, P.F.; Zhang, H.R. Analysis of quality factors for Rayleigh channel waves. *Appl. Geophys.* **2014**, *11*, 107–114. [CrossRef]
8. Yang, S.T.; Wei, J.C.; Cheng, J.L.; Shi, L.Q.; Wen, Z.J. Numerical simulations of full-wave fields and analysis of channel wave characteristics in 3-D coal mine roadway models. *Appl. Geophys.* **2016**, *13*, 621–630. [CrossRef]
9. Pei, X.D. Signal acquisition method for 3D seismic exploration in high density coal mining area. *Arab. J. Geosci.* **2020**, *13*, 17–40.
10. Zhu, M.B.; Cheng, J.Y.; Cui, W.X.; Yue, H. Comprehensive prediction of coal seam thickness by using in-seam seismic surveys and Bayesian kriging. *Acta Geophys.* **2019**, *67*, 825–836.
11. Zhang, L.Y.; Li, A.; Yang, J.G. Data processing of a wide-azimuth, broadband, high-density 3D seismic survey using a low-frequency vibroseis: A case study from Northeast China. *Explor. Geophys.* **2020**, *51*, 652–666. [CrossRef]
12. Ji, G.Z.; Zhang, P.S.; Guo, L.Q. Characteristics of dispersion curves for Love channel waves in transversely isotropic media. *Appl. Geophys.* **2020**, *17*, 243–252. [CrossRef]
13. Wei, W.; Gao, X.; Wu, Y.H. A Study on the Imaging Method for the Channel Wave Dispersion Curve Variability Function. *Minerals* **2023**, *13*, 50.
14. Wang, Y.H. Stable Q analysis on vertical seismic profiling data. *Geophysics* **2014**, *79*, 217–225. [CrossRef]
15. Wu, Y.H.; Zhu, G.W.; Wang, W.; Gao, Z. Quantitative Evaluation of Faults by Combined Channel Wave Seismic Transmission-Reflection Detection Method. *Minerals.* **2022**, *12*, 1022. [CrossRef]
16. Zhang, P.S.; Ou, Y.C.; Li, S.L. Development quo-status and thinking of mine geophysical prospecting technology and equipment in China. *Coal Sci. Technol.* **2021**, *49*, 1–15.
17. Teng, J.W.; Li, S.Y.; Jia, M.K.; Liu, H.; Liu, G.; Wang, W.; Volker, S.; Feng, L.; Yao, X.; Wang, K.; et al. Research and Application of In-seam Seismic Survey Technology for Disaster-causing Potential Geology Anomalous Body in Coal Seam. *Acta Geol. Sin. (Engl. Ed.)* **2020**, *94*, 10–26. [CrossRef]
18. Wu, Y.; Zhu, G.; Wang, W. Precise prediction of the collapse column based on channel wave spectral disparity characteristics and velocity tomography imaging. *J. Geophys. Eng.* **2022**, *19*, 326–335. [CrossRef]
19. Wang, R.; Li, Q. Application research of attribute fusion technology based on principal component analysis in fracture identification. *IOP Conf. Ser. Earth Environ. Sci.* **2021**, *671*, 012026. [CrossRef]
20. Trugman, D.T.; Ross, Z.E.; Johnson, P.A. Imaging Stress and Faulting Complexity Through Earthquake Waveform Similarity. *Geophys. Res. Lett.* **2020**, *47*, 1–8. [CrossRef]
21. Gao, D.H. The Seismic Kinetic Character and Attribute Analysis of the Marine Strata of Paleozeic Erathem in the South Yellow Sea. Master's Thesis, Shandong University of Science and Technology, Qingdao, China, 2012.
22. Pimienta, L.; Schubnel, A.; Violay, M.; Fortin, J.; Guéguen, Y.; Lyon-Caen, H. Anomalous V_p/V_s ratios at seismic frequencies might evidence highly damaged rocks in subduction zones. *Geophys. Res. Lett.* **2018**, *45*, 12210–12217. [CrossRef]
23. Hu, Z.A.; Cao, L.K.; Wu, R.X.; Ji, G. Estimation method of coal channel Q value based on frequency shift phenomenon of transmitting channel wave. *Explor. Geophys.* **2023**, *54*, 79–87. [CrossRef]
24. Kim, J.K.; Wee, S.H.; Yoo, S.H.; Kim, K.H.; Noh, J.S.; Kwon, Y.J. S-wave velocity structures at Yedang Reservoir Dam inferred from amplification characteristics determined using H/V spectral ratios with background noise. *Explor. Geophys.* **2021**, *52*, 590–599. [CrossRef]
25. Heit, B.; Mancilla, F.D.L.; Yuan, X.; Morales, J.; Stich, D.; Martín, R.; Molina-Aguilera, A. Tearing of the mantle lithosphere along the intermediate-depth seismicity zone beneath the Gibraltar Arc: The onset of lithospheric delamination. *Geophys. Res. Lett.* **2017**, *44*, 4027–4035. [CrossRef]
26. Li, X.; Ji, G.; Guan, B.; Du, Z.; Han, C.; Cheng, Q. Elimination of seismic characteristics of solid-filled in ultra-deep fractured-vuggy reservoirs. *Explor. Geophys.* **2024**, *55*, 246–262. [CrossRef]
27. Kennett, B.L.N.; Engdahl, E.R.; Buland, R. Constraints on seismic velocities in the Earth from traveltimes. *Geophys. J. Int.* **1995**, *122*, 108–124. [CrossRef]

28. Waqas, A.; Shazia, N.; Muyyassar, H. Accentuating the Reservoir Potentials of Balkassar Field by Delineating the Fracture Trends and Hydrocarbon Prospects with the Aid of Seismic Attributes. *ACS Omega* **2024**, *9*, 27047–27064. [CrossRef]
29. Wu, Y.H.; Wang, W.; Zhu, G.W.; Wang, P. Application of seismic multiattribute machine learning to determine coal strata thickness. *J. Geophys. Eng.* **2021**, *18*, 834–844. [CrossRef]
30. Barnes, A.E. The calculation of instantaneous frequency and instantaneous bandwidth. *Geophysics* **1992**, *57*, 1520–1524. [CrossRef]

water

MDPI

Article

Assessment of Groundwater Quality through Hydrochemistry Using Principal Components Analysis (PCA) and Water Quality Index (WQI) in Kızılırmak Delta, Turkey

Sema Arıman [1,*], Nazire Göksu Soydan-Oksal [2], Neslihan Beden [1] and Hayatullah Ahmadzai [3]

1 Department of Meteorological Engineering, Özdemir Bayraktar Faculty of Aeronautics and Astronautics, University of Samsun, 55420 Samsun, Turkey; neslihan.beden@samsun.edu.tr

2 Department of Civil Engineering, Faculty of Engineering, Mersin University, 33000 Mersin, Turkey; goksusoydan@mersin.edu.tr

3 Department of Environmental Engineering, Ondokuz Mayıs University, 55000 Samsun, Turkey; hayatahmadzai419@gmail.com

* Correspondence: sema.ariman@samsun.edu.tr

Abstract: This study aimed to characterize the chemical composition and spatial distribution of groundwater in the Kızılırmak Delta of Turkey and to evaluate the suitability of groundwater in the Kızılırmak Delta for drinking water use through a Water Quality Index (WQI) assessment. Eleven water parameters, including nitrate (NO_3^-), calcium (Ca^{2+}), magnesium (Mg^{2+}), sodium (Na^+), chloride (Cl^-), potassium (K^+), bicarbonate (HCO_3^-), sulfate (SO_4^{2-}), hardness (measured as $CaCO_3$), electrical conductivity (EC), and pH were analyzed to determine the water quality of each groundwater sample. The WQI was determined using the weighted arithmetic index method and the method specified by the Canadian Council of Ministers of the Environment (CCME). The spatial distribution of the result for all observation wells was plotted. Principal Component Analysis (PCA) was generated utilizing the analytical data from eleven selected samples. As a result of the study, according to the calculated WQI values, the water in most of the wells was not suitable for drinking purposes. The minimum Ca^{2+} concentration in the study area was 108,817 mg/L, and the maximum was 692,382 mg/L, which showed that the samples in all wells exceeded the WHO limit. The same situation is valid for Mg^{2+}, and the values vary between 100.383 and 5183.026 mg/L. From the spatial distribution of the water quality parameters it has been understood that the eastern part of the region is more suitable than the western part for drinking purposes. The results from correlation analysis showed the strongest positive correlation between Mg^{2+} and Na^+ and Na^+ and EC as 0.989. The present study shows that the groundwater of the delta, which has deteriorating water quality, should be treated before it is used for drinking water and protected from contamination hazards.

Keywords: groundwater quality; hydrogeochemistry; Water Quality Index (WQI); principal component analysis

1. Introduction

Natural resources are indispensable for a country's development in every sense, and groundwater has a significant place among these resources regarding the water budget. Agricultural activities, the most essential tool for growth in many countries, lead to rapid pollution and even groundwater consumption [1,2]. Due to the gradual decrease in the availability and quality of groundwater resources recently, regular observation and hydrochemical evaluation of groundwater quality is necessary for sustainable development and strategic planning under this semi-confined aquifer.

Nowadays, stress on groundwater is increasing for many reasons, the most important of which are rapid population growth, unplanned industrialization, climate change, and climate variability. Groundwater sources are endangered by overuse, pollution, and inadequate development. This issue is more noticeable in regions where surface water sources are under stress and surface groundwater is used simultaneously [3,4]. For instance, water-intensive agricultural practices exist [5,6] water demand is rising [7], and surface water pollution is common in residential areas with inadequate water supply systems [8]. When groundwater is polluted, restoring and maintaining proper quality by removing contaminants from the source to protect public health and the environment is challenging. Therefore, it becomes necessary to regularly monitor groundwater quality and find ways to defend it from pollution [9].

Various natural and human-induced factors can detrimentally impact groundwater quality by introducing metals and nutrients. Natural sources of contamination include geothermal systems [10] and geological formations [11], while human activities such as mining, agriculture, and industrial or urban pollution also contribute to groundwater pollution [12]. Geogenic pollutants in groundwater primarily originate from natural geological processes such as the weathering of rocks and minerals, leaching of trace elements from soil and sediment, and interactions between water and geological formations [13]. Examples of geogenic pollutants include heavy metals like arsenic, lead, mercury, and fluoride; salts and ions like chloride, sulfate, and bicarbonate; hydrocarbons such as methane and petroleum compounds; and radionuclides such as radium and uranium, which can naturally occur in geological formations and be dissolved into groundwater [14]. The hydrochemical composition of groundwater is influenced by chemical interactions between water and minerals along its flow path, the mixing of different water sources, the chemical properties of recharged water, and the geochemical characteristics of the surrounding formations. These processes involve complex interactions between water and the geological materials through which it passes, shaping the overall chemical composition of groundwater [15]. These geogenic pollutants can pose risks to human health and the environment if present in groundwater at elevated concentrations [16].

Nowadays, hydrogeochemical processes have garnered increased attention [17] due to their valuable insights into the interaction between groundwater and the environment. The groundwater's hydrochemical properties, including both anions and cations, are predominantly dictated by the surrounding area's geology and the aquifer's characteristics [18,19]. Anions such as chloride (Cl^-), sulfate (SO_4^{2-}), and bicarbonate (HCO_3^-) may derive from the dissolution of minerals rich in these ions, such as halite, gypsum, and calcite, respectively. Similarly, cations such as Ca^{2+}, magnesium (Mg^{2+}), sodium (Na^+), and potassium (K^+) are released into groundwater through the weathering and dissolution of minerals containing these elements, such as calcite, dolomite, feldspar, and mica [20]. These geogenic processes play a fundamental role in shaping the hydrochemical composition of groundwater.

In recent years, many researchers have paid great attention to studies on groundwater quality. An evaluation of the aquifer water quality due to seawater intrusion in the coastal region of the current study area was previously carried out by Fırat Ersoy et al. [21] and classified as non-consumable based on the groundwater permeability index, sodium adsorption rate, Na%, and Kelly Ratio (KR) values. Karakuş [22] studied the water quality of the Kızılırmak River and stated that portions near the Sivas city center and in the south of the province did not meet the standards for drinking water purposes. Samsunlu et al. [23] comparatively investigated the past and present status of the water quality of the Kızılırmak Delta. Demirci et al. [24] investigated the spatial distribution of saltwater intrusion on the northeastern coast of the Kızılırmak Delta.

Water quality management only readily provides an accurate overview of temporal and spatial trends in the global water quality in an aquifer, except for a few attempts to establish a comprehensive methodology that significantly consolidates data sources and makes them practical information. The WQI is a widely accepted indicator of water resource management and the quality assignment and classification of ground and surface waters [25,26]. Furthermore, the WQI combines the chemical and physical factors used to measure water quality into a single parameter [27,28]. The WQI was first developed by

Horton [29] based on a weighted arithmetical calculation. Afterward, other WQIs were created, such as the Oregon WQI [30], the Canadian WQI [31], the National Sanitation Foundation WQI [32], and the Weighted Arithmetic WQI Method (WAWQI) [33]. Combining the WQI with the Geographical Information System (GIS) to illustrate spatial variation in water quality and identify vulnerable sites assists in decision-making for sustainable groundwater management [34]. To demonstrate that groundwater is suitable for drinking, a WQI is utilized [34–38].

The simplicity of the computation and utilization of water quality indices such as the WQI has led to their widespread application in evaluating groundwater quality across a broad spectrum of landscapes and land use patterns. However, this method is a dimensionless numerical value that gives accurate results depending on multiple WQ parameters in more than one location. Many researchers [28,36,38] developed various WQI models based on the weighting and rating of different water quality parameters derived from the weighted arithmetic method. In recent years, groundwater quality assessment and monitoring have been done using the GIS and interpolation methods. It has proven a powerful tool for evaluating and analyzing spatial information in water resources [2,39]. Rabeiy [40] evaluated the quality of 812 groundwater samples from the central region of Upper Egypt (Sohag Governorate) regarding drinking and irrigation. The WQI converted the water parameter into a single indicator value representing the water quality level. The spatial distribution of the estimated values of each groundwater parameter is spatially modeled using the GIS. In the study by Singh and Noori [41] on 35 groundwater samples in Kabul, Afghanistan, water types, geochemical properties, quality, and the feasibility of the samples as drinking water were analyzed. Additionally, they used the GIS to model the spatial distribution of water quality with the parameters examined in the WQI calculation.

The Kızılırmak Delta is a RAMSAR site with international protection status and is an important wetland. In addition, agriculture and animal husbandry, the region's most important sources of income, are carried out in a significant part of the study area. For this reason, the delta's water resources are vital for economic and social activities. However, the fact that socio-economic activities, water use, and water quality directly affect each other reveals that a study on water quality is necessary for the region. Hence, conducting comprehensive studies to evaluate the groundwater quality and effectively communicating the findings to policymakers, managers, and water users is imperative. Also, it holds significant importance to analyze the hydrogeochemistry of the groundwater, assess the water quality, and understand the influencing factors in a given area to promote the sustainable utilization of groundwater reservoirs. From this point of view, a study, which is thought to contribute to the literature, was conducted for the following purposes in the Kızılırmak Delta, Turkey, which has the potential to be affected by salinity: (i) the chemical composition and spatial distribution of the groundwater are characterized, (ii) the sources of major ions present in the groundwater are identified, and (iii) the suitability of the groundwater in the Kızılırmak Delta for drinking purposes is evaluated, through a Water Quality Index (WQI) assessment. The study's findings provide further information on water quality assessment in the Kızılırmak Delta, and local governments can benefit from the study's results to develop sustainable groundwater management.

2. Materials and Methods

2.1. Study Area

This study was carried out in the Kızılırmak Delta located in the downstream area of the Kızılırmak Basin, between latitudes 41°44′ and 41°27′ N and longitudes 35°34′ and 36° 7′, with a total area of 103,537 hectares (ha) (Figure 1). Different climatic characteristics prevail in the interior and coastal areas downstream of Kızılırmak. In the coastal area, a typical Black Sea climate is observed. The average annual temperature in the study area is 13.7 °C; the hottest months are July and August, with 22.8 °C. The month with the lowest average temperature is January, with 5.7 °C. The annual average relative humidity is 74.9%. Total annual precipitation is 791.8 mm. The month with the highest precipitation is December (102.1 mm), and the month with the lowest precipitation is July (29.8 mm). The study area has fertile soil rich in content through the organic matter carried by the Kızılırmak River [42].

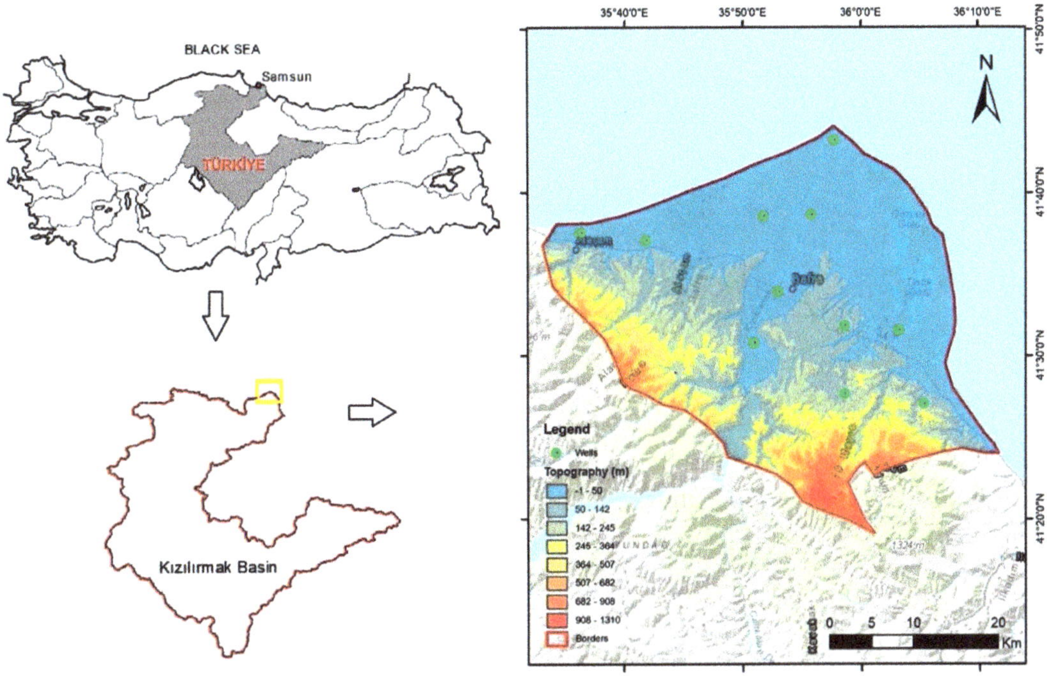

Figure 1. The location of the study area and observation wells used for groundwater samples.

Groundwater samples were obtained from eleven observation wells and analyzed for anions and cations. Locations of the wells in the WGS84 projection system for water quality monitoring are displayed in Table 1 and Figure 1. The samples were collected during different seasons and dates, and for the present study, the average values of each parameter were used.

Table 1. Water quality range [33].

WQI Level	Water Quality Status	Grade
0–25	Excellent water quality	A
26–50	Good water quality	B
51–75	Poor water quality	C
76–100	Very poor water quality	D
>100	Unsuitable for drinking	E

2.2. Geology and Hydrogeology of the Study Area

The Kızılırmak Delta was formed by the accumulation of fertile sediments carried by the Kızılırmak River. From the Black Sea to the inner regions, the area, which is primarily flat, gradually rises in steps. The formations seen in the delta are Mesozoic and Cenozoic (Eocene, Neogene) age volcanic, metamorphic, and fluvial deposits and Quaternary alluviums. The Upper Cretaceous flysch series are the oldest deposits widely distributed in the delta and represent the Mesozoic period. The delta is a plain consisting of a mixture of clay, silt, and sand spread over a wide area, following the Kızılırmak bed from the sea. Plio-Quaternary deposits primarily composed of clay, sand, and gravel are encountered on the slopes with the rise of the plain areas. The volcanics in the Flysch series consist of tuff layers and andesites [42,43].

The Kızılırmak Delta coastal region has many wells drilled in alluviums. The delta's primary aquifer is the alluvium unit on the river's banks and plains [43]. Groundwater is found near the surface in the alluvium, which is composed of silt, clay, gravel, and sand. These wells range in depth from 50 to 100 m and in flow rate from 1 to 60 L/s. The transmissivity coefficient in the coastal region of the Kızılırmak Delta varies between 18 and 8320 m^2/day, indicating that it is an unconfined aquifer. In the aquifer, groundwater flows in three directions: north, northeast, and northwest [21].

2.3. Data Acquisition

To evaluate the water quality, data on eleven observation wells (January–December 2016) (Figure 1) representing the study area's unconfined aquifer system were obtained from the General Directorate of State Hydraulic Works. Each well was pumped for a minimum of 15 min, after which the samples were collected and stored at 4 °C until the analysis. A portable multiparameter meter measured EC (electrical conductivity) and pH values. Hydrochemical parameters, including NO_3^-, Ca^{2+}, Mg^{2+}, Na^+, Cl^-, K^+, HCO_3^-, SO_4^{2-}, and hardness (measured as $CaCO_3$), were analyzed using ion chromatography. Distilled water (used as blank samples) underwent parallel analysis to ensure quality control, and all measurements were duplicated. Method accuracy and quality control were verified by examining Standard Reference Materials (SRM). The suitability of the study area's groundwater for drinking water supply was evaluated using the ten groundwater quality parameters to reveal the groundwater quality index. Also, adding more parameters to the groundwater quality model would needlessly increase uncertainty, which is another reason to focus only on the most significant water quality parameters.

2.3.1. Determination of the Water Quality Index for Groundwater

The Weighted Arithmetic Index Method (WAIM) [33] was used to calculate the WQI. The calculation of the WQI utilized the weighted arithmetic method, which has been employed in several earlier studies [35,44–46].

This method is based on the following steps.

In the first step, for each parameter, a unit weight factor was calculated using Equation (1):

$$W_n = \frac{K}{S_n} \tag{1}$$

W_n is the unit weight factor, K is the constant for proportionality, and S_n is the standard desirable value of the nth parameter.

The following equation can calculate the K factor (Equation (2)):

$$K = \frac{1}{\frac{1}{S_1} + \frac{1}{S_2} + \frac{1}{S_3} + \ldots + \frac{1}{S_n}} = \frac{1}{\sum \frac{1}{S_n}} \tag{2}$$

The sum of the weight factor calculation must equal 1.

Equation (3) calculates the sub-index (Q_n) value in the second step.

$$Q_n = \frac{V_n - V_0}{S_n - V_0} \times 100 \tag{3}$$

Q_n is the sub-index value on the parameter, the mean concentration of the nth parameter, S_n is the standard desirable value of the nth parameter, and V_0 is the actual value of the parameter in pure water (generally V_0 = 0 for most of the parameters except pH and Dissolved Oxygen (DO).

For the pH, Q_n must be calculated according to Equation (4).

$$Q_{pH} = \frac{V_{pH}}{s\tan dart_{pH} - 7} \times 100 \quad (4)$$

The third step involves computing the overall WQI using Equation (5). Calculating the overall water quality index involves a linear combination of the quality rating and the unit weight.

$$OverallWQI = \frac{\sum W_n \times Q_n}{\sum W_n} \quad (5)$$

The WQI values obtained were grouped into five categories based on the calculated WQI: excellent water quality (WQI < 25); good water quality (WQI 25–50); poor water quality (WQI 50–75); very poor water quality (WQI 75–100); and water unsuitable for drinking (WQI > 100) [33]. Table 1 shows all the categories of the WQI based on Brown, McClelland, Deininger, and O'Connor [33].

2.3.2. Canadian Council of Ministers of the Environment Water Quality Index (CCME WQI)

The WQI, developed by the Canadian Council of Ministers of the Environment (CCME) in 2001, serves to simplify the presentation of intricate and technical water quality information. The WQI comprises three distinct factors, and comprehensive documentation is available (47).

Factor 1: F1 (Scope)

The scope evaluation determines the degree to which water quality guidelines are not met within the selected timeframe. This involves calculating the number of parameters for which the objective limits are not achieved. This methodology has been adapted from the British Columbia Water Quality Index. Equation (6) can find factor 1.

$$F_1 = \frac{Number\,of\,field\,\text{variables}}{Total\,number\,of\,\text{variables}} \times 100 \quad (6)$$

The number of field variables refers to parameters that exceed specified limits, while the total number of variables indicates the overall count of parameters studied.

Factor 2: F2 (Frequency)

The frequency of non-compliance (i.e., the number of times the measured or observed value falls outside the acceptable limits) indicates the proportion of individual tests that do not meet the objectives (termed "failed tests"). The formulation of this factor is taken directly from the British Columbia Water Quality Index. This value can be calculated using Equation (7).

$$F_2 = \frac{Number\,of\,field\,tests}{Total\,number\,of\,tests} \times 100 \quad (7)$$

Factor 3: F3 (Amplitude)

The amplitude reflects the extent to which the objectives are not achieved, representing the deviation of failed test values from their intended objectives. Its computation involves three steps. In the first step the "excursion" is calculated. An "excursion" refers to the number of instances in which an individual concentration exceeds (or falls below, in the case of a minimum objective) the intended objective and can be computed as follows: if the test value should not exceed the objective, the excursion is the difference between the test value and the objective; and can be calculated using Equation (8).

$$Excursion_i = \frac{Field\,test\,value_i}{Objective_i} - 1 \quad (8)$$

In instances where the test value should not be lower than the objective, the excursion can be calculated by using Equation (9).

$$Excursion_i = \frac{Objective_i}{Field\,test\,value_i} - 1 \tag{9}$$

In the second step to determine the total amount of non-compliance across individual tests, the excursions of each test from their respective objectives are summed up and divided by the total number of tests. This value is known as the normalized sum of excursions (NSE) and can be calculated using the following formula:

$$NSE = \frac{\sum_{i=1}^{n} excursion_i}{Total\,number\,of\,tests} \tag{10}$$

In the third step, F3 is calculated using an asymptotic function that scales the NSE from objectives to yield a value between 0 and 100.

$$F_3 = \frac{NSE}{0.01NSE + 0.01} \tag{11}$$

Finally, the CCME WQI can be found using the following formula: variations in each of the three components directly correspond to changes in the index.

$$CCMEWQI = 100 - \left(\frac{\sqrt{F_1^2 + F_2^2 + F_3^2}}{1.732}\right) \tag{12}$$

where 1732 serves as a scaling factor and normalizes the obtained values to fall between 0 and 100, with 100 denoting the highest water quality and 0 indicating the lowest level of quality [47]. The classification of the CCMEWQI ranges is clearly stated in Table 2.

Table 2. Classification of CCME WQI ranges with descriptions Source [47].

Water Quality	CCMEWQI Value	Description
Excellent	95–100	Water quality is protected with a virtual absence of threat or impairment, conditions very close to natural or pristine levels.
Very good	89–94	Water quality is protected with a slight threat or impairment, conditions close to natural or pristine levels.
Good	80–88	Water quality is protected with only a minor threat or impairment; conditions rarely depart from natural or desirable levels.
Fair	65–79	Water quality is usually protected but occasionally threatened or impaired; conditions sometimes depart from natural or desirable.
Poor (Marginal)	45–64	Water quality is frequently threatened or impaired; conditions often depart from natural or desirable levels.
Poor	0–44	Water quality is almost always threatened or impaired; conditions usually depart from natural or desirable levels.

2.4. Principal Component Analysis (PCA)

The principal components present uncorrelated linear combinations of the original variables and explain the total variance of the original data [48]. In PCA, the correlation between all components is zero because the method generates all components to be orthogonal to each other. The first principal component explains most of the variation present in the original data [49].

The number of parameters is reduced by applying PCA, assuming a complete dataset containing k variables $x_1, x_2, \ldots, x_k$ measured on n variables. The component analysis is performed on the correlation matrix, so it can be assumed that each variable has a sample mean of 0 and variance of 1 without loss of generality. For a complete data set, any principal component is a linear combination, and all parameter variables can be written as in Equation (13):

$$y = a_1x_1 + a_2x_2 + \ldots + a_kx_k \quad (13)$$

The a_i (i = 1 to k) are constants [50].

The PCA procedure is implemented as follows:

i. First, the covariance/correlation matrix is calculated (here, the most appropriate data set to perform PCA is the one with the highest correlation between the individual indicators).
ii. Determine the number of principal components to be considered according to the percentage of variance they explain.
iii. Rotate the factors to improve their interpretation (this is done by maximizing the loading of individual indicators on individual factors) [46,51].

2.5. Hierarchical Cluster Analysis (CA)

Clustering is a statistical approach that facilitates the grouping of similar locations in various clusters based on distance criteria that characterize the parameters to be classified [52]. Principal Component Analysis (PCA) was the basis for performing Hierarchical Cluster Analyses (HCA), which quantified the degree of similarity between the samples. Factor scores from PCA were utilized in HCA as variables for statistical grouping. The aim of HCA is to find homogeneous subgroups of instances within a set of groups that maximize variation between groups while minimizing within-group variance. In the study area, wells with comparable chemical properties were categorized using HCA using the Ward method and Square Euclidean distance.

3. Results and Discussion

3.1. Hydrogeochemical Characteristics

Groundwater, essential for many uses, including agriculture and drinking water, is experiencing a decline in quality attributed to a mix of natural phenomena and human activities. For this reason, the assessment of groundwater quality is of immense importance to society [53]. In the study area discussed in this article, GW is used for drinking and irrigation purposes, which is why investigation into the water quality of this area is essential. All the parameters were compared according to concentrations in WHO drinking water standards. Table 3 shows the WQI values of the parameters used in this study for the eleven different wells considered and the value of each parameter according to WHO drinking water quality standards. The statistical analyses of groundwater quality parameters are shown in Table 4.

As seen from Table 3, except pH, most of the parameters of each well in the study area indicate that the GW in these wells is unsuitable for drinking water according to WHO standards. A necessary component for any application is the water's pH. According to the guidelines, a pH range of 6–9 is suitable for practically all applications [54].

The EC values of the samples examined fell within the range of 652 to 6112 µS/cm (with an average of 1762 µS/cm), suggesting that most of the groundwater in the area is more mineralized. Nearly all of the groundwater samples taken in the region were found to be not the acceptable limit of 500 µS/cm, as defined by the WHO [55]. It is known that sodium salts are found in all food and drinking water [40]. However, WHO has stated that concentrations above 200 mg/L taste unpleasant and are unacceptable. Among the samples obtained from the study area wells, the lowest sodium concentration value was determined to be 14.484 mg/L and the highest value was 1334.799 mg/L. It is seen that the sodium value obtained from five wells exceeds WHO standards. Most of these areas are located east of the studied area, in the region where the sea enters the delta (Figure 1).

Potassium is released into groundwater by weathering potassium-containing minerals in rocks and soils. Common minerals like feldspar, mica, and clay minerals are primary contributors to potassium in groundwater [56]. Agricultural activities are the predominant source of increased potassium levels in groundwater [44]. The maximum and minimum potassium values in the eleven wells in the study area were 25.02 mg/L and 1.95 mg/L, respectively, while the average value was 7.36 mg/L. It is seen that both the average potassium value and three of the wells exceed the WHO limit. It was determined that the potassium value was higher in the wells in the north of the delta than in other regions. Agricultural practices and the natural weathering of silicate minerals are likely the primary factors contributing to elevated potassium levels in the groundwater of the study area. Conversely, the lower potassium concentrations in the groundwater can be attributed to its binding with clay minerals and the greater durability of potassium-bearing minerals against weathering processes [57].

Calcium is released into groundwater through the weathering of calcium-containing minerals present in rocks and soils [58]. In addition, calcium-containing fertilizers, such as calcium carbonate (lime), can also be applied to agricultural fields to adjust the soil pH and improve the soil structure. However, overuse of these fertilizers may lead to Ca^{2+} ions leaching into groundwater. The minimum calcium concentration in the study area was 108.817 mg/L, and the maximum was 692.382 mg/L. According to the WHO limit, all wells in the study area exceeded the limit value. The increase in the calcium concentration in the groundwater can be attributed to both natural geological processes and human activities. Like the calcium concentration, the sodium concentration is more intense in the eastern part of the study area. The ion exchange reactions involving sodium and calcium frequently occur when freshwater mixes with saltwater. In cases where saline water infiltrates fresh groundwater, calcium is released from solid exchange surfaces in exchange for sodium in the groundwater. Conversely, when flushing out a saline aquifer with fresh groundwater, the opposite processes occur. While magnesium and potassium may also undergo exchange for sodium, their contributions are generally less significant. Cation exchange between exchange surfaces and groundwater thus regulates the proportions of sodium and calcium in saline groundwater [59].

Table 3. Groundwater water quality parameters of some wells in the Kızılırmak Delta.

								Parameters					
Stations	**X (m)**	**Y(m)**	**pH**	**EC**	**Na^+ (mg/L)**	**K^+ (mg/L)**	**Ca^{2+} (mg/L)**	**Mg^{2+} (mg/L)**	**HCO_3^- (mg/L)**	**Cl^- (mg/L)**	**SO_4^{2-} (mg/L)**	**Hardness**	**NO_3^- (mg/L)**
Well 1	41°31′51″	35°58′37″	7.12	1015	142.768	3.128	136.42	386.771	546.739	24.106	37.463	37.700	3.782
Well 2	41°27′41″	35°58′37″	8.5	1005	14.484	5.474	162.925	318.865	399.071	3.191	9.606	51.400	2.790
Well 3	41°30′48″	35°50′59″	7.7	1162	50.118	11.339	122.044	308.531	233.097	6.027	13.929	40.900	2.604
Well 4	41°33′55″	35°53′0″	7.6	1260	135.181	4.692	111.422	534.394	549.18	134.71	67.722	45.910	16.120
Well 5	41°37′31″	35°56′23″	7.4	672	19.771	2.346	129.859	128.432	369.781	24.106	39.865	36.710	<0.01
Well 6	41°38′34″	35°51′46″	7.5	1546	207.599	5.083	151.903	615.586	375.273	243.896	340.533	58.730	26.600
Well 7	41°42′53″	35°55′59″	7.5	2401	348.299	25.024	108.817	1778.851	685.865	383.215	475.497	87.410	31.000
Well 8	41°38′39″	35°55′47″	7.3	1950	229.67	7.429	205.811	863.592	487.55	311.96	400.09	80.600	50.900
Well 9	41°31′33″	35°3′16″	7	6112	1334.799	9.384	692.382	5183.026	732.85	638.1	1474.521	348.280	52.950
Well 10	41°37′5″	35°41′51″	7.4	652	17.243	1.955	118.036	100.383	371.002	28.006	37.944	32.890	17.110
Well 11	41°38′34″	35°51′46″	7.28	1606	187.598	5.083	130.26	1198.695	745.054	216.6	113.351	73.100	27.160
Average			7.5	1761	244.321	7.358	188.171	1037.921	499.587	183.083	273.684	81.239	23.102
WHO standard [55,60].			7.5	500	200	10	75	50	500	250	250	100	50

Table 4. Statistical analysis of groundwater quality parameters.

Parameters	Min	Max	Mean	Median	SD
pH	7	8.5	7.5	7.4	0.39
EC (μS/cm)	652	6112	1761	1403	1472
Na^+ (mg/L)	14.484	1334.799	244.321	165.100	359.281
K^+ (mg/L)	1.955	25.024	7.358	5.242	21.799
Ca^{2+} (mg/L)	108.817	692.382	188.171	130.030	171.762
Mg^{2+} (mg/L)	100.383	5183.026	1037.921	574.900	1394.753
HCO_3^- (mg/L)	233.097	745.054	499.587	488.700	160.746
Cl^- (mg/L)	3.191	638.100	183.083	81.350	199.711
SO_4^{2-} (mg/L)	9.606	1474.521	273.684	53.750	420.049
Hardness (mg/L)	32.890	348.280	81.239	55.065	86.441
NO_3^- (mg/L)	<0.01	52.950	23.102	16.616	18.845

The magnesium concentration was identified as one of the significant parameters among the major cationic constituents, with samples ranging from 100.383 to 5183.026 mg/L. It was noticed that all samples exceeded the desirable limit of 50 mg/L for drinking water, as set by the WHO [60]. Aquifers with naturally high concentrations of magnesium-containing minerals may also contribute to elevated magnesium levels in the groundwater. The high levels of Ca^{2+} and Mg^{2+} can be attributed to the lithological composition of the area and the presence of minerals such as calcite and dolomite, which are fundamental constituents of siltstone, a prevalent geological feature in the study region [61].

Sulfate in groundwater typically occurs in the form of soluble Ca^{2+}, Mg^{2+}, and Na^+ salts. There is a notable variation in the SO_4^{2-} concentration over time, particularly during rainfall infiltration and groundwater recharge [62]. Using sulfate-containing fertilizers such as ammonium SO_4^{2-} or potassium sulfate can also contribute to high SO_4^{2-} levels in groundwater. Over the past few decades, atmospheric deposition has emerged as a significant contributor of sulfate to soil, eventually permeating groundwater. Given the mobility of sulfate within soil, its introduction into the soil can consequently affect shallow aquifers as well [63]. The sulfate content in the groundwater samples ranged from 9.606 to 1474.521 mg/L, exceeding the WHO permissible limit of 250 mg/L at some stations (wells 6, 7, 8, and 9). The fact that these stations are within the region's borders where intensive agriculture is practiced may have increased the SO_4^{2-} value.

Bicarbonate in groundwater primarily originates from dissolved carbonate minerals like calcite and dolomite found in rocks and soil. As rainwater or surface water seeps through these carbonate-rich formations, it reacts with the minerals, releasing HCO_3^- ions into the groundwater [64]. Bicarbonate in groundwater is mainly derived from the carbonate minerals and CO_2 in the atmosphere and soil. At neutral pH, bicarbonate is the dominant ion [65]. An alkaline environment is created in the groundwater due to the reaction between the carbon dioxide in the soil and the minerals that form rocks. The HCO_3^- concentrations vary from 233.097 and 745.054 mg/L and exceed the permitted limit of 500 mg/L in some stations (wells 1, 4, 7, 9, and 11) [60].

Chloride originates from sodium chloride, which dissolves in water from rocks and soil. The presence of sodium chloride in water generally has minimal impact on its suitability unless it reaches concentrations that render the water non-potable or corrosive [61]. The chloride concentration of groundwater samples taken from wells in the study area varies between 3.191 and 638.100 mg/L. The maximum allowable concentration of Cl^- ions in drinking water is 250 mg/L. Water with chloride ion concentrations exceeding 250 mg/L tastes salty. A higher chloride concentration indicates a higher degree of organic contaminants, so chloride is an important parameter in assessing groundwater quality [66]. The spatial distribution of Cl^- is created via Arc-GIS 10.8.2 and given in Figures 2 and 3. Furthermore, it can be seen that this parameter is high in the eastern part of the study area. Sulfate is dissolved and leached from rocks containing gypsum, iron sulfides, and other sulfur-bearing compounds [9]. The groundwater hardness levels in the region varied between 32.890 and 348.280 mg/L (as $CaCO_3$), with an average value of 81.239 mg/L. Approximately 90% of the wells had hardness levels within the desirable 100 mg/L limit. The anionic and catatonic content of the groundwater was found to be $Mg^{2+} > HCO_3^- > SO_4^{2-} > Cl^- > Na^+$, as shown in Table 3.

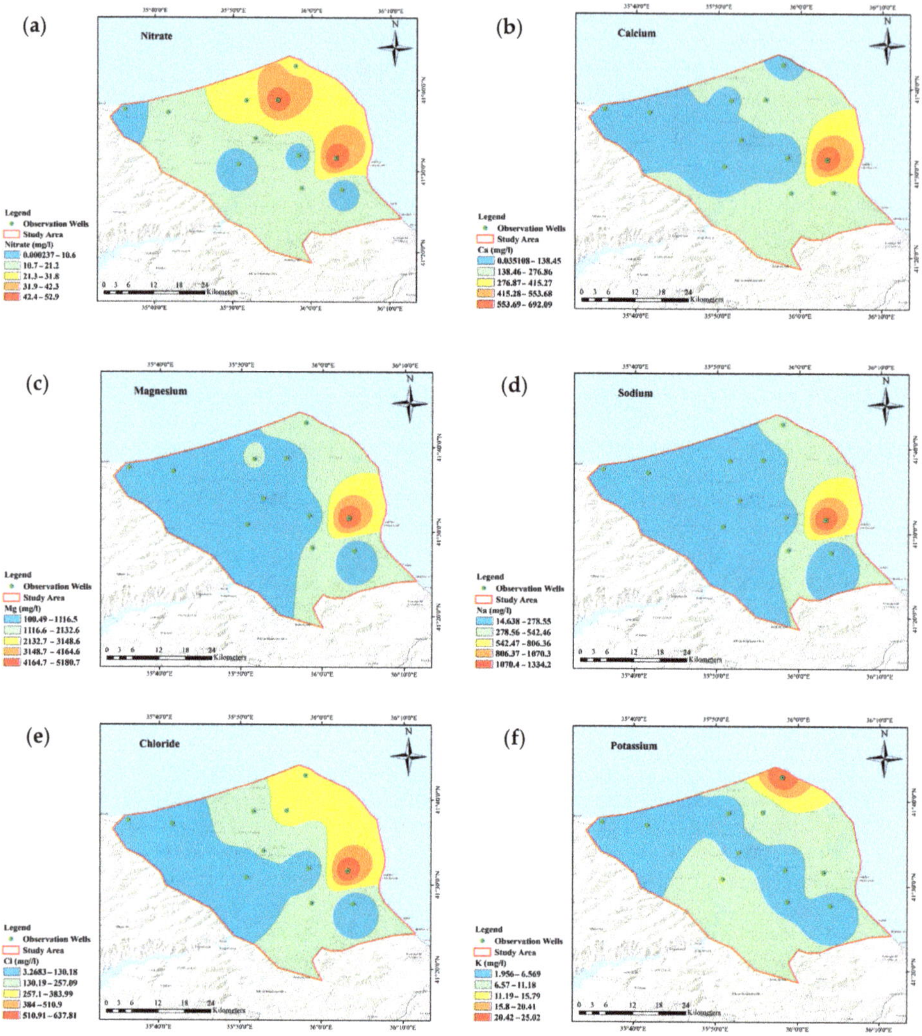

Figure 2. The thematic layer for each parameter of groundwater in the study area. (**a**) Nitrate (NO_3), (**b**) calcium (Ca^{2+}), (**c**) magnesium (Mg^{2+}), (**d**) sodium (Na^+), (**e**) chloride (Cl^-), (**f**) potassium (K^+).

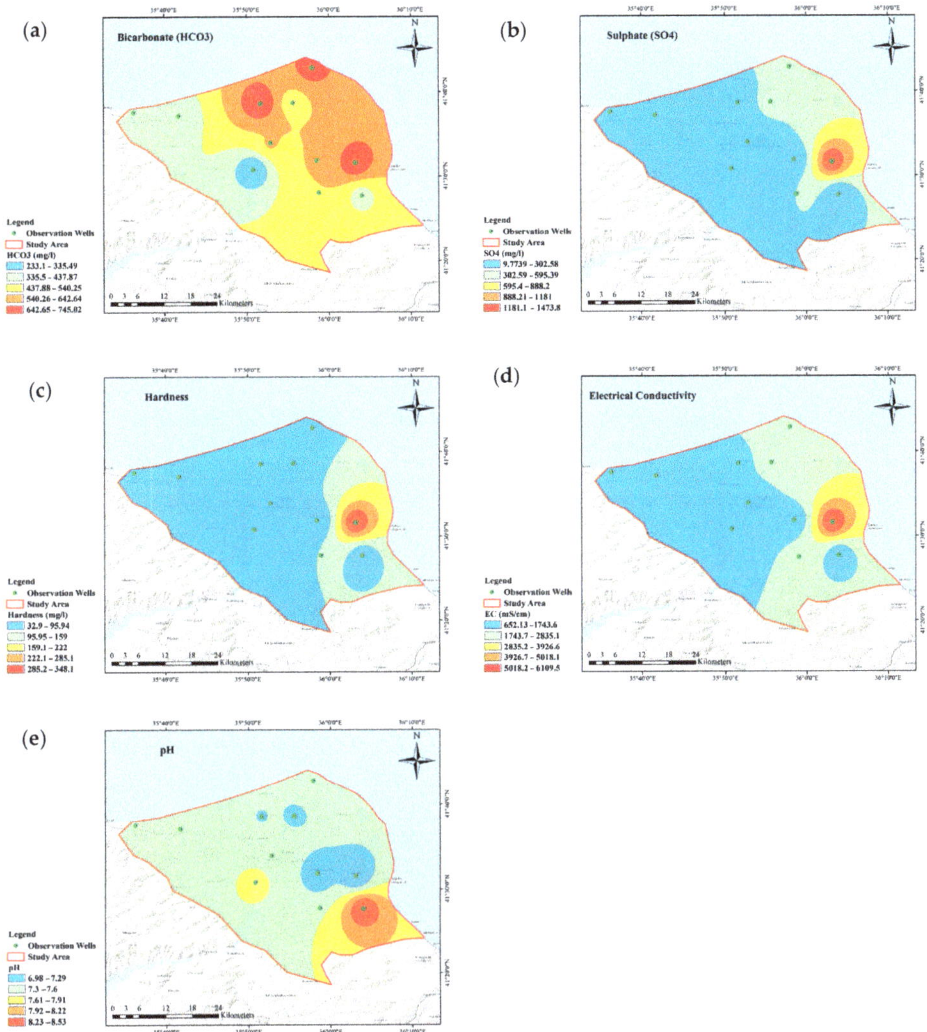

Figure 3. The thematic layer for each groundwater parameter in the study area. (**a**) Bicarbonate (HCO_3^-), (**b**) sulphate (SO_4^{2-}), (**c**) hardness ($CaCO_3$), (**d**) electrical conductivity (EC), (**e**) pH.

Large quantities of nitrates are also generated from organic waste produced by agricultural livestock and urban sewage, potentially reaching groundwater reservoirs. Particularly in areas with intensive feedlots, livestock waste significantly contributes to excess nutrients entering the environment. While concerns regarding the pollution of both surface waters and aquifers due to increased use of mineral fertilizers are supported by observed correlations linking fertilizer application to nitrate leaching [67], pinpointing the direct impact of fertilizer application on water nitrate content is challenging. Nitrate concentrations in groundwater can vary significantly across different locations, even when uniform farm management practices are employed [68]. The study area is a RAMSAR area with international protection status and is an important wetland. Agriculture and animal husbandry are carried out in this study area. Animal waste, plants, and remaining animals go through ammonification in the soil, generating ammonia that is then converted to nitrate by Nitrosomonas and Nitrobacter bacteria [9]. Nitrogen released through weathering has a significant impact on soil and water quality. Additionally, denitrification significantly impacts how much nitrogen is released during the weathering process of bedrock, which affects the amount of nitrate in groundwater [9,69]. According to the study by [70], approximately 25% of the nitrogen applied to agricultural ecosystems leaches into water resources. However, there are significant uncertainties about the time of transfer of agriculturally applied nitrogen from topsoil to groundwater. This uncertainty arises from a deficient mechanistic understanding of the impact of different nitrogen transformations in the soil [71].

Nitrate pollution is a critical anthropogenic variable impacting groundwater quality [72]. Their overabundance or accumulation due to human activities can harm water quality, health, and environmental sustainability. When it exceeds the WHO's allowable limit of 50 mg/L for nitrate, its higher concentration generally poses health risks [73,74]. The study area's concentration ranges from <0.01 to 52.498 mg/L (Table 3). Overall, the low nitrate concentration in the assessed area is likely a result of a combination of factors, including organic agricultural activities, natural processes, land management practices, and geological conditions. Since this region has international protection status, attempts are made to protect an important wetland and ecosystem health. For this reason, projects are being carried out by the Ministry of Agriculture and Forest regarding organic farming. In addition, ecosystem-based solutions are being implemented to remove nitrate and nitrite pollution from soil in this region. Therefore, except for two wells (wells 8 and 9), nitrate concentration was found below WHO limit values (50 mg/L).

3.2. Groundwater Quality Mapping/Spatial Distribution Pattern

This study evaluated the quality and pollution status of water samples collected from eleven underground wells in the Kızılırmak Delta. The results of the chemical parameters of the groundwater samples taken from the Kızılırmak Delta are given in Table 5. The result of the analysis parameters was plotted on the thematic map for all observation wells using ArcGIS 10.8.2, as shown in Figures 2 and 3. The present study utilized the Inverse Distance Weighted (IDW) interpolation technique, which has proven to be a proficient method for producing spatial distribution maps by spatially interpolating groundwater quality parameters. The closest specified locations were considered when calculating the weights, allocated to different parameters at each location based on distance. As seen in Figures 2 and 3, the contour maps of the groundwater quality parameters were created with a spatial distribution pattern. The pH distribution pattern indicates the presence of alkaline groundwater in the northeast region of the Kızılırmak Delta, the NE (Figure 2a). The groundwater's Ca^{2+} and Mg^{2+} ions came from the rock formations in the study area, which contains Ca and Mg leaching (Figures 2 and 3). Weathering and erosion of these rocks release calcium and magnesium into soil and water systems. The concentration of calcium and magnesium in water can vary depending on geological factors, such as the composition of underlying rocks and soils. They are naturally present in dissolved form, contributing to water hardness.

Table 5. GIS-based WQI index values of each monitoring well according to Brown et al. [33].

Stations	GIS-Based WQI	Remarks
Well 1	69.14688	Poor
Well 2	200.0549	Unsuitable
Well 3	144.0467	Unsuitable
Well 4	143.5819	Unsuitable
Well 5	66.03841	Poor
Well 6	157.0077	Unsuitable
Well 7	367.2081	Unsuitable
Well 8	184.5786	Unsuitable
Well 9	772.5291	Unsuitable
Well 10	63.94814	Poor
Well 11	211.5382	Unsuitable

To summarize, it is obvious from Figures 2 and 3 that the water quality in the west of the region is insufficient for drinking water and irrigation purposes. When the figures are examined, it can be clearly seen that high levels were also obtained for HCO_3^- and NO_3^- in the wells in the north of the region. It can be said that the samples obtained from the east of the region are useful for drinking water and irrigation.

3.3. GIS-Based Groundwater Quality Index

3.3.1. GIS-Based Groundwater Quality Index by Brown et al. [33]

Eleven groundwater parameters, NO_3^-, Ca^{2+}, Mg^{2+}, Na^+, Cl^-, K^+, HCO_3^-, SO_4^{2-}, hardness ($CaCO_3$), EC, and pH, were analyzed to evaluate the groundwater quality of eleven different wells in the study area for drinking and irrigation purposes. These wells are monitoring wells, and these parameters are measured during different seasons of the year. Two types of GIS-based WQI calculators were applied to the study area data, and each method was discussed in the method section of Brown et al. [33]. The results shown in Table 6 show that most of the wells have low-quality water for drinking purposes.

The calculated result of the WQI for each well shows that most of the water of the well is not suitable for drinking purposes. It exceeds the water quality rating as per Brown, McClelland, Deininger, and O'Connor [33]. Among these monitoring wells, well 10 has a low water quality index of 63.948, but all other wells have E-grade water quality, which is unsuitable for drinking purposes. These exceeded the limit of the WQI for these wells due to the high concentration of different parameters in the groundwater; For instance, in Well 1, Na^+ and Mg^{2+} concentrations were higher than the WHO standards for drinking water. Well, 9 had the highest GIS-based WQI (772.529), indicating the lowest water quality at the site, which is attributed to elevated levels of Na^+, Ca^{2+}, Mg^{2+}, SO_4^{2-}, and $CaCO_3$, as detailed in Table 5. Figure 4 shows the plotted map of the WQI for all eleven sites based on the Brown et al. [33] WQI method.

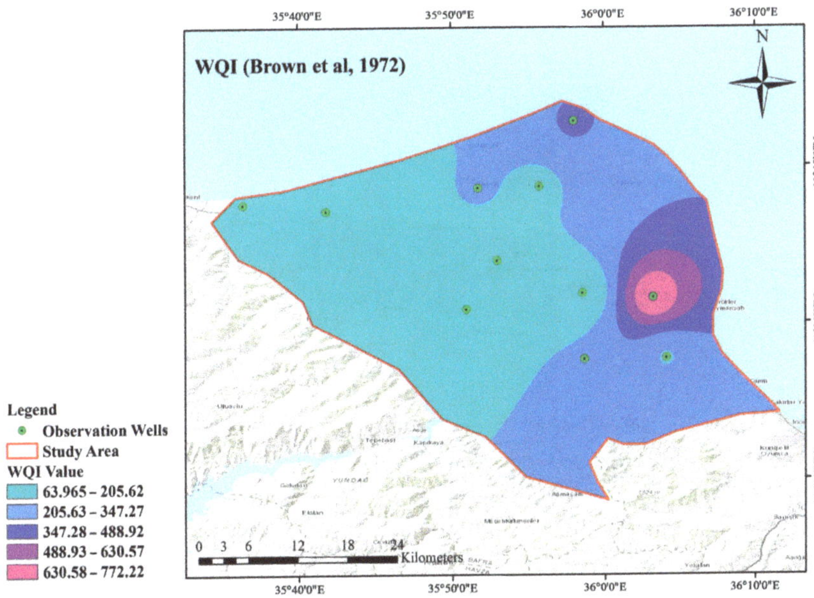

Figure 4. Plotted map of WQI for eleven wells based on WQI method [33].

Table 6. CCME WQI index values of each monitoring well developed by the Canadian Council of Ministers of the Environment (2001).

Stations	CCME WQI	Remarks
Well 1	77.728	Fair
Well 2	70.306	Fair
Well 3	62.884	Fair
Well 4	62.880	Fair
Well 5	77.731	Fair
Well 6	62.878	Fair
Well 7	40.577	Poor
Well 8	48.027	Marginal
Well 9	32.887	Poor
Well 10	77.728	Fair
Well 11	70.278	Fair

3.3.2. CCME WQI

As indicated in Figure 5, the northeast side of the Kızılırmak Delta site has lower water quality than the southwest side. The northeast part of the study area is influenced by the high concentrations of Na^{+}, Mg^{2+}, Ca^{2+}, SO_4^{2-}, and $CaCO_3$, which have decreased the water quality and made it unsuitable for drinking purposes. On the other hand, the CCME WQI method is applied for the same groundwater parameters in the study area.

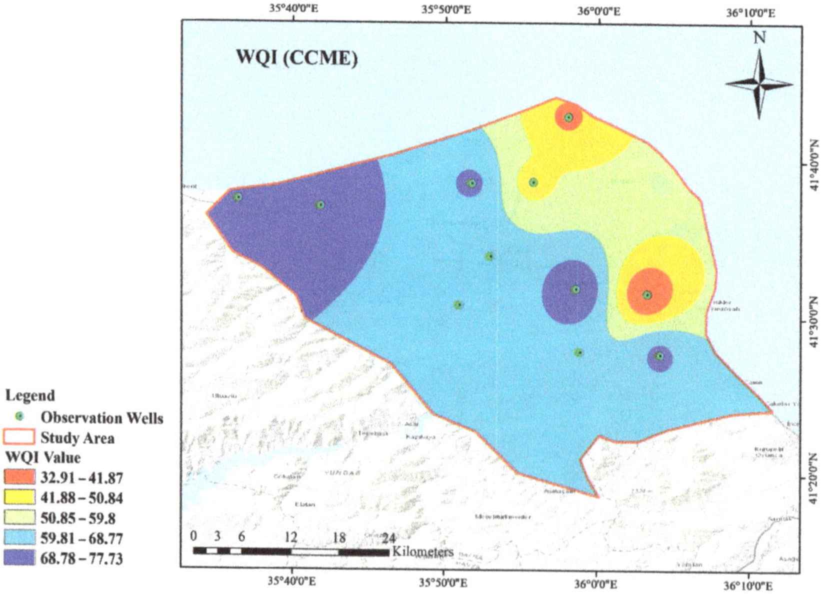

Figure 5. The water quality index map for the study area, based on the CCME method.

As indicated in Table 6, most sites have fair or lower than fair water quality or poor or marginal water quality, which is unsuitable for drinking purposes. As shown in Table 6, these types of water quality in the CCME method are usually or almost always threatened or impaired and cannot be suitable for drinking purposes. As indicated in Figure 5, only two sites have low water quality, which is in the range of 32.9 to 41.87, defined as poor grade, and other sites have a fair and marginal grade, which are the lowest water quality grades for drinking purposes.

Additionally, some studies [9,75–78] have used various WQIs to draw attention to the variations in the spatial-temporal classification of water sources. This context also includes calculations using a variety of arithmetic and logarithmic indexes. According to the findings, the indices are generally successful in accurately classifying water quality levels.

3.4. Principle Component Analysis (PCA)

Principal Component Analysis (PCA) is a statistical technique for reducing the number of multi-index data [52,79]. PCA and Cluster Analysis (CA) are the most widely used multivariate statistical methods of environmental samples. PCA is used for data reduction and reveals a few representative factors in analyzing relationships between identified variables. This study used PCA and CA to assess the accuracy of analytical data and analyses. The component analysis technique performs classification by comparing grouped data by measuring either similarity or distance between them. The objectives of CA analysis are to group similar components [80]. A groundwater quality study was performed in 2016 using PCA with varimax rotation and Kaiser normalization component values. PCA was generated utilizing the analytical data from 11 selected samples. It is a method of statistical analysis used to reduce the size of a multi-index data set. Eleven parameters were used, and two components with eigenvalues greater than one were identified (Figure 6). To comprehend the groundwater parameter structure, Figure 6 shows screen plots utilized to calculate the number of PCs. For the groundwater quality data, the overall variation for PC1 and PC2 was 70.50% and 11.49%, respectively. A highly positive loading of the K^+ physicochemical characteristic was indicated by the factor (PC2) in the data set. All of the physicochemical characteristics were low positively loaded (except pH), according to the factor (PC1) in the data sets. The varimax rotated method factor loadings from the PCA results are shown in Table 7. PCA indicates whether a positive or negative relationship exists. A value approaching 1 indicates a significant relationship between the parameter and the PC. A value greater than 0.75 indicates a strong correlation. It was found that the values in the range of 0.5 to 0.74 were closely related [52]. Table 7 illustrates the processes that the factor loadings have been used to interpret.

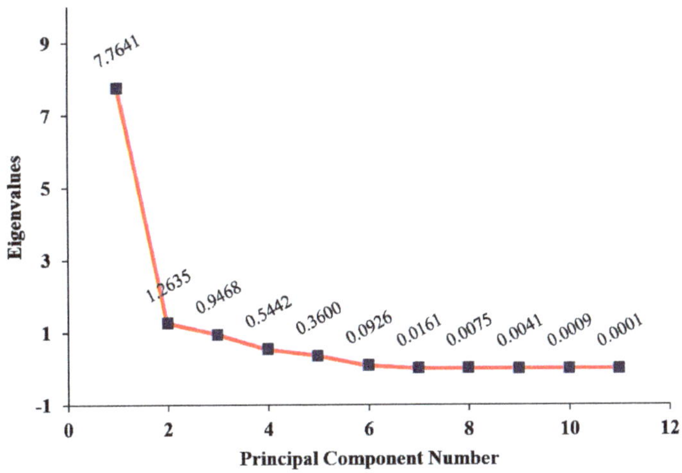

Figure 6. Screen plot of the characteristic eigenvalues used in principal component analysis.

Table 7. Varimax rotated method factor loadings from PCA.

Variables	Component Matrix	
	PC1	PC2
pH	−0.18876	0.15442
EC	0.35053	0.16871
Na^+	0.35206	0.09999
K^+	0.02028	0.79085
Ca^{2+}	0.31866	0.23649
Mg^{2+}	0.35237	0.09261
HCO_3^-	0.24332	−0.11207
Cl^-	0.3326	−0.23585
SO_4^{2-}	0.34852	−0.0826
$CaCO_3$	0.34591	0.15731
NO_3^-	0.28371	−0.38469
Eigenvalues	7.764	1.26353
Variability (%)	70.50	11.49
Cumulative (%)	70.50	82.07

Figure 7 illustrates the relationships between PC1 and PC2, contributing the most to the overall variance. The variables given in Figure 7 are all represented by a vector, and the vector's direction and length indicate each variable's contribution to the two principal components. It can be said that PC1, on the horizontal axis, has positive coefficients for K^+, Ca^{2+}, EC, hardness, Mg^{2+}, Na^+, SO_4^{2-}, HCO_3^-, Cl^-, and NO_3^- but negative coefficients for pH. For PC1, it can be said that vectors directed to the right side of the axis have a positive effect, while those directed to the left side have a negative effect. PC2, on the vertical axis, has negative coefficients for SO_4^{2-}, HCO_3^-, Cl^-, and NO_3^- and positive coefficients for pH, K^+, Ca^{2+}, EC, hardness, Mg^{2+}, and Na^+ (Table 8).

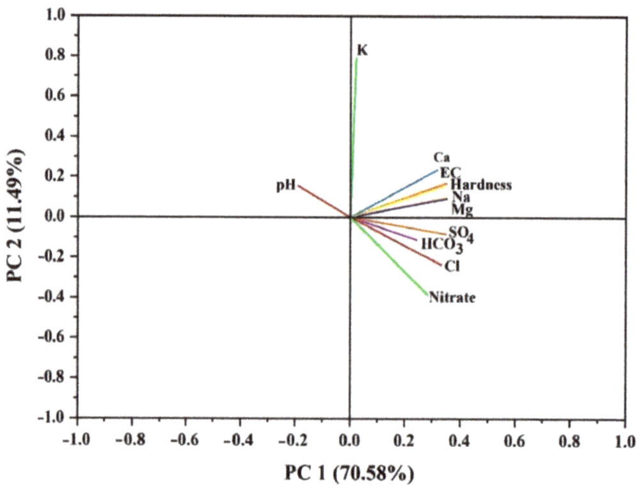

Figure 7. The 2D plot of PC1 and PC2.

Table 8. Correlation matrix of groundwater quality parameters.

	pH	EC	Na^+	K^+	Ca^{2+}	Mg^{2+}	HCO_3^-	Cl^-	SO_4^{2-}	$CaCO_3$	NO_3^-
pH	1	−0.436	−0.488	−0.054	−0.296	−0.443	−0.468	−0.470	−0.448	−0.411	−0.469
EC		1	**0.989**	0.197	**0.928**	**0.987**	0.605	0.854	**0.944**	**0.985**	0.678
Na^+			1	0.099	**0.918**	**0.989**	0.610	0.855	**0.960**	**0.988**	0.673
K^+				1	0.164	0.101	0.078	−0.116	−0.080	0.129	−0.210
Ca^{2+}					1	**0.902**	0.383	0.711	0.873	**0.962**	0.577
Mg^{2+}						1	0.669	0.871	**0.950**	**0.982**	0.679
HCO_3^-							1	0.680	0.547	0.563	0.566
Cl^-								1	**0.933**	0.810	0.911
SO_4^{2-}									1	**0.939**	0.793
$CaCO_3$										1	0.636
NO_3^-											1

3.5. Hierarchical Cluster Analysis (HCA)

Hierarchical cluster analysis used groundwater samples from different classes. The dendrograms have several clusters, each with one or more variables. Clusters were selected based on the diagram of the dendrogram in Figure 8 to facilitate understanding. They can be divided into two main groups based on the geochemical parameters of the variables as well as the outputs of the cluster tree (Figure 8). Cluster analysis and the derived dendrogram based on eleven parameters illustrated that sampling sites are ordinated into five groups: group 1 included pH, EC, Na^+; group 2 included Mg^{2+}, hardness; group 3 included SO_4^{2-}, Ca^{2+}, Cl^-; group 4 included Na^+, HCO_3^-; and Group 5 included K^+. Extreme values in the data sets are sensitive to the clusters. The similarity between PCA factors and HCA clusters supports the PCA-suggested dominating processes; the similarities among the parameters and their distribution facilitated an understanding of the impact of water quality. According to CA analysis, the most representative variable was EC, while the least representative variable was pH.

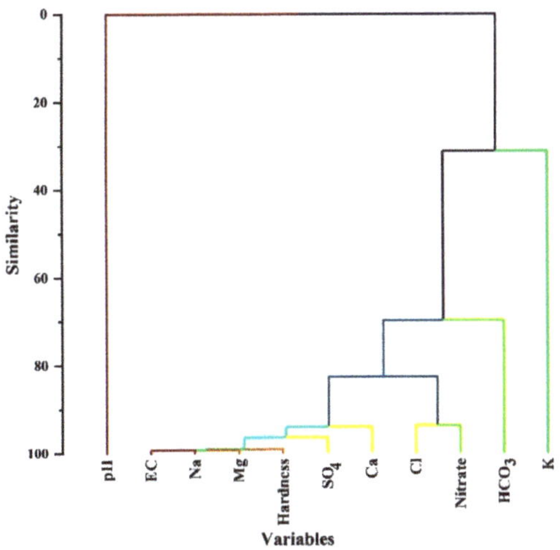

Figure 8. Dendrogram resulting from the hierarchical cluster analysis of groundwater samples.

The clusters impact extreme values in the data sets. The similarity between PCA factors and HCA clusters supports the dominating processes suggested by PCA. Therefore, incorporating physicochemical variables into the PCA to evaluate groundwater quality is a flexible and useful method that offers fresh insights and a remarkable performance.

3.6. Correlation Coefficient Matrix Analysis

The relationship between two variables is usually determined using the correlation coefficient. In this study, the relationship of eleven parameters with each other was determined via the correlation coefficient and is given in Table 8. In the correlation analysis, the Pearson correlation method was used. The correlation values between the parameters were within the 95% confidence interval. The bold prominences when r values were more than +0.75 are strongly correlated. A correlation coefficient close to 1 or −1 indicates a strong relationship between two parameters, while a correlation coefficient of zero indicates no relationship [52]. The results show that a highly positive correlation is observed between Na^+ and EC (0.98), Ca^{2+} and EC (0.92), Mg^{2+} and EC (0.98), $SO_4{}^{2-}$ and EC (0.94), $CaCO_3$ and EC (0.98), Ca^{2+} and Na^+ (0.91), Mg^{2+} and Na^+ (0.98), $SO_4{}^{2-}$ and Na^+ (0.95), $CaCO_3$ and Na^+ (0.98), Mg^{2+} and Ca^{2+} (0.90), $CaCO_3$ and Ca^{2+} (0.96), $SO_4{}^{2-}$ and Mg^{2+} (0.95), $CaCO_3$ and Mg^{2+} (0.98), $SO_4{}^{2-}$ and Cl^- (0.93), and $CaCO_3$ and $SO_4{}^{2-}$ (0.93). It was observed that while K^+ showed a low correlation with all other parameters, pH showed both a negative and low correlation.

4. Conclusions

This study aimed to analyze the chemical composition and spatial distribution of groundwater within the Kızılırmak Delta of Turkey. It also sought to assess the suitability of the groundwater in the Kızılırmak Delta for drinking water purposes by conducting a WQI evaluation. Samples were collected from eleven wells for analysis. Eleven key water parameters, comprising $NO_3{}^-$, Ca^{2+}, Mg^{2+}, Na^+, Cl^-, K^+, $HCO_3{}^-$, $SO_4{}^{2-}$, hardness (measured as $CaCO_3$), EC, and pH, were scrutinized to gauge the quality of each groundwater sample. The WQI was computed using the weighted arithmetic index and the CCME methods. In this study, the chemical composition and spatial distribution of the groundwater in the Kızılırmak Delta of Turkey were characterized, the WQI was determined, and the relationships between the parameters were analyzed using the PCA method.

- The WQI value was high due to the high values of Ca^{2+}, Mg^{2+}, and $SO_4{}^{2-}$ in some wells, indicating that the groundwater has the potential for salinization.
- A low nitrate concentration was observed in this region. The reason is a combination of factors, including suitable agricultural activities, natural processes, land management practices, and geological conditions. In addition, since this region has international protection status, an important wetland and ecosystem health are also being protected.
- Approximately 90% of wells had hardness levels within the desirable 100 mg/L limit. The low hardness may be because the hardness level decreases when mixed with groundwater and surface water or other sources with lower hardness. Another reason is that groundwater with a short residence time may need more contact with minerals in the aquifer to accumulate significant hardness.
- According to the WQI values, most wells' water is unsuitable for drinking and use. On the other hand, the CCME WQI method indicated that most sites have fair or lower than fair water quality or poor or marginal water quality, which is also unsuitable for drinking purposes.
- Based on the spatial distribution of the water quality, it is estimated that the region's western part is inadequate for drinking water and irrigation in the Kızılırmak Delta.
- The correlation coefficient determined the relationships between the groundwater parameters, and the highest correlation was between Mg^{2+} and $CaCO_3$.

The present study shows that the delta's groundwater, which has deteriorating water quality, should be treated before it is used for drinking water and protected from contamination hazards.

Author Contributions: Conceptualization, S.A. and N.G.S.-O.; methodology, N.B., N.G.S.-O. and S.A.; software, N.B. and formal analysis, N.B. and H.A.; investigation, S.A.; resources, N.B., N.G.S.-O., S.A. and H.A.; data curation, N.B., N.G.S.-O. and S.A.; writing—original draft preparation, N.G.S.-O. and S.A.; writing—review and editing, N.B.; visualization, H.A.; supervision, S.A.; project administration, N.B. All authors have read and agreed to the published version of the manuscript.

Funding: This work was supported by the project SCORE (Smart Control of the Climate Resilience in European Coastal Cities), funded by the European Commission's Horizon 2020 research and innovation programme under grant agreement no.101003534.

Data Availability Statement: The data that support the findings of this study are available from the corresponding author upon reasonable request.

Acknowledgments: We thank the General Directorate of State Hydraulic Works for the groundwater quality data used in this study.

Conflicts of Interest: The authors declare no conflicts of interest.

References

1. Siebert, S.; Burke, J.; Faures, J.-M.; Frenken, K.; Hoogeveen, J.; Döll, P.; Portmann, F.T. Groundwater use for irrigation-a global inventory. *Hydrol. Earth Syst. Sci.* **2010**, *14*, 1863–1880. [CrossRef]
2. Eid, M.H.; Elbagory, M.; Tamma, A.A.; Gad, M.; Elsayed, S.; Hussein, H.; Moghanm, F.S.; Omara, A.E.-D.; Kovács, A.; Péter, S. Evaluation of groundwater quality for irrigation in deep aquifers using multiple graphical and indexing approaches supported with machine learning models and GIS techniques, Souf Valley, Algeria. *Water* **2023**, *15*, 182. [CrossRef]
3. Taşan, S. Estimation of groundwater quality using an integration of water quality index, artificial intelligence methods and GIS: Case study, Central Mediterranean Region of Turkey. *Appl. Water Sci.* **2023**, *13*, 15. [CrossRef]
4. Husna, A.; Akmalia, R.; Rohmat, F.I.W.; Rohmat, F.I.W.; Rohmat, D.; Wijayasari, W.; Alvando, P.V.; Wijaya, A. Groundwater Sustainability Assessment against the Population Growth Modelling in Bima City, Indonesia. *Water* **2023**, *15*, 4262. [CrossRef]
5. Singh, A. Groundwater modelling for the assessment of water management alternatives. *J. Hydrol.* **2013**, *481*, 220–229. [CrossRef]
6. Rohmat, F.I.; Labadie, J.W.; Gates, T.K. Deep learning for compute-efficient modeling of BMP impacts on stream-aquifer exchange and water law compliance in an irrigated river basin. *Environ. Model. Softw.* **2019**, *122*, 104529. [CrossRef]
7. Barbieri, M.; Barberio, M.D.; Banzato, F.; Billi, A.; Boschetti, T.; Franchini, S.; Gori, F.; Petitta, M. Climate change and its effect on groundwater quality. *Environ. Geochem. Health* **2023**, *45*, 1133–1144. [CrossRef]
8. Foster, S.; Hirata, R.; Eichholz, M.; Alam, M.-F. Urban self-supply from groundwater—An analysis of management aspects and policy needs. *Water* **2022**, *14*, 575. [CrossRef]
9. Ram, A.; Tiwari, S.; Pandey, H.; Chaurasia, A.K.; Singh, S.; Singh, Y. Groundwater quality assessment using water quality index (WQI) under GIS framework. *Appl. Water Sci.* **2021**, *11*, 46. [CrossRef]
10. Goodsell, T.; Carling, G.; Aanderud, Z.; Nelson, S.; Fernandez, D.; Tingey, D. Thermal groundwater contributions of arsenic and other trace elements to the middle Provo River, Utah, USA. *Environ. Earth Sci.* **2017**, *76*, 268. [CrossRef]
11. Schreiber, M.; Gotkowitz, M.; Simo, J.; Freiberg, P. Mechanisms of arsenic release to ground water from naturally occurring sources, eastern Wisconsin. In *Arsenic in Ground Water: Geochemistry and Occurrence*; Springer: Boston, MA, USA, 2003; pp. 259–280.
12. Selck, B.J.; Carling, G.T.; Kirby, S.M.; Hansen, N.C.; Bickmore, B.R.; Tingey, D.G.; Rey, K.; Wallace, J.; Jordan, J.L. Investigating anthropogenic and geogenic sources of groundwater contamination in a semi-arid alluvial basin, Goshen Valley, UT, USA. *Water Air Soil Pollut.* **2018**, *229*, 186. [CrossRef]
13. Kushawaha, J.; Aithani, D. Geogenic pollutants in groundwater and their removal techniques. In *Groundwater Geochemistry: Pollution and Remediation Methods*; John Wiley & Sons Ltd.: Hoboken, NJ, USA, 2021; pp. 1–21.
14. Yuan, L.; Wang, K.; Zhao, Q.; Yang, L.; Wang, G.; Jiang, M.; Li, L. An overview of in situ remediation for groundwater co-contaminated with heavy metals and petroleum hydrocarbons. *J. Environ. Manag.* **2024**, *349*, 119342. [CrossRef] [PubMed]
15. Singh, C.K.; Kumar, A.; Shashtri, S.; Kumar, A.; Kumar, P.; Mallick, J. Multivariate statistical analysis and geochemical modeling for geochemical assessment of groundwater of Delhi, India. *J. Geochem. Explor.* **2017**, *175*, 59–71. [CrossRef]
16. Karunanidhi, D.; Subramani, T.; Roy, P.D.; Li, H. Impact of groundwater contamination on human health. *Environ. Geochem. Health* **2021**, *43*, 643–647. [CrossRef] [PubMed]
17. Cao, X.; Lu, Y.; Wang, C.; Zhang, M.; Yuan, J.; Zhang, A.; Song, S.; Baninla, Y.; Khan, K.; Wang, Y. Hydrogeochemistry and quality of surface water and groundwater in the drinking water source area of an urbanizing region. *Ecotoxicol. Environ. Saf.* **2019**, *186*, 109628. [CrossRef] [PubMed]
18. Mallick, J.; Singh, C.K.; AlMesfer, M.K.; Kumar, A.; Khan, R.A.; Islam, S.; Rahman, A. Hydro-geochemical assessment of groundwater quality in Aseer Region, Saudi Arabia. *Water* **2018**, *10*, 1847. [CrossRef]
19. Gautam, A.; Rai, S.; Rai, S.; Ram, K. Impact of anthropogenic and geological factors on groundwater hydrochemistry in the unconfined aquifers of Indo-Gangetic plain. *Phys. Chem. Earth Parts A/B/C* **2022**, *126*, 103109. [CrossRef]
20. Ruiz-Pico, Á.; Pérez-Cuenca, Á.; Serrano-Agila, R.; Maza-Criollo, D.; Leiva-Piedra, J.; Salazar-Campos, J. Hydrochemical characterization of groundwater in the Loja Basin (Ecuador). *Appl. Geochem.* **2019**, *104*, 1–9. [CrossRef]

21. Fırat-Ersoy, A.; Turan-Ayyıldız, N.; Arslan, H.; Kuleyin, A. Assessment of seawater intrusion in Kızılırmak delta coastal area (North Turkey) using hydrochemical and isotopic data. *Environ. Earth Sci.* **2021**, *80*, 400. [CrossRef]
22. Karakuş, C.B. Evaluation of water quality of Kızılırmak River (Sivas/Turkey) using geo-statistical and multivariable statistical approaches. *Environ. Dev. Sustain.* **2020**, *22*, 4735–4769. [CrossRef]
23. Samsunlu, A.; Akca, L.; Kinaci, C.; Findik, N.; Tanik, A. Significance of wetlands in water quality management-past and present situation of Kizilirmak Delta, Turkey. *Water Sci. Technol.* **2002**, *46*, 145–152. [CrossRef] [PubMed]
24. Demirci, I.; Gündoğdu, N.Y.; Candansayar, M.E.; Soupios, P.; Vafidis, A.; Arslan, H. Determination and evaluation of saltwater intrusion on bafra plain: Joint interpretation of geophysical, hydrogeological and hydrochemical data. *Pure Appl. Geophys.* **2020**, *177*, 5621–5640. [CrossRef]
25. Mohebbi, M.R.; Saeedi, R.; Montazeri, A.; Vaghefi, K.A.; Labbafi, S.; Oktaie, S.; Abtahi, M.; Mohagheghian, A. Assessment of water quality in groundwater resources of Iran using a modified drinking water quality index (DWQI). *Ecol. Indic.* **2013**, *30*, 28–34. [CrossRef]
26. Asadollah, S.B.H.S.; Sharafati, A.; Motta, D.; Yaseen, Z.M. River water quality index prediction and uncertainty analysis: A comparative study of machine learning models. *J. Environ. Chem. Eng.* **2021**, *9*, 104599. [CrossRef]
27. Sánchez, E.; Colmenarejo, M.F.; Vicente, J.; Rubio, A.; García, M.G.; Travieso, L.; Borja, R. Use of the water quality index and dissolved oxygen deficit as simple indicators of watersheds pollution. *Ecol. Indic.* **2007**, *7*, 315–328. [CrossRef]
28. Chauhan, S.S.; Trivedi, M.K. Artificial neural network-based assessment of water quality index (WQI) of surface water in Gwalior-Chambal region. *Int. J. Energy Environ. Eng.* **2023**, *14*, 47–61. [CrossRef]
29. Horton, R.K. An index number system for rating water quality. *J. Water Pollut. Control. Fed.* **1965**, *37*, 300–306.
30. Lumb, A.; Sharma, T.; Bibeault, J.-F. A review of genesis and evolution of water quality index (WQI) and some future directions. *Water Qual. Expo. Health* **2011**, *3*, 11–24. [CrossRef]
31. Davies, J.-M. Application and tests of the Canadian water quality index for assessing changes in water quality in lakes and rivers of central North America. *Lake Reserv. Manag.* **2006**, *22*, 308–320. [CrossRef]
32. Wills, M.; Irvine, K.N. Application of the national sanitation foundation water quality index in Cazenovia Creek, NY, pilot watershed management project. *Middle States Geogr.* **1996**, *1996*, 95–104.
33. Brown, R.M.; McClelland, N.I.; Deininger, R.A.; O'Connor, M.F. A water quality index—Crashing the psychological barrier. In *Indicators of Environmental Quality: Proceedings of a Symposium Held during the AAAS Meeting, Philadelphia, Pennsylvania, 26–31 December 1971*; Springer: Boston, MA, USA, 1972; pp. 173–182.
34. Hamidi, M.D.; Kissane, S.; Bogush, A.A.; Karim, A.Q.; Sagintayev, J.; Towers, S.; Greenwell, H.C. Spatial estimation of groundwater quality, hydrogeochemical investigation, and health impacts of shallow groundwater in Kabul city, Afghanistan. *Sustain. Water Resour. Manag.* **2023**, *9*, 20. [CrossRef]
35. Udeshani, W.; Dissanayake, H.; Gunatilake, S.; Chandrajith, R. Assessment of groundwater quality using water quality index (WQI): A case study of a hard rock terrain in Sri Lanka. *Groundw. Sustain. Dev.* **2020**, *11*, 100421. [CrossRef]
36. Krishan, G.; Kumar, M.; Rao, M.S.; Garg, R.; Yadav, B.K.; Kansal, M.; Singh, S.; Bradley, A.; Muste, M.; Sharma, L. Integrated approach for the investigation of groundwater quality through hydrochemistry and water quality index (WQI). *Urban Clim.* **2023**, *47*, 101383. [CrossRef]
37. Krishnamoorthy, N.; Thirumalai, R.; Sundar, M.L.; Anusuya, M.; Kumar, P.M.; Hemalatha, E.; Prasad, M.M.; Munjal, N. Assessment of underground water quality and water quality index across the Noyyal River basin of Tirupur District in South India. *Urban Clim.* **2023**, *49*, 101436. [CrossRef]
38. El Yousfi, Y.; Himi, M.; El Ouarghi, H.; Aqnouy, M.; Benyoussef, S.; Gueddari, H.; Ait Hmeid, H.; Alitane, A.; Chaibi, M.; Zahid, M. Assessment and Prediction of the Water Quality Index for the Groundwater of the Ghiss-Nekkor (Al Hoceima, Northeastern Morocco). *Sustainability* **2022**, *15*, 402. [CrossRef]
39. Soujanya Kamble, B.; Saxena, P.R.; Kurakalva, R.M.; Shankar, K. Evaluation of seasonal and temporal variations of groundwater quality around Jawaharnagar municipal solid waste dumpsite of Hyderabad city, India. *SN Appl. Sci.* **2020**, *2*, 498. [CrossRef]
40. Rabeiy, R.E. Assessment and modeling of groundwater quality using WQI and GIS in Upper Egypt area. *Environ. Sci. Pollut. Res.* **2018**, *25*, 30808–30817. [CrossRef] [PubMed]
41. Singh, S.; Noori, A.R. Groundwater quality assessment and modeling utilizing water quality index and GIS in Kabul Basin, Afghanistan. *Environ. Monit. Assess.* **2022**, *194*, 673. [CrossRef]
42. Şener, E.; Şener, Ş.; Varol, S. Evaluation of irrigation water quality using GIS-based analytic hierarchy process (AHP) in Kızılırmak Delta (Turkey). *Arab. J. Geosci.* **2022**, *15*, 678. [CrossRef]
43. SMM. *Determination of the Kızılırmak Delta Water Footprint, Sub-Project Report*; SMM: Samsun, Turkey, 2018; 594p.
44. Chowdhury, R.M.; Muntasir, S.Y.; Hossain, M.M. Water quality index of water bodies along Faridpur-Barisal road in Bangladesh. *Glob. J. Eng. Technol.* **2012**, *2*, 1–8.
45. Ponsadailakshmi, S.; Sankari, S.G.; Prasanna, S.M.; Madhurambal, G. Evaluation of water quality suitability for drinking using drinking water quality index in Nagapattinam district, Tamil Nadu in Southern India. *Groundw. Sustain. Dev.* **2018**, *6*, 43–49. [CrossRef]
46. Tripathi, M.; Singal, S.K. Use of principal component analysis for parameter selection for development of a novel water quality index: A case study of river Ganga India. *Ecol. Indic.* **2019**, *96*, 430–436. [CrossRef]

47. Canadian Council of Ministers of the Environment. *Canadian Environmental Quality Guidelines*; Canadian Council of Ministers of the Environment: Winnipeg, MB, Canada, 2002; Volume 2.
48. Jolliffe, I.T. Discarding variables in a principal component analysis. I: Artificial data. *J. R. Stat. Soc. Ser. C Appl. Stat.* **1972**, *21*, 160–173. [CrossRef]
49. Mahapatra, S.; Sahu, M.; Patel, R.; Panda, B.N. Prediction of water quality using principal component analysis. *Water Qual. Expo. Health* **2012**, *4*, 93–104. [CrossRef]
50. Jolliffe, I.T. Discarding variables in a principal component analysis. II: Real data. *J. R. Stat. Soc. Ser. C Appl. Stat.* **1973**, *22*, 21–31. [CrossRef]
51. Commission, J.R.C.-E. *Handbook on Constructing Composite Indicators: Methodology and User Guide*; OECD Publishing: Paris, France, 2008.
52. Samtio, M.S.; Jahangir, T.M.; Mastoi, A.S.; Lanjwani, M.F.; Rajper, R.H.; Lashari, R.A.; Agheem, M.H.; Noonari, M.W. Impact of rock-water interaction on hydrogeochemical characteristics of groundwater: Using multivariate statistical, water quality index and irrigation indices of chachro sub-district, thar desert, sindh, Pakistan. *Groundw. Sustain. Dev.* **2023**, *20*, 100878. [CrossRef]
53. Doerr, S.H.; Shakesby, R.; Walsh, R. Soil water repellency: Its causes, characteristics and hydro-geomorphological significance. *Earth-Sci. Rev.* **2000**, *51*, 33–65. [CrossRef]
54. Gazette, R. Regulation on Surface Water Quality. Annex-5 (Amended: RG-10/8/2016-29797) Environmental QUALITY Standards for Some Parameters in Surface Water Bodies and Purposes of Use. (28483 Repeated), 30, 2012. Available online: https://www.mevzuat.gov.tr/mevzuat?MevzuatNo=16806&MevzuatTur=7&MevzuatTertip=5 (accessed on 27 March 2024).
55. WHO. *Guidelines for Drinking-Water Quality*; World Health Organization: Geneva, Switzerland, 2011; Volume 38, pp. 104–108.
56. Jena, S.K. A review on potash recovery from different rock and mineral sources. *Min. Metall. Explor.* **2021**, *38*, 47–68. [CrossRef]
57. Rajmohan, N.; Elango, L. Nutrient chemistry of groundwater in an intensively irrigated region of southern India. *Environ. Geol.* **2005**, *47*, 820–830. [CrossRef]
58. Laghrib, F.; Bahaj, T.; El Kasmi, S.; Hilali, M.; Kacimi, I.; Nouayti, N.; Dakak, H.; Bouzekraoui, M.; El Fatni, O.; Hammani, O. Hydrogeochemical study of groundwater in arid and semi-arid regions of the Infracenomanian aquifers (Cretaceous Errachidia basin, Southeastern Morocco). Using hydrochemical modeling and multivariate statistical analysis. *J. Afr. Earth Sci.* **2024**, *209*, 105132. [CrossRef]
59. Li, C.; Gao, X.; Li, S.; Bundschuh, J. A review of the distribution, sources, genesis, and environmental concerns of salinity in groundwater. *Environ. Sci. Pollut. Res.* **2020**, *27*, 41157–41174. [CrossRef] [PubMed]
60. WHO. *Use of Glycated Hemoglobin (HbA1c) in Diagnosis of Diabetes Mellitus: Abbreviated Report of a WHO Consultation*; World Health Organization: Geneva, Switzerland, 2011.
61. Haritash, A.; Kaushik, C.P.; Kaushik, A.; Kansal, A.; Yadav, A.K. Suitability assessment of groundwater for drinking, irrigation and industrial use in some North Indian villages. *Environ. Monit. Assess.* **2008**, *145*, 397–406. [CrossRef] [PubMed]
62. Jain, C.; Bandyopadhyay, A.; Bhadra, A. Assessment of ground water quality for drinking purpose, District Nainital, Uttarakhand, India. *Environ. Monit. Assess.* **2010**, *166*, 663–676. [CrossRef] [PubMed]
63. Sharma, M.; Kumar, M. Sulphate contamination in groundwater and its remediation: An overview. *Environ. Monit. Assess.* **2020**, *192*, 1–10. [CrossRef] [PubMed]
64. Hanor, J.S.; Wendeborn, F.C. Origin of sodium bicarbonate groundwaters, Southern Hills Aquifer System, USA by silicate hydrolysis. *Appl. Geochem.* **2023**, *148*, 105512. [CrossRef]
65. Lin, G.-F.; Wu, M.-C. An RBF network with a two-step learning algorithm for developing a reservoir inflow forecasting model. *J. Hydrol.* **2011**, *405*, 439–450. [CrossRef]
66. Singh, P.; Khan, I. Ground water quality assessment of Dhankawadi ward of Pune by using GIS. *Int. J. Geomat. Geosci.* **2011**, *2*, 688–703.
67. Craswell, E. Fertilizers and nitrate pollution of surface and ground water: An increasingly pervasive global problem. *SN Appl. Sci.* **2021**, *3*, 518.
68. Liu, G.; Wu, W.; Zhang, J. Regional differentiation of non-point source pollution of agriculture-derived nitrate nitrogen in groundwater in northern China. *Agric. Ecosyst. Environ.* **2005**, *107*, 211–220. [CrossRef]
69. McCray, J.E.; Kirkland, S.L.; Siegrist, R.L.; Thyne, G.D. Model parameters for simulating fate and transport of on-site wastewater nutrients. *Groundwater* **2005**, *43*, 628–639. [CrossRef]
70. Galloway, J.N.; Dentener, F.J.; Capone, D.G.; Boyer, E.W.; Howarth, R.W.; Seitzinger, S.P.; Asner, G.P.; Cleveland, C.C.; Green, P.; Holland, E.A. Nitrogen cycles: Past, present, and future. *Biogeochemistry* **2004**, *70*, 153–226. [CrossRef]
71. Galloway, J.N.; Aber, J.D.; Erisman, J.W.; Seitzinger, S.P.; Howarth, R.W.; Cowling, E.B.; Cosby, B.J. The nitrogen cascade. *Bioscience* **2003**, *53*, 341–356. [CrossRef]
72. Panneerselvam, B.; Muniraj, K.; Thomas, M.; Ravichandran, N.; Bidorn, B. Identifying influencing groundwater parameter on human health associate with irrigation indices using the Automatic Linear Model (ALM) in a semi-arid region in India. *Environ. Res.* **2021**, *202*, 111778. [CrossRef] [PubMed]
73. Kumar, M.; Singh, D.; Prabha, R.; Sharma, A.K. Role of cyanobacteria in nutrient cycle and use efficiency in the soil. *Nutr. Use Effic. Basics Adv.* **2015**, 163–171.
74. WHO. *One Health*; World Health Organization: Geneva, Switzerland, 2017; p. 736.

75. Chaurasia, A.K.; Pandey, H.; Tiwari, S.; Pandey, P.; Ram, A. Groundwater vulnerability assessment using water quality index (WQI) under geographic information system (GIS) framework in parts of Uttar Pradesh, India. *Sustain. Water Resour. Manag.* **2021**, *7*, 40. [CrossRef]
76. Changsheng, H.; Akram, W.; Rashid, A.; Ullah, Z.; Shah, M.; Alrefaei, A.F.; Kamel, M.; Aleya, L.; Abdel-Daim, M.M. Quality Assessment of Groundwater Based on Geochemical Modelling and Water Quality Index (WQI). *Water* **2022**, *14*, 3888. [CrossRef]
77. Saqib, N.; Rai, P.K.; Kanga, S.; Kumar, D.; Đurin, B.; Singh, S.K. Assessment of Ground Water Quality of Lucknow City under GIS Framework Using Water Quality Index (WQI). *Water* **2023**, *15*, 3048. [CrossRef]
78. Gaagai, A.; Aouissi, H.A.; Bencedira, S.; Hinge, G.; Athamena, A.; Heddam, S.; Gad, M.; Elsherbiny, O.; Elsayed, S.; Eid, M.H. Application of water quality indices, machine learning approaches, and GIS to identify groundwater quality for irrigation purposes: A case study of Sahara Aquifer, Doucen Plain, Algeria. *Water* **2023**, *15*, 289. [CrossRef]
79. Howladar, M.F.; Chakma, E.; Koley, N.J.; Islam, S.; Al Numanbakth, M.A.; Ahmed, Z.; Chowdhury, T.R.; Akter, S. The water quality and pollution sources assessment of Surma river, Bangladesh using, hydrochemical, multivariate statistical and water quality index methods. *Groundw. Sustain. Dev.* **2021**, *12*, 100523. [CrossRef]
80. Frías, S.; Conde, J.E.; Rodríguez-Bencomo, J.J.; García-Montelongo, F.; Pérez-Trujillo, J.P. Classification of commercial wines from the Canary Islands (Spain) by chemometric techniques using metallic contents. *Talanta* **2003**, *59*, 335–344. [CrossRef]

water

MDPI

Article

Assessing the Relationship Between Groundwater Availability, Access, and Contamination Risk in Arizona's Drinking Water Sources

Simone A. Williams [1,2,*], Adriana A. Zuniga-Teran [3,4], Sharon B. Megdal [2], David M. Quanrud [5] and Gary Christopherson [3]

1 Arid Lands Resource Sciences, The University of Arizona, Tucson, AZ 85721, USA
2 Water Resources Research Center, The University of Arizona, Tucson, AZ 85721, USA; smegdal@arizona.edu
3 School of Geography, Development & Environment, The University of Arizona, Tucson, AZ 85721, USA; aazuniga@arizona.edu (A.A.Z.-T.); gchristo@arizona.edu (G.C.)
4 The Udall Center for Studies in Public Policy, The University of Arizona, Tucson, AZ 85721, USA
5 School of Natural Resources and the Environment, The University of Arizona, Tucson, AZ 85721, USA; quanrud@arizona.edu
* Correspondence: sawillms@arizona.edu

Abstract: Groundwater is a critical drinking water source in arid regions globally, where reliance on groundwater is highest. However, disparities in groundwater availability, access, and quality pose challenges to water security. This case study employs geostatistical tools, multivariate regression, and clustering analysis to examine the intersection of groundwater level changes (availability), socioeconomic and regulatory factors (access), and nitrate and arsenic contamination (quality) across 1881 groundwater-supplied drinking water service areas in Arizona. Groundwater availability declined over 20-year and 10-year periods, particularly outside designated management areas, with mean annual decline rates ranging from −15.97 to −0.003 m/year. In contrast, increases (0.003 to 13.41 m/year) were concentrated in urban and managed areas. Karst aquifers show long-term resilience but short-term vulnerability. Non-designated areas exhibit mixed effects, reflecting variable management effectiveness. Disparities in groundwater access emerge along various socioeconomic and regulatory lines. Communities with higher Black populations are twice as likely (OR = 2.01, $p < 0.001$) to experience groundwater declines, while Hispanic/Latino communities have lower depletion risks (OR = 0.92, $p < 0.001$). Tribal oversight significantly reduces groundwater decline risk (OR = 0.62, $p < 0.001$), whereas state–primacy areas show mixed effects. Higher female populations correlate with increased groundwater declines, while older populations (65+) experience greater stability. Married-family households and institutional housing are associated with greater declines. Migrant worker housing shows protective effects in long-term models. Rising groundwater levels are associated with higher nitrate and arsenic detection, reinforcing recharge-driven contaminant mobilization. Nitrate exceedance (OR = 1.05) responds more to short-term groundwater changes, while arsenic exceedance persists over longer timescales (OR = 1.01–1.05), reflecting their distinct hydrogeochemical behaviors. Community water systems show higher pollutant detection rates than domestic well areas, suggesting monitoring and infrastructure differences influence contamination patterns. Tribal primacy areas experience lower groundwater declines but show mixed effects on water quality, with reduced nitrate exceedance probabilities; yet they show variable arsenic contamination patterns, suggesting that governance influences availability and contamination dynamics. These findings advance groundwater sustainability research by quantifying disparities across multiple timescales and socio-hydrogeological drivers of groundwater vulnerability. The results underscore the need for expanded managed aquifer recharge, targeted regulatory interventions, and strengthened Tribal water governance to reduce inequities in availability, access, and contamination risk to support equitable and sustainable groundwater management.

Keywords: groundwater access; water level trends; public water systems; arsenic and nitrate contamination; groundwater depletion; drinking water security; water equity

1. Introduction

Although the availability of safe drinking water is essential to human well-being, 2.2 billion people worldwide lacked access to safe drinking water in 2022 [1]. There is a well-established relationship between water resource development and economic development [2]. Countries with the capacity to invest in water infrastructure and management not only achieve higher development levels but also provide safer drinking water, enhancing their populations' quality of life. However, challenges remain, even in developed countries like the United States. The crises in Flint, Michigan, and Newark, New Jersey, highlight persistent inequities in water safety and availability [3–5]. Additionally, hundreds of rural communities across the U.S. face water quality issues, inadequate infrastructure, and unreliable service [6–9].

Low-income and minority communities in the U.S. often face disproportionate risks from drinking water contamination. Systems serving areas with higher material deprivation, lower median incomes, and larger populations of Hispanic, Native American, and other non-white residents frequently show elevated levels of contaminants like arsenic, nitrate, and lead [3,10–14]. These disparities are particularly pronounced in small rural water systems and Tribal communities—Indigenous nations in the U.S. with self-governing authority—where limited resources, aging infrastructure, and historical contamination contribute to persistent environmental injustices [15–17].

The concept of water security, encompassing availability, quality, safety, reliability, access, and sustainability, has been widely used to examine these challenges [18]. Scott, et al. [19] define water security as "the sustainable availability of adequate quantities and qualities of water for resilient societies and ecosystems in the face of uncertain global change". The international community has advanced water security through initiatives like the UN Sustainable Development Goals (SDGs), particularly Goal 6, which aims to ensure universal access to water and sanitation by 2030 [1].

Groundwater is crucial for achieving water security due to its wide accessibility, large storage capacity, and resilience to climate variability [1,20–23]. It supports domestic, agricultural, industrial, and municipal needs, especially in arid regions where it often serves as the primary water source [2,24]. In areas where groundwater levels are stable, even the most dispersed and economically disadvantaged communities can benefit from reliable water access through shallow aquifers [1].

However, rapid urbanization, intensive agriculture, industrialization, and other changes in land use have accelerated groundwater extraction and have altered recharge patterns, with negative implications for watershed sustainability [24,25]. Excessive pumping depletes aquifers, causing land subsidence, loss of riparian vegetation, species loss, and reduced soil moisture, which, in turn, increases irrigation demands, compounding groundwater scarcity [26]. The interplay between groundwater quantity and quality is critical, as overexploitation can degrade water quality, posing risks to both human and ecosystem health [24].

This study builds on previous research that mapped groundwater vulnerability in Arizona [27] and assessed equity in groundwater vulnerability and risk [28]. Building on this work, we examine how groundwater availability, access, and contamination risk interact through a case study of Arizona. By integrating spatial and temporal models with equity-focused variables, the analysis explores how hydrogeological conditions, service area characteristics, and policy factors shape groundwater vulnerability. The results show significant relationships between groundwater availability, access, and quality across regulatory frameworks, infrastructure types, environmental settings, and community socioeconomic contexts, with implications for sustainable groundwater management.

1.1. Groundwater Sustainability Through a Trivalent Framework

This study employs an integrated framework to explore the connections between groundwater availability, access, and contamination risk, providing a comprehensive approach to understanding groundwater sustainability. Availability refers to the physical condition and quantity of groundwater shaped by factors such as aquifer depletion and recharge rates. Access encompasses the socioeconomic, regulatory, and infrastructural factors that influence communities' ability to obtain and use groundwater for drinking and other essential purposes. Contamination risk (quality) focuses on the presence of harmful contaminants, such as nitrate and arsenic, which impact groundwater safety and usability. These trivalent dimensions are interconnected, and this study explores how groundwater level changes, socioeconomic factors, and contamination risks converge within Arizona's groundwater-dependent drinking water systems. This analysis highlights the complex interplay between factors influencing groundwater sustainability.

Assessing groundwater resources and progress toward water security remains challenging. Various modeling approaches, including physical, conceptual, numerical, statistical, and hybrid models, are used to evaluate groundwater availability, access, and quality [26,27,29–34]. In addition, indices have been developed to enhance assessment efforts and track global progress. For instance, the SDGs include water-related indicators such as Indicator 6.1.1, which monitors the proportion of the population using safely managed drinking water services and emphasizes the importance of access, availability, and reliability [1]. However, global indices are often inadequate for local-scale assessments, prompting interdisciplinary researchers to create comparative frameworks such as the Water Poverty Index, the Arab Water Sustainability Index (AWSI), and the European Sustainable Water Management Index [1].

Despite their utility, most indices remain qualitative and do not adequately account for the impacts of aridification and groundwater overexploitation. Empirical assessments of groundwater availability are further complicated by inconsistent and incomplete data, hindering the ability to track trends over time. In Arizona, a state often recognized for its groundwater management efforts [35], research has primarily focused on regulated areas that have more reliable and cohesive data. In contrast, unregulated regions rely on indirect indicators, such as streamflow reduction or vegetation decline [25,26], making groundwater depletion more difficult to quantify. The continued decline in groundwater levels poses a growing risk to freshwater resources and could threaten the achievement of SDGs, increasing the likelihood of a future water crisis.

1.2. Background—Groundwater Dynamics Through an Equity Lens

Groundwater resources are deeply interconnected with surface water and human systems, making equity a critical factor in groundwater sustainability. Hydrologically, groundwater and surface water interact through natural recharge and discharge processes, while socially, access to groundwater sources is shaped by infrastructure, pumping, and regulatory frameworks [25]. Aquifers are naturally replenished through rainfall and percolation, but recharge is influenced by soil composition, geomorphology, land use, and aquifer material [24]. In arid and semi-arid regions like Arizona, where natural recharge is low, managed aquifer recharge programs play a crucial role in increasing groundwater storage, particularly in areas supplied by the Central Arizona Project, which diverts surface water from the Colorado River [25].

Effective groundwater governance is essential for sustainability and equitable access. Scholars highlight four key governance factors: (1) access to information, (2) institutional setting, (3) civil society participation, and (4) policy frameworks, all of which influence groundwater equity [36,37]. For instance, Arizona's managed aquifer recharge programs, regulated by the Arizona Water Banking Authority, help mitigate overdrafts in areas served by the Central Arizona Project. However, regulation alone is insufficient in regions with minimal natural recharge. Unregulated areas remain due to continued groundwater extraction with minimal monitoring and enforcement, jeopardizing availability, access, and long-term sustainability [26]. Even with regulated pumping and managed recharge, long-term monitoring, and early warning systems are crucial for preventing resource depletion [2,24]. Thus, regulation alone is insufficient—it must be complemented by environmental, socioeconomic, and political considerations to ensure groundwater sustainability [24].

Arizona's groundwater governance faces structural gaps that affect availability, access, and quality. Weak regulatory oversight and the lack of integration between land use policies and groundwater management further exacerbate sustainability challenges [28,38]. Fragmented policies fail to integrate groundwater-surface water connections and provide differential protection of karst and alluvial aquifers. While policies primarily focus on alluvial aquifers in urban areas, karst systems—which cover nearly half of Arizona—lack tailored protections despite their heightened vulnerability to contamination from land use changes and climate impacts [28,39,40]. The differential treatment of these aquifers highlights broader governance deficiencies, particularly in rural and Indigenous communities that depend on karst systems [26].

Groundwater access and equity are also influenced by aquifer levels, as shallow aquifers tend to be more accessible to economically disadvantaged populations with limited technological resources [2]. However, limited hydrogeological characterization, inadequate groundwater infrastructure and maintenance, and insufficient technical capacity—alongside broader socioeconomic and governance challenges—impede the sustainable use of local groundwater resources [1]. While the *Groundwater Management Act* (GMA) has advanced groundwater security, it falls short in addressing unregulated pumping, rising water demands, and critical hydrological linkages [26,41,42]. Effective management requires strategies that account for the distinct hydrogeological dynamics of these aquifer types [28].

Sociodemographic factors play a critical role in shaping groundwater access and quality, with contamination risks disproportionately affecting low-income and minority communities. Pace and colleagues [5] found that relative to community water system users, households relying on domestic wells experience heightened risks from arsenic, nitrate, and hexavalent chromium contamination, with racial and economic disparities in water quality outcomes. Similarly, Balazs and colleagues (2012) documented that nitrate contamination disproportionately affects renters, lower-income households, and smaller water systems in California's San Joaquin Valley [13]. Additional studies across the U.S. consistently link groundwater contamination risks to race, income, and service infrastructure across various spatial and temporal scales, reinforcing broader environmental justice concerns [3,7,10,11,43–49].

Groundwater access is influenced by aquifer depth, infrastructure reliability, service costs, and regulatory oversight, especially in rural areas [1,9]. Rural communities—both in Arizona and nationally—face heightened vulnerabilities due to small, under-resourced water systems that struggle with monitoring, reporting, and compliance [4,10,12,50]. In these areas, while shallow aquifers are more accessible, private domestic wells, which often serve low-income populations, remain largely unregulated, increasing the risk of undetected contamination [5,28,51].

Research consistently shows that small water systems serving vulnerable populations experience more frequent water quality violations, raising significant environmental justice concerns [5,13,14]. In Arizona, groundwater contamination risks are influenced by socio-hydrogeological factors, such as aquifer type, water system type, groundwater policies, and socioeconomic conditions. Rural domestic wells and Tribal systems show the highest vulnerabilities [28]. Studies in Arizona and California reveal that water quality violations are more common in low-income communities and smaller systems serving renters and communities of color [14,28]. Additionally, communities with higher proportions of institutionalized groups, and migrant workers face elevated contamination levels, underscoring persistent disparities in groundwater safety [5,28].

Communities outside designated groundwater management areas face greater risks due to weak regulations that allow over-extraction. Declining water levels, inadequate infrastructure, and high water costs create additional burdens for low-income and Indigenous populations [26,52,53]. Contamination from agricultural runoff and legacy pollutants further exacerbates risks in these vulnerable communities [5,43,54–56].

Legacies of inequity in groundwater access and quality reflect long-standing disparities in water governance, infrastructure investment, and regulatory enforcement. Municipal expansion and selective annexation determine who receives regulated water services, leaving many disadvantaged communities without access to safe drinking water. These inequities stem from multiple factors, including the quality of infrastructure (whether natural, built, or managerial), the costs associated with extending and maintaining water services to marginalized areas, and poorly enforced regulations (building codes and public health) [9,11,49,57–63]. Together, these factors affect both the quality of water resources and the reliability of water service delivery and cost (low ability to pay for the service) [9]. Empirical research highlights that disparities in drinking water safety arise from the interaction of natural, built, sociopolitical, and environmental systems, disproportionately affecting low-income, rural, and communities of color [5,36,64]. Despite clear evidence of equity issues in groundwater quality and quantity across the Western U.S., no study has comprehensively examined the intersection of groundwater availability, access, and contamination risk from a multidimensional equity perspective.

1.3. Research Questions and Hypothesis

This study expands prior research [5,9,27,28], by applying a trivalent framework—integrating groundwater availability, access, and contamination risk—to examine the interconnections among these dimensions in Arizona's drinking water systems. Specifically, it investigates how access factors—such as aquifer type, policy designations, and socioeconomic conditions—and water quality indicators (nitrate and arsenic contamination) influence groundwater availability, and how these relationships vary across different regulatory, hydrogeological, spatial, and temporal contexts. We hypothesize that groundwater availability, measured by changes in groundwater levels, is shaped by access factors and water quality, with greater declines expected in areas with unregulated water supplies, weaker regulatory oversight, vulnerable aquifers, and disadvantaged communities. These influences are expected to vary spatially across different regions and temporally over time due to changes in groundwater management practices, climate conditions, and land use. By exploring the interactions among these dimensions, this research offers an equity-based perspective on groundwater sustainability and the factors driving disparities in safe drinking water availability, access, and quality.

2. Material and Methods

Figure 1 presents the Groundwater Vulnerability Assessment Trivalent Framework, outlining the stepwise methodology. It integrates raw data, preprocessing, and key variable stratification into groundwater availability, access, and quality. These elements inform statistical and sensitivity analyses of groundwater levels, regulatory and socioeconomic influences, and contamination risk. Later sections provide details on each component.

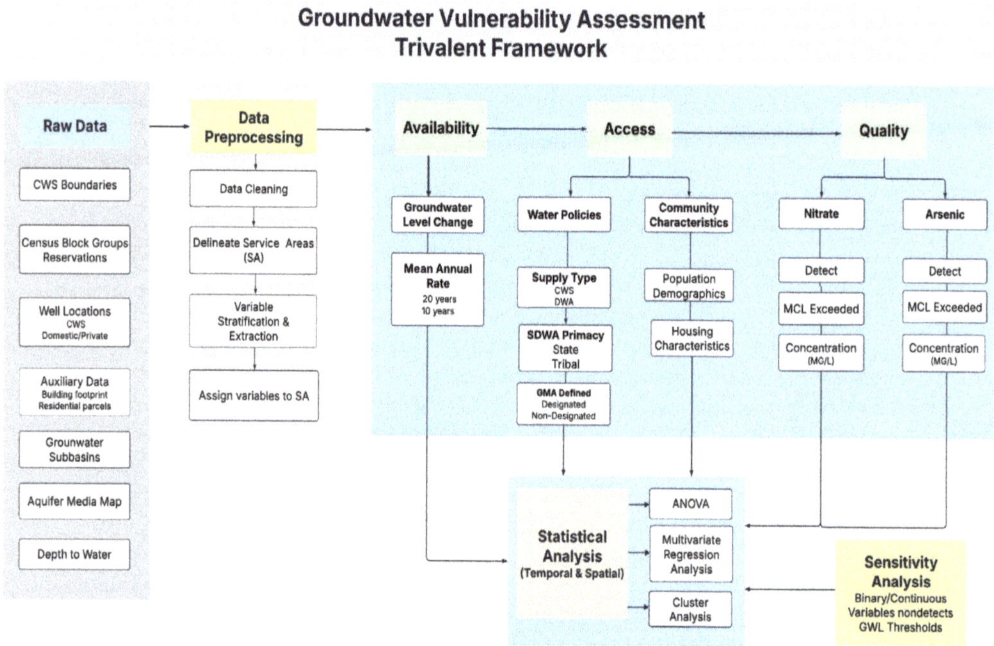

Figure 1. Flowchart of trivalent framework applied in this study.

2.1. Study Area

Arizona, located in the southwestern United States, spans approximately 113,990 mi^2 (or 295,232 km^2) and features diverse hydrogeological conditions, including extensive alluvial plains and complex karst systems. Groundwater is a critical drinking water source for both urban and rural populations, delivered through community water systems (CWS) or domestic water areas (DWAs) representing an areal cluster of individual private wells and managed through varied regulatory frameworks across state and Tribal jurisdictions.

The 1980 Groundwater Management Act (GMA) established Active Management Areas (AMAs) to regulate groundwater in regions with significant depletion. AMAs cover over 80% of Arizona's population and are governed by strict regulations to balance annual groundwater withdrawals with natural or artificial recharge. In comparison, Irrigation Non-Expansion Areas (INAs) and the remaining regions, referred to as "Non-Designated Areas" (NDAs), have minimal or no regulatory oversight, leading to disparities in groundwater availability and access, particularly in rural and underserved regions [42,65,66]. Groundwater governance also differs between state and Tribal jurisdictions. Tribal lands, such as the Navajo Nation, manage water under sovereign authority. While the *Safe Drinking Water Act*

(SDWA) applies to public water systems on Tribal lands, enforcement and infrastructure support vary, contributing to gaps in monitoring, regulation, and water access. These issues are compounded by unresolved water rights and underfunded systems. See Figure 2.

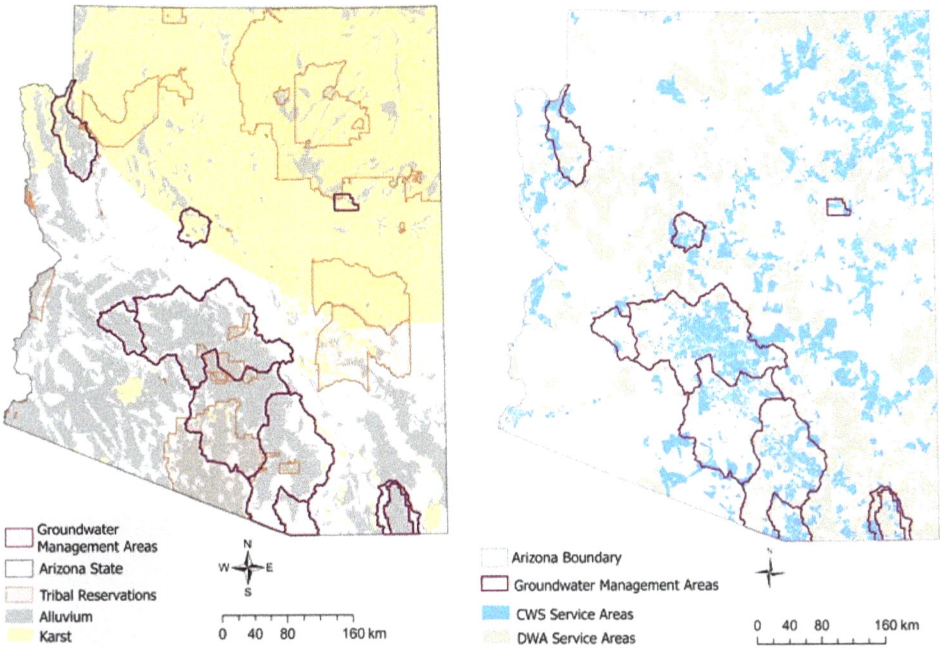

Figure 2. Map of study area. The map shows binary delineations of aquifer type, regulatory jurisdiction (Tribal reservations versus the State), designated management areas, and supply type (CWS—community water system; DWA—domestic well area).

Groundwater availability, reflected in water level changes, varies across aquifer types. Alluvial aquifers in regulated urban areas face stricter oversight, while karst systems in rural and Tribal regions are more vulnerable to contamination and depletion due to their complex hydrogeology and limited regulation [28]. Arizona faces growing water security challenges due to groundwater depletion, rising demand, and contamination threats. Climate change, rapid urbanization, and agricultural intensification exacerbate these risks, impacting both human and ecosystem health [20,67–69]. Despite recent efforts to address governance gaps such as rural groundwater management [70], Arizona's groundwater governance, primarily focused on quantity, lacks integrated approaches to address availability, quality, and access disparities [26]. This highlights the need for comprehensive policies that address the dynamic interplay between availability, access, and contamination risk to ensure sustainable and equitable groundwater management.

2.2. Data Input and Sources

This study utilizes multiple data sources to assess groundwater availability, access, and contamination risk, integrating information on hydrogeological characteristics, water supply systems, community demographics, and water quality indicators. Table 1 summarizes the data categories and their respective sources.

Table 1. Summary of data sources.

Category	Description	Data Sources
Availability	Changes in groundwater levels in subbasins are used as indicators of water access variability over time.	Arizona Department of Water Resources (ADWR) [71–73]
Access	Water supply type, service area boundaries, locations of public and private wells, and regulatory compliance data.	ADWR, Environmental Protection Agency (EPA) including a Freedom of Information Act request [74–78], Navajo Nation data via direct request
	Population demographic and socioeconomic data to evaluate the communities that rely on different water supply systems.	U.S. Census Bureau [79]
	Policy Attributes: aquifer types, policy designations, and SDWA jurisdictions.	ADWR, U.S. Geological Survey (USGS), EPA
Quality	Nitrate (NO_3) and arsenic (As) concentrations in groundwater; derived detection and exceedance variables.	Arizona Department of Environmental Quality (ADEQ), EPA, USGS [74–76,80]
Auxiliary Data	Infrastructure-related data, including building footprints, residential parcels, and pipeline networks.	County governments, online platforms, and direct records requests.

2.3. Measures for Assessing Groundwater Availability, Access, and Quality

This study employs a spatial ecological design to examine the relationship between groundwater access and quality while accounting for policy, service area characteristics, and socioeconomic attributes. The design facilitates robust statistical analyses that mitigate spatial dependence, data aggregation biases, and spatial correlation effects, ensuring the findings are contextually specific and reliable. Measures for evaluating access, vulnerability, and equity were selected based on prior research in environmental equity and groundwater risk, as well as insights from a companion study focused on groundwater quality [26–28].

Water access varies by hydrogeology, socioeconomic factors, and regulatory frameworks. Socioeconomic factors, including housing, population demographics, and institutional facilities, served as proxies for community resilience, following established methodologies [3,7,11,46]. Hydrogeological variables, such as aquifer type and well location, were selected for their known associations with contaminant behavior and recharge dynamics [8,10,81–86]. Dominant aquifer properties (karst versus alluvial) were derived from USGS datasets to assess water access variability across the study area.

Service areas were delineated using spatial data on exempt domestic wells, water system boundaries, residential parcels, building footprints, and census block groups [5,26,28,49,87]. Water systems in these areas were categorized as either community water systems (CWS), regulated under the *Safe Drinking Water Act* (SDWA), or domestic well areas (DWAs), which primarily rely on unregulated private domestic wells [3,5,10]. Of the 1881 mapped service areas, 55.3% were designated as CWS and 44.7% were designated as DWAs. Additionally, equity considerations included regulatory distinctions between designated and undesignated groundwater management areas, as well as Tribal and State jurisdictions. These distinctions align with research on regulatory frameworks and environmental justice outcomes [3–5,11,17,49,62,88,89].

Groundwater availability trends in subbasins containing drinking water sources were evaluated using rates of water level change as proxies for available quantities. Groundwater level changes over 20-year (2001–2020) and 10-year (2011–2020) periods were calculated using ADWR index well data. All groundwater level changes are reported in feet per year (ft/year), with metric equivalents in meters per year (m/year) provided in parentheses. The long-term period captures the cumulative impacts of Arizona's groundwater management policies, prolonged drought conditions, and evolving water conservation practices. The short-term period reflects recent developments, including the implementation of the 2019 Drought Contingency Plan, increased groundwater reliance due to Colorado River shortages, and the expansion of groundwater pumping in unregulated areas. This temporal framework allows for the identification of both long-term trends and more recent shifts in groundwater dynamics.

Two contaminants, arsenic and nitrate, were used as primary water quality indicators due to their prevalence, toxicity, and regulation under drinking water standards. These contaminants present significant health risks, originating from natural sources and human activities, respectively. Data were sourced from agency databases, *Freedom of Information Act* (FOIA) requests, and published literature. Compliance monitoring data for CWS were obtained from the EPA, while private well data were provided by the Arizona Department of Environmental Quality (ADEQ), the U.S. Geological Survey (USGS), and previous studies [90]. The dataset encompasses 27,334 arsenic observations and 47,783 nitrate observations collected between 1998 and 2019 across state and Tribal water systems in Arizona. To handle non-detection observations, multiple imputation methods and sensitivity analyses were employed [3,11]. Using the Amelia package in R, contaminant concentrations below detection limits were replaced with values based on the EPA's MDL or MRL divided by the square root of two, ensuring that imputed values remained within defined bounds [46,91]. In cases where both MDL and MRL were provided, we prioritized MDL to minimize potential bias, as MRLs can overestimate contaminant levels. Studies have shown that MRL-based imputation may introduce conservative bias. Sensitivity analyses in our study confirmed that imputation choices did not significantly affect the results. Records below the MDL accounted for 97.8% of arsenic and 54.5% of nitrate observations, highlighting the challenges of measuring contaminants at low concentrations. Water quality variables were analyzed as continuous measures of contaminant concentrations (*ValueImp1*) and binary indicators for detection (*Detect*) and regulatory exceedance (*MCLExceeded*). This dual framework enables a comprehensive evaluation of both contamination extent and its regulatory implications across groundwater systems.

2.4. Statistical and Spatiotemporal Analysis

This study applies the methods used in Williams, et al. [28], utilizing areal apportionment and spatial overlay techniques to link demographic, socioeconomic, and hydrogeologic data to service areas while expanding the analysis to include groundwater availability and developing new models to examine the relationship between availability, access, and quality. We employed geostatistical tools, ANOVA, and multivariate regression models to analyze relationships between groundwater level changes and explanatory variables. These methods assessed mean differences and associations among groundwater level change, water quality indicators, policy frameworks, service area characteristics, and socioeconomic factors. Integrated datasets included groundwater levels, hydrogeologic attributes, water service boundaries, and regulatory designations, with all analyses conducted at the service area level.

The hypothesis posited that groundwater availability, measured by changes in groundwater levels, is influenced by access factors and water quality, with greater declines expected in areas with unregulated water supplies, weaker regulatory oversight, vulnerable aquifers, and disadvantaged communities. ANOVA and pairwise comparisons evaluated mean differences in groundwater level changes, while mixed-effects regression models estimated associations, incorporating interaction terms to examine combined effects of demographic, hydrogeologic, and policy variables [9,10]. The logistic regression models estimated prevalence ratios for binary outcomes [5,10,92]. Variables were stratified into subgroups based on water quality (continuous and binary for nitrate and arsenic detection and exceedance), hydrogeologic characteristics, and policy designations [5,10]. Regression models controlled for service area characteristics, aquifer types, and regulatory designations to explore associations with groundwater level changes.

Hierarchical and k-means clustering in ArcGIS grouped service areas by groundwater level change patterns and shared characteristics, identifying spatial clusters with significant declines and associated predictors. These clusters highlighted high-risk areas, offering insights for targeted policy and management interventions [46,93,94].

Sensitivity analyses assessed the robustness of findings by testing various groundwater level thresholds, comparing models, and stratifying by aquifer type, supply type, policy designation, and regulatory primacy [10]. Water quality variables were analyzed as continuous (measured concentrations) and binary (detection and regulatory exceedance), and two contaminants, nitrate and arsenic, were included to ensure robustness. These methods accounted for spatial and temporal variability, integrating predictors such as aquifer types, policy designations, and demographic characteristics [26,28].

3. Results

Understanding the relationship between groundwater availability, access, and quality is critical for ensuring the sustainability of Arizona's drinking water resources. This study examines how changes in groundwater levels (availability) intersect with patterns of contaminant occurrence (quality) and socio-hydrogeological factors that influence water distribution (access). The findings identify areas at heightened risk of contamination and limited access by analyzing spatial and temporal variations in groundwater level fluctuations alongside hydrogeology, supply service area characteristics, and regulatory oversight. These insights support the development of targeted management strategies to promote equitable access to safe drinking water and ensure the long-term sustainability of groundwater resources.

3.1. *Temporal Patterns in Rates of Groundwater Level Change*

The distribution of groundwater availability in Arizona's drinking water supply service areas exhibits significant variability over 20-year and 10-year periods (Figure 3). The 20-year groundwater level change map reveals widespread declines ranging from −52.40 to −0.01 ft/year (−15.97 to −0.003 m/year), particularly outside designated groundwater management areas (AMAs), highlighting long-term overuse and insufficient regulation. In contrast, positive changes (0.01 to 44.00 ft/year or 0.003 to 13.41 m/year) are localized, primarily within AMAs and urban centers, indicating the success of targeted recharge efforts and stricter management practices.

The 10-year groundwater level change map reflects similar patterns but shows fewer areas of decline, suggesting potential stabilization in groundwater levels due to improved management strategies and conservation efforts. Positive trends are more pronounced in managed regions, reinforcing the impact of recharge projects and policy interventions. However, the long-term map underscores cumulative impacts, with persistent declines in unmanaged areas and positive changes concentrated in AMAs and INAs with active groundwater management. Tribal reservations show limited improvements, with many areas experiencing declines or lacking data (represented in gray). The prevalence of no-data areas across Tribal lands and unmanaged regions presents a barrier to comprehensive assessment and sustainable groundwater management.

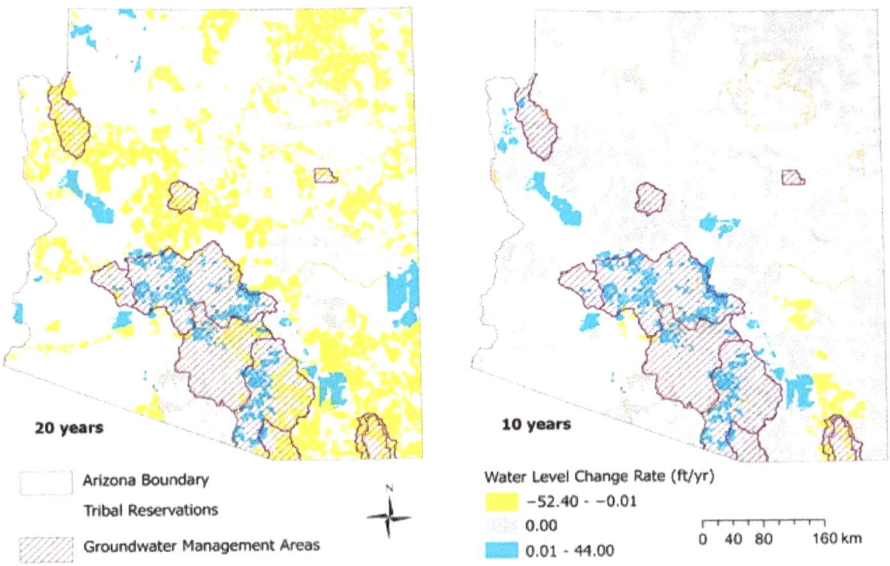

Figure 3. Groundwater level change patterns. Mean annual groundwater level change over 20 years (**left**) and 10 years (**right**) in ft/year. Note: 1 ft/year = 0.3048 m/year.

3.2. Spatial Distribution of Groundwater Level Changes

The spatial analysis of mean rates of groundwater level change (Figure 4) revealed significant clustering patterns. The spatial cluster analysis of groundwater level changes reveals distinct patterns over the long-term (2001–2020) and short-term (2011–2020) periods. In the 20-year analysis, extensive high-high clusters (red) indicate widespread groundwater declines, particularly in central Arizona (Pinal and Maricopa counties), southeastern Arizona, and parts of northern Arizona, reflecting long-term cumulative impacts of intensive groundwater use for agriculture, urbanization, and industrial activities. In contrast, the 10-year period shows a notable reduction in red clusters and an increase in low-low clusters (blue), especially within AMAs and regions benefiting from managed aquifer recharge projects. This shift suggests improvements in groundwater conditions in specific areas, potentially due to recent policy interventions, including the Drought Contingency Plan (2019) and expanded water conservation programs. Despite these improvements, persistent red clusters in rural and agricultural zones highlight areas of continued groundwater stress, particularly outside regulated AMAs.

Moran's I results confirm statistically significant clustering ($p < 0.01$) in both periods, with the 10-year period showing more balanced clustering, suggesting divergent trends influenced by policy and environmental factors. The Getis-Ord Gi* statistic further identifies significant clusters, confirming areas of persistent decline and regions with improved conditions, reflecting the impact of groundwater management efforts and environmental changes across Arizona.

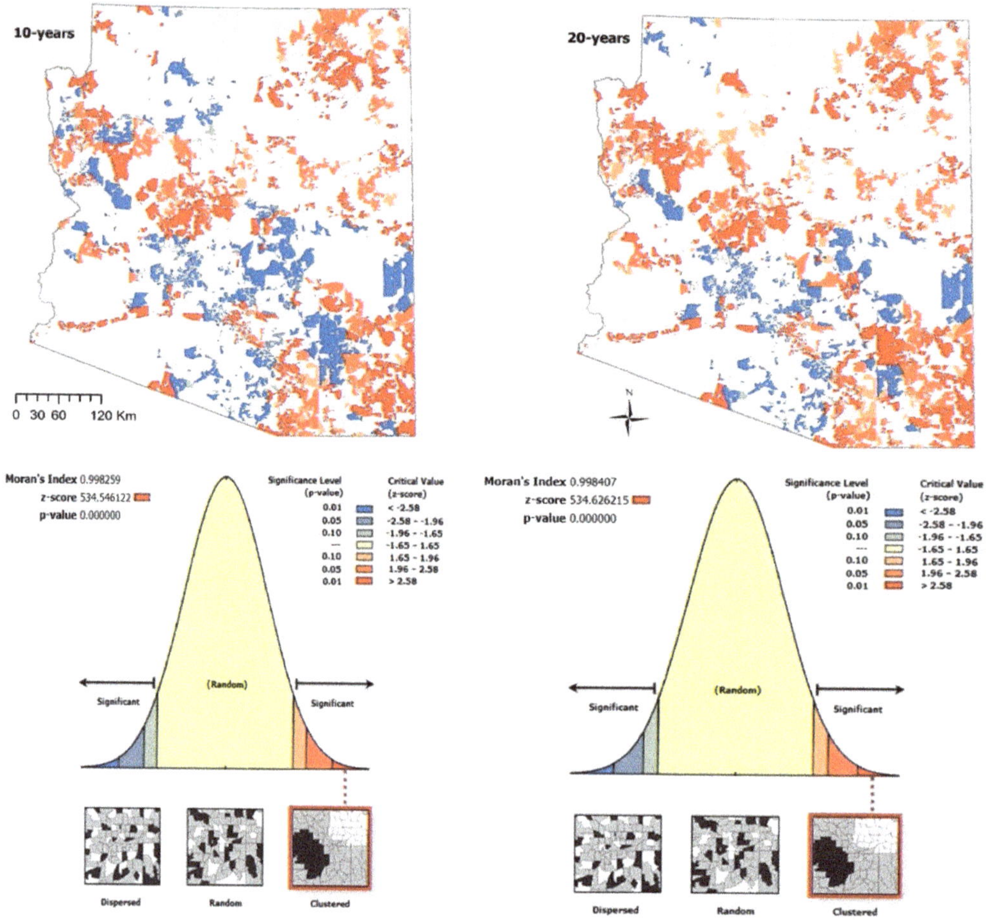

Figure 4. Spatial clusters of mean groundwater decline rates in Arizona (2001–2020). This figure shows spatial clusters of groundwater level changes over 10-year (2011–2020) and 20-year (2001–2020) periods in Arizona, based on Moran's Index of spatial autocorrelation analysis. High-high (red) and low-low (blue) clusters indicate significant groundwater decline and recovery, respectively. Moran's I is a unitless measure of spatial clustering relative to a random distribution. The shaded illustrations at the bottom use light and dark tones to show how similar values are distributed across space. Spatial pattern types: Dispersed (left) shows values spread apart, Random (middle) shows no pattern, and Clustered (right) shows similar values grouped together.

3.3. *Groundwater Access and Pollution*

3.3.1. Groundwater Level Change and Contaminant Detection

Examination of the rates of change in groundwater levels with arsenic and nitrate detection across groundwater sources revealed notable disparities and significant associations with predictors (Appendix A—Tables A1 and A2). Specifically, areas experiencing higher rates of groundwater level recharge over 20 years (*Mean_20*) are linked to an increased probability of arsenic detection ($\beta = 0.016$ $p < 0.001$) and nitrate detection ($\beta = 0.0044$, $p < 0.001$). As shown in Figure 5, drinking water sources located in groundwater subbasins experiencing higher rates of decline in groundwater levels exhibited lower rates of arsenic and nitrate detection, while those with higher rates of groundwater level recovery tended to show higher arsenic and nitrate detections. Moreover, systems under Tribal primacy show a notably lower likelihood of arsenic and nitrate detection compared to State-regulated systems. Aquifer type also matters, with karst aquifer-dependent systems displaying a lower likelihood of arsenic detection than alluvial aquifers. Additionally, DWA sources exhibit a lower likelihood of arsenic detection and a higher likelihood of nitrate detection than CWS.

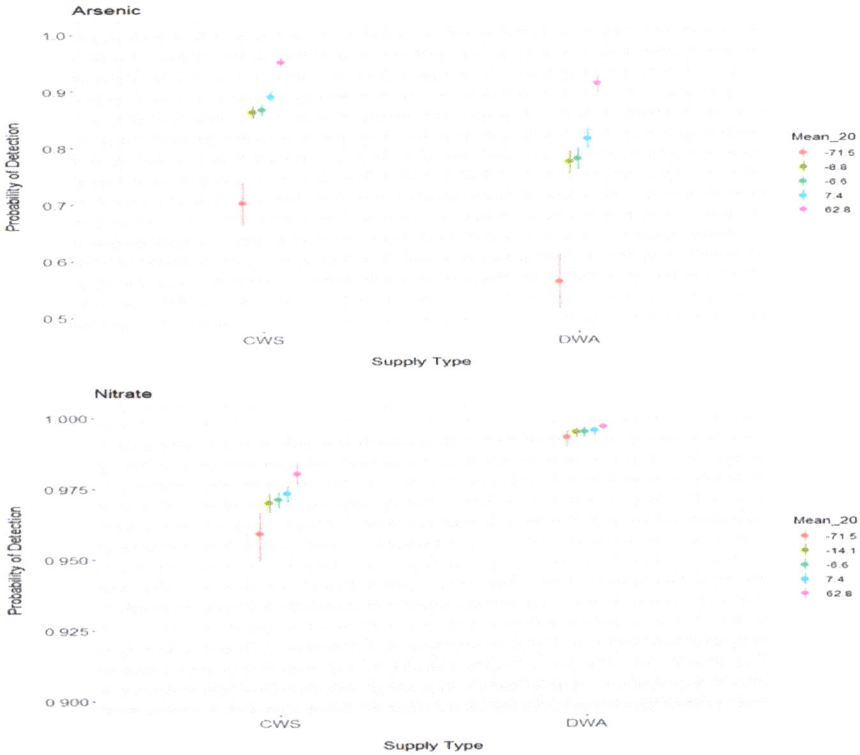

Figure 5. Mean difference in arsenic and nitrate detection relative to groundwater level change. This figure illustrates the probability of arsenic (**top**) and nitrate (**bottom**) detection in different water supplies as a function of the mean annual rate of groundwater level change (ft/year) over 20 years (*Mean_20*). Each colored point represents a different rate, with the bar showing the range of probability for each rate. The x-axis denotes the supply type, while the y-axis shows the probability of detection. Note: 1 ft/year = 0.3048 m/year.

3.3.2. Temporal Changes in Groundwater Availability and MCL Exceedance

The decade spanning 2011 to 2020 witnessed a notable correlation between the rate of groundwater level change (*Mean_10*) and the probability of arsenic and nitrate MCL exceedances in groundwater sources of drinking water (Figure 6). Both arsenic and nitrate MCL exceedances are positively related to groundwater level changes. For arsenic, the relationship is linear, suggesting a steady increase in exceedance probability with rising groundwater levels. For nitrate, the relationship is exponential, indicating a rapid increase in exceedance probability as groundwater levels rise more significantly. These findings highlight a linkage between increased groundwater access and contamination risk in arid regions, pointing to the importance of considering groundwater level changes as a critical factor in assessing the risk of contaminant exceedances in drinking water supplies.

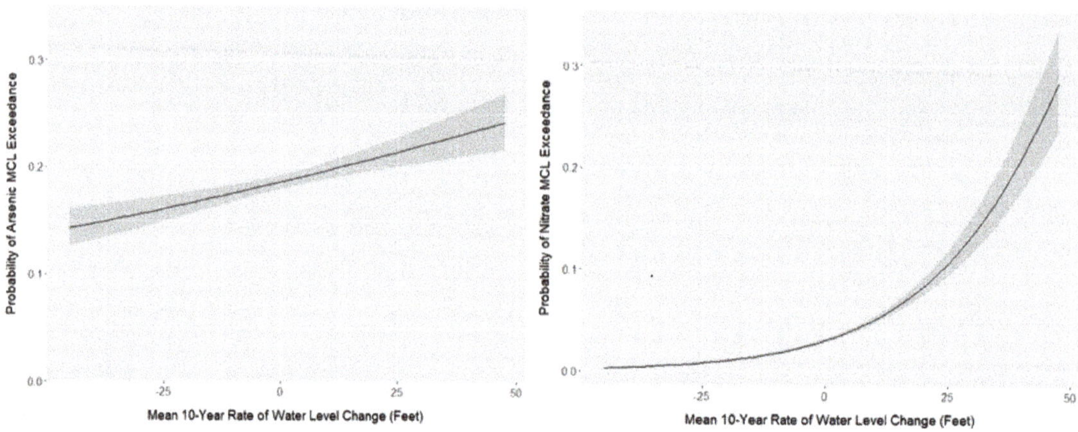

Figure 6. Comparison of mean rates of change in groundwater levels and MCL exceedances. This graph shows that the probability of MCL exceedance for arsenic (**left**) and nitrate (**right**) increases with the mean annual rate of groundwater level change over 10 years (*Mean_10*) (ft/year). The shaded ribbons/error bars represent the 95% confidence intervals of predictions. Note: 1 ft/year = 0.3048 m/year.

3.4. Temporal Associations: Groundwater Level Change and Arsenic MCL Exceedance

Figure 7 illustrates mean differences in arsenic MCL exceedances in groundwater sources, comparing CWS and DWA over 20-year (*Mean_20*) and 10-year (*Mean_10*) periods. Notably, DWA exhibits higher mean arsenic MCL exceedances than CWS across both periods. As changes in groundwater level increase, MCL exceedances also rise, with the highest exceedances observed in DWA, particularly at a mean change rate of 62.8 feet over 20 years and an average change rate of 47.4 feet over 10 years.

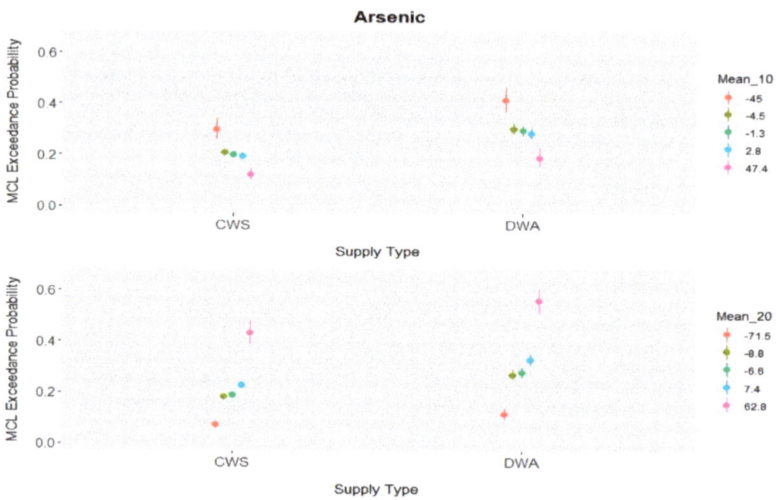

Figure 7. Mean annual rates of groundwater level change (ft/yr) and arsenic MCL exceedances. This figure shows the mean annual groundwater level change (ft/yr) over 20 years (**bottom**) and 10 years (**top**), along with arsenic MCL exceedance probabilities in community water systems (CWS) and domestic well areas (DWAs). For reference, 1 ft/yr = 0.3048 m/yr. *Mean_20* and *Mean_10* represent groundwater level change over 20-year and 10-year periods, respectively.

The temporal analysis of arsenic exceedances in groundwater over a 20-year period and a 10-year period reveals contrasting trends. Over the 20-year span, higher groundwater recharge rates were associated with increased arsenic exceedances, indicating a potential link between recharge events and arsenic contamination. In contrast, during the more recent 10-year period, higher recharge rates were correlated with lower nitrate exceedances. These findings suggest that temporal shifts in recharge dynamics, water policies, and land use practices may influence nitrate contamination patterns in groundwater.

3.5. Temporal Relationship Between Groundwater Availability and Predictors

The analysis of groundwater level changes over 20 years reveals notable relationships between water quality indicators, water policy variables, demographic factors, and housing characteristics. For *Detect* (arsenic presence) and *MCLExceeded* (regulatory exceedance), higher odds ratios were observed in the 20-year decline model compared to the 10-year model (*Detect*: OR = 0.92, 0.94; *MCLExceeded*: OR = 0.97, 0.98). These results indicate that arsenic detection and exceedance of MCLs are more strongly associated with groundwater declines in the longer-term model.

For arsenic, higher rates of annual groundwater level change from 2011 to 2020 were positively associated with MCL exceedance (Estimate = 0.007, $p < 0.001$, OR = 1.01). Similarly, higher rates of groundwater level change increased the likelihood of nitrate MCL exceedance (Estimate = 0.049, $p < 0.001$, OR = 1.05). Non-Designated Areas showed differing associations: positive for arsenic (Estimate = 0.101, $p = 0.004$, OR = 1.11) and negative for nitrate (Estimate = −0.733, $p < 0.001$, OR = 0.48).

Water policy variables, such as Non-designated Areas (NDAs) and Tribal jurisdictions, showed slightly higher detection probability in the 20-year model compared to the 10-year model (*NDA*: OR = 1.54, 1.52; *PRIMACY(Tribal)*: OR = 0.62, 0.69), emphasizing their consistent influence on groundwater declines. In contrast, karst aquifers exhibited marginally lower detection probability in the 10-year model (OR = 1.18, 1.16), indicating a nuanced temporal dynamic in these regions.

The interaction between tribal governance and karst aquifers demonstrated reduced detection probability in the short-term mode (OR = 1.66, 1.49), highlighting the evolving impact of Tribal management in karst regions over time. Demographic and housing factors, including the percent of Black and Hispanic/Latino population in service areas, as well as housing types such as migrant worker housing and rental units, were consistently associated with detection probability across both periods, reflecting stable relationships with groundwater availability.

3.6. Main and Interaction Effects of Predictors on Groundwater Decline

This section examines the relationship between groundwater decline, water quality, and access. Logistic regression analysis identified significant associations between groundwater level changes in drinking water supply service areas and key water access and quality predictor variables (Appendix A—Tables A1 and A2).

3.6.1. Water Quality Variables

The regression models examining groundwater level (GWL) decline reveal distinct patterns between binary (detection and MCL exceedance) and continuous (measured concentrations) definitions of nitrate and arsenic contamination, highlighting variations in their associations and temporal effects.

Binary Variables: Nitrate and arsenic detection and MCL exceedance are significantly and negatively associated with the probability of GWL decline across both the 20-year and 10-year periods (Table 2), suggesting that areas with detectable contamination are less likely to experience declining groundwater levels. This negative association suggests that areas with stable or increasing groundwater levels are more likely to exhibit contaminant detection and exceedances, indicative of long-term recharge or stabilization processes that mobilize contaminants into aquifers. For nitrate detection, the 20-year model shows a weaker effect compared to the 10-year model, indicating more immediate stabilization effects in the shorter term. While significantly and negatively associated with GWL decline, the rate of nitrate MCL Exceedance over 20 years also has a slightly weaker influence than the 10-year rate of nitrate MCL Exceedance, pointing to more immediate interactions between nitrate contamination and groundwater dynamics, where recharge stabilizes water levels while mobilizing nitrates.

Table 2. Regression results for groundwater decline models with binary water quality predictors.

	Model: 20-Year GWL					Model: 10-Year GWL				
Variables	**est.**	**Std Error**	**Statistic**	***p* Value**	**Log Odds**	**est.**	**Std Error**	**Statistic**	***p* Value**	**Log Odds**
Nitrate (binary)										
Detect	−0.075	0.006	−11.67	<0.001	0.93	−0.114	0.007	−15.458	<0.001	0.89
MCL Exceeded	−0.089	0.009	−9.919	<0.001	0.91	−0.102	0.010	−9.817	<0.001	0.90
Water Policy										
GMA Designation (NDA)	0.300	0.007	43.109	<0.001	1.35	0.178	0.008	22.100	<0.001	1.20
SDWA Primacy (Tribal)	−0.173	0.017	−10.289	<0.001	0.84	−0.112	0.02	−5.725	<0.001	0.89
Aquifer Type (Karst)	0.121	0.006	20.829	<0.001	1.13	−0.003	0.007	−0.507	0.612	1.00
Supply Type (DWA)	0.033	0.006	5.758	<0.001	1.03	0.014	0.007	2.180	0.029	1.01
Arsenic (binary)										
Detect	−0.086	0.006	−13.954	<0.001	0.92	−0.059	0.007	−8.832	<0.001	0.94
MCL Exceeded	−0.028	0.006	−4.931	<0.001	0.97	−0.023	0.006	−3.766	<0.001	0.98
Water Policy										
GMA Designation (NDA)	0.431	0.008	51.122	<0.001	1.54	0.418	0.009	45.877	<0.001	1.52
SDWA Primacy (Tribal)	−0.484	0.008	−57.086	<0.001	0.62	−0.367	0.009	−40.059	<0.001	0.69
Aquifer Type (Karst)	0.163	0.007	22.692	<0.001	1.18	0.148	0.008	19.075	<0.001	1.16
Supply Type (DWA)	0.014	0.007	1.895	0.058	1.01	−0.024	0.008	−2.996	0.003	0.98

Response Variable: mean annual rate of water level decline over 20 yrs and 10 yrs − binary variable, 1 = decline.

Similarly, arsenic detection demonstrates a stronger relationship with groundwater stabilization in the 20-year model (estimate = −0.086, $p < 0.001$, OR = 0.92) relative to the 10-year model (estimate = −0.059, $p < 0.001$, OR = 0.94). Arsenic MCL exceedance is associated with a reduced probability of groundwater declines, with odds ratios of 0.97 (estimate = −0.028, $p < 0.001$) and 0.98 (estimate = −0.023, $p < 0.001$) for the 20-year and 10-year models, respectively. These results suggest groundwater sources with detectable or elevated arsenic levels were less likely to experience significant GWL declines, potentially due to limited usage or reliance on these sources. A complementary reciprocal groundwater recharge model supports these findings, showing that areas with detectable nitrate and arsenic are more likely to have stable or rising groundwater levels, highlighting the role of recharge in both water level stabilization and contaminant mobilization within aquifers.

Continuous Variables: Measured nitrate and arsenic concentrations exhibit strong, positive associations with mean GWL decline (Table 3), suggesting that higher contaminant levels are more likely to occur in areas of intensive extraction or localized contamination hotspots. For nitrate concentration, the association with groundwater decline is stronger over the 20-year period than in the more recent 10-year period. A similar trend is observed for arsenic concentration, with a stronger effect in the 20-year model. The 20-year model showed an exceptionally high odds ratio, indicating a substantial long-term association between arsenic levels and GWL decline. The 10-year model also reflected a positive and statistically significant relationship, though the magnitude of the association was notably reduced compared to the longer period. These results suggest that areas with elevated arsenic levels likely experience persistent reliance on groundwater, particularly in regions with limited alternative water sources.

Table 3. The regression results for groundwater decline models with continuous water quality predictors.

	Model: 20-Year GWL					Model: 10-Year GWL				
Variables	**est.**	**Std Error**	**Statistic**	***p* Value**	**Log Odds**	**est.**	**Std Error**	**Statistic**	***p* Value**	**Log Odds**
Water Quality										
Nitrate (NO_3)	0.228	0.012	18.792	<0.001	1.26	0.105	0.007	15.499	<0.001	1.11
Water Policy										
GMA Designation (NDA)	−5.989	0.286	−20.96	<0.001	0.00	−1.770	0.160	−11.035	<0.001	0.17
SDWA Primacy (Tribal)	6.621	0.684	9.684	<0.001	751.01	6.012	0.384	15.664	<0.001	408.10
Aquifer Type (Karst)	−7.171	0.236	−30.407	<0.001	0.00	0.243	0.132	1.839	0.066	1.28
Supply Type (DWA)	−3.151	0.229	−13.774	<0.001	0.04	−0.885	0.128	−6.889	<0.001	0.41
Water Quality										
Arsenic (As)	91.637	6.129	14.951	<0.001	6.27E+39	38.494	3.942	9.764	<0.001	5.22E+16
Water Policy										
GMA Designation (NDA)	−10.46	0.330	−31.702	<0.001	0.00	−4.169	0.212	−19.642	<0.001	0.02
SDWA Primacy (Tribal)	8.103	0.330	24.536	<0.001	3305.44	0.139	0.212	0.653	0.514	1.15
Aquifer Type (Karst)	−1.465	0.281	−5.215	<0.001	0.23	2.390	0.181	13.223	<0.001	10.91
Supply Type (DWA)	−5.435	0.286	−18.985	<0.001	0.00	−4.951	0.184	−26.889	<0.001	0.01

Response Variable: mean annual rate of water level decline over 20 years − continuous variable.

3.6.2. Water Policy

The analysis of groundwater management policies, regulatory oversight, aquifer type, and water supply type reveals significant differences in magnitude, direction, and temporal trends. Non-designated groundwater management areas are less likely to experience groundwater level declines than regulated areas, with the strongest protective effects observed in long-term models. However, these effects weaken in the 10-year models, suggesting increasing short-term groundwater decline pressures in non-designated areas over the last decade.

Tribal oversight strongly mitigates groundwater level declines, particularly over the long term, with significant protective effects observed for both nitrate and arsenic. The positive estimates and extremely high odds ratios indicate that service areas under tribal governance are substantially less likely to experience GWL declines than those under state oversight. However, in the short term, tribal oversight remains an important factor for mitigating nitrate-related groundwater issues, whereas its effect on arsenic becomes statistically insignificant. This suggests that the influence of tribal governance varies depending on the type of contaminant, the timeframe, and whether contamination sources are geogenic or anthropogenic.

Karst aquifers exhibit contrasting patterns in water availability and contamination risk compared to alluvial aquifers. Long-term models indicate greater stability in groundwater storage, as karst aquifers show a lower likelihood of groundwater level declines. However, in short-term models, they become more vulnerable to instability, particularly in arsenic-contaminated areas, where groundwater level declines are more pronounced. These trends may be influenced by recent extraction pressures, climate variability, or localized contamination dynamics.

Domestic well areas consistently show a higher probability of GWL declines than community water systems. The strongest association is observed in the long-term arsenic model, where groundwater declines are most pronounced. In short-term models, the risks persist but are slightly reduced, indicating ongoing yet moderated groundwater vulnerabilities in unregulated areas.

3.6.3. Service Area Socioeconomic Characteristics

Racial composition of drinking water service areas shows significant associations with groundwater availability, as reflected in groundwater level trends (Appendix A—Tables A1 and A2). Service areas with higher percentages of Black populations experience reduced groundwater availability across all models, with the strongest effects in the 20-year models. Although slightly weaker in the 10-year models, the association remains consistent, indicating persistent groundwater access challenges for these communities. Conversely, areas with higher Hispanic/Latino populations are associated with greater groundwater availability, as indicated by positive associations across all models. For Native American populations, the associations vary; while the 20-year nitrate model shows no significant relationship, the 10-year nitrate and arsenic models suggest moderate improvements in groundwater availability, possibly reflecting localized management efforts or contextual factors.

Gender and age characteristics show significant associations with water availability. Service areas with higher female populations are associated with reduced groundwater availability, with the strongest associations observed over 20 years. These associations remain significant but are slightly weaker in the 10-year models, indicating persistent but moderating groundwater challenges for these areas. Service areas with a higher percentage of children under five show weak positive associations with groundwater declines, suggesting minor but consistent effects on groundwater availability. In contrast, areas with a larger older population (over 65 years) consistently exhibit improved groundwater availability, with the strongest effects seen over longer periods.

Housing characteristics also display distinct relationships with groundwater availability. Higher percentages of married-couple households are associated with reduced groundwater levels across all models, with the strongest effects observed over 20 years. Conversely, renter-occupied housing is consistently associated with improved groundwater availability, showing the most pronounced effects in long-term arsenic models and short-term nitrate models.

Seasonal housing exhibits a stronger impact on changing groundwater levels over longer timescales (Appendix A—Tables A1 and A2), suggesting a cumulative effect of periodic high water demand. Institutional housing types, such as correctional facilities, show a consistent but diminishing probability of groundwater decline over time. Non-institutional student housing has the strongest association with groundwater declines across both models, reflecting significant concentrated water use. Similar trends in arsenic models are observed for seasonal and institutional housing types, reinforcing their role in shaping groundwater demand and availability.

3.6.4. Groundwater Sustainability: Interaction Dynamics and Asymmetrical Drivers

The sensitivity analysis reveals notable patterns of symmetry and asymmetry in how different factors influence groundwater level declines and stabilization. Symmetrical relationships indicate that variables driving groundwater depletion also affect recharge, reinforcing a reciprocal relationship between groundwater decline and recovery. For example, arsenic detection and regulatory exceedance consistently show inverse relationships—areas with elevated arsenic levels are less likely to experience groundwater declines and more likely to stabilize. This pattern suggests that recharge and stable aquifer conditions may mobilize contaminants, linking water quality to groundwater availability dynamics.

In contrast, asymmetrical relationships were observed for certain hydrogeological and policy-related variables. Karst aquifers showed a higher likelihood of groundwater declines but a lower likelihood of stabilization, likely due to rapid recharge and depletion cycles that increase vulnerability to extraction pressures while offering limited long-term storage stability. Similarly, the interaction between tribal primacy and karst aquifers revealed contrasting effects between the decline and stabilization models. This suggests that tribal management influences groundwater dynamics differently depending on whether the system is depleting or recovering, particularly in vulnerable aquifers.

Significant interaction effects were observed across the 20-year and 10-year models (Appendix A—Table A1). The interaction between tribal primacy and karst aquifers showed a strong protective effect in both models, with greater significance in the 10-year period, indicating an increasing role of tribal governance in short-term groundwater management. Similarly, non-designated groundwater areas in rural regions showed protective effects against groundwater declines, though the effect was more pronounced over 10 years.

Several variables exhibited non-symmetrical effects, meaning their influence on groundwater declines did not mirror their effect on stabilization. Karst aquifers were positively associated with groundwater declines in the 20-year model but showed no significant effect in the 10-year model, possibly due to short-term variability in recharge dynamics. Migrant worker housing areas had a stronger protective effect in the 20-year model compared to the 10-year model, suggesting that long-term groundwater management practices may benefit transient worker communities more than short-term interventions.

Some variables were statistically significant but had limited practical impact, as indicated by odds ratios close to 1. In the continuous models, nitrate levels had minimal influence in the 10-year model but a much stronger effect over 20 years, reinforcing the cumulative impact of contamination and recharge processes on groundwater availability. Similarly, demographic factors such as younger populations and housing types like rental and non-institutional housing had minor effects, while group housing variables exhibited localized water use patterns without broader impacts on groundwater trends. In the binary models, nitrate and arsenic exceedance were statistically significant but had little influence on groundwater level changes. Other demographic factors, such as Native American population percentages and rural housing, were also statistically significant but had negligible effects, suggesting that while these variables contribute to groundwater dynamics, they are not primary drivers of availability shifts.

4. Discussion

Groundwater sustainability is shaped by three interconnected dimensions: availability, influenced by depletion and recharge; access, determined by socioeconomic, regulatory, and infrastructural factors; and quality, impacted by contaminants like nitrate and arsenic. This study examines how groundwater level changes, socioeconomic disparities, and contamination risks intersect in Arizona's groundwater-dependent drinking water systems, highlighting the complex interplay of environmental, social, and policy factors. The findings reveal significant spatial and temporal disparities in groundwater availability, access, and quality, with implications for resource management and policymaking. The following section examines these results in the context of existing research, regulatory frameworks, and potential policy interventions.

4.1. Spatial and Temporal Shifts in Groundwater Availability

Temporal analysis of groundwater levels over 20-year (2001–2020) and 10-year (2011–2020) periods reveal evolving relationships between groundwater availability, access, and quality. Short-term models show stronger associations with water quality variables, such as nitrate and arsenic detection and exceedance, reflecting the effect of recent recharge and contamination dynamics. In contrast, long-term models indicate cumulative effects, where recharge stabilizes groundwater levels but mobilizes contaminants over time. These findings align with Schaider, et al. [3], who reported persistent nitrate contamination disparities in U.S. drinking water linked to agricultural intensification. However, unlike Schaider et al., our study suggests that Arizona's long-term groundwater management practices have moderated some disparities, emphasizing sustained policy interventions.

The shift from high-high to low-low clusters (Figure 4) in the short term suggests stabilization due to management strategies, including the Groundwater Management Act, the Drought Contingency Plan, and expanded Managed Aquifer Recharge (MAR) projects. Similar improvements were noted by Tadych, et al. [25] following MAR implementation, though our findings highlight additional benefits of Arizona-specific regulations and variations across aquifer types, regulatory jurisdictions, and water supply systems.

Persistent declines in rural, Tribal, and unregulated areas expose policy gaps, especially outside AMAs where groundwater use is largely unregulated. Targeted interventions and adaptive management practices can mitigate these vulnerabilities. Equity concerns persist as marginalized communities face disproportionate risks related to groundwater access and quality, reinforcing Balazs and Ray's [9] emphasis on equity-focused water governance. Our findings show that inequities endure despite broad regulatory efforts, particularly in unregulated areas.

In arid regions like Arizona, spatial and temporal variations in groundwater dynamics are also influenced by climate variability, extraction practices, land use changes, and human activities. Climate variability influences fluctuations in precipitation, affecting recharge rates and long-term water availability. Intensive extraction disrupts flow patterns, accelerates depletion, and reduces storage capacity. Urban expansion and agriculture modify recharge zones and increase demand, while mining contributes to groundwater loss through dewatering, aquifer compaction, and disrupted recharge. Mondal [24] reported similar climate-driven groundwater declines, but Arizona's land use policies and water banking programs have mitigated some depletion risks, influencing regional groundwater trends.

4.2. Water Quality

The relationship between groundwater level changes and contaminant concentration, detection, and MCL exceedance reveals the dual impact of recharge processes on water levels and contamination. Consistent negative associations between contaminant detection, MCL exceedances, and groundwater decline suggest that recharge mobilizes nitrate and arsenic, stabilizing water levels while increasing contamination risks. These patterns align with Ward, et al. [54], who reported that recharge events enhance nitrate leaching in agricultural regions, leading to replenishment and heightened contamination.

Stronger associations in the 10-year models indicate more immediate interactions, likely driven by recent changes in land use, agricultural practices, or recharge infrastructure. This aligns with Orellana-Macías and Perles Roselló [56], who found that nitrate concentrations respond rapidly to land use changes, especially in intensive agricultural areas. In contrast, weaker associations in the 20-year models suggest nitrate contamination persists as a marker for groundwater stabilization over time, influenced by historical land management and prolonged agricultural activities.

Arsenic-related variables exhibit weaker associations compared to nitrate, particularly in the 10-year models. Minor associations for arsenic detection and exceedance imply that arsenic dynamics are less sensitive to recharge processes or short-term hydrological changes. This observation is supported by Hoover, et al. [16], who noted that arsenic mobilization often results from localized geochemical conditions, such as redox fluctuations and mineral interactions, rather than broad hydrological shifts.

For both nitrate and arsenic, stronger associations in the 10-year models highlight GWL changes' sensitivity to short-term contamination patterns. The weakening of these associations in the 20-year models suggests broader influences, including regional management practices, climatic trends, and aquifer characteristics. Banerjee, et al. [86] emphasized that vulnerability assessments should integrate both short- and long-term hydrological and geochemical data to capture comprehensive risks. By incorporating both temporal scales and variable formats, this study provides a more comprehensive understanding of how groundwater availability interacts with quality and access, offering valuable insights for groundwater management and policy development.

Comparing continuous (*As* and NO_3) and binary (*Detect* and *MCLExceeded*) water quality variables highlights key insights into GWL changes. Continuous models reveal larger effect sizes, reflecting the stronger influence of actual contaminant concentrations. This finding aligns with Nemčić-Jurec, et al. [48], who reported that continuous metrics offer detailed insights into groundwater quality dynamics compared to binary regulatory thresholds. In contrast, binary models provide simpler interpretations with smaller effect sizes, which are suitable for compliance assessments and broad patterns. These results support Pennino, et al. [47], who emphasized the complementary roles of continuous and binary metrics in environmental monitoring. These approaches underscore the role of recharge in contaminant mobilization and its stabilizing effect on groundwater levels.

The analysis also reveals that areas with rising groundwater levels show higher arsenic and nitrate detection rates due to the reintroduction of water via recharge. This process can cause desorption driven by changes in pH and redox conditions, as well as the presence of competing ions like sulfate (SO_4) and chloride (Cl), which release arsenic from sediments into the water. Conversely, subbasins with declining groundwater levels exhibit lower arsenic and nitrate detection rates, potentially due to reduced mobilization under compacted aquifer conditions, limiting arsenic release.

Additionally, Community Water Systems (CWS) in Arizona show higher detection rates for arsenic and nitrate compared to Domestic Well Areas (DWA). This discrepancy may be attributed to stricter monitoring and testing requirements for CWS and differences in hydrological characteristics. Regulated extraction and structured groundwater management in designated areas further explain variations in detection rates.

4.3. Water Access

4.3.1. Policy

Policy variables such as *GMADefinedNDA* and *PRIMACYTribal* consistently influence groundwater level trends across both short- and long-term periods. Stronger long-term associations highlight the cumulative effects of governance, particularly through Tribal oversight. Interaction effects, like *PRIMACYTribal*: *AquiferTypekarst*, underscore evolving Tribal management strategies in karst aquifers, where improved recharge and governance reduce groundwater risks over time. This aligns with Balazs, et al. [13], who found governance structures significantly impact contamination risks, though our study reveals a more pronounced effect of Tribal governance in Arizona due to region-specific water rights and co-management policies.

Non-designated areas (*GMADefinedNDA*) show strong protective effects against GWL declines, particularly in the 20-year models for nitrate and arsenic, reflecting long-term management benefits. These findings are consistent with Megdal, et al. [41], highlighting the effectiveness of groundwater management in arid regions. However, weaker impacts in the 10-year models suggest that immediate regulatory interventions are needed to address short-term pressures, such as increased groundwater extraction—a trend also noted by Engel [42].

Tribal oversight (*PRIMACYTribal*) effectively mitigates GWL declines, with significant protective effects in long-term models. This supports Hoover, et al. [16], who reported that Tribal-led governance fosters groundwater resilience. However, the lack of significance in the 10-year arsenic model suggests emerging challenges linked to localized contamination or resource limitations, emphasizing the need for continued investment in Tribal infrastructure, as noted by Zuniga-Teran, et al. [38].

Karst aquifers (*AquiferTypekarst*) show temporal variability. Long-term models suggest resilience due to effective recharge processes, while short-term models reveal increased vulnerabilities, likely driven by extraction pressures or climate variability. Kalhor, et al. [33] similarly emphasized the susceptibility of karst systems to hydrological changes. Domestic Well Areas (*SupplyTypeDWA*) face persistent risks, reflecting challenges of unregulated groundwater use, underscoring the call for integrated policies to promote sustainable access by Johnson, et al. [51].

4.3.2. Community Socioeconomic Characteristics

Demographic and housing variables significantly impact groundwater level changes, shaping localized disparities in access. Transient populations and housing pressures contribute to spatial variability, extending Cid-Escobar et al.'s [1] findings on socio-technical barriers in rural groundwater access. Our results reveal compounded racial and socioeconomic disparities, particularly in marginalized communities.

Areas with higher Black populations face greater probabilities of groundwater level decline, reflecting systemic inequities in groundwater access, consistent with Switzer and Teodoro [7]. This trend has been more pronounced over the past 20 years, suggesting gradual improvements in governance or infrastructure. Conversely, Hispanic/Latino populations show protective effects, potentially due to cultural practices or infrastructure advantages, similar to Delpla, et al. [12]. Native American/Alaska Native populations also show strong protective effects in short-term models, indicating shifts in groundwater management practices, aligning with Hoover, et al. [16] on the benefits of Tribal water governance.

Gender and age dynamics further reveal disparities. Areas with higher female populations exhibit greater groundwater level declines, indicating reduced groundwater access, consistent with McDonald and Grineski [60]. This suggests that gender disparities in resource allocation and infrastructure access may contribute to groundwater availability challenges. In contrast, areas with higher proportions of older populations (65+ years) are associated with lower probabilities of groundwater level decline ($p < 0.001$, OR = 0.99 across models), suggesting relatively better groundwater availability. This trend may reflect lower extraction pressures, reduced water demand, or stable infrastructure in older communities. While these findings challenge assumptions about aging populations facing increased water insecurity, further research is needed to assess how governance and infrastructure adaptation influence groundwater sustainability for different demographic groups.

Housing characteristics significantly influence groundwater access and contamination risks, with distinct patterns by contaminant type and timeframe. Long-term trends show that agricultural groundwater reliance heightens nitrate contamination and depletion risks, supporting Fischer, et al. [50]. In contrast, arsenic models suggest natural recharge processes mitigate long-term risks, with weaker associations in short-term models suggesting these contamination dynamics evolve gradually.

Household structure also plays a role. Married-couple households experience reduced groundwater access over time, likely reflecting cumulative water demands. Conversely, renter-occupied housing areas benefit from improved access, possibly due to urban infrastructure efficiencies, as noted by Wallace and York [93]. Student housing has a significant long-term influence on groundwater availability, while seasonal and institutional housing show modest impacts.

These findings underscore the importance of integrating socioeconomic considerations into groundwater policies to address disparities and improve water-use practices, reinforcing, and echoing Mueller and Gasteyer's [59] emphasis on environmental justice in water governance.

4.3.3. Implications of Interactions and Asymmetrical Drivers in Groundwater Dynamics

Understanding temporal dynamics and interaction effects in groundwater management is crucial for developing targeted, sustainable interventions that address diverse environmental and community needs. The interaction between Tribal oversight and karst aquifers (*AquiferTypekarst*) highlights the sustained effectiveness of Tribal water governance in mitigating groundwater depletion in karst aquifers, supporting findings by Hoover et al. [16]. Similarly, the interaction between non-designated groundwater management areas (*GMADefinedNDA*) and rural areas (*Rural_pcnt*) reveals localized regulatory impacts and evolving rural dynamics, consistent with Fischer, et al. [50]. Karst aquifers show stronger associations with Native populations in the 10-year nitrate model, underlining the need for targeted management strategies to address short-term contamination risks. This aligns with Zuniga-Teran, et al. [38], who identified groundwater vulnerabilities in Native communities. Tribal oversight in karst regions reinforces the importance of sustained, long-term management strategies for these sensitive aquifers.

In contrast, non-symmetric effects for karst aquifers suggest their influence on groundwater levels is more pronounced over longer timescales, likely due to slower hydrological or geophysical processes, consistent with Kalhor, et al. [33]. Housing-related variables show greater cumulative impacts over extended periods, while short-term effects are less significant. As discussed by Banerjee, et al. [86], symmetric effects often indicate stable groundwater access patterns, while asymmetry highlights unique vulnerabilities or localized dynamics.

The results highlight consistent statistical associations for several variables but with limited explanatory power for groundwater level changes. For example, while nitrate and arsenic exceedance variables were statistically significant, their odds ratios near 1 indicate minimal practical influence on groundwater dynamics, echoing Pennino, et al. [47], who noted such indicators often mark localized conditions rather than drive hydrological trends. Similarly, demographic variables such as *Indian_Alaska_pcnt* and housing-related factors like *Rural_pcnt* and group housing types demonstrated consistent statistical significance without notable impacts on GWL trends.

These findings illustrate the importance of distinguishing between statistical significance and practical implications in groundwater management. Variables with negligible practical effects may reflect sample size influences or underlying mechanisms not captured in the models. Nemčić-Jurec, et al. [48] stressed the importance of nuanced groundwater quality interpretations, noting statistical associations do not always equate to hydrological relevance. Future research should explore indirect or contextual effects and investigate potential drivers that may contribute to groundwater level changes beyond the observed statistical patterns. Understanding these nuances is critical for developing policies that address both ecological and hydrological processes comprehensively.

5. Conclusions

This study provides a comprehensive analysis of groundwater availability, access, and quality in Arizona's groundwater-dependent drinking water systems. By examining groundwater level changes over both 20-year and 10-year periods, we identified persistent vulnerabilities shaped by infrastructural, environmental, socioeconomic, and policy factors. Our findings highlight that while long-term groundwater management strategies, such as the Groundwater Management Act and MAR projects, have contributed to stabilizing groundwater levels within designated groundwater management areas, significant challenges persist in rural, Tribal, and unregulated regions where oversight and resources are limited.

Groundwater availability is influenced by the cumulative effects of climate variability, land use changes, and extraction pressures, with adaptive management practices demonstrating the potential to mitigate depletion risks. Recharge plays a dual role—stabilizing groundwater levels while mobilizing contaminants like nitrate and arsenic, linking availability directly to water quality challenges. Stronger short-term associations between groundwater level changes and contamination risks reflect the immediate effects of land use and policy interventions, while long-term patterns highlight the influence of sustained governance and environmental conditions.

Access to groundwater is shaped by socioeconomic disparities, governance structures, and infrastructural factors. Tribal governance has emerged as a critical factor in mitigating groundwater declines, particularly in karst aquifers, demonstrating the effectiveness of localized management strategies. However, disparities persist, with marginalized communities facing disproportionate risks related to groundwater access and quality. These findings align with broader evidence that integrated, equity-focused, and data-driven policies are essential for sustainable groundwater governance.

To advance equitable and sustainable groundwater management, this study identifies five policy priorities. First, expand regulatory oversight beyond Active Management Areas to protect rural and non-designated areas where domestic wells and small systems remain vulnerable. Second, strengthen Tribal water governance through continued investment in infrastructure, technical capacity, and collaborative partnerships, reinforcing self-determined and resilient water management. Third, expand and tailor MAR programs—particularly in vulnerable aquifers like karst systems—using climate projections to enhance long-term effectiveness. Fourth, integrate equity into groundwater policy by leveraging socioeconomic data to guide planning and address the disproportionate risks faced by marginalized communities. Finally, invest in data-driven management through robust monitoring and accessible data-sharing platforms to support real-time, adaptive decision-making. These recommendations align with the study's findings and emphasize the need for integrated, equity-focused, and evidence-based strategies across Arizona and other arid regions.

In conclusion, ensuring groundwater sustainability in Arizona and similar arid regions requires a holistic approach that integrates environmental management, equitable access, and water quality protection. Policy interventions must address regulatory gaps, particularly outside designated groundwater management areas, enhance support for Tribal and rural water governance, and promote adaptive strategies that consider both short- and long-term hydrological dynamics. By framing groundwater management within the interconnected dimensions of availability, access, and quality, this study offers actionable insights to inform more resilient and equitable water resource policies.

Author Contributions: Conceptualization, S.A.W.; methodology, S.A.W., D.M.Q., A.A.Z.-T. and G.C.; software, S.A.W.; validation, S.A.W., S.B.M., D.M.Q. and A.A.Z.-T.; formal analysis, S.A.W.; investigation, S.A.W.; resources, S.A.W., S.B.M. and D.M.Q.; data curation, S.A.W.; writing—original draft preparation, S.A.W. and A.A.Z.-T.; writing—review and editing, S.A.W., S.B.M., A.A.Z.-T., D.M.Q. and G.C.; visualization, S.A.W., G.C. and D.M.Q.; project administration, S.A.W.; funding acquisition, S.A.W., S.B.M. and D.M.Q. All authors have read and agreed to the published version of the manuscript.

Funding: This research was supported by a Babbitt Center for Land and Water Policy Dissertation Fellowship and fellowships from the University of Arizona Graduate College through the Arid Lands Resource Sciences Graduate Interdisciplinary Program.

Data Availability Statement: Processed data supporting the findings of this study are included in the article. Raw data are available from the corresponding author upon reasonable request. Access to some third-party datasets may require prior permission.

Acknowledgments: We extend our sincere gratitude to the Navajo Nation for sharing spatial data and well information for this study. We remain committed to upholding Tribal data sovereignty and privacy. In accordance with our agreement, the research findings will be shared to support their groundwater management initiatives. We also express our appreciation to Eric Scott for providing technical guidance in developing the software codes utilized in this analysis.

Conflicts of Interest: The authors confirm there are no conflicts of interest. The funding entities had no involvement in the study's design, data collection, analysis, interpretation, manuscript preparation, or the decision to publish the findings.

Appendix A

Table A1. Logistic Regression Results for Nitrate Levels, Socioeconomic Predictors, and Interaction Terms.

	Model: 20-Year GWL and Nitrate					Model: 10-Year GWL and Nitrate				
Variables	Estimate	Std Error	Statistic	*p* Value	Log Odds	Estimate	Std Error	Statistic	*p* Value	Log Odds
Water Quality										
Nitrate (NO_3)	0.228	0.012	18.79	<0.001	1.26	0.105	0.007	15.499	<0.001	1.11
Race (%)										
Black	0.698	0.03	23.5	<0.001	2.01	0.237	0.017	14.25	<0.001	1.27
Hispanic/Latino	−0.085	0.004	−19.36	<0.001	0.92	−0.054	0.002	−21.931	<0.001	0.95
Native American	−0.003	0.009	−0.285	0.776	1.00	−0.097	0.005	−18.844	<0.001	0.91
Gender and Age										
Female	0.122	0.01	12.81	<0.001	1.13	0.088	0.005	16.453	<0.001	1.09
Age < 5	0.032	0.005	6.445	<0.001	1.03	0.021	0.003	7.384	<0.001	1.02
Age 65+	−0.009	0.001	−14.88	<0.001	0.99	−0.006	0.000	−17.777	<0.001	0.99
Housing Type (%)										
Rural	0.047	0.002	19.14	<0.001	1.05	0.001	0.001	0.426	0.67	1.00
Married w/Children	−0.100	0.009	−10.83	<0.001	0.91	−0.072	0.005	−13.910	<0.001	0.93
Family Households	0.165	0.009	18.25	<0.001	1.18	0.090	0.005	17.691	<0.001	1.09
Migrant Workers	−0.254	0.049	−5.202	<0.001	0.78	−0.356	0.027	−13.010	<0.001	0.70
Rental Units	−0.385	0.025	−15.15	<0.001	0.68	−0.102	0.014	−7.174	<0.001	0.90
Fully Owned Homes	−0.206	0.006	−35	<0.001	0.81	−0.076	0.003	−23.011	<0.001	0.93
Renter−Occupied	−0.154	0.006	−24.29	<0.001	0.86	−0.074	0.004	−20.765	<0.001	0.93
Group Housing (%)										
Seasonal Housing	−0.088	0.006	−15.16	<0.001	0.92	−0.011	0.003	−3.243	0.001	0.99
Non−Institutional Housing	0.008	0.002	3.436	0.001	1.01	−0.004	0.001	−2.951	0.003	1.00
Correctional Facilities	−0.061	0.004	−15.65	<0.001	0.94	−0.055	0.002	−25.074	<0.001	0.95

Table A1. *Cont.*

Variables	Model: 20-Year GWL and Nitrate					Model: 10-Year GWL and Nitrate				
	Estimate	Std Error	Statistic	*p* Value	Log Odds	Estimate	Std Error	Statistic	*p* Value	Log Odds
Juvenile Facilities	−0.042	0.006	−6.883	<0.001	0.96	−0.012	0.003	−3.660	<0.001	0.99
Nursing Homes	−0.054	0.003	−16.15	<0.001	0.95	−0.011	0.002	−5.988	<0.001	0.99
Student Housing	−0.188	0.005	−35.87	<0.001	0.83	−0.050	0.003	−16.865	<0.001	0.95
Military Housing	−0.013	0.01	−1.261	0.207	0.99	0.004	0.006	0.638	0.524	1.00
Variable Interaction Terms										
Rural × BIPOC (%)	0.000	0.000	−8.258	<0.001	1.00	0.000	0.000	0.496	0.62	1.00
Karst × Native American (%)	0.070	0.016	4.408	<0.001	1.07	0.106	0.009	11.796	<0.001	1.11
Tribal Primacy × Karst Aquifer	−3.898	1.436	−2.714	0.007	0.02	−6.644	0.806	−8.242	<0.001	0.00
NDA × Rural (%)	−0.056	0.003	−16.28	<0.001	0.95	−0.036	0.002	−18.573	<0.001	0.96

Response Variable: mean 20 yr water level decline rate − binary variable, 1 = decline.

Table A2. Logistic Regression Results for Arsenic Levels, Socioeconomic Predictors, and Interaction Terms.

Variables	Model: 20-Year GWL and Arsenic					Model: 10-Year GWL and Arsenic				
	Estimate	Std Error	Statistic	*p* Value	Log Odds	Estimate	Std Error	Statistic	*p* Value	Log Odds
Water Quality										
Arsenic (As)	91.637	6.129	14.951	<0.001	6E+39	38.494	3.942	9.764	<0.001	5E+16
Race (%)										
Black	0.63	0.039	16.041	<0.001	1.88	0.309	0.025	12.232	<0.001	1.36
Hispanic/Latino	−0.085	0.005	−16.05	<0.001	0.92	−0.029	0.003	−8.399	<0.001	0.97
Native American	−0.064	0.006	−11.37	<0.001	0.94	−0.031	0.004	−8.565	<0.001	0.97
Gender and Age										
Female	0.239	0.011	20.869	<0.001	1.27	0.097	0.007	13.206	<0.001	1.10
Age < 5	0.054	0.01	5.309	<0.001	1.06	0.009	0.006	1.314	0.189	1.01
Age 65+	−0.010	0.001	−18.02	<0.001	0.99	−0.006	0.000	−15.27	<0.001	0.99
Housing Type (%)										
Rural	−0.015	0.003	−4.814	<0.001	0.98	−0.02	0.002	−9.684	<0.001	0.98
Married w/Children	−0.216	0.01	−21.34	<0.001	0.81	−0.119	0.007	−18.363	<0.001	0.89
Family Households	0.172	0.01	17.111	<0.001	1.19	0.093	0.006	14.47	<0.001	1.10
Migrant Workers	−0.543	0.057	−9.46	<0.001	0.58	−0.722	0.037	−19.559	<0.001	0.49
Rental Units	−0.318	0.033	−9.76	<0.001	0.73	−0.173	0.021	−8.253	<0.001	0.84
Fully Owned Homes	−0.169	0.007	−23.03	<0.001	0.84	−0.081	0.005	−17.14	<0.001	0.92
Renter−Occupied	−0.240	0.008	−30.45	<0.001	0.79	−0.132	0.005	−26.09	<0.001	0.88
Group Housing (%)										
Seasonal Housing	−0.161	0.008	−21.02	<0.001	0.85	−0.056	0.005	−11.35	<0.001	0.95
Non−Institutional Housing	−0.001	0.003	−0.349	0.727	1.00	−0.007	0.002	−4.095	<0.001	0.99
Correctional Facilities	−0.075	0.005	−16.25	<0.001	0.93	−0.031	0.003	−10.55	<0.001	0.97
Juvenile Facilities	−0.012	0.008	−1.399	0.162	0.99	0.019	0.005	3.493	<0.001	1.02
Nursing Homes	−0.002	0.004	−0.564	0.573	1.00	0.001	0.003	0.19	0.849	1.00
Student Housing	−0.056	0.007	−7.778	<0.001	0.95	−0.014	0.005	−3.054	0.002	0.99
Military Housing	0.119	0.013	8.794	<0.001	1.13	0.066	0.009	7.614	<0.001	1.07
Variable Interaction Terms										
Rural × BIPOC (%)	0.000	0.000	−7.578	<0.001	1.00	0.000	0.000	−6.932	<0.001	1.00
Karst × Native American (%)	−0.014	0.008	−1.803	0.071	0.99	0.040	0.005	7.968	<0.001	1.04
Tribal Primacy × Karst Aquifer	−6.087	0.582	−10.46	<0.001	0.00	−1.568	0.374	−4.188	<0.001	0.21
NDA × Rural (%)	0.013	0.004	3.263	0.001	1.01	−0.001	0.003	−0.529	0.597	1.00

Response Variable: mean 20 yr water level decline rate − binary variable, 1 = decline.

References

1. Cid-Escobar, D.; Folch, A.; Ferrer, N.; Katuva, J.; Sanchez-Vila, X. An assessment tool to improve rural groundwater access: Integrating hydrogeological modelling with socio-technical factors. *Sci. Total Environ.* **2024**, *912*, 168864. [CrossRef]
2. Ciampi, L.; Plumpton, H.J.; Osbahr, H.; Cornforth, R.J.; Petty, C. Building resilience through improving groundwater management for sustainable agricultural intensification in African Sahel. *CABI Agric. Biosci.* **2022**, *3*, 63. [CrossRef]
3. Schaider, L.A.; Swetschinski, L.; Campbell, C.; Rudel, R.A. Environmental justice and drinking water quality: Are there socioeconomic disparities in nitrate levels in U.S. drinking water? *Environ. Health* **2019**, *18*, 3. [CrossRef]
4. Allaire, M.; Wu, H.; Lall, U. National trends in drinking water quality violations. *Proc. Natl. Acad. Sci. USA* **2018**, *115*, 2078–2083. [CrossRef] [PubMed]
5. Pace, C.; Balazs, C.; Bangia, K.; Depsky, N.; Renteria, A.; Morello-Frosch, R.; Cushing, L.J. Inequities in Drinking Water Quality Among Domestic Well Communities and Community Water Systems, California, 2011–2019. *Am. J. Public Health* **2022**, *112*, 88–97. [CrossRef] [PubMed]
6. Thiros, S.A.; Paul, A.P.; Bexfield, L.M.; Anning, D.W. Water Quality in Basin-Fill Aquifers of the Southwestern United States: Arizona, California, Colorado, Nevada, New Mexico, and Utah, 1993–2009. *U.S. Geol. Surv. Circ. 1358* **2014**. [CrossRef]
7. Switzer, D.; Teodoro, M.P. The Color of Drinking Water. *J. (Am. Water Work. Assoc.)* **2017**, *109*, 40–45. [CrossRef]
8. Vanderslice, J. Drinking Water Infrastructure and Environmental Disparities: Evidence and Methodological Considerations. *Am. J. Public Health* **2011**, *101*, S109–S114. [CrossRef]
9. Balazs, C.L.; Ray, I. The drinking water disparities framework: On the origins and persistence of inequities in exposure. *Am. J. Public Health* **2014**, *104*, 603–611. [CrossRef]
10. Spaur, M.; Lombard, M.A.; Ayotte, J.D.; Harvey, D.E.; Bostick, B.C.; Chillrud, S.N.; Navas-Acien, A.; Nigra, A.E. Associations between private well water and community water supply arsenic concentrations in the conterminous United States. *Sci. Total Environ.* **2021**, *787*, 147555. [CrossRef]
11. Nigra, A.E.; Chen, Q.; Chillrud, S.N.; Wang, L.; Harvey, D.; Mailloux, B.; Factor-Litvak, P.; Navas-Acien, A. Inequalities in Public Water Arsenic Concentrations in Counties and Community Water Systems across the United States, 2006–2011. *Environ. Health Perspect.* **2020**, *128*, 127001. [CrossRef] [PubMed]
12. Delpla, I.; Benmarhnia, T.; Lebel, A.; Levallois, P.; Rodriguez, M.J. Investigating social inequalities in exposure to drinking water contaminants in rural areas. *Environ. Pollut.* **2015**, *207*, 88–96.
13. Balazs, C.L.; Morello-Frosch, R.; Hubbard, A.E.; Ray, I. Environmental justice implications of arsenic contamination in California's San Joaquin Valley: A cross-sectional, cluster-design examining exposure and compliance in community drinking water systems. *Environ. Health Glob. Access Sci. Sour.* **2012**, *11*, 84. [CrossRef]
14. Balazs, C.; Morello-Frosch, R.; Hubbard, A.; Ray, I. Social disparities in nitrate-contaminated drinking water in California's San Joaquin Valley. *Environ. Health Perspect.* **2011**, *119*, 1272–1278. [CrossRef]
15. Nigra, A.E.; Navas-Acien, A. Arsenic in US correctional facility drinking water, 2006–2011. *Environ. Res.* **2020**, *188*, 109768. [CrossRef] [PubMed]
16. Hoover, J.; Gonzales, M.; Shuey, C.; Barney, Y.; Lewis, J. Elevated Arsenic and Uranium Concentrations in Unregulated Water Sources on the Navajo Nation, USA. *Expos. Health* **2017**, *9*, 113–124. [CrossRef]
17. Foster, S.A.; Pennino, M.J.; Compton, J.E.; Leibowitz, S.G.; Kile, M.L. Arsenic Drinking Water Violations Decreased across the United States Following Revision of the Maximum Contaminant Level. *Environ. Sci. Technol.* **2019**, *53*, 11478–11485. [CrossRef] [PubMed]
18. Gerlak, A.K.; House-Peters, L.; Varady, R.G.; Albrecht, T.; Zúñiga-Terán, A.; De Grenade, R.R.; Cook, C.; Scott, C.A. Water security: A review of place-based research. *Environ. Sci. Policy* **2018**, *82*, 79–89. [CrossRef]
19. Scott, C.A.; Meza, F.J.; Varady, R.G.; Tiessen, H.; McEvoy, J.; Garfin, G.M.; Wilder, M.; Farfán, L.M.; Pablos, N.P.; Montaña, E. Water Security and Adaptive Management in the Arid Americas. *Ann. Assoc. Am. Geogr.* **2013**, *103*, 280–289. [CrossRef]
20. Aslam, R.A.; Shrestha, S.; Pandey, V.P. Groundwater vulnerability to climate change: A review of the assessment methodology. *Sci. Total Environ.* **2018**, *612*, 853–875. [CrossRef]
21. WHO/UNICEF. *Progress on Sanitation and Drinking Water: 2015 Update and MDG Assessment*; World Health Organization: Geneva, Switzerland, 2015.
22. Kumar, D.M.; Shah, T. *Groundwater Pollution and Contamination in India: The Emerging Challenge*; International Water Management Institute: New Delhi, India, 2006.
23. UN-Water. *Groundwater: Making the Invisible Visible*; United Nations Educational, Scientific and Cultural Organization: Paris, France, 2022; p. 246.
24. Mondal, M. Risk of groundwater depletion in Jaipur district, India: A prediction of groundwater for 2028 using artificial neural network. *Rend. Lincei. Sci. Fis. E Nat.* **2024**, *35*, 971–984. [CrossRef]
25. Tadych, D.E.; Ford, M.; Colby, B.G.; Condon, L.E. Historical patterns of well drilling and groundwater depth in Arizona considering groundwater regulation and surface water access. *JAWRA J. Am. Water Resour. Assoc.* **2024**, *60*, 1193–1208. [CrossRef]

26. Williams, S.A. Enhancing Assessment of Groundwater Vulnerability and Contamination Risk in Arid Lands. Ph.D. Thesis, The University of Arizona, Tucson, AZ, USA, 2024.
27. Williams, S.A.; Megdal, S.B.; Zuniga-Teran, A.A.; Quanrud, D.M.; Christopherson, G. Mapping Groundwater Vulnerability in Arid Regions: A Comparative Risk Assessment Using Modified DRASTIC Models, Land Use, and Climate Change Factors. *Land* **2024**, *14*, 58. [CrossRef]
28. Williams, S.A.; Megdal, S.B.; Zuniga-Teran, A.A.; Quanrud, D.M.; Christopherson, G. Equity Assessment of Groundwater Vulnerability and Risk in Drinking Water Supplies in Arid Regions. *Water* **2024**, *16*, 3520. [CrossRef]
29. Tan, M.L.; Gassman, P.W.; Yang, X.; Haywood, J. A review of SWAT applications, performance and future needs for simulation of hydro-climatic extremes. *Adv. Water Resour.* **2020**, *143*, 103662. [CrossRef]
30. Taghavi, N.; Niven, R.K.; Paull, D.J.; Kramer, M. Groundwater vulnerability assessment: A review including new statistical and hybrid methods. *Sci. Total Environ.* **2022**, *822*, 153486. [CrossRef]
31. Mussa, K.R.; Mjemah, I.C.; Muzuka, A.N.N. A review on the state of knowledge, conceptual and theoretical contentions of major theories and principles governing groundwater flow modeling. *Appl. Water Sci.* **2020**, *10*, 146. [CrossRef]
32. Machiwal, D.; Cloutier, V.; Güler, C.; Kazakis, N. A review of GIS-integrated statistical techniques for groundwater quality evaluation and protection. *Environ. Earth Sci.* **2018**, *77*, 681. [CrossRef]
33. Kalhor, K.; Ghasemizadeh, R.; Rajic, L.; Alshawabkeh, A. Assessment of groundwater quality and remediation in karst aquifers: A review. *Groundw. Sustain. Dev.* **2019**, *8*, 104–121. [CrossRef]
34. Chaudhry, A.K.; Kumar, K.; Alam, M.A. Groundwater flow and contaminant transport models–a short review. *Desalin. Water Treat.* **2021**, *211*, 80–91. [CrossRef]
35. Megdal, S.B.; Gerlak, A.K.; Varady, R.G.; Huang, L.-Y. Groundwater Governance in the United States: Common Priorities and Challenges. *Groundwater* **2015**, *53*, 677–684. [CrossRef]
36. Williams, S.A.; Eden, S.; Megdal, S.B.; Joe-Gaddy, V. Diversity, Equity, Inclusion, and Justice in Water Dialogues: A Review and Conceptualization. *J. Contemp. Water Res. Educ.* **2023**, *177*, 113–139. [CrossRef]
37. Varady, R.G.; Zuniga-Teran, A.A.; Gerlak, A.K.; Megdal, S.B. Modes and approaches of groundwater governance: A survey of lessons learned from selected cases across the globe. *Water* **2016**, *8*, 417. [CrossRef]
38. Zuniga-Teran, A.A.; Mussetta, P.C.; Lutz Ley, A.N.; Díaz-Caravantes, R.E.; Gerlak, A.K. Analyzing water policy impacts on vulnerability: Cases across the rural-urban continuum in the arid Americas. *Environ. Dev.* **2021**, *38*, 100552. [CrossRef]
39. Weary, D.J.; Doctor, D.H. *Karst in the United States: A Digital Map Compilation and Database. 2014–1156*; U.S. Geological Survey: Reston, VA, USA, 2014; p. 23.
40. Elshall, A.S.; Arik, A.D.; El-Kadi, A.I.; Pierce, S.; Ye, M.; Burnett, K.M.; Wada, C.A.; Bremer, L.L.; Chun, G. Groundwater sustainability: A review of the interactions between science and policy. *Environ. Res. Lett.* **2020**, *15*, 093004. [CrossRef]
41. Megdal, S.B.; Dillon, P.; Seasholes, K. Water banks: Using managed aquifer recharge to meet water policy objectives. *Water* **2014**, *6*, 1500–1514. [CrossRef]
42. Engel, K.H.; Loiseleur, E.; Drilhon, E. Arizona's Groundwater Management Act at Forty: Tackling Unfinished Business. *Arizona J. Environ. Law Policy* **2020**, *10*, 187.
43. Zheng, Y.; Flanagan, S.V. The case for universal screening of private well water quality in the US and testing requirements to achieve it: Evidence from arsenic. *Environ. Health Perspect.* **2017**, *125*, 085002. [CrossRef]
44. Schaider, L.A.; Rudel, R.A.; Ackerman, J.M.; Dunagan, S.C.; Brody, J.G. Pharmaceuticals, perfluorosurfactants, and other organic wastewater compounds in public drinking water wells in a shallow sand and gravel aquifer. *Sci. Total Environ.* **2014**, *468–469*, 384–393. [CrossRef]
45. Nigra, A.E.; Sanchez, T.R.; Nachman, K.E.; Harvey, D.E.; Chillrud, S.N.; Graziano, J.H.; Navas-Acien, A. The effect of the Environmental Protection Agency maximum contaminant level on arsenic exposure in the USA from 2003 to 2014: An analysis of the National Health and Nutrition Examination Survey (NHANES). *Lancet Public Health* **2017**, *2*, e513–e521. [CrossRef]
46. Ravalli, F.; Yu, Y.; Bostick, B.C.; Chillrud, S.N.; Schilling, K.; Basu, A.; Navas-Acien, A.; Nigra, A.E. Sociodemographic inequalities in uranium and other metals in community water systems across the USA, 2006–2011: A cross-sectional study. *Lancet Planet. Health* **2022**, *6*, e320–e330. [CrossRef]
47. Pennino, M.J.; Compton, J.E.; Leibowitz, S.G. Trends in Drinking Water Nitrate Violations Across the United States. *Environ. Sci. Technol.* **2017**, *51*, 13450–13460. [CrossRef] [PubMed]
48. Nemčić-Jurec, J.; Ruk, D.; Oreščanin, V.; Kovač, I.; Ujević Bošnjak, M.; Kinsela, A.S. Groundwater contamination in public water supply wells: Risk assessment, evaluation of trends and impact of rainfall on groundwater quality. *Appl. Water Sci.* **2022**, *12*, 172. [CrossRef]
49. Konisky, D.M.; Reenock, C.; Conley, S. Environmental injustice in Clean Water Act enforcement: Racial and income disparities in inspection time. *Environ. Res. Lett.* **2021**, *16*, 084020. [CrossRef]
50. Fischer, C.; Aubron, C.; Trouvé, A.; Sekhar, M.; Ruiz, L. Groundwater irrigation reduces overall poverty but increases socioeconomic vulnerability in a semiarid region of southern India. *Sci. Rep.* **2022**, *12*, 8850. [CrossRef]

51. Johnson, T.D.; Belitz, K.; Lombard, M.A. Estimating domestic well locations and populations served in the contiguous U.S. for years 2000 and 2010. *Sci. Total Environ.* **2019**, *687*, 1261–1273. [CrossRef]
52. McGreal, B.; Eden, S. Arizona Groundwater Management: Past, Present and Future. 2021. Available online: https://wrrc.arizona.edu/publications/arroyo/arroyo-2021-arizona-groundwater-management-past-present-and-future (accessed on 12 March 2024).
53. GWAICC. Issue Brief: Exempt Wells. 2020. Available online: https://www.azwater.gov//sites/default/files/2022-08/ISSUE%20BRIEF%20-%20Exempt%20Wells%20-%20Final.pdf (accessed on 30 June 2024).
54. Ward, M.H.; Jones, R.R.; Brender, J.D.; De Kok, T.M.; Weyer, P.J.; Nolan, B.T.; Villanueva, C.M.; Van Breda, S.G. Drinking water nitrate and human health: An updated review. *Int. J. Environ. Res. Public Health* **2018**, *15*, 1557. [CrossRef]
55. Bayabil, H.K.; Teshome, F.T.; Li, Y.C. Emerging Contaminants in Soil and Water. *Front. Environ. Sci.* **2022**, *10*, 873499. [CrossRef]
56. Orellana-Macías, J.M.; Perles Roselló, M.J. Assessment of Risk and Social Impact on Groundwater Pollution by Nitrates. Implementation in the Gallocanta Groundwater Body (NE Spain). *Water* **2022**, *14*, 202. [CrossRef]
57. Schneider, F.; Buser, T.; Graefe, O. Scales of justice in water governance: Hydropower controversies in Switzerland. *Water Policy* **2014**, *16*, 137–154. [CrossRef]
58. Neal, M.J.; Lukasiewicz, A.; Syme, G.J. Why justice matters in water governance: Some ideas for a 'water justice framework'. *Water Policy* **2014**, *16*, 1–18. [CrossRef]
59. Mueller, J.T.; Gasteyer, S. The widespread and unjust drinking water and clean water crisis in the United States. *Nat. Commun.* **2021**, *12*, 3544. [CrossRef] [PubMed]
60. McDonald, Y.J.; Grineski, S.E. Disparities in access to residential plumbing: A binational comparison of environmental injustice in El Paso and Ciudad Juárez. *Popul. Environ.* **2012**, *34*, 194–216. [CrossRef]
61. Jaffee, D.; Case, R.A. Draining us dry: Scarcity discourses in contention over bottled water extraction. *Local Environ.* **2018**, *23*, 485–501. [CrossRef]
62. Cory, D.C.; Rahman, T. Environmental justice and enforcement of the safe drinking water act: The Arizona arsenic experience. *Ecol. Econ.* **2009**, *68*, 1825–1837. [CrossRef]
63. Balazs, C.L. Just Water? Social Disparities and Drinking Water Quality in California's San Joaquin Valley. Ph.D. Thesis, University of California, Berkeley, CA, USA, 2011.
64. Zwarteveen, M.Z.; Boelens, R. Defining, researching and struggling for water justice: Some conceptual building blocks for research and action. *Water Int.* **2014**, *39*, 143–158. [CrossRef]
65. Ferris, K.; Porter, S. *The Elusive Concept of an Assured Water Supply: The Role of the CAGRD and Replenishment*; Kyl Center for Water Policy at Morrison Institute, Arizona State University: Tempe, AZ, USA, 2019.
66. GWAICC. Issue Brief: AMA Management Structure. 2021. Available online: www.azwater.gov//sites/default/files/2022-08/Issue%20-%20Management%20Structure-01052021_Final.pdf (accessed on 20 March 2021).
67. Al Atawneh, D.; Cartwright, N.; Bertone, E. Climate change and its impact on the projected values of groundwater recharge: A review. *J. Hydrol.* **2021**, *601*, 126602. [CrossRef]
68. Wilder, M.; Liverman, D.; Bellante, L.; Osborne, T. Southwest climate gap: Poverty and environmental justice in the US Southwest. *Local Environ.* **2016**, *21*, 1332–1353. [CrossRef]
69. Silber-Coats, N.; Eden, S. Arizona Water Banking, Recharge, and Recovery. 2017. Available online: https://wrrc.arizona.edu/publications/arroyo/arroyo-2017-arizona-water-banking-recharge-and-recovery (accessed on 2 March 2022).
70. GWPC. Rural Groundwater Committee Principles, Objectives, and Proposals. 2023. Available online: https://www.azwater.gov/sites/default/files/2023-07/20230712_Rural_GW_Committee_Proposals_All.pdf (accessed on 2 March 2022).
71. ADWR. *Statewide Groundwater Level Changes in Arizona Water Years 1997 to 2017, 2007 to 2017 and 2016 to 2017*; ADWR: Phoenix, AZ, USA, 2020.
72. ADWR. *Statewide Groundwater Level Changes in Arizona Water Years 2017 to 2018, 2008 to 2018, and 1998 to 2018*; ADWR: Phoenix, AZ, USA, 2021.
73. ADWR. *Statewide Groundwater Level Changes in Arizona Water Years 2019 to 2020, 2010 to 2020, and 1990 to 2020*; ADWR: Phoenix, AZ, USA, 2023.
74. EPA, Six-Year Review 2 Contaminant Occurrence Data (1998–2005). Available online: https://www.epa.gov/dwsixyearreview/six-year-review-2-contaminant-occurrence-data-1998-2005 (accessed on 1 January 2024).
75. EPA, Six-Year Review 3 Compliance Monitoring Data (2006–2011). Available online: https://www.epa.gov/dwsixyearreview/six-year-review-3-compliance-monitoring-data-2006-2011 (accessed on 1 January 2024).
76. EPA. Six-Year Review 4 Compliance Monitoring Data for Arizona (2012–2019). Data Obtained via FOIA Request EPA-2023-002267. Available online: https://www.epa.gov/dwsixyearreview/six-year-review-4-compliance-monitoring-data-2012-2019 (accessed on 18 May 2023).
77. ADWR. Annual Water Supply and Demand Dashboard. Available online: https://www.azwater.gov/ama/ama-data (accessed on 1 August 2024).

78. ADWR. GIS Data and Maps Datasets. Available online: https://www.azwater.gov/gis-data-and-maps (accessed on 15 January 2024).
79. U.S. Census Bureau. 2020 Decennial Census of Population and Housing. Available online: https://www.census.gov/programs-surveys/decennial-census/decade/2020.html (accessed on 15 January 2024).
80. ADEQ. Water Quality Databases. Available online: https://waterdata.azdeq.gov/AZWQDB/Pages/Public/ (accessed on 15 February 2024).
81. Smerdon, B.D. A synopsis of climate change effects on groundwater recharge. *J. Hydrol.* **2017**, *555*, 125–128. [CrossRef]
82. Owuor, S.O.; Butterbach-Bahl, K.; Guzha, A.C.; Rufino, M.C.; Pelster, D.E.; Díaz-Pinés, E.; Breuer, L. Groundwater recharge rates and surface runoff response to land use and land cover changes in semi-arid environments. *Ecol. Process.* **2016**, *5*, 16. [CrossRef]
83. Machiwal, D.; Jha, M.K.; Singh, V.P.; Mohan, C. Assessment and mapping of groundwater vulnerability to pollution: Current status and challenges. *Earth Sci. Rev.* **2018**, *185*, 901–927.
84. Fakhreddine, S.; Prommer, H.; Scanlon, B.R.; Ying, S.C.; Nicot, J.-P. Mobilization of arsenic and other naturally occurring contaminants during managed aquifer recharge: A critical review. *Environ. Sci. Technol.* **2021**, *55*, 2208–2223.
85. Becker, B.; Reichel, F.; Bachmann, D.; Schinke, R. High groundwater levels: Processes, consequences, and management. *WIREs Water* **2022**, *9*, e1605. [CrossRef]
86. Banerjee, A.; Creedon, L.; Jones, N.; Gill, L.; Gharbia, S. Dynamic Groundwater Contamination Vulnerability Assessment Techniques: A Systematic Review. *Hydrology* **2023**, *10*, 182. [CrossRef]
87. Statman-Weil, Z.; Nanus, L.; Wilkinson, N. Disparities in community water system compliance with the Safe Drinking Water Act. *Appl. Geogr.* **2020**, *121*, 102264. [CrossRef]
88. Hoover, J.H.; Coker, E.; Barney, Y.; Shuey, C.; Lewis, J. Spatial clustering of metal and metalloid mixtures in unregulated water sources on the Navajo Nation—Arizona, New Mexico, and Utah, USA. *Sci. Total Environ.* **2018**, *633*, 1667–1678. [CrossRef]
89. USEPA. *Providing Safe Drinking Water in America: 2013 National Public Water Systems Compliance Report*; US Environmental Protection Agency Office of Enforcement and Compliance Assurance: Washington, DC, USA, 2015.
90. Baker, P.; Friel, S.; Kay, A.; Baum, F.; Strazdins, L.; Mackean, T. What Enables and Constrains the Inclusion of the Social Determinants of Health Inequities in Government Policy Agendas? A Narrative Review. *Int. J. Health Policy Manag.* **2017**, *7*, 101–111. [CrossRef]
91. Hu, X.C.; Tokranov, A.K.; Liddie, J.; Zhang, X.; Grandjean, P.; Hart, J.E.; Laden, F.; Sun, Q.; Yeung, L.W.Y.; Sunderland, E.M. Tap Water Contributions to Plasma Concentrations of Poly- and Perfluoroalkyl Substances (PFAS) in a Nationwide Prospective Cohort of U.S. Women. *Environ. Health Perspect.* **2019**, *127*, 067006. [CrossRef]
92. USEPA. *EPA Annual Environmental Justice Progress Report FY 2020*; Environmental Protection Agency: Washington, DC, USA, 2020; p. 52, EPA Publication Number: 230R20002.
93. Wallace, K.J.; York, J.M. A systems change framework for evaluating academic equity and inclusion in an Ecology and Evolution Graduate Program. *Ecol. Evol.* **2020**, *10*, 10922–10929. [CrossRef]
94. Wallace, M.; Sharfstein, J.M.; Kaminsky, J.; Lessler, J. Comparison of US county-level public health performance rankings with county cluster and national rankings: Assessment based on prevalence rates of smoking and obesity and motor vehicle crash death rates. *JAMA Netw. Open* **2019**, *2*, e186816.

water

MDPI

Article

Harnessing Explainable AI for Sustainable Agriculture: SHAP-Based Feature Selection in Multi-Model Evaluation of Irrigation Water Quality Indices

Enas E. Hussein [1], Bilel Zerouali [2,3,*], Nadjem Bailek [4,5], Abdessamed Derdour [6], Sherif S. M. Ghoneim [7], Celso Augusto Guimarães Santos [8] and Mofreh A. Hashim [1]

1 National Water Research Center, Shubra El-Kheima 13411, Egypt; enas_el-sayed@nwrc.gov.eg (E.E.H.); mofrehhashim@yahoo.com (M.A.H.)
2 Laboratory of Architecture, Cities and Environment, Faculty of Civil Engineering and Architecture, Department of Hydraulic, Hassiba Benbouali University of Chlef, B.P. 78C, Ouled Fares 02180, Algeria
3 Vegetal Chemistry-Water-Energy Laboratory, Faculty of Civil Engineering and Architecture, Department of Hydraulic, Hassiba Benbouali University of Chlef, B.P. 78C, Ouled Fares 02180, Algeria
4 Laboratory of Mathematics Modeling and Applications, Department of Mathematics and Computer Science, Faculty of Sciences and Technology, Ahmed Draia University of Adrar, Adrar 01000, Algeria; bailek.nadjem@univ-adrar.edu.dz
5 MEU Research Unit, Middle East University, Amman 11831, Jordan
6 Laboratory for the Sustainable Management of Natural Resources in Arid and Semi-Arid Zones, University Center Salhi Ahmed Naama, P.O. Box 66, Naama 45000, Algeria; derdour@cuniv-naama.dz
7 Department of Electrical Engineering, College of Engineering, Taif University, Taif 21944, Saudi Arabia; s.ghoneim@tu.edu.sa
8 Department of Civil and Environmental Engineering, Federal University of Paraíba, João Pessoa 58051-900, Brazil; celso@ct.ufpb.br
* Correspondence: b.zerouali@univ-chlef.dz

Abstract: Irrigation water quality is crucial for sustainable agriculture and environmental health, influencing crop productivity and ecosystem balance globally. This study evaluates the performance of multiple deep learning models in classifying the Irrigation Water Quality Index (IWQI), addressing the challenge of accurate water quality prediction by examining the impact of increasing input complexity, particularly through chemical ions and derived quality indices. The models tested include convolutional neural networks (CNN), CNN-Long Short-Term Memory networks (CNN-LSTM), CNN-bidirectional Long Short-Term Memory networks (CNN-BiLSTM), and CNN-bidirectional Gated Recurrent Unit networks (CNN-BiGRUs). Feature selection via SHapley Additive exPlanations (SHAP) provided insights into individual feature contributions to the model predictions. The objectives were to compare the performance of 16 models and identify the most effective approach for accurate IWQI classification. This study utilized data from 166 wells in Algeria's Naama region, with 70% of the data for training and 30% for testing. Results indicate that the CNN-BiLSTM model outperformed others, achieving an accuracy of 0.94 and an area under the curve (AUC) of 0.994. While CNN models effectively capture spatial features, they struggle with temporal dependencies—a limitation addressed by LSTM and BiGRU layers, which were further enhanced through bidirectional processing in the CNN-BiLSTM model. Feature importance analysis revealed that the quality index (qi) qi-Na was the most significant predictor in both Model 15 (0.68) and Model 16 (0.67). The quality index qi-EC showed a slight decrease in importance, from 0.19 to 0.18 between the models, while qi-SAR and qi-Cl maintained similar importance levels. Notably, Model 16 included qi-HCO_3 with a minor importance score of 0.02. Overall, these findings underscore the critical role of sodium levels in water quality predictions and suggest areas for enhancing model performance. Despite the computational demands of the CNN-BiLSTM model, the results contribute to the development of robust models for effective water quality management, thereby promoting agricultural sustainability.

Keywords: CNN-BiLSTM; SHAP; multi-deep-learning; water quality; chemical ions; agricultural sustainability

1. Introduction

Currently, sustaining agricultural systems and protecting water bodies from contamination are key priorities in the collection and management of water resources [1,2]. Surface water bodies, including rivers, lakes, and reservoirs, hold particular significance due to their critical roles in irrigation, domestic consumption, and industrial activities [3]. However, these sources are highly susceptible to various forms of pollution, making it imperative to monitor and assess their water quality regularly [4].

Surface waters are influenced by an interplay of physical, chemical, and biological factors. Physical parameters, such as temperature, turbidity, and color, affect not only aquatic ecosystems but also the efficiency of water treatment processes [5]. Chemical constituents include nutrients such as nitrogen and phosphorus, as well as xenobiotic compounds [6,7]. The concentrations of these pollutants are critical for predicting the extent of environmental damage to water resources and assessing their suitability for specific applications [8–10]. Lastly, biological constituents, including pathogenic bacteria, algae, and other aquatic organisms, are essential indicators for evaluating water quality [11].

Water Quality Indices (WQI) have been developed using various methodologies, as extensively documented in the literature [12–15]. Most WQIs generate a single index by aggregating multiple sub-indices, typically through weighting methods that combine several water quality indicators into a final value [16]. In weighted approaches, each factor is assigned a weight reflecting its relative impact on overall water quality [17]. These methods generally involve standardizing all values to a common scale before summing them for final analysis. Alternatively, multi-criteria approaches employ a comprehensive index that integrates multiple parameters, often utilizing advanced analytical techniques to account for complex parameter interactions [18].

The advent of Artificial Intelligence (AI) has significantly advanced the efficiency and accuracy of water quality modeling by enabling the development of sophisticated models capable of representing complex relationships and processing large datasets [19]. In this technological era, deep neural networks—particularly Convolutional Neural Networks (CNNs) and Recurrent Neural Networks (RNNs), including Long Short-Term Memory (LSTM) and Gated Recurrent Unit (GRU) architectures—have become increasingly prevalent for water quality analysis and forecasting [20–22]. CNNs, known for their ability to extract features from spatial data, are particularly effective in processing and analyzing multi-dimensional datasets [23,24].

In addition to neural networks, other nonlinear methods, such as Random Forest (RF) and Support Vector Regression (SVR), have gained prominence due to their capability to handle complex interactions among variables and produce accurate forecasts [25]. RF is an ensemble learning method that constructs multiple decision trees, each making predictions based on specific decision rules, and combines these predictions into a single output through aggregation [26]. Conversely, SVR generates a hyperplane designed to minimize prediction errors within a defined margin [27]. Both methods are highly effective in modeling the intricate interrelationships among water quality parameters, making significant contributions to the assessment of overall water health.

Gaussian Process Regression (GPR) has emerged as a powerful approach for modeling water quality time-series data. In addition to generating predictions, GPR offers the unique advantage of quantifying uncertainty, which is particularly valuable in water management decisions where uncertainty plays a critical role [28]. Boosting methods, such as Gradient Boosting Machines (GBM) and XGBoost, further enhance model accuracy by iteratively adding observations specifically targeted to correct residual prediction errors. These techniques systematically address the model's shortcomings, resulting in refined predictions and improved overall accuracy [29].

Support Vector Machine (SVM) has been identified as a highly suitable model for predicting river water quality, particularly when compared to Artificial Neural Networks (ANN) and Decision Tree (DT) models. This approach yielded statistically significant results in the Langat River Basin [30]. Additionally, the XGBoost classifier demonstrated exceptional performance in drinking water quality classification, achieving an accuracy of 95.12% [31]. Numerous studies have also investigated various machine learning models—including k-Nearest Neighbors, Naive Bayes, SVM, DT, and RF—for water quality classification, underscoring the effectiveness of incorporating multiple parameters to enhance WQI predictions [32].

In some studies, the combination of RF and DT models achieved 100% accuracy in assessing the water quality of Taal Lake [33]. Similarly, the XGBoost algorithm demonstrated superior performance in coastal water quality classification compared to other algorithms [34]. For river water quality prediction, research has indicated that effective classification can be achieved with a minimal set of parameters, highlighting the potential of supervised machine learning techniques for environmental monitoring [35]. Advances in Artificial Intelligence (AI) have significantly enhanced water quality classification, resulting in substantial improvements in predictive accuracy. Hybrid architectures, such as the SVR-XGBoost model, have proven highly effective, achieving a remarkable 99.4% accuracy and F1-score in binary water quality classification [36]. These integrations underscore the critical role of machine learning algorithms in water quality assessment, providing robust tools for environmental monitoring and global resource management.

Similar successes have been observed in predicting and classifying the Water Quality Index (WQI) of river water. For instance, the MLP classifier achieved an accuracy of 81% [37], while the Gradient Boosting model attained 99.50% accuracy in Water Quality Classification (WQC) [38]. In contrast, some predictions yielded variable WQC values, with accuracies as low as 50% [38]. Other studies have effectively integrated Kernel Principal Component Analysis (KPCA) with DT for feature extraction and classification, achieving overall accuracies of up to 99%, including 68% specifically in water quality classification [39].

Moreover, the use of Transformers, equipped with an attention mechanism, offers enhanced versatility for modeling dependencies in time-series data [40,41]. This capability is particularly valuable for detecting subtle changes in various water attributes that may indicate broader environmental shifts. Additionally, recent studies utilizing hyperparameter-tuned deep learning models for water quality monitoring have shown improved performance in handling sequential data and identifying key factors influencing water quality [42].

The integration of advanced AI technologies has transformed the landscape of water quality modeling, making it both more accurate and comprehensive. This study seeks to evaluate how these advancements enhance the understanding of water quality fluctuations, ultimately improving water resource management. As these technologies continue to evolve, they offer effective solutions to contemporary environmental challenges related to water resource prediction, management, and conservation. The Wilaya of Naama, characterized by its diverse landscape of high plateaus and steppe plains, provides a unique case study to examine the effects of arid climatic conditions on water resource management. Understanding how these geographical and climatic features influence water quality can strengthen predictive modeling efforts and guide the development of effective environmental management strategies. The database and methodology for calculating the Irrigation Water Quality Index (IWQI) employed in this study are grounded in the work of Hussein et al. [25]. A central hypothesis of this research is that the integration of advanced machine learning algorithms will significantly enhance the accuracy of IWQI predictions. Hussein et al.'s foundational contributions defined the database structure and established a robust framework for IWQI estimation, supporting both quality assurance and forecasting.

This study focuses on calculating the Irrigation Water Quality Index (IWQI) using advanced machine learning algorithms, including CNN, CNN-LSTM, CNN-BiGRU, and CNN-BiLSTM. It is hypothesized that integrating SHapley Additive exPlanations (SHAP) will enhance model interpretability by providing a clearer understanding of the contributions of individual features to predictions. This research seeks to improve predictive accuracy and facilitate algorithm fine-tuning while leveraging SHAP to elucidate feature impact. Ultimately, the study aims to advance the precision and reliability of IWQI predictions.

2. Materials and Methods

2.1. Study Area and Dataset Description

The Wilaya of Naama, located in northwestern Algeria, spans approximately 29,950 km^2, making it one of the largest regions in the country [25]. The area is characterized by an arid climate and is geographically positioned within UTM Zone 30 N, with Easting coordinates ranging from 600,000 to 800,000 m and Northing coordinates from 3,550,000 to 3,800,000 m (Figure 1). Situated within the Saharan Atlas, Naama's landscape features high plateaus and expansive steppe plains [43]. It is bordered by the Sahara Desert to the south and the Tell Atlas mountains to the north, with elevations reaching up to 2000 m above sea level [44]. Administratively, the Wilaya is divided into 12 municipalities, including major ones such as Naama, Mecheria, and Ain Sefra, with a total population of approximately 298,045, according to the 2018 census. Water resources in Naama are primarily derived from underground aquifers, notably the Continental Intercalary (Albian) and the Terminal Complex, which are vital for both agricultural and domestic purposes. The region also contains several ephemeral wadis that flow intermittently following seasonal rainfall.

Despite the region's challenging climatic conditions, agriculture remains a cornerstone of Naama's economy. Key crops, including cereals, olives, and dates, are cultivated using traditional irrigation techniques, while livestock farming—primarily sheep and goats—is widespread [45]. However, these activities are heavily dependent on the availability of water resources. The region faces significant threats from desertification, driven by overgrazing, deforestation, and climate change. The fragile steppe ecosystem is particularly vulnerable, with soil erosion and vegetation loss emerging as critical concerns. In response, various initiatives have been implemented to combat desertification, including reforestation, sustainable land management practices, and water conservation techniques. Algeria's recent strategy seeks to enhance agricultural productivity by expanding arable land and mitigating desertification through reforestation efforts, such as the revival of the Green Wall project. Additionally, the strategy emphasizes fodder cultivation to support livestock farming, particularly in southeastern regions, with a strong focus on the Wilaya of Naama.

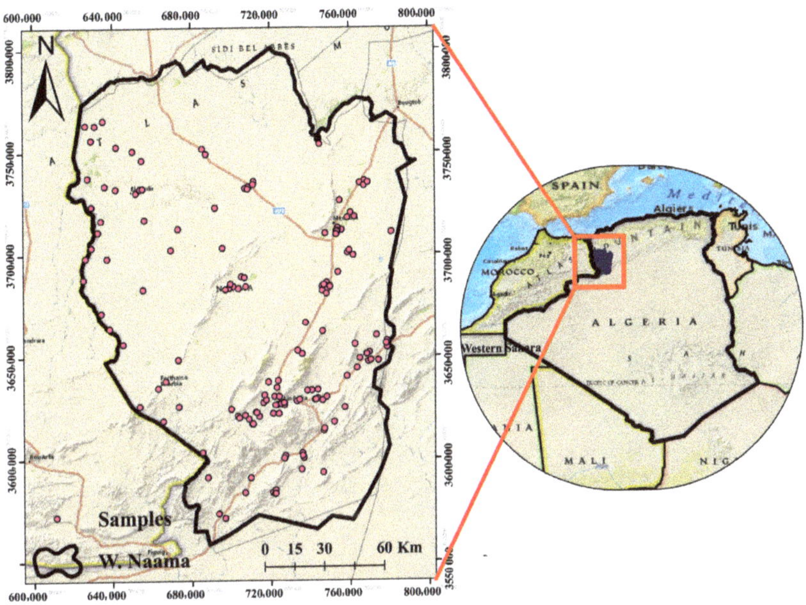

Figure 1. Geographic location of the study area and sample point locations used in this study.

A total of 166 samples were collected from the study area to analyze ten elements commonly assessed in water quality: Na^+, K^+, Mg^{2+}, Cl^-, Ca^{2+}, NO_3^-, HCO_3^-, SO_4^{2-}, pH, and electrical conductivity (EC). In situ measurements of pH, EC, and temperature were performed using a multiparameter portable quality meter (HANNA HI98194, Hanna Instruments, Woonsocket, RI, USA) [46]. To determine the concentrations of cations and anions, samples were transported to the Laboratory for Sustainable Management of Natural Resources in the Arid and Semi-arid Region at Naama and stored at 4 °C. Chemical analyses were conducted following established guidelines for water and waste analysis (EPA-600/4-79-020) [47].

Ions analyzed include magnesium (Mg), calcium (Ca), and bicarbonate (HCO_3^-), using titration methods. Sodium and potassium concentrations were determined through atomic absorption spectrometry (PerkinElmer, Waltham, MA, USA), while chloride content was measured using the Mohr method. A UV-Vis spectrophotometer (Thermo Fisher Scientific, Waltham, MA, USA) was employed to estimate nitrate (NO_3^-) and sulfate (SO_4^{2-}) concentrations. The FAO's global standards were applied to evaluate the suitability of groundwater from Naama for irrigation purposes.

The IWQ parameter (qi) is calculated using Equation (1), which depends on the tolerance limits of the parameters shown in Table 1:

$$q_i = qi_{max} - \frac{\left[\left(x_{ij} - x_{i\,inf}\right) * qi_{amp}\right]}{X_{amp}} \quad (1)$$

In this equation, qi represents the quality of each parameter, qi_{max} indicates the maximum value of qi for each class, x_{ij} refers to the observed value of each parameter, $x_{i\,inf}$ is the value associated with the lower limit class of the parameter, qi_{amp} signifies the amplitude of the quality measurement class, and X_{amp} denotes the amplitude class.

Finally, the IWQI was calculated using Equation (2). The relative weight of each parameter was applied as outlined by Meireles et al. [48]:

$$IWQI = \sum_{i=1}^{n} q_i w_i \quad (2)$$

Table 1. Limiting values for parameters used in quality assessments (qi).

qi Range	EC (uS cm^{-1})	SAR	Na^+	Cl^-	HCO_3^-
0–35	EC < 750or EC ≥ 3000	SAR < 2or SAR ≥ 12	Na < 2or Na ≥ 12	Cl < 1or Cl ≥ 10	HCO_3^- < 1or HCO_3^- ≥ 8.5
35–60	1500 ≤ EC < 3000	6 ≤ SAR < 12	6 ≤ Na < 12	7 ≤ Cl < 10	4.5 ≤ HCO_3^- < 8.5
60–85	750 ≤ EC < 1500	3 ≤ SAR < 6	3 ≤ Na < 6	4 ≤ Cl < 7	1.5 ≤ HCO_3^- < 4.5
85–100	200 ≤ EC < 750	2 ≤ SAR < 3	2 ≤ Na < 3	1 ≤ Cl < 4	1 ≤ HCO_3^- < 1.5

Figure 2 illustrates the distribution of the model input variables and the target variable, generated using Python 3.12 software. This study builds upon the work of Hussein et al. [25], who conducted the first detailed analysis of IWQI indices in the Naama region of Algeria. Hussein and colleagues quantified various water quality parameters with high precision, providing a robust foundation for evaluating water resource quality in the region.

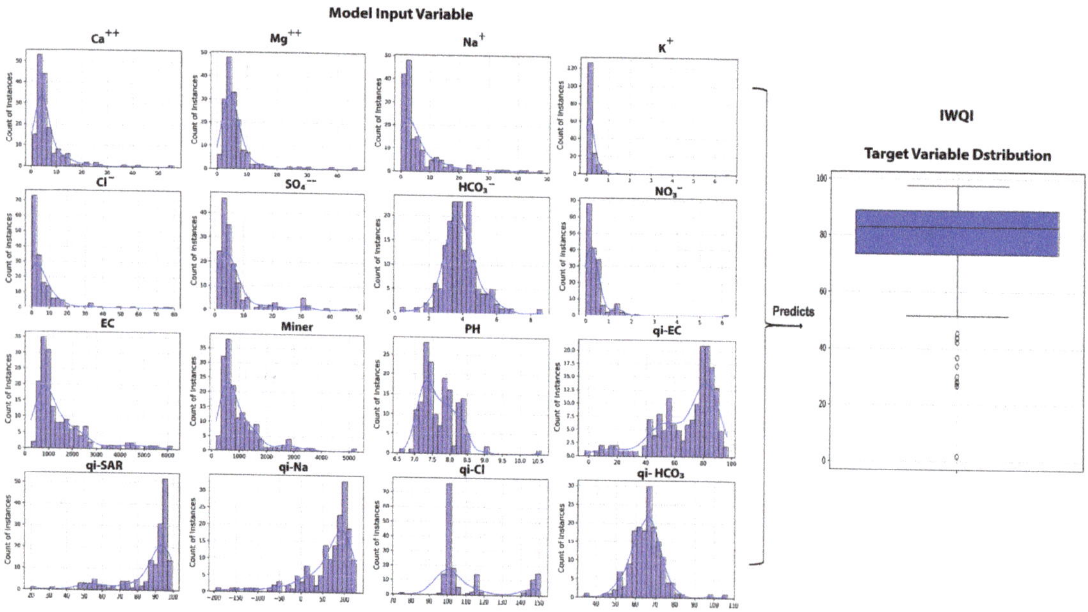

Figure 2. Distribution of model input variables and target variable analysis.

The coefficients of the parameters in Table 2 reveal notable distribution characteristics and variability, as outlined in the findings below. Calcium (Ca^{2+}) and magnesium (Mg^{2+}) exhibit high positive skewness values of 3.73 and 4.07, respectively, along with high kurtosis values of 18, indicating distributions with significant outliers. Sodium (Na^+) also shows positive skewness (2.49) and kurtosis (6.60), suggesting higher values, though less pronounced than those of Ca^{2+} and Mg^{2+}. Potassium (K^+) displays exceptionally high skewness (10.58) and kurtosis (123.18), indicating a strongly right-skewed distribution with numerous extreme values. Chloride (Cl^-) and sulfate (SO_4^{2-}) also exhibit positive skewness values of 3.73 and 2.65, respectively, and kurtosis values of 15.58 and 7.05, reflecting right-skewed distributions due to several high-value observations.

Table 2. Descriptive statistics for the IWQI dataset.

Parameter	Max	Min	Mean ± STD	Skewness	Kurtosis
Ca^{++}	56.10	0.60	7.01 ± 7.05	3.73	18.04
Mg^{++}	46.67	0.25	6.29 ± 5.76	4.07	20.86
Na^{+}	48.48	0.22	6.81 ± 8.65	2.49	6.60
K^{+}	6.69	0.03	0.25 ± 0.54	10.58	123.18
Cl^{-}	79.41	0.28	7.87 ± 12.39	3.73	15.58
SO_4^{--}	49.38	0.79	7.64 ± 8.99	2.65	7.05
HCO_3^{-}	8.67	0.33	3.90 ± 1.04	0.62	2.88
NO_3^{-}	6.29	0.02	0.44 ± 0.60	6.16	54.13
EC	6200	290	1464 ± 1100	2.08	4.39
Miner.	5270	186	1016.5 ± 753	2.30	6.99
PH	10.60	6.58	7.71 ± 0.51	1.32	4.78
qi-EC	97.55	−2.33	68.01 ± 21.18	−1.15	0.79
qi-SAR	99.41	20.06	86.20 ± 15.11	−1.86	2.92
qi-Na	126.74	−195.29	63.07 ± 59.28	−1.94	4.05
qi-Cl	150.77	73.59	111.07 ± 18.74	1.25	0.02
qi-HCO_3	106.72	33.92	65.48 ± 8.45	0.55	4.47
IWQI	97.64	1.81	78.15 ± 15.94	−1.87	4.15

In contrast, bicarbonate (HCO_3^-) exhibits a skewness of 0.62 and a kurtosis of 2.88, indicating a distribution closer to normal with low skewness and kurtosis. Nitrate (NO_3^-), on the other hand, displays a high positive skewness of 6.16 and a kurtosis of 54.13, reflecting a highly right-skewed distribution with pronounced peaks in the tails. Electrical conductivity (EC) is similarly right-skewed, with a skewness of 2.08 and a moderate kurtosis of 4.39. Extreme mineral content values show positive skewness (2.30) and higher kurtosis (6.99), emphasizing the presence of outliers in the distribution. pH levels demonstrate a slightly right-skewed distribution, with a skewness of 1.32 and a kurtosis of 4.78, suggesting a tendency toward higher values within the observed range.

The quality indices exhibit varied distributions. The quality index for EC (qi-EC) has a negative skewness (−1.15) and low kurtosis (0.79), indicating a left-leaning distribution with few outliers. Similarly, the quality index for SAR (qi-SAR) is left-skewed (−1.86) with moderate kurtosis (2.92). The quality index for Na^+ (qi-Na) shows a strong left skew (−1.94) with high kurtosis (4.05), suggesting the presence of numerous extreme values on the left side of the distribution. In contrast, the quality index for Cl^- (qi-Cl^-) is right-skewed (1.25) with nearly no kurtosis (0.02), indicating a distribution with a normal-like tail. The quality index for HCO_3^- (qi-HCO_3^-) has a right skew (0.55) and notable kurtosis (4.47), reflecting a generally normal spread with a few outliers. Lastly, the IWQI is left-skewed (−1.87) with moderate kurtosis (4.15), suggesting the presence of some extreme low values.

To categorize the dataset's water quality, key parameters such as calcium, magnesium, sodium, chloride, bicarbonate, and electrical conductivity were selected. The data were normalized to ensure consistency across different units. A Pearson correlation analysis was conducted to identify relationships among the parameters, followed by the establishment of quality categories based on established guidelines, such as "Excellent", "Good", "Fair", "Poor", and "Unsuitable". Each water sample was classified according to these thresholds, and the categorization was validated against recognized standards. The results were visualized to facilitate interpretation and effective communication of the findings. The Pearson correlation analysis of the dataset revealed notable interactions among the parameters, providing valuable insights into their interactions. Figure 3 presents the Pearson correlation coefficients between the input parameters and IWQI indices. The data indicate a significant correlation between calcium and chloride concentrations, with a correlation coefficient of 0.779, suggesting that both ions may originate from the same source or mechanism. Similarly, the strong correlation of 0.863 between Na^+ and Cl^- suggests a significant relationship, indicating the likelihood of their co-occurrence or mutual influence. Additionally, given that mineral content typically increases with electrical conductivity, the very high positive correlation between EC and mineral concentration (0.978) highlights a strong relationship between electrical conductivity and mineral levels.

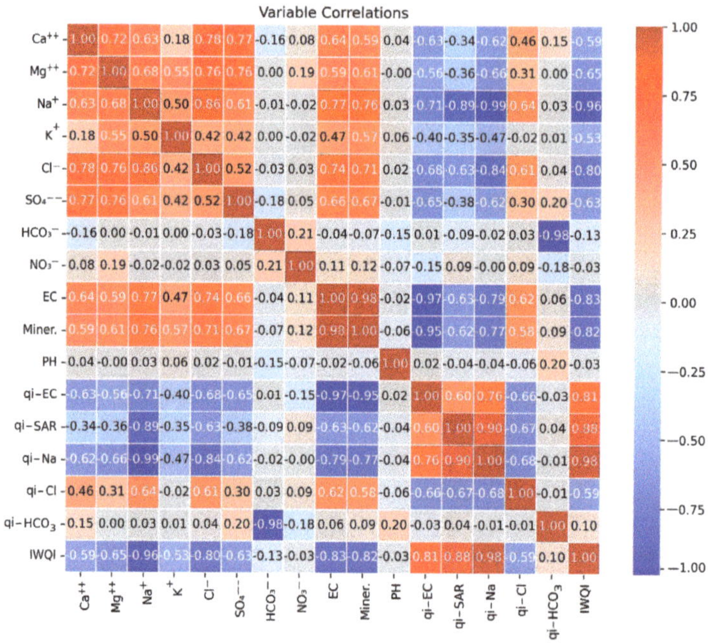

Figure 3. Pearson correlation analysis between input parameters and IWQI indices.

In contrast, the very high negative correlation between qi-Na and Na^+(−0.990) indicates a significant inverse relationship, implying that as the qi-Na index increases, sodium concentration decreases. This finding suggests the influence of distinct mechanisms or measurement factors affecting the relationship between the quality index and sodium levels. Similarly, the strong negative correlation between qi-EC and EC (−0.972) demonstrates an inverse relationship, where higher values of the qi-EC index correspond to lower electrical conductivity (Figure 3).

The analysis further highlights relationships among different ions, illustrating the complex chemistry of irrigation water. A moderate to strong positive correlation exists between Ca^{2+} and Mg^{2+} (0.724), as well as between Mg^{++} and Na^+ (0.675), indicating that increases in calcium and magnesium, and magnesium and sodium, are somewhat interrelated. Additionally, the high positive correlation between qi-SAR and qi-Na (0.903) strongly suggests a relationship between the SAR index and sodium concentration. In contrast, weak or negligible correlations, such as that between pH and most other variables, and the very weak positive correlation between HCO_3^- and qi-EC (0.008), indicate minimal relationships in these cases (Figure 3).

The IWQI is negatively correlated with key parameters, specifically Ca^{2+} (−0.591), Mg^{2+} (−0.652), Na^+ (−0.961), and Cl^- (−0.801) (Figure 3). These high negative correlation values indicate that as the concentrations of calcium, magnesium, sodium, and chloride ions increase, the IWQI decreases. This inverse relationship suggests that higher concentrations of these ions reflect poorer irrigation water quality, as indexed by the IWQ. The IWQI also exhibits strong negative correlations with qi-Na and electrical conductivity, with correlation values of −0.976 and −0.834, respectively. This indicates that increased sodium adsorption ratio and electrical conductivity are associated with a decline in IWQI, likely due to salinity or ion imbalances adversely affecting water quality. Conversely, the IWQI shows a high positive correlation with qi-SAR (0.885) and qi-Na (0.976), implying that higher values of these indices correspond to better irrigation water quality, as measured by IWQI. The correlation with qi-HCO_3 is low and positive (0.099), suggesting a minor influence of bicarbonate concentration on the IWQI. These correlations underscore the importance of monitoring specific ions and indices for effective irrigation water quality management. The strong negative correlations with Ca^{2+}, Mg^{++}, Na^{+}, and Cl^{-} further suggest that controlling these ions could significantly improve the IWQI (Figure 3).

As shown in Figure 4, the dataset's water quality distribution exhibits variability across five quality categories. A review of the data samples reveals that most samples fall into the "Excellent" or "Very Good" categories, with 75 samples classified as "Excellent" and 57 as "Very Good", together accounting for 45.18% of the dataset. This distribution aligns with the correlation analysis findings, as many samples had ion concentrations below the critical limits, which are less likely to adversely affect water quality. In contrast, eleven samples (6.63%) were classified as "Good", sixteen samples (96.64%) as "Satisfactory", and seven samples (4.22%) as "Unsuitable" (Figure 5). The relatively low percentage of samples in the lower quality categories supports the conclusions from the correlation analysis that samples with high concentrations of negatively correlated ions—such as calcium, magnesium, sodium, and chloride—are less frequent in the dataset.

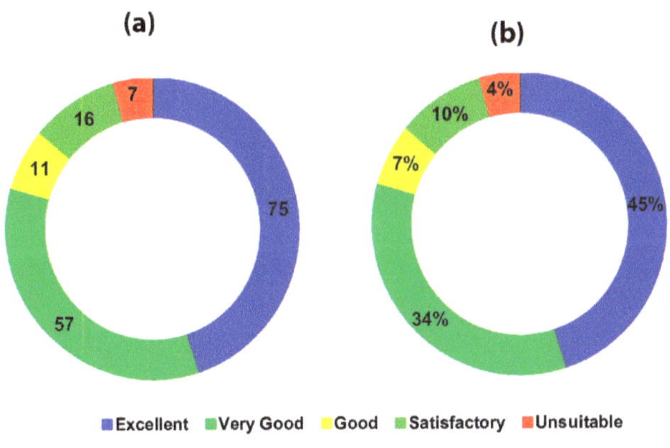

Figure 4. Distribution of IWQI data samples in (**a**) number and (**b**) percentage.

To further investigate the complex relationships identified in the correlation analysis and their impact on water quality classification, a series of models were developed. These models, summarized in Table 3, incorporate various ions and quality indices to represent the intricate interactions observed. The models range from simple single-ion evaluations to comprehensive multi-parameter assessments, enabling a systematic exploration of how different combinations of ions and indices contribute to water quality classification.

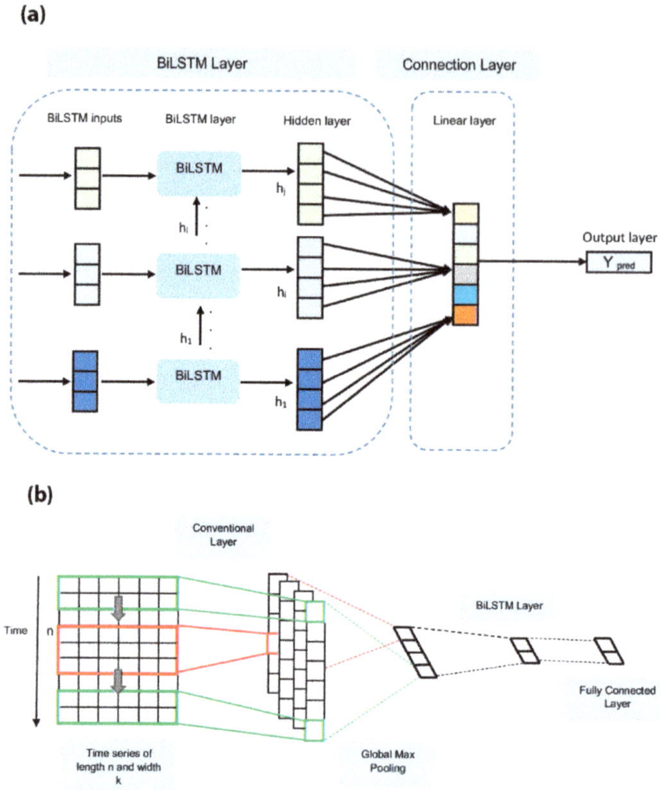

Figure 5. Representation of (**a**) BiLSTM and (**b**) hybrid CNN-BiLSTM architecture models.

Table 3. Overview of water quality classification models.

Models	Features	Description
Model 1	Ca^{++}	Focuses solely on calcium ion concentration, serving as a simple model to identify the effect of a single ion on water quality classification.
Model 2	Ca^{++}, Mg^{++}	Adds magnesium ion to assess how the presence of a second ion influences classification performance.
Model 3	Ca^{++}, Mg^{++}, Na^{+}	Extends the feature set with sodium ion to evaluate its contribution to classification accuracy.
Model 4	Ca^{++}, Mg^{++}, Na^{+}, K^{+}	Includes potassium, examining the effect of adding this ion to the existing set.
Model 5	Ca^{++}, Mg^{++}, Na^{+}, K^{+}, Cl^{-}	Adds chloride ion concentration to the previous group to analyze its impact on classification.
Model 6	Ca^{++}, Mg^{++}, Na^{+}, K^{+}, Cl^{-}, SO_4^{--}	Introduces sulfate ion concentration, testing its contribution to model complexity and performance.
Model 7	Ca^{++}, Mg^{++}, Na^{+}, K^{+}, Cl^{-}, SO_4^{--}, HCO_3^{-}	Adds bicarbonate ion concentration, assessing its role in water quality classification.

Table 3. *Cont.*

Models	Features	Description
Model 8	Ca^{++}, Mg^{++}, Na^{+}, K^{+}, Cl^{-}, SO_4^{--}, HCO_3^{-}, NO_3^{-}	Includes nitrate ion concentration to evaluate its influence on classification.
Model 9	Ca^{++}, Mg^{++}, Na^{+}, K^{+}, Cl^{-}, SO_4^{--}, HCO_3^{-}, NO_3^{-}, EC	Adds electrical conductivity to capture overall ion concentration, crucial for water quality assessment.
Model 10	Ca^{++}, Mg^{++}, Na^{+}, K^{+}, Cl^{-}, SO_4^{--}, HCO_3^{-}, NO_3^{-}, EC, Miner	Introduces a mineral index, aggregating information from multiple ions to enhance mineral-related quality assessment.
Model 11	Ca^{++}, Mg^{++}, Na^{+}, K^{+}, Cl^{-}, SO_4^{--}, HCO_3^{-}, NO_3^{-}, EC, Miner, PH	Adds pH, a critical parameter for water quality assessment, affecting various chemical processes.
Model 12	qi-EC	Focuses on the quality index for electrical conductivity, using a standardized metric to streamline the feature set.
Model 13	qi-EC, qi-SAR	Adds the sodium adsorption ratio index to evaluate interactions among different quality metrics.
Model 14	qi-EC, qi-SAR, qi-Na	Includes the quality index for sodium, testing the impact of multiple quality indices on classifier accuracy.
Model 15	qi-EC, qi-SAR, qi-Na, qi-Cl	Adds the chloride quality index, testing the effect of multiple indices on classification accuracy.
Model 16	qi-EC, qi-SAR, qi-Na, qi-Cl, qi-HCO_3	Combines the bicarbonate quality index with other indices to evaluate the impact of a comprehensive range of quality indices.

Figure 5 illustrates the BiLSTM and hybrid CNN-BiLSTM architecture models employed to achieve the objectives of this study, utilizing data collected from 166 groundwater wells, as presented by Hussein et al. [25].

2.2. Convolutional Neural Networks (CNNs)

CNNs are a specialized type of deep learning model designed for processing grid-like data formats, such as images and time-series [49]. CNNs are highly efficient at representing spatial hierarchies and local structures in data through the use of convolutional layers, making them particularly valuable for applications that require temporal recognition, such as time-series classification with the IWQI. The key components of CNNs are as follows:

Convolutional layers: These layers form the foundational elements of CNNs and perform convolution operations on the input data. Convolution is achieved by applying a filter (or kernel) across the input, generating feature maps that emphasize significant patterns. Filters can capture local features such as edges or textures in image data, as well as trends and cycles in time-series data. The convolution operation is represented as follows:

$$(f * x)(t) = \sum_{i=1}^{k} f(i) \cdot x(t - i + 1) \quad (3)$$

where f is the filter (kernel) containing learnable weights. For one-dimensional convolutions, the filter is a one-dimensional array; x represents the input, and k is the filter size.

Convolution operation: The filter slides over the input data, computing the weighted sum of the input, and the filter at each position; f(i) is the i-th weight of the filter; and $x(t - i + 1)$ is the value of the input sequence at the position shifted by $i - 1$. The result of this convolution operation is a new sequence, known as a feature map (or matrix for 2D data), which highlights specific patterns in the input data as detected by the filter.

Activation functions: After the convolution operation, nonlinearity is introduced into the model, typically using the Rectified Linear Unit (ReLU) activation function. This nonlinearity enables the network to capture complex relationships within the data, improving its predictive capabilities. The ReLU function is defined as

$$\mathrm{ReLU}(z) = \max(0, z) \tag{4}$$

Pooling layers: Pooling layers reduce the size of feature maps by discarding low-level features, thus optimizing computation. For instance, max pooling may reduce a 5 × 5 feature map to a 3 × 3 map, conserving computational resources. Pooling enhances the model's classification functionality by retaining essential features while reducing spatial dimensions [36]. This operation can be represented as

$$y(t) = \max_{i=1,\ldots,k} x(t + i - 1) \tag{5}$$

Fully connected layers: After several layers of convolution and pooling, the data are passed to fully connected layers where the high-level reasoning occurs. Each neuron in a fully connected layer is linked to all neurons in the subsequent layer, allowing the model to make predictions based on features extracted by previous layers [49]. The fully connected layer operation is expressed as

$$z = W \cdot x + b \tag{6}$$

where W represents the weight matrix, x is the input vector, and b is the bias vector. This layer combines features from the convolutional and pooling layers to perform the final classification [50,51].

Dropout layers: To prevent overfitting, dropout layers randomly drop a proportion of neurons during training. This technique enhances model robustness by preventing reliance on specific neurons, thus improving generalization to unseen data.

2.3. CNN-Long Short-Term Memory (CNN-LSTM)

The CNN-LSTM architecture combines the feature-extracting capabilities of CNN with the sequential processing abilities of LSTM networks, making it well-suited for tasks involving spatiotemporal data [52]. The CNN layers, described above in equations 1–4, extract spatial features from the time-series input data. The LSTM layers continue the feature extraction process by capturing sequential patterns and temporal dependencies in the data. This is achieved through the following equations [53].

$$f_t = \sigma(W_f \cdot [h_{t-1}, x_t] + b_f) \tag{7}$$

$$i_t = \sigma(W_i \cdot [h_{t-1}, x_t] + b_i) \tag{8}$$

$$\tilde{C}_t = \tanh(W_i \cdot [W_C, x_t] + b_C) \tag{9}$$

$$C_t = f_t * C_{t-1} + i_t * \tilde{C}_t \tag{10}$$

$$ot = \sigma(W_0 \cdot [h_{t-1}, x_t] + b_0) \tag{11}$$

$$h_t = o_t * \tanh(C_t) \tag{12}$$

where f_t, i_t, and o_t represent the forget, input, and output gates, respectively. C_t is the cell state, and h_t is the hidden state. These equations enable the LSTM layers to manage information flow, capturing relevant temporal patterns and dependencies in the sequential data.

2.4. CNN-Bidirectional Long Short-Term Memory (CNN-BiLSTM)

The CNN-BiLSTM architecture is a development over CNN-LSTM, where regular LSTM layers are replaced with bidirectional LSTM (BiLSTM) layers. Unlike the standard LSTM, which only updates information in a forward direction, BiLSTM enhances the representation of temporal dependencies by processing data in both forward and backward passes [54]. The CNN-BiLSTM architecture consists of two main components: (i) convolutional layers: These layers extract spatial features from the input data. Using convolution operations, CNNs effectively identify and represent spatial hierarchies within the data; and (ii) bidirectional LSTM layers: After spatial feature extraction, the data are passed through BiLSTM layers, which handle the temporal aspects of the sequence through bidirectional processing. This approach enables the model to capture dependencies from both past and future contexts, providing a more comprehensive understanding of temporal relationships within the data [55,56]. This bidirectional processing allows the CNN-BiLSTM model to capture a broader range of temporal dependencies, enhancing its performance on tasks that require contextual understanding in both forward and backward directions.

For the forward pass,

$$\overrightarrow{h_t} = \mathrm{LSTM}\left(x_t, \overrightarrow{h_{t-1}}\right) \tag{13}$$

The forward LSTM processes the input sequence from the beginning to the end, maintaining a hidden state $\overrightarrow{h_t}$.

For the backward pass,

$$\left(\overleftarrow{h_t}\right) = \mathrm{LSTM}\left(x_t, \overleftarrow{h_{t-1}}\right) \tag{14}$$

The backward LSTM processes the input sequence from the end to the beginning, maintaining a hidden state $\overleftarrow{h_t}$.

Combined hidden state:

$$h_t = \left[\overrightarrow{h_t}, \ \overleftarrow{h_t}\right] \tag{15}$$

The forward and backward outputs are then concatenated, providing a comprehensive representation of the temporal dependencies within the input sequence [57]. This bidirectional processing enables the model to recognize dependencies from both past and future contexts, enhancing its understanding of the temporal dynamics in the IWQI data. In classification tasks, the CNN-BiLSTM model is particularly useful for analyzing temporal dependencies in both directions, making it suitable for applications such as sentiment analysis in textual data, complex time-series forecasting, and advanced video clustering [58]. Figure 5 illustrates the CNN-BiLSTM and CNN-BiGRU architecture models proposed in this paper.

2.5. CNN-Bidirectional Gated Recurrent Unit (CNN-BiGRU)

The CNN-bidirectional Gated Recurrent Unit (CNN-BiGRU) model integrates CNNs with BiGRUs to extract both the spatial and temporal features within sequential data streams [59]. This architecture leverages the strengths of both components to enhance model performance on time-series and sequence-based tasks. Issues such as vanishing and exploding gradients, as well as the representation of long-term dependencies, are effectively addressed in GRUs, offering improvements over traditional RNNs [60,61].

Bidirectional processing enables the model to read input sequences in both directions, enhancing its contextual understanding. The CNN-BiGRU architecture consists of two main components [49,50,61,62]:

Convolutional layers: These CNN layers extract spatial features from the input data by convolving over local structures to identify patterns and hierarchical structures, making them well suited for data in grids or tabular formats.

Bidirectional Gated Recurrent Units (BiGRUs): For temporal dependency modeling, BiGRU layers are applied to the spatial features extracted by the CNN layers. Unlike conventional GRUs or vanilla RNNs, BiGRUs process data in both forward and backward directions, allowing the model to capture time-dependent features from past and future contexts.

For the forward pass:

$$\vec{h_t} = GRU_{forward}\left(x_t, \overrightarrow{h_{t-1}}\right) \tag{16}$$

For the backward pass:

$$\widehat{h_t} = GRU_{backward}\left(x_t, \widehat{h_{t+1}}\right) \tag{17}$$

where $\vec{h_t}$ and $\widehat{h_t}$ represent the hidden states from the forward and backward GRUs, respectively. The $GRU_{forward}$ processes the input sequence from the beginning to the end, while the $GRU_{backward}$ processes it from the end to the beginning.

Combined hidden state:

$$h_t = \left[\vec{h_t}, \ \widehat{h_t}\right] \tag{18}$$

The combined hidden state h_t is formed by concatenating the forward and backward hidden states at each time step. This representation captures dependencies from both directions, providing a richer temporal context [62,63].

The implementation of the CNN, CNN-LSTM, CNN-BiLSTM, and CNN-BiGRU models was carried out using Python version 3.10, which offers enhanced features and optimizations for machine learning applications. These models were developed within the Conda Python 3.12 software environment, which provides a robust platform for managing dependencies and package installations, ensuring that all required libraries, such as TensorFlow v2.16.1 and Keras 3.7.0., were properly configured. Additionally, Google Colab was utilized for its cloud-based computing capabilities, offering scalable resources and enabling collaborative development. With Colab's GPU acceleration, the training process was significantly expedited. The integration of the latest Python version, Conda's environment management, and Colab's computational power ensured an efficient implementation and experimentation process for these advanced neural network architectures. Python and Colab were also employed for generating the figures presented in this study.

2.6. CNN-BiLSTM Model Architecture

In the CNN-BiLSTM model architecture, the input layer has a shape of (None, 5, 1), meaning it accepts sequences of five time steps with one feature per step. This layer does not contain trainable parameters and serves as a conduit for data entry, allowing new data to pass through to higher layers without interaction, as shown in Table 3.

This convolutional layer (conv1d) has an output size of (None, 5, 64) with 128 parameters, applying 64 filters of kernel size 1. This configuration helps capture local features and temporal windows within the data, adding value to the visualization process by highlighting the location of significant patterns. The Max Pooling Layer (max_pooling1d) maintains a batch output shape of (None, 5, 64) without adding parameters, as shown in Table 4. This layer downsamples the data along the time axis, reducing its dimensionality while retaining essential spatial features. This approach conserves computational resources and mitigates overfitting by simplifying the data representation.

Table 4. CNN-BiLSTM model architecture used in this study.

Layer (Type)	Output Shape	Parameters
input_layer (InputLayer)	(None, 5, 1)	0
conv1d (Conv1D)	(None, 5, 64)	128
max_pooling1d (MaxPooling1D)	(None, 5, 64)	0
bidirectional (Bidirectional)	(None, 5, 100)	46,000
bidirectional_1 (Bidirectional)	(None, 100)	60,000
dense (Dense)	(None, 64)	6464
dense_1 (Dense)	(None, 32)	2000
dense_2 (Dense)	(None, 5)	165

The first bidirectional LSTM layer has an output shape of (None, 5, 100) and contains a total of 46,000 parameters. This layer processes the input in both forward and backward directions, capturing temporal dependencies from both past and future contexts. This bidirectional processing strengthens the model's capacity to understand and utilize sequential patterns in the data. The second bidirectional LSTM layer, referred to as bidirectional_1, is expected to output a dimension of (None, 100). As shown in Table 4, this layer has a total of 60,000 parameters. This in turn improves the temporal patterns elicited from the first bidirectional LSTM. It acts as a feature summarization layer, refining the temporal patterns extracted from the first bidirectional LSTM layer, thereby providing a more nuanced representation of sequential dependencies and patterns in the data.

Following the previous operations, the feature space is reduced to 64 units by the dense layer with an output shape of (None, 64) and a total of 6464 parameters. This layer uses the ReLU activation function, introducing non-linear transformations that enable the model to learn complex relationships in the data. Next, an additional dense layer, named dense_1, has 2000 parameters and outputs a shape of (None, 32), further condensing the feature space 32 units using ReLU activation. This step refines the features, making them smaller but highly representative before the final classification layer.

Output Layer: The last layer, dense_2, has an output shape of (None, 5) with 165 parameters, and it produces the classification result across five classes, as shown in Table 3. Using the softmax activation function, this layer outputs a probability distribution across the classes, making it suitable for multi-class classification.

2.7. CNN-BiLSTM Model Hyperparameters

In configuring the hyperparameters for the CNN-BiLSTM model, the hyperparameter 'Conv1D Filters' was set to 64, defining the number of filters applied in the convolution layer. This parameter determines the depth of the output volume; a higher number of filters enables the model to capture a wider range of features, allowing it to recognize more intricate patterns in the data. For the Conv1D kernel size, a value of 1 was chosen, enabling the model to learn features from each individual time step. This configuration is particularly suitable for sequential data, where each time step holds unique importance.

In the bidirectional LSTM layer, each LSTM unit has 50 units. This hyperparameter controls the number of outputs generated by the layer, thereby influencing the model's capacity to capture complex temporal patterns in the data. This unit size provides an effective balance between complexity and performance (Table 5).

Table 5. Hyperparameters of CNN-BiLSTM model.

Hyperparameter	Value
Conv1D Filters	64
Conv1D Kernel Size	1
LSTM Units	50
Dense Layer 1 Units	64
Dense Layer 2 Units	32
Optimizer	adam
Loss	sparse categorical crossentropy
Batch Size	32
Epochs	100

The Dense Layer 1 dimensions are set to 64, indicating the number of neurons in the first dense layer. This value provides an optimal balance between model complexity and performance, effectively enhancing the feature space without overcomplicating it. Similarly, the Dense Layer 2 units are set to 32, meaning the second dense layer contains 32 neurons. This further reduces the feature space, refining the features before they are passed to the output layer for final classification.

The optimizer used is Adam, one of the most popular optimization algorithms, which automatically adjusts the learning rate during training [64]. Adam is derived from AdaGrad and RMSProp and is well-suited to a variety of problems due to its ability to converge efficiently and stably [65]. For a multi-class classification problem, the ideal loss function is the sparse categorical cross-entropy, which calculates the difference between the predicted probabilities and the true class labels to formulate the optimization loss.

The batch size is set to 32, which is the number of samples the model processes in each session before updating the model parameters. This batch size strikes a balance between training stability and speed. Epochs are set to 100, representing the total number of complete passes through the training data (Table 5). Each epoch allows the model to iteratively adjust parameters with the aim of minimizing the loss function and improving accuracy.

Feature selection is one of the most crucial steps in the machine learning model development process. It involves identifying and selecting the most relevant features from a pool of candidate features. The goal of feature selection is to improve model performance and generalization while reducing overfitting by simplifying the model.

2.8. SHapley Additive Explanations (SHAP) for Feature Selection

SHAP is a widely used method for explaining how individual features influence model outcomes [66,67]. In SHAP, a value is assigned to each feature, representing its contribution to the final prediction. This approach has become increasingly popular in machine learning as it provides practitioners with a clear understanding of how each feature impacts model behavior and outputs [68,69]. SHAP-based feature selection is particularly useful for identifying which features have the greatest influence on a model, allowing for streamlined, transparent models. SHAP values are rooted in concepts from game theory, where a fair allocation of rewards among players is possible. In this analogy, the features are the "players", and the model's prediction is the "reward". SHAP values reveal each feature's contribution to the final prediction. To achieve a fair measure of feature importance, SHAP considers all possible feature combinations and estimates the change in the model output when adding or excluding a feature from a combination [70]. The Shapley value represents the average importance of a feature in making a prediction, accounting for interactions among all features. This ensures an equitable distribution of importance across features, considering all possible feature combinations.

While the concept of SHAP is rooted in game theory, SHAP values calculate the marginal influence of each feature with respect to all possible feature coalitions. The mathematical expression for calculating the Shapley value, ϕ_i, of a feature i within a set of features S can be expressed as follows:

$$\phi_i(f,\ x) = \sum_{S\subset N\setminus\{i\}} \frac{|S|!(M-|S|-1)!}{M!}[f_x(S) - f_x(S)\{i\}] \tag{19}$$

where $\phi_i(f,\ x)$ represents the contribution of feature i for model f at record x, $\frac{|S|!(M-|S|-1)!}{M!}$ is the weighting function that reflects the impact of each subset of features, $[f_x(S) - f_x(S)\{i\}]$ denotes the impact of removing feature i from subset S, N is the full set of features, and $f_x(S)$ represents the model's prediction when only the features in subset S are used. This formulation allows SHAP values to fairly allocate the contribution of each feature by considering its effect across all possible feature combinations.

SHAP analysis was conducted within the Conda environment and on Google Colab, enabling efficient dependency management and the utilization of cloud-based resources. This approach provided deeper insights into the contributions of individual features to model predictions for water quality classification.

2.9. Model Evaluation

In analyzing groundwater suitability and IWQI, a set of performance metrics was employed to evaluate the classification models. Accuracy provided an initial measure of the overall correctness of model predictions. To address the imbalanced nature of the data, micro-averaged metrics such as micro precision, micro recall, and micro F1 score were used. These metrics aggregate the contributions of all classes to provide a global performance measure. Additionally, macro-averaged metrics like macro precision, macro recall, and macro F1 score were implemented to evaluate model performance on a per-class basis, treating each class equally [71–75]. These performance metrics, along with their respective formulas, were summarized as follows.

$$\text{Accuracy} = \frac{\text{Number of Correct Predictions}}{\text{Total Number of Predictions}} \tag{20}$$

$$\text{Micro Precision} = \frac{\text{Total True Positives}}{\text{Total True Positives} + \text{Total False Positives}} \tag{21}$$

$$\text{Micro Recall} = \frac{\text{Total True Positives}}{\text{Total True Positives} + \text{Total False Negatives}} \tag{22}$$

$$\text{Micro F1 Score} = 2 \times \frac{\text{Micro Precision} \times \text{Micro Recall}}{\text{Micro Precision} + \text{Micro Recall}} \tag{23}$$

$$\text{Macro Precision} = \frac{1}{\text{N}}\sum_{\text{i}=1}^{\text{N}} \text{Precision}_{\text{i}} \tag{24}$$

$$\text{Macro Recall} = \frac{1}{\text{N}}\sum_{\text{i}=1}^{\text{N}} \text{Recall}_{\text{i}} \tag{25}$$

$$\text{Macro F1 Score} = 2 \times \frac{\text{Macro Precision} \times \text{Macro Recall}}{\text{Macro Precision} + \text{Macro Recall}} \tag{26}$$

For model evaluation, a 70/30% train-test split was used, a commonly employed methodology in the literature. The training set, comprising 70% of the data, allowed the models to learn the underlying patterns and relationships between various water quality parameters—such as chemical ions and derived quality indices—and the target variable representing water quality classification. The testing set, containing the remaining 30% of the data, was used to evaluate how well the models generalize to new, unseen data.

3. Results

The performance metrics for the CNN, CNN-LSTM, CNN-BiLSTM, and CNN-BiGRU models provide insight into how varying model complexities and feature sets influence the classification of IWQ (Figure 6). Among the CNN models, there was a clear trend of performance improvements correlating with increased model complexity (Figure 6a). The baseline accuracy for Model 1, which used only calcium ion concentration, was 0.62, demonstrating that a single feature significantly limits the model's ability to accurately classify water quality. Each subsequent model incorporated additional features, such as magnesium and sodium, resulting in a steady increase in accuracy. Model 16, which included the full set of water quality indices, achieved an accuracy of 0.90. This substantial improvement highlights the importance of using multiple parameters to enhance the model's classification capability. Similarly, micro precision, micro recall, and micro F1 score followed this upward trend, increasing from 0.62 for the simplest model to 0.90 for Model 16, indicating consistent performance across different water quality types and suggesting that these models improve not only overall accuracy but also their classification performance. Macro metrics also started from lower values and showed significant improvement with the addition of features, particularly from Model 12 onward. For instance, macro precision rose from 0.444 in Model 1 to 0.867 in Model 16 (Figure 6a). This indicates that more complex models perform better at balancing classification performance across various water quality categories.

For the CNN-LSTM models, performance metrics showed similar improvement trends. Starting with an accuracy of 0.62 for Model 1, accuracy rose substantially to 0.92 for Model 16, underscoring the value of combining CNN and LSTM architectures (Figure 6b). This integration effectively captures both spatial and temporal dependencies within the data, which are crucial for accurate water quality classification. Alongside accuracy gains, micro metrics (micro precision, micro recall, and micro F1 score) increased from 0.62 to 0.88, indicating that the model's ability to handle complex, imbalanced data improved significantly with increased complexity. Macro metrics displayed parallel trends, with macro precision rising from 0.384 to 0.826 and macro recall from 0.371 to 0.839. These improvements reflect the model's enhanced capacity to detect instances across all water quality classes, particularly those with lower frequencies. In the CNN-BiLSTM models, accuracy ranged from 0.62 to 0.94. Initially, accuracy was lower, but it grew steadily with the addition of features, peaking at 0.94 for Model 16. Micro accuracy also showed robust performance, increasing from 0.62 to 0.94, which again highlights the model's balanced classification capability across the different classes (Figure 6c). Macro metrics showed substantial improvement as well, with macro precision rising from 0.384 to 0.925 and macro recall from 0.371 to 0.908, indicating that the model's ability to correctly identify positive instances increased significantly as more features were incorporated.

The CNN-BiGRU models played a crucial role in extracting key features necessary for IWQ classification (Figure 6d). Starting from a baseline of 0.60 for Model 1, which relied solely on calcium ion concentration, accuracy improved as more ions were incorporated into the model, such as magnesium, sodium, and potassium. Notably, Model 3 achieved an accuracy of 0.74, demonstrating that including a variety of ions allows the model to capture finer variations in water quality, thereby enhancing classification accuracy. Models 4 through 7 showed an accuracy range of 0.66 to 0.74, with gradual improvements as additional ions were introduced. A more significant performance boost occurred in

Models 8 to 10 with the inclusion of nitrate and electrical conductivity (EC). While Model 8 maintained an accuracy of 0.74, the inclusion of EC in Model 9 resulted in a slight increase to 0.70 (Figure 6d), highlighting the relevance of EC, as it represents a cumulative measure of ion concentration, a vital indicator of overall water quality. Model 10, which included the mineral index, achieved an accuracy of 0.72, suggesting that this aggregated metric may enhance the model's classification capabilities by summarizing multiple ion-related metrics. More pronounced improvements were observed with Models 12 through 16, which used quality indices for electrical conductivity (qi-EC), sodium adsorption ratio (qi-SAR), sodium (qi-Na), chloride (qi-Cl), and bicarbonate (qi-HCO_3). These sophisticated models leverage derived indices to encapsulate complex ion interactions as standardized measures. Model 12 reached an accuracy of 0.82, with peak accuracies of 0.90 and 0.92 in Models 15 and 16, respectively (Figure 6d). This sharp improvement underscores the value of quality indices, which likely capture critical aspects of water quality not represented by individual ion concentrations. The upward trend in accuracy was consistently reflected in precision, recall, and F1 scores for CNN-BiGRU models. Although the macro precision, recall, and F1 scores showed some variability, they exhibited a clear upward trend as model complexity increased.

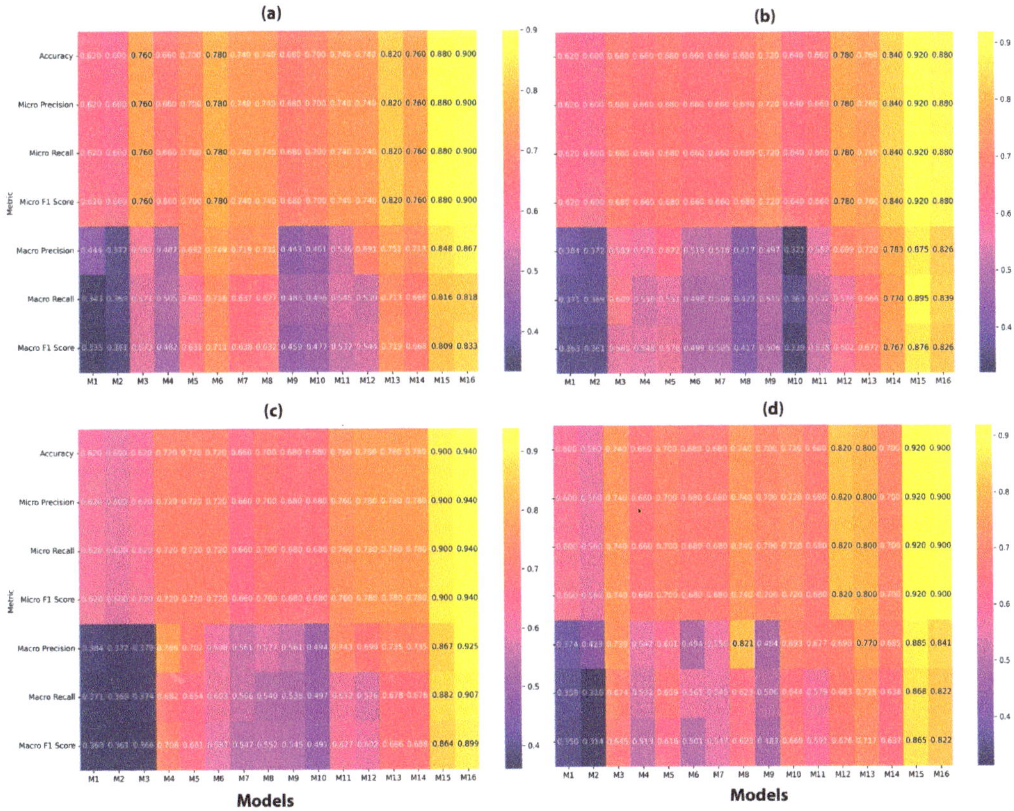

Figure 6. Summary of performance metrics for the (**a**) CNN, (**b**) CNN-LSTM, (**c**) CNN-BiLSTM, and (**d**) CNN-BiGRU models across different levels of complexity during the test phase.

The assessment results for all models reveal an intricate understanding of their effectiveness in classifying IWQ, with each method offering distinct advantages and limitations

(Figures 7–9). The CNN model M16 achieved a notable test accuracy of 0.84 and a test loss of 0.334, indicating strong classification capability with minimal predictive error (Figure 7a). The low test loss suggests that the model's predictions closely align with actual labels, reflecting effective learning from the training data. The CNN model exhibits strong performance in terms of specificity, with high values across all classes, ranging from 0.9118 to 1.0, indicating the model's effectiveness at identifying negative cases. However, sensitivity varies more significantly, with values between 0.6 and 0.9375. Sensitivity is particularly low for Class 4, suggesting that the model could benefit from improvements in detecting positive cases for this class (Figure 8a). The AUC score of 0.9819 reflects excellent overall classification performance, emphasizing the model's capability to differentiate between classes effectively (Figure 9a). In summary, while the CNN model demonstrates robustness in specificity and overall classification performance, enhancing sensitivity for Class 4 could further improve its effectiveness.

The CNN-LSTM model, M15, achieved a notable test accuracy of 0.92 with a test loss of 0.493, demonstrating strong classification ability (Figure 7b). The CNN-LSTM model performs impressively across multiple metrics. Specificity is high across all classes, ranging from 0.9565 to 1.0, indicating excellent accuracy in identifying negative cases. Sensitivity varies, with Class 0 achieving perfect sensitivity at 1.0, while Classes 1 and 2 have lower sensitivities of 0.81 and 0.75, respectively. Classes 3 and 4 also perform well, with sensitivity values of 0.8 and 1.0. The AUC score of 0.9730 highlights the model's strong capability to distinguish between classes effectively. While the CNN-LSTM model demonstrates robust classification performance overall, increasing sensitivity for Classes 2 and 3 could further enhance its effectiveness (Figure 8b). The AUC score of 0.973 also underscores the model's ability to rank predictions accurately, showcasing a high level of proficiency in class separation (Figure 9b).

The CNN-BiLSTM model (M16) achieved a test accuracy of 0.94 and a notably low test loss of 0.184, indicating high reliability and minimal deviation between predicted versus actual values (Figure 7c). This model demonstrates exceptional performance across all metrics. Specificity remains consistently high, ranging from 0.9565 to 1.0, reflecting strong accuracy in identifying negative cases across all classes. Sensitivity varies slightly; Classes 0, 2, and 4 achieve perfect sensitivity at 1.0, while Class 3 shows a lower sensitivity of 0.6, suggesting an area for potential improvement in detecting positive cases (Figure 8c). The model's impressive AUC score of 0.994 highlights its excellent capability in distinguishing between different water quality classes, making it a valuable tool for practical applications (Figure 9c).

The best-performing model was the CNN-BiGRU model on M15, with a test accuracy of 92% and a test loss of 0.4998, indicating effective generalization from the training set to new examples (Figure 7d). The model displayed high specificity, especially for Classes 1 and 4, achieving perfect scores in minimizing false positives for these classes. However, sensitivity for Class 4 was notably lower at 0.6667, suggesting that the model occasionally missed instances of this class. This issue may stem from class imbalance or intrinsic challenges in distinguishing Class 4 from other classes (Figure 8d). The model achieved an AUC of 0.9743, underscoring its strong capability in class differentiation and its effective ranking of predictions within closely related classes (Figure 9d). Overall, while each model had distinct advantages—CNN-BiLSTM demonstrated the highest accuracy and AUC, and CNN showed the best specificity—each also presented opportunities for improvement, particularly in enhancing sensitivity for specific classes.

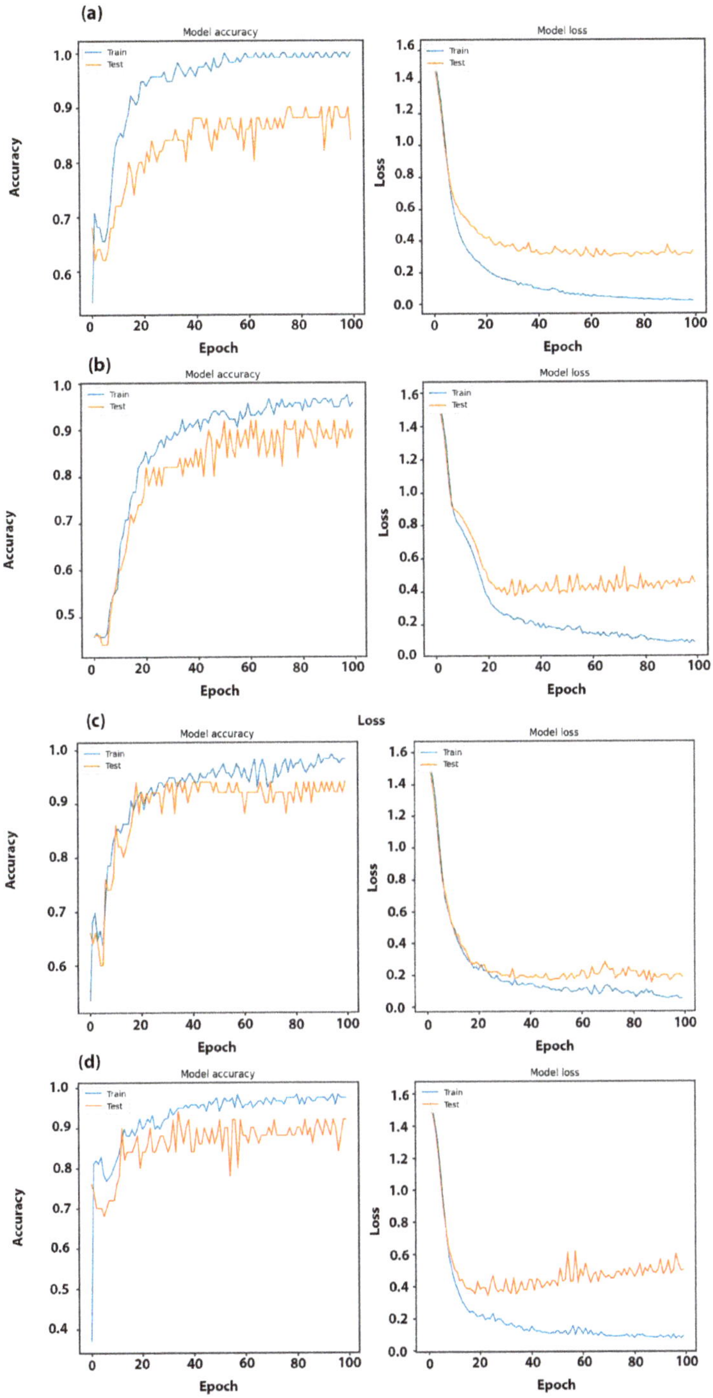

Figure 7. Best accuracy and loss for the (**a**) CNN (M16), (**b**) CNN-LSTM (M15), (**c**) CNN-BiLSTM (M16), and (**d**) CNN-BiGRU (M15) models.

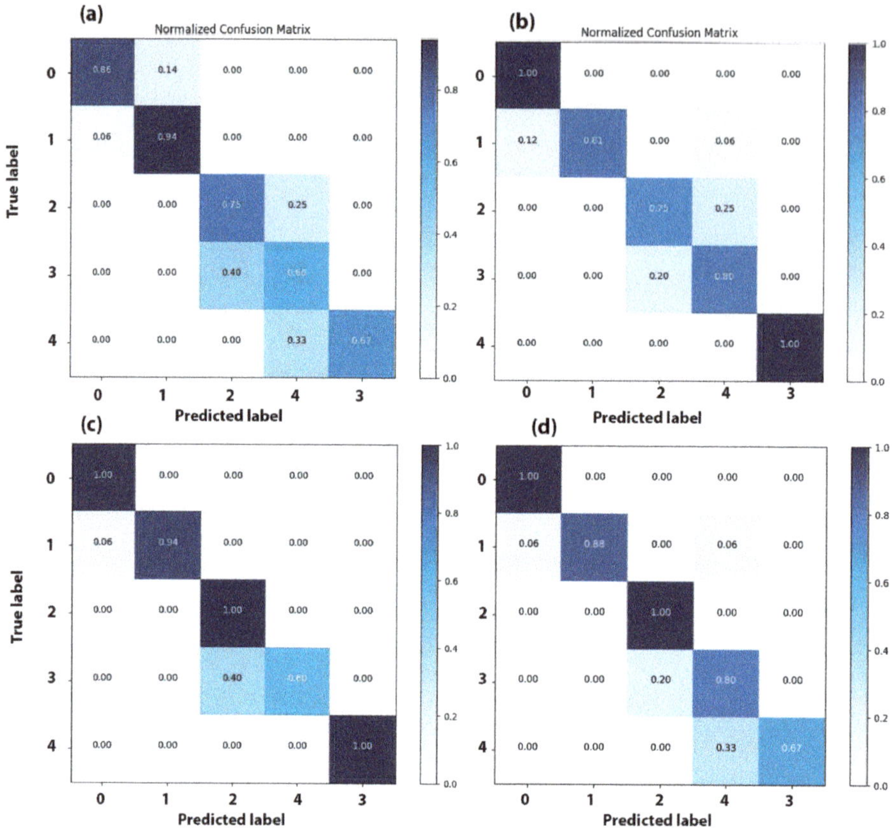

Figure 8. Confusion matrix for best-performing models: (**a**) CNN (M16), (**b**) CNN-LSTM (M15), (**c**) CNN-BiLSTM (M16), and (**d**) CNN-BiGRU (M15).

These models have demonstrated a robust ability to classify irrigation water quality efficiently, presenting significant potential for refinement and improved predictive accuracy. With additional calibration and the incorporation of advanced techniques such as class balancing, these models could further enhance their precision, reliability, and overall effectiveness in real-world applications.

Complementary analyses using the receiver operating characteristic (ROC) curve offer comprehensive insights into each model's classification performance. The CNN model demonstrates outstanding performance for Class 0, with an AUC of 0.99 (Figure 9a), confirming its ability to accurately distinguish this class from others. Class 1 also achieves an AUC of 0.99, indicating strong predictive reliability. Class 2 scores well with an AUC of 0.97, showing solid classification but slightly less confidence than the first two classes. Class 3 receives an AUC of 0.96, suggesting that while performance is strong, there remains room for improvement. Notably, Class 4 achieves a perfect AUC of 1.00, indicating that the model can accurately identify when this class is absent. The CNN-LSTM model follows a similar trend, with an AUC of 0.99 for Class 0 and 0.94 for Class 1. Class 2 performs impressively with an AUC of 0.98, while Class 3 scores adequately at 0.96, similar to the CNN model. Once again, Class 4 achieves a perfect AUC of 1.00, attesting to the model's robustness for this class (Figure 9b). The CNN-BiLSTM model further enhances classification performance, achieving perfect AUC scores of 1.0 for both Class 0 and Class 1, indicating exceptionally strong classification capabilities. Class 2 scores 0.98, Class 3 scores 0.97, and Class 4 maintains a perfect AUC of 1.00 (Figure 9c). The CNN-BiGRU model shows comparable trends: Class 0 achieves an AUC of 0.99, Class 1 scores 0.94, Class 2 performs well with 0.98, Class 3 scores satisfactorily at 0.96, and Class 4 again achieves a perfect AUC of 1.00 (Figure 9d). The ROC curve analysis reinforces the model's effectiveness in classifying IWQ, especially for Classes 0 and 4. These slight variations in AUC scores across classes highlight areas for potential improvement, particularly in enhancing sensitivity for classes with relatively lower performance rates.

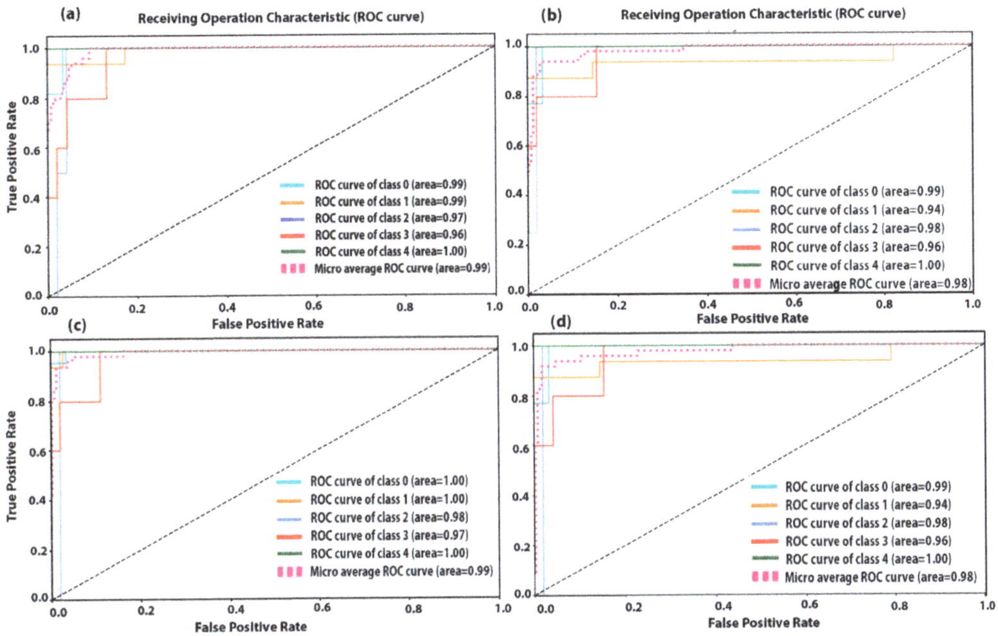

Figure 9. ROC AUC of best-performing models: (**a**) CNN (M16), (**b**) CNN-LSTM (M15), (**c**) CNN-BiLSTM (M16), and (**d**) CNN-BiGRU (M15).

The CNN-BiLSTM model demonstrates outstanding performance, with a test accuracy of 0.94 and an AUC of 0.994, underscoring its superior ability to classify water quality types accurately (Figure 10). The low test loss of 0.184 indicates precise predictions with minimal error. High specificity values across classes confirm the model's effectiveness in minimizing false positives, while high sensitivity values for most classes highlight its strong detection capabilities. Although sensitivity for Class 4 is somewhat lower, indicating a potential area for further improvement, the model's overall performance remains robust and highly effective for water quality classification tasks.

CNNs are highly effective at extracting spatial features from discrete time-series data, making them suitable for grid-based or tabular data analysis. However, a key limitation of CNNs is their inability to represent temporal dependencies, which limits their effectiveness for sequential or time-series analysis. CNN-LSTM networks address this limitation by combining CNN's spatial feature extraction capabilities with the LSTM's ability to model temporal patterns, creating a powerful model for spatiotemporal data. Although effective, CNN-LSTM networks are more complex, resulting in longer training times and higher computational costs. CNN-BiLSTM models further enhance temporal pattern recognition by processing data in both forward and backward directions, enabling more accurate predictions. However, this bidirectional approach increases computational demands and often requires larger datasets for optimal performance. CNN-BiGRU models provide a good balance, capturing bidirectional temporal dependencies with less complexity than CNN-BiLSTM, thus offering greater efficiency. Nonetheless, they still require more computational resources than traditional CNNs and need careful tuning to prevent overfitting. Table 6 summarizes the advantages and disadvantages of CNN, CNN-LSTM, CNN-BiLSTM, and CNN-BiGRU models in terms of accuracy and AUC.

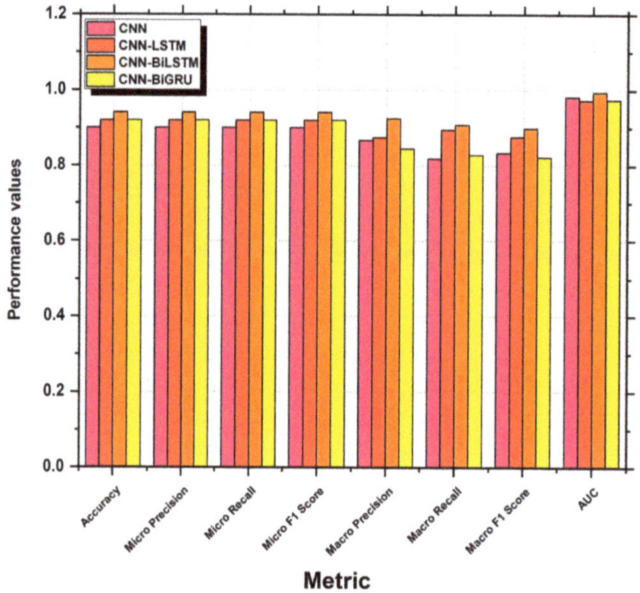

Figure 10. Performance metrics for the best CNN, CNN-LSTM, and CNN-BiLSTM models.

Table 6. Summary of advantages and disadvantages of CNN, CNN-LSTM, CNN-BiLSTM, and CNN-BiGRU models.

Model	Advantages	Disadvantages	Performance
CNN	- Effective in extracting spatial features and patterns	- Limited in capturing temporal dependencies	Accuracy: 0.9 AUC: 0.9819
	- Computationally efficient and faster to train	- May not perform well on sequential data	
	- Suitable for tasks with spatial hierarchies (e.g., images)	- Requires large datasets for optimal performance	
	- Lower complexity compared with LSTM-based models	- Can struggle with long-range dependencies	
CNN-LSTM	- Combines spatial feature extraction with temporal sequence modeling	- More complex and computationally intensive than standalone CNN	Accuracy: 0.92 AUC: 0.973
	- Effective in handling spatiotemporal data	- Longer training times compared with CNN	
	- Capture long-range dependencies and sequential patterns	- Requires careful tuning of hyperparameters	
	- Versatile for tasks involving both spatial and temporal features	- Potentially prone to overfitting without proper regularization	

Table 6. *Cont.*

Model	Advantages	Disadvantages	Performance
CNN-BiLSTM	- Captures bidirectional temporal dependencies (past and future context)	- Highest computational complexity among the models	Accuracy: 0.94 AUC: 0.994
	- Effective in modeling complex spatiotemporal relationships	- Longer training times and higher memory usage	
	- Provides comprehensive understanding of sequential data	- Requires significant computational resources	
	- Improved performance for tasks requiring contextual understanding	- May be prone to overfitting without proper regularization	
CNN-BiGRU	- Captures bidirectional temporal dependencies with reduced complexity compared with CNN-BiLSTM	- Still more complex and computationally intensive than standalone CNN	Accuracy: 0.92 AUC: 0. 0.974
	- Effective in modeling spatiotemporal relationships with fewer resources	- Training times and memory usage higher than CNN but lower than CNN-BiLSTM	
	- Balance performance and computational efficiency	- May require careful tuning to avoid overfitting	

This section examines feature importance for the top-performing models, Models 15 and 16. Each value represents the expected contribution of a feature toward the prediction of a given instance. Positive SHAP values indicate a tendency toward higher predictions, reflecting a favorable outcome, while negative SHAP values suggest a tendency towards lower predictions or less favorable outcomes (Figures 11 and 12).

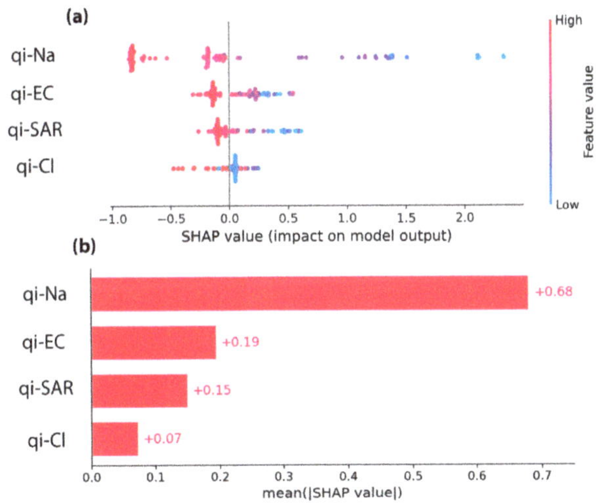

Figure 11. (**a**) Feature importance assessed by SHAP values and (**b**) predictor importance based on SHAP values for Model 15.

For Model 15, the feature, qi-EC, shows a marked contrast in its contributions as indicated by the SHAP values. The maximum SHAP value of 0.5372 reflects a strong positive contribution in certain instances, while the minimum value of −0.3136 shows that in some cases, this feature contributes to negative predictions. With a mean SHAP value of 0.0509, qi-EC has an overall marginally positive impact, suggesting that it generally contributes positively but with variable importance across different cases (Figure 11a).

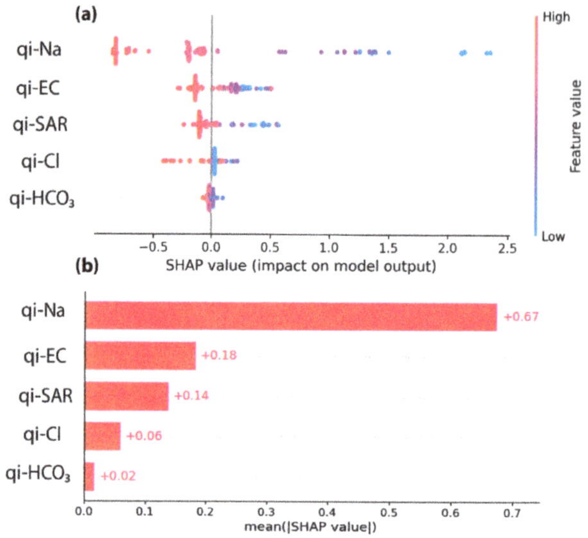

Figure 12. (**a**) Feature importance assessed by SHAP values and (**b**) predictor importance based on SHAP values for Model 16.

For qi-SAR, the maximum SHAP value is 0.6063, and the minimum is −0.2689, with an average of 0. 0180, indicating a generally non-significant impact on the model's predictions. In contrast, qi-Na exhibits a more varied influence on model predictions. The maximum SHAP value of 2.3342 reflects an extremely high positive contribution in certain instances, while the minimum of −0.8574 shows a negative impact in others (Figure 11a). The average SHAP value of −0.0882 suggests a slightly negative overall contribution, indicating that while qi-Na can have large positive contributions, it tends to detract from performance on average. The feature qi-Cl also shows nuanced impacts on the model's predictions. With a maximum SHAP value of 0.2430, qi-Cl provides a relatively positive contribution in some cases. However, its minimum SHAP value of −0.4811 suggests that it may hinder model performance in certain instances (Figure 11a). The average SHAP value for qi-Cl is 0.0193, indicating a slight positive impact overall. This suggests that qi-Cl generally contributes positively, albeit minimally, and its effect may be overshadowed by more influential features (Figure 11a).

The SHAP values for qi-Na exhibit a multivariate relationship with water quality predictions, generally showing a negative impact. The two most prominent negative values, −0.8441 and −0.8276 (Figure 11a), suggest that increasing sodium levels often decrease the likelihood of positive water quality outcomes. This strong negative effect underscores the importance of monitoring sodium levels, as higher concentrations may lead to adverse results in model predictions. However, certain instances show a positive SHAP value for qi-Na, such as 2.3342, indicating that, under specific conditions, high sodium levels can positively influence water quality predictions. Figures 11 and 12 illustrate the complexity of this relationship, highlighting the need for further research to better understand sodium's interactions with other variables. With an importance score of 0.68 (Figure 11b), qi-Na is the most influential predictor in this model, making sodium management a priority for stakeholders aiming to improve water quality outcomes. The SHAP values for qi-EC range from negative to positive, indicating varied effects on predictions depending on specific instances. For example, values like −0.7454 and −0.3136 reflect a significant negative contribution in some cases, while a positive value of 0.5352 shows a strong positive contribution in others. The maximum SHAP value for qi-EC is 0.5372, and the minimum is −0.3136, with an average of 0.0509, indicating an overall modest positive impact (Figure 11b). With an importance score of 0.19 (Figure 11b), qi-EC ranks as the second most significant feature after qi-Na. Although it contributes meaningfully to predictions, qi-EC is less influential than sodium. Nevertheless, understanding the dynamics of qi-EC may help refine model predictions and improve comprehensive water quality evaluations.

Regarding the SHAP values for qi-SAR, most values are negative, indicating that higher qi-SAR levels generally reduce the likelihood of favorable outcomes. Occasionally, however, positive SHAP values—such as 0.6063—suggest that qi-SAR can contribute positively in certain, albeit less frequent, cases (Figure 11a). Following this, qi-EC has an importance score of 0.19, indicating a moderate contribution, while qi-SAR has an importance of 0.15, signifying a slightly lower but still relevant impact. Finally, qi-Cl has the lowest importance score at 0.07, suggesting a minimal influence on the model's predictions. Together, these values highlight the varying degrees of impact each feature has on overall model performance, with qi-Cl variations unlikely to significantly affect predictions (Figure 11b).

Comparing Model 15 to Model 16, qi-Na remains the most influential predictor in both models, with importance scores of 0.68 and 0.67, respectively, making it crucial for accurate water quality predictions. The importance of qi-EC decreases slightly, from an importance score of 0.19 in Model 15 to 0.18 in Model 16, indicating a somewhat reduced role in Model 16. qi-SAR and qi-Cl maintain similar levels of importance, with scores of 15 and 0.07, respectively (Figure 12a). Notably, Model 16 introduces qi-HCO_3, albeit with a low importance score of 0.02 (Figure 12b), suggesting that while additional features were incorporated, they did not significantly affect model performance. Despite these minor shifts in feature importance, qi-Na consistently remains the central feature in both models.

The SHAP explanatory force plot is a powerful tool for understanding model outcomes, as it illustrates the role of specific elements in the prediction process. The base value represents a model output without feature impact, serving as a comparison point. Each feature's SHAP value reflects its contribution to moving the prediction away from this base value: positive SHAP values push the prediction higher, while negative values pull it lower. Typically, the plot uses color coding—blue for features contributing positively and red for features with a negative impact. The length of each bar corresponds to the size of the SHAP values, indicating which features have the greatest influence on prediction.

The force plot for Model 15 (Figure 13a) shows a model output value of 1.03 and visualizes how different features support this prediction. Starting from a base value of approximately 94.13, the contributions from individual features increase the value to 99.42, suggesting that the features contribute positively to the model's prediction by pushing it well above the base value. Some features show a significant contribution to this increase, underscoring their importance in the model's decision-making. The magnitude of the shift from base to output highlights the interaction between these features, emphasizing their critical role in explaining the model's behavior and the drivers behind its predictions.

The force plot for Model 16 (Figure 13b) displays a slightly higher output value of 1.06, beginning from the same base value of 94.1303 as in plot (Figure 13a). This shared base value allows for direct comparison of feature contributions between the two outputs. In this plot, the cumulative contributions from individual features result in a similarly elevated output, approaching a final value of 99.4231. The rise in output demonstrates that these features reinforce one another, adding up to produce a more confident prediction from the model. This plot highlights the strong features that enhance model performance while also illustrating how various contributions can subtly shift the overall prediction.

By comparing these two plots, it becomes evident that model predictions are dynamically influenced by feature interactions.

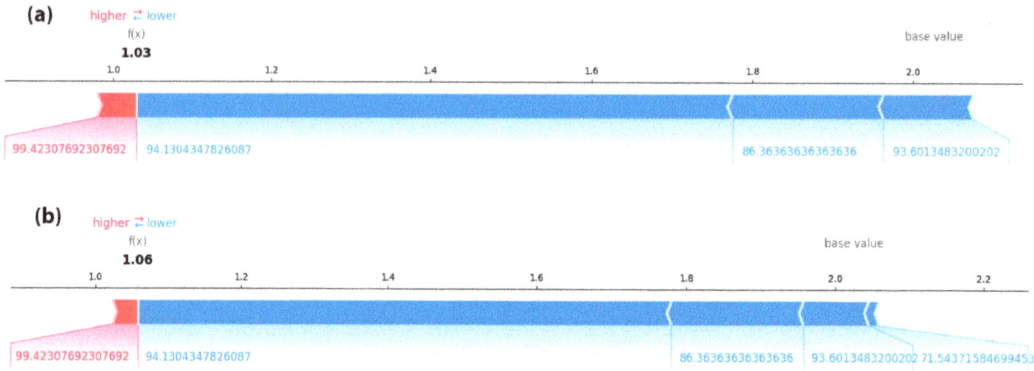

Figure 13. Force plot illustrating the contribution to the model's prediction for a specific observation in (**a**) Model 15 and (**b**) Model 16.

In Model M15 (Figure 13a), the base value is close to 1, with a normalization value of 0.3 representing the model's projected output before considering the effects of individual features. The distribution of SHAP values demonstrates how these features impact the prediction. For instance, a feature with a SHAP value of 99 adds significantly to the prediction, while features with SHAP values under 1 contribute more modestly. A SHAP value of 42 also positively influences the predicted outcome, and another feature with a SHAP value of 94.13 increases the forecast, albeit to a lesser degree than the feature with a value of 12. The overall prediction, calculated using logistic regression, yields f(x) = 1.03, achieved by adding the base value to the cumulative contributions of these features, which collectively produce a positive effect. Results from M15 and M16 show that the positive contributions of features to model predictions are substantial, while the base value remains relatively constant. In Figure 13b, Model M16 illustrates this further by extending the model's range with additional features, which influences the output based on input variations. Examining these contributions provides valuable insights into model performance, highlighting which features most significantly impact predictions.

4. Discussion

The evaluation of model performance across different architectures—CNN, CNN-LSTM, CNN-BiGRU, and CNN-BiLSTM—shows that the CNN-BiLSTM model performs best. This architecture achieved an accuracy of 0.94 for Model 15, using quality indices qi-EC, qi-SAR, qi-Na, and qi-Cl. Additionally, its micro precision, recall, and F1 scores all reached 0.94, outperforming CNN, CNN-LSTM, and CNN-BiGRU, which recorded accuracy values of 0.90 and 0.92, respectively. Further supporting CNN-BiLSTM's superior performance are the increases in macro precision to 0.925, macro recall to 0.908, and macro f1 score to 0.899, indicating a better balance across classes and improved handling of class imbalance. SHAP analysis underscores qi-Na as the most impactful predictor, highlighting the importance of sodium monitoring, as the SHAP values for qi Na tend to be strongly negative and play a crucial role in water quality prediction. Although qi EC shows moderate importance, other features, such as qi-SAR and qi-Cl, enhance the model's predictive capability by augmenting the contribution of significant attributes. Model 16, which includes qi-HCO_3, further strengthens the CNN-BiLSTM model's predictive accuracy without affecting its 0.94 accuracy score while increasing AUC to an impressive

0.994. This high AUC suggests an exceptional ability to distinguish between different water quality levels. The CNN-BiLSTM's strength lies in its capacity to capture both spatial and temporal patterns in the data, making it particularly effective for complex classification tasks. SHAP analysis reveals that, although qi-HCO_3 has a relatively low importance score, its contribution may enhance model robustness by capturing additional variability in water quality assessments. While CNN and CNN-LSTM also performed well, they did not achieve optimal results, particularly in AUC and macro metrics. Overall, the CNN-BiLSTM model stands out in water quality classification, especially for samples containing qi-HCO_3, and demonstrated strong predictability and resilience across diverse datasets. SHAP analysis further supports these findings, underscoring the prioritization of key features such as qi-Na in water quality management.

The comparative analysis of water quality classification methods across various studies reveals insightful trends in model performance and applicability (Table 7). In this study, the CNN-BiLSTM model applied to the Naama region of Algeria achieved an impressive accuracy of 0.94 and an AUC of 0.994, underscoring its robustness and ability to distinguish between water quality categories effectively. This performance is comparable to that of Lap et al. [76], who used the RF model in the Kim Hai system of Vietnam, also achieving an accuracy of 0.94, although AUC was not reported. In Adrar, Algeria, Derdour et al. [77] demonstrated high efficacy with an SVM model, obtaining a slightly higher accuracy of 94.2 and a perfect AUC of 1.0, indicative of flawless classification capabilities. Similarly, Hussein et al. [78] achieved notable results using SVM in Sindh, Pakistan, with an accuracy of 93.33 and an AUC of 0.98, despite working with a smaller sample size of 80. Nadiri et al. [79] reported an AUC of 0.99 for RF in Gulfepe-Zarinabad, Iran, based on just 28 samples, highlighting the model's potential to handle limited data effectively.

In contrast, Abu El-Magd et al. [80] used an SVM model in Ismalia, Egypt, achieving an accuracy of 90.6 and an AUC of 0.906, reflecting solid performance but slightly lower than other studies. Zheng et al. [81] applied XGboost to a landfill site in Yopurga, China, achieving an AUC of 0.90 with 28 samples, indicating a moderate yet reliable performance. The study by Ahn et al. [82] in South Korea, employing an artificial neural network (ANN) model, stands out with a high accuracy of 98.6, derived from a large sample size of 4774, showcasing the model's robustness in handling extensive datasets. In Northern Iran, Sahour et al. [83] and Mosavi et al. [84] in the Dezekord-Kamfiruz watershed, both using RF, reported accuracies of 0.92 and 0.86, respectively, with Sahour et al. [83] also achieving an AUC of 0.95, underscoring strong model reliability. Shamsuddin et al. [30] in Malaysia, using SVM with 40 samples, reported a high accuracy of 0.96, while Abbas et al. [29], using XGboost in Mirpurkhas, Pakistan, achieved an AUC of 0.96 with 422 samples, indicating effective model application across diverse regions. Lastly, Mohseni et al. [16] in India and Bedi et al. [85] in the Midcontinental United States achieved AUCs of 0.9048 and an accuracy of 92.0, respectively, using XGboost and ANN, demonstrating reliable performance on moderate-sized datasets.

In summary, while CNN-BiLSTM models achieve high accuracy and AUC in the Naama region, other models such as SVM, RF, and XGboost also demonstrate competitive performance across various geographical regions and sample sizes (Table 7). The choice of model often depends on dataset characteristics, computational resources, and the desired balance between accuracy and interpretability. ANN, SVM, and RF models consistently deliver strong results, making them versatile options for water quality classification tasks.

Table 7. Performance comparison of the CNN-BiLSTM model with other predictive models from the literature, based on data volume, accuracy, and AUC.

Authors	Methods	Case Study	Number of Samples	Accuracy	AUC
This study	CNN-BiLSTM Based-SHAP	Naama region (Algeria)	166	0.94	0.994
Lap et al. [76]	RF	Kim Hai system (Vietnam)	-	0.94	-
Hussein et al. [78]	SVM	Sindh, (Pakistan)	80	93.33	0.98
Nadiri et al. [79]	RF	Gulfepe-Zarinabad (Iran)	28	-	0.99
Abu El-Magd et al. [80]	SVM	Ismalia, (Egypt)	67	90.6	0.906
Zheng et al. [81]	XGboost	Yopurga landfill (China)	28	-	0.90
Sahour et al. [83]	RF	Northern Iran	248	0.92	0.95
Shamsuddin et al. [30]	SVM	Langat River Basin (Malaysia)	40	0.96	-
Abbas et al. [29]	XGboost	Mirpurkhas (Pakistan)	422	-	0.96
Mohseni et al. [16]	XGboost	Madhya Pradesh (India)	54	-	0.9048
Mosavi et al. [84]	RF	Dezekord-Kamfiruz watershed (Iran)	339	0.86	-
Bedi et al. [85]	ANN	Midcontinental United States	303	92.0	-

5. Conclusions and Future Recommendations

This study utilized CNN, CNN-LSTM, CNN-BiGRU, and CNN-BiLSTM models to classify the Irrigation Water Quality Indices (IWQIs) in the Naama region of northwestern Algeria. The results highlight that incorporating temporal dependencies significantly improves the performance of machine learning models in this context. While CNNs are highly effective for image analysis, they are limited in capturing temporal patterns, presenting challenges for dynamic datasets. Among the tested models, CNN-BiLSTM achieved the highest accuracy and AUC values, showcasing its outstanding predictive capabilities. However, this model also has high computational demands and requires substantial training data.

Specifically, in Model 16—which includes qi-EC, qi-SAR, qi-Na, qi-Cl, and qi-HCO_3—the CNN-BiLSTM model achieved an accuracy of 0.94, with micro precision, recall, and F1 scores of 0.94. The model also attained a macro precision of 0.925, a macro recall of 0.908, a macro F1 score of 0.899, and an AUC of 0.994. These results demonstrate that the CNN-BiLSTM model outperforms other models, establishing it as the most effective choice for this specific classification task. The SHAP analysis further emphasizes the importance of key features, particularly qi-Na, in the model's predictive performance. Future studies should focus on improving the efficiency of these models to achieve high accuracy with reduced computational cost. Expanding this study to include a broader sampling area across different zones and regions of the country could help generalize the findings. Additionally, exploring hybrid approaches that leverage the strengths of various architectures and incorporating techniques like attention mechanisms could enhance model performance. Collaborating with domain experts to integrate specialized knowledge into the models may also improve predictive accuracy. Together, these advancements could lead to better practices in water quality management, ultimately supporting agricultural productivity and sustainability.

Author Contributions: E.E.H.: Contributed to the conception and design of the study and conducted data analysis. B.Z.: Led data acquisition and contributed to the interpretation of results. N.B.: Assisted in methodology design and provided critical input on data analysis. A.D.: Conducted literature review and contributed to data acquisition. S.S.M.G.: Analyzed statistical data and contributed to the discussion section. C.A.G.S.: Provided subject matter expertise and contributed to the study design. M.A.H.: Supervised the research project and contributed to the interpretation of findings. All authors have read and agreed to the published version of the manuscript.

Funding: This research was funded by Taif University, Taif, Saudi Arabia (TU-DSPP-2024-14).

Data Availability Statement: The data that support the findings of this study are available on request from the corresponding author, [B.Z.].

Acknowledgments: The authors gratefully acknowledge Taif University, Saudi Arabia, for supporting this work under project number TU-DSPP-2024-14.

Conflicts of Interest: The authors declare no conflicts of interest.

References

1. Khouri, L.; Sallom, A. The Impact of Spatial and Temporal Shifts on Orontes River Water Quality Parameters. *DYSONA Appl. Sci.* **2023**, *4*, 35–41.
2. Zerouali, B.; Bailek, N.; Islam, A.R.M.T.; Katipoğlu, O.M.; Ayek, A.A.E.; Santos, C.A.G.; Rajput, J.; Wong, Y.J.; Abda, Z.; Chettih, M.; et al. Enhancing Groundwater Potential Zone Mapping with a Hybrid Analytical Method: The Case of Semiarid Basin. *Groundw. Sustain. Dev.* **2024**, *26*, 101261. [CrossRef]
3. World Health Organization. *Protecting Surface Water for Health: Identifying, Assessing and Managing Drinking-Water Quality Risks in Surface-Water Catchments*; World Health Organization: Geneva, Switzerland, 2016.
4. Döndü, M.; Özdemir, N.; Demirak, A.; Doğan, H.M.; Dincer, N.G.; Keskin, F. Seasonal Assessment of the Impact of Fresh Waters Feeding the Bay of Gökova with Water Quality Index (WQI) and Comprehensive Pollution Index (CPI). *Environ. Forensics* **2024**, *25*, 68–80. [CrossRef]
5. Aqel, H.; Sannan, N.; Al-Hunaiti, A.; Fodah, R. Integrated Water Quality Dynamics in Wadi Hanifah: Physical, Chemical, and Biological Perspectives. *PLoS ONE* **2024**, *19*, e0298200. [CrossRef]
6. Derdour, A.; Belam, N.; Chebab, W. Traditional Irrigation System and Methods of Water Harvesting in the Oasis of Sfissifa Ksour Mountains-Algeria. *LARHYSS J.* **2022**, *23*, 17–35.
7. de Almeida, M.B.G.; de Jesus, A.M.D.; Pereira, A.S.; Fiore, F.A. Evaluating Centrifuged Water Treatment Plant Sludge as an Adsorbent for Nutrients, Microorganisms, and Heavy Metals Removal from Wastewater. *J. Clean. Prod.* **2024**, *468*, 142975. [CrossRef]
8. Maqsoom, A.; Aslam, B.; Alwetaishi, M.; Awais, M.; Hassan, U.; Maqsoom, S.; Alaloul, W.S.; Musarat, M.A.; Zerouali, B.; Hussein, E.E. A Gis-Based Groundwater Contamination Assessment Using Modified Drastic Geospatial Technique. *Water* **2021**, *13*, 2868. [CrossRef]
9. Aldris, B.; Farhoud, N. Wastewater Treatment Efficiency of an Experimental MBBR System under Different Influent Concentrations. *DYSONA Appl. Sci.* **2020**, *1*, 20–28.
10. Chaïeb, A.; Arabi, M.; Gamagami, O.; Benyoussef, S.; Aknaf, A.; Moumen, A. Assessing the Environmental Impacts of Treated Wastewater Reuse on Water-Soil-Plant Ecosystems in Oued Bou Naim, Eastern Morocco. *Ecol. Eng. Environ. Technol.* **2024**, *25*, 208–219. [CrossRef]
11. Sahoo, S.K.; Sahu, A.K.; Baitharu, I. Environmental and Health Risks Assessment of Chemical Pollutants in Drinking Water and Wastewaters. In *Emerging Innovative Trends in the Application of Biological Processes for Industrial Wastewater Treatment*; Elsevier: Amsterdam, The Netherlands, 2024; pp. 291–311.
12. Mladenović-Ranisavljević, I.I.; Žerajić, S.A. Comparison of Different Models of Water Quality Index in the Assessment of Surface Water Quality. *Int. J. Environ. Sci. Technol.* **2018**, *15*, 665–674. [CrossRef]
13. Uddin, M.G.; Nash, S.; Olbert, A.I. A Review of Water Quality Index Models and Their Use for Assessing Surface Water Quality. *Ecol. Indic.* **2021**, *122*, 107218. [CrossRef]
14. Kachroud, M.; Trolard, F.; Kefi, M.; Jebari, S.; Bourrié, G. Water Quality Indices: Challenges and Application Limits in the Literature. *Water* **2019**, *11*, 361. [CrossRef]
15. Kumar, D.; Kumar, R.; Sharma, M.; Awasthi, A.; Kumar, M. Global Water Quality Indices: Development, Implications, and Limitations. *Total Environ. Adv.* **2024**, *9*, 200095. [CrossRef]

16. Mohseni, U.; Pande, C.B.; Pal, S.C.; Alshehri, F. Prediction of Weighted Arithmetic Water Quality Index for Urban Water Quality Using Ensemble Machine Learning Model. *Chemosphere* **2024**, *352*, 141393. [CrossRef]
17. Han, X.; Tang, F.; Liu, A.-L. Drinking Water Quality Evaluation in Supply Systems in Wuhan, China: Application of Entropy Weight Water Quality Index and Multivariate Statistical Analysis. *Environ. Sci. Pollut. Res.* **2024**, *31*, 280–292. [CrossRef]
18. Mishra, A.P.; Singh, S.; Jani, M.; Singh, K.A.; Pande, C.B.; Varade, A.M. Assessment of Water Quality Index Using Analytic Hierarchy Process (AHP) and GIS: A Case Study of a Struggling Asan River. *Int. J. Environ. Anal. Chem.* **2024**, *104*, 1159–1171. [CrossRef]
19. Kushwaha, N.L.; Kudnar, N.S.; Vishwakarma, D.K.; Subeesh, A.; Jatav, M.S.; Gaddikeri, V.; Ahmed, A.A.; Abdelaty, I. Stacked Hybridization to Enhance the Performance of Artificial Neural Networks (ANN) for Prediction of Water Quality Index in the Bagh River Basin, India. *Heliyon* **2024**, *10*, e31085. [CrossRef]
20. Yang, S.; Zhong, S.; Chen, K. W-WaveNet: A Multi-Site Water Quality Prediction Model Incorporating Adaptive Graph Convolution and CNN-LSTM. *PLoS ONE* **2024**, *19*, e0276155. [CrossRef]
21. Zamani, M.G.; Nikoo, M.R.; Al-Rawas, G.; Nazari, R.; Rastad, D.; Gandomi, A.H. Hybrid WT–CNN–GRU-Based Model for the Estimation of Reservoir Water Quality Variables Considering Spatio-Temporal Features. *J. Environ. Manag.* **2024**, *358*, 120756. [CrossRef]
22. Anand, M.V.; Sohitha, C.; Saraswathi, G.N.; Lavanya, G.V. Water Quality Prediction Using CNN. *J. Phys. Conf. Ser.* **2023**, *2484*, 12051. [CrossRef]
23. Tian, Q.; Luo, W.; Guo, L. Water Quality Prediction in the Yellow River Source Area Based on the DeepTCN-GRU Model. *J. Water Process Eng.* **2024**, *59*, 105052. [CrossRef]
24. Wang, Z.; Wang, Q.; Wu, T. A Novel Hybrid Model for Water Quality Prediction Based on VMD and IGOA Optimized for LSTM. *Front. Environ. Sci. Eng.* **2023**, *17*, 88. [CrossRef]
25. Hussein, E.E.; Derdour, A.; Zerouali, B.; Almaliki, A.; Wong, Y.J.; Ballesta-de los Santos, M.; Minh Ngoc, P.; Hashim, M.A.; Elbeltagi, A. Groundwater Quality Assessment and Irrigation Water Quality Index Prediction Using Machine Learning Algorithms. *Water* **2024**, *16*, 264. [CrossRef]
26. Abda, Z.; Zerouali, B.; Chettih, M.; Guimarães Santos, C.A.; de Farias, C.A.S.; Elbeltagi, A. Assessing Machine Learning Models for Streamflow Estimation: A Case Study in Oued Sebaou Watershed (Northern Algeria). *Hydrol. Sci. J.* **2022**, *67*, 1328–1341. [CrossRef]
27. Ren, J.; Cui, J.; Dong, W.; Xiao, Y.; Xu, M.; Liu, S.; Wan, J.; Li, Z.; Zhang, J. Remote Sensing Inversion of Typical Offshore Water Quality Parameter Concentration Based on Improved SVR Algorithm. *Remote Sens.* **2023**, *15*, 2104. [CrossRef]
28. Hamada, M.S.; Zaqoot, H.A.; Sethar, W.A. Using a Supervised Machine Learning Approach to Predict Water Quality at the Gaza Wastewater Treatment Plant. *Environ. Sci. Adv.* **2024**, *3*, 132–144. [CrossRef]
29. Abbas, F.; Cai, Z.; Shoaib, M.; Iqbal, J.; Ismail, M.; Alrefaei, A.F.; Albeshr, M.F. Machine Learning Models for Water Quality Prediction: A Comprehensive Analysis and Uncertainty Assessment in Mirpurkhas, Sindh, Pakistan. *Water* **2024**, *16*, 941. [CrossRef]
30. Shamsuddin, I.I.S.; Othman, Z.; Sani, N.S. Water Quality Index Classification Based on Machine Learning: A Case from the Langat River Basin Model. *Water* **2022**, *14*, 2939. [CrossRef]
31. Kirui, J. Machine Learning Models for Drinking Water Quality Classification. In Proceedings of the 2024 International Conference on Control, Automation and Diagnosis (ICCAD), Paris, France, 15–17 May 2024; p. 1.
32. Ravindra, N.S.; Munnu, P.N.; Bhagwat, P.G.; Sankalp, B.; Yogiraj, A. Review of Machine Learning and Flask for Water Quality Classification. *Int. J. Res. Publ. Rev.* **2024**, *5*, 3032–3037. [CrossRef]
33. Tanega, M.C.; Fajardo, A.C.; Limbago, J.S.; Pabalate, F.M. Performance Analysis of Machine Learning Classification Model and Ensemble Learning for Water Quality Index of Taal Lake. In Proceedings of the 2023 International Workshop on Artificial Intelligence and Image Processing (IWAIIP), Yogyakarta, Indonesia, 1–2 December 2023; pp. 10–15.
34. Uddin, M.G.; Nash, S.; Rahman, A.; Olbert, A.I. Performance Analysis of the Water Quality Index Model for Predicting Water State Using Machine Learning Techniques. *Process Saf. Environ. Prot.* **2023**, *169*, 808–828. [CrossRef]
35. Fernández del Castillo, A.; Yebra-Montes, C.; Verduzco Garibay, M.; de Anda, J.; Garcia-Gonzalez, A.; Gradilla-Hernández, M.S. Simple Prediction of an Ecosystem-Specific Water Quality Index and the Water Quality Classification of a Highly Polluted River through Supervised Machine Learning. *Water* **2022**, *14*, 1235. [CrossRef]
36. Yurtsever, M.; Emeç, M. Potable Water Quality Prediction Using Artificial Intelligence and Machine Learning Algorithms for Better Sustainability. *Ege Acad. Rev.* **2023**, *23*, 265–278. [CrossRef]
37. Nair, J.P.; Vijaya, M.S. River Water Quality Prediction and Index Classification Using Machine Learning. *In J. Phys. Conf. Ser.* **2022**, *2325*, 12011. [CrossRef]
38. Shams, M.Y.; Elshewey, A.M.; El-Kenawy, E.-S.M.; Ibrahim, A.; Talaat, F.M.; Tarek, Z. Water Quality Prediction Using Machine Learning Models Based on Grid Search Method. *Multimed. Tools Appl.* **2024**, *83*, 35307–35334. [CrossRef]

39. Dilmi, S. A Combined Water Quality Classification Model Based on Kernel Principal Component Analysis and Machine Learning Techniques. *Desalin. Water Treat.* **2022**, *279*, 61–67. [CrossRef]
40. Zerouali, B.; Bailek, N.; Tariq, A.; Kuriqi, A.; Guermoui, M.; Alharbi, A.H.; Khafaga, D.S.; El-Kenawy, E.-S.M. Enhancing Deep Learning-Based Slope Stability Classification Using a Novel Metaheuristic Optimization Algorithm for Feature Selection. *Sci. Rep.* **2024**, *14*, 21812. [CrossRef]
41. Belletreche, M.; Bailek, N.; Abotaleb, M.; Bouchouicha, K.; Zerouali, B.; Guermoui, M.; Kuriqi, A.; Alharbi, A.H.; Khafaga, D.S.; El-Shimy, M. Hybrid Attention-Based Deep Neural Networks for Short-Term Wind Power Forecasting Using Meteorological Data in Desert Regions. *Sci. Rep.* **2024**, *14*, 21842. [CrossRef]
42. VM, S.P.; Subramanian, P.; Soniya, M.; Pugalenthi, R. Water Quality Index Prediction and Classification Using Hyperparameter Tuned Deep Learning Approach. *Glob. NEST J.* **2024**, *26*, 05821.
43. Abdessamed, D.; Jodar-Abellan, A.; Ghoneim, S.S.M.; Almaliki, A.; Hussein, E.E.; Pardo, M.Á. Groundwater Quality Assessment for Sustainable Human Consumption in Arid Areas Based on GIS and Water Quality Index in the Watershed of Ain Sefra (SW of Algeria). *Environ. Earth Sci.* **2023**, *82*, 510. [CrossRef]
44. Bouarfa, S.; Derdour, A.; Okkacha, Y.; Almaliki, A.H.; Jodar-Abellan, A.; Hussein, E.E. Sedimentological Investigation of the Potential Origin and Provenance of Sand Deposits in an Arid Area: A Case Study of the Ksour Mountains Region in Algeria. *Arab. J. Geosci.* **2022**, *15*, 1460. [CrossRef]
45. Derdour, A.; Hitache, R.; Mebarki, A. Assessment of Water Resources for Livestock in the Wilaya of Naama, Southwest Algeria: A GIS-Based Approach. *Environ. Monit. Assess.* **2023**, *195*, 1419. [CrossRef] [PubMed]
46. Guo, Y.; Zhang, Z.; Yao, B.; Chai, J.; Zhang, S.; Liu, J.; Zhao, Z.; Xue, C. Fabrication and Performance of a Ta2O5 Thin Film PH Sensor Manufactured Using MEMS Processes. *Sensors* **2023**, *23*, 6061. [CrossRef] [PubMed]
47. De Giglio, O.; Diella, G.; Lopuzzo, M.; Triggiano, F.; Calia, C.; Pousis, C.; Fasano, F.; Calabrese, G.; Rafaschieri, V.; Carpagnano, L.F. Management of Microbiological Contamination of the Water Network of a Newly Built Hospital Pavilion. *Pathogens* **2021**, *10*, 75. [CrossRef]
48. Meireles, A.C.M.; de Andrade, E.M.; Chaves, L.C.G.; Frischkorn, H.; Crisostomo, L.A. A New Proposal of the Classification of Irrigation Water. *Rev. Ciência Agronômica* **2010**, *41*, 349–357. [CrossRef]
49. Zhao, B.; Lu, H.; Chen, S.; Liu, J.; Wu, D. Convolutional Neural Networks for Time Series Classification. *J. Syst. Eng. Electron.* **2017**, *28*, 162–169. [CrossRef]
50. Müller, P.N.; Müller, A.J.; Achenbach, P.; Göbel, S. Imu-Based Fitness Activity Recognition Using Cnns for Time Series Classification. *Sensors* **2024**, *24*, 742. [CrossRef]
51. Sayadi, H.; Gao, Y.; Mohammadi Makrani, H.; Lin, J.; Costa, P.C.; Rafatirad, S.; Homayoun, H. Towards Accurate Run-Time Hardware-Assisted Stealthy Malware Detection: A Lightweight, yet Effective Time Series CNN-Based Approach. *Cryptography* **2021**, *5*, 28. [CrossRef]
52. Tovar, M.; Robles, M.; Rashid, F. PV Power Prediction, Using CNN-LSTM Hybrid Neural Network Model. Case of Study: Temixco-Morelos, México. *Energies* **2020**, *13*, 6512. [CrossRef]
53. Liu, T.; Bao, J.; Wang, J.; Zhang, Y. A Hybrid CNN–LSTM Algorithm for Online Defect Recognition of CO2 Welding. *Sensors* **2018**, *18*, 4369. [CrossRef]
54. Cui, F.; Wang, R.; Ding, W.; Chen, Y.; Huang, L. A Novel DE-CNN-BiLSTM Multi-Fusion Model for EEG Emotion Recognition. *Mathematics* **2022**, *10*, 582. [CrossRef]
55. Staffini, A. A CNN–BiLSTM Architecture for Macroeconomic Time Series Forecasting. *Eng. Proc.* **2023**, *39*, 33. [CrossRef]
56. Rhanoui, M.; Mikram, M.; Yousfi, S.; Barzali, S. A CNN-BiLSTM Model for Document-Level Sentiment Analysis. *Mach. Learn. Knowl. Extr.* **2019**, *1*, 832–847. [CrossRef]
57. Nie, L.; Zhang, L.; Xu, S.; Cai, W.; Yang, H. Remaining Useful Life Prediction of Milling Cutters Based on CNN-BiLSTM and Attention Mechanism. *Symmetry* **2022**, *14*, 2243. [CrossRef]
58. Zhang, J.; Ye, L.; Lai, Y. Stock Price Prediction Using CNN-BiLSTM-Attention Model. *Mathematics* **2023**, *11*, 1985. [CrossRef]
59. Tang, C.; Zhang, D.; Tian, Q. Convolutional Neural Network–Bidirectional Gated Recurrent Unit Facial Expression Recognition Method Fused with Attention Mechanism. *Appl. Sci.* **2023**, *13*, 12418. [CrossRef]
60. Mohsen, S. Recognition of Human Activity Using GRU Deep Learning Algorithm. *Multimed. Tools Appl.* **2023**, *82*, 47733–47749. [CrossRef]
61. Abumohsen, M.; Owda, A.Y.; Owda, M. Electrical Load Forecasting Using LSTM, GRU, and RNN Algorithms. *Energies* **2023**, *16*, 2283. [CrossRef]
62. Liu, J.; Yang, Y.; Lv, S.; Wang, J.; Chen, H. Attention-Based BiGRU-CNN for Chinese Question Classification. *J. Ambient Intell. Humaniz. Comput.* **2019**, 1–12. [CrossRef]
63. Bao, T.; Ren, N.; Luo, R.; Wang, B.; Shen, G.; Guo, T. A Bert-Based Hybrid Short Text Classification Model Incorporating Cnn and Attention-Based Bigru. *J. Organ. End User Comput.* **2021**, *33*, 1–21. [CrossRef]

64. Zerouali, B.; Santos, C.A.G.; de Farias, C.A.S.; Muniz, R.S.; Difi, S.; Abda, Z.; Chettih, M.; Heddam, S.; Anwar, S.A.; Elbeltagi, A. Artificial Intelligent Systems Optimized by Metaheuristic Algorithms and Teleconnection Indices for Rainfall Modeling: The Case of a Humid Region in the Mediterranean Basin. *Heliyon* **2023**, *9*, e15355. [CrossRef]
65. Elbeltagi, A.; Seifi, A.; Ehteram, M.; Zerouali, B.; Vishwakarma, D.K.; Pandey, K. GLUE Analysis of Meteorological-Based Crop Coefficient Predictions to Derive the Explicit Equation. *Neural Comput. Appl.* **2023**, *35*, 14799–14824. [CrossRef]
66. Anjum, M.; Khan, K.; Ahmad, W.; Ahmad, A.; Amin, M.N.; Nafees, A. New SHapley Additive ExPlanations (SHAP) Approach to Evaluate the Raw Materials Interactions of Steel-Fiber-Reinforced Concrete. *Materials* **2022**, *15*, 6261. [CrossRef] [PubMed]
67. Dong, S.; Khattak, A.; Ullah, I.; Zhou, J.; Hussain, A. Predicting and Analyzing Road Traffic Injury Severity Using Boosting-Based Ensemble Learning Models with SHAPley Additive ExPlanations. *Int. J. Environ. Res. Public Health* **2022**, *19*, 2925. [CrossRef] [PubMed]
68. Ekanayake, I.U.; Meddage, D.P.P.; Rathnayake, U. A Novel Approach to Explain the Black-Box Nature of Machine Learning in Compressive Strength Predictions of Concrete Using Shapley Additive Explanations (SHAP). *Case Stud. Constr. Mater.* **2022**, *16*, e01059. [CrossRef]
69. Mangalathu, S.; Hwang, S.-H.; Jeon, J.-S. Failure Mode and Effects Analysis of RC Members Based on Machine-Learning-Based SHapley Additive ExPlanations (SHAP) Approach. *Eng. Struct.* **2020**, *219*, 110927. [CrossRef]
70. Ibrahim, B.; Ewusi, A.; Ahenkorah, I. Assessing the Suitability of Boosting Machine-Learning Algorithms for Classifying Arsenic-Contaminated Waters: A Novel Model-Explainable Approach Using Shapley Additive Explanations. *Water* **2022**, *14*, 3509. [CrossRef]
71. Suhaimi, N.S.; Othman, Z.; Yaakub, M.R. Comparative Analysis between Macro and Micro-Accuracy in Imbalance Dataset for Movie Review Classification. In Proceedings of the Seventh International Congress on Information and Communication Technology: ICICT 2022, London, UK, 21–24 February 2022; Springer: Berlin/Heidelberg, Germany; Volume 3, p. 83.
72. Opitz, J.; Burst, S. Macro F1 and Macro F1. *arXiv* **2019**, arXiv:1911.03347.
73. Takahashi, K.; Yamamoto, K.; Kuchiba, A.; Koyama, T. Confidence Interval for Micro-Averaged F 1 and Macro-Averaged F 1 Scores. *Appl. Intell.* **2022**, *52*, 4961–4972. [CrossRef]
74. El-kenawy, E.-S.M.; Albalawi, F.; Ward, S.A.; Ghoneim, S.S.M.; Eid, M.M.; Abdelhamid, A.A.; Bailek, N.; Ibrahim, A. Feature Selection and Classification of Transformer Faults Based on Novel Meta-Heuristic Algorithm. *Mathematics* **2022**, *10*, 3144. [CrossRef]
75. Nasir, N.; Kansal, A.; Alshaltone, O.; Barneih, F.; Sameer, M.; Shanableh, A.; Al-Shamma'a, A. Water Quality Classification Using Machine Learning Algorithms. *J. Water Process Eng* **2022**, *48*, 102920. [CrossRef]
76. Lap, B.Q.; Du Nguyen, H.; Hang, P.T.; Phi, N.Q.; Hoang, V.T.; Linh, P.G.; Hang, B.T.T. Predicting Water Quality Index (WQI) by Feature Selection and Machine Learning: A Case Study of An Kim Hai Irrigation System. *Ecol. Inform.* **2023**, *74*, 101991. [CrossRef]
77. Derdour, A.; Abdo, H.G.; Almohamad, H.; Alodah, A.; Al Dughairi, A.A.; Ghoneim, S.S.M.; Ali, E. Prediction of Groundwater Quality Index Using Classification Techniques in Arid Environments. *Sustainability* **2023**, *15*, 9687. [CrossRef]
78. Hussein, E.E.; Jat Baloch, M.Y.; Nigar, A.; Abualkhair, H.F.; Aldawood, F.K.; Tageldin, E. Machine Learning Algorithms for Predicting the Water Quality Index. *Water* **2023**, *15*, 3540. [CrossRef]
79. Nadiri, A.A.; Barzegar, R.; Sadeghfam, S.; Rostami, A.A. Developing a Data-Fused Water Quality Index Based on Artificial Intelligence Models to Mitigate Conflicts between GQI and GWQI. *Water* **2022**, *14*, 3185. [CrossRef]
80. Abu El-Magd, S.A.; Ismael, I.S.; El-Sabri, M.A.S.; Abdo, M.S.; Farhat, H.I. Integrated Machine Learning–Based Model and WQI for Groundwater Quality Assessment: ML, Geospatial, and Hydro-Index Approaches. *Environ. Sci. Pollut. Res.* **2023**, *30*, 53862–53875. [CrossRef]
81. Zheng, H.; Hou, S.; Liu, J.; Xiong, Y.; Wang, Y. Advanced Machine Learning and Water Quality Index (WQI) Assessment: Evaluating Groundwater Quality at the Yopurga Landfill. *Water* **2024**, *16*, 1666. [CrossRef]
82. Ahn, S.H.; Jeong, D.H.; Kim, M.; Lee, T.K.; Kim, H.-K. Prediction of Groundwater Quality Index to Assess Suitability for Drinking Purpose Using Averaged Neural Network and Geospatial Analysis. *Ecotoxicol. Environ. Saf.* **2023**, *265*, 115485. [CrossRef]
83. Sahour, S.; Khanbeyki, M.; Gholami, V.; Sahour, H.; Kahvazade, I.; Karimi, H. Evaluation of Machine Learning Algorithms for Groundwater Quality Modeling. *Environ. Sci. Pollut. Res.* **2023**, *30*, 46004–46021. [CrossRef]
84. Mosavi, A.; Sajedi Hosseini, F.; Choubin, B.; Goodarzi, M.; Dineva, A.A.; Rafiei Sardooi, E. Ensemble Boosting and Bagging Based Machine Learning Models for Groundwater Potential Prediction. *Water Resour. Manag.* **2021**, *35*, 23–37. [CrossRef]
85. Bedi, S.; Samal, A.; Ray, C.; Snow, D. Comparative Evaluation of Machine Learning Models for Groundwater Quality Assessment. *Environ. Monit. Assess.* **2020**, *192*, 776. [CrossRef]

water

MDPI

Article

Integrated Geospatial and Analytical Hierarchy Process Approach for Assessing Sustainable Management of Groundwater Recharge Potential in Barind Tract

Md. Zahed Hossain [1,2], Sajal Kumar Adhikary [2], Hrithik Nath [2,3], Abdulla Al Kafy [4], Hamad Ahmed Altuwaijri [5] and Muhammad Tauhidur Rahman [6,*]

1 Department of Civil Engineering, Bangladesh University of Business and Technology (BUBT), Dhaka 1216, Bangladesh; zahedhossain@bubt.edu.bd
2 Department of Civil Engineering, Khulna University Engineering & Technology (KUET), Khulna 9203, Bangladesh; sajal@ce.kuet.ac.bd (S.K.A.); hrithik@uctc.edu.bd (H.N.)
3 Department of Civil Engineering, University of Creative Technology Chittagong (UCTC), Chattogram 4212, Bangladesh
4 Department of Geography & The Environment, The University of Texas at Austin, 1 University Station A3100, Austin, TX 78712, USA; abdullaalkafy@utexas.edu
5 Department of Geography, College of Humanities and Social Sciences, King Saud University, Riyadh 11451, Saudi Arabia; haaltuwaijri@ksu.edu.sa
6 Geospatial Information Sciences Program, School of Economic, Political and Policy Sciences, University of Texas at Dallas, 800 Campbell Road, Richardson, TX 75080, USA
* Correspondence: mtr@utdallas.edu

Abstract: Groundwater depletion in Bangladesh's Barind tract poses significant challenges for sustainable water management. This study aims to delineate groundwater recharge potential zones in this region using an integrated geospatial and Analytical Hierarchy Process (AHP) approach. The methodology combines remote-sensing data with GIS analysis, considering seven factors influencing groundwater recharge: rainfall, soil type, geology, slope, lineament density, land use/land cover, and drainage density. The AHP method was employed to assess the variability of groundwater recharge potential within the 7586 km^2 study area. Thematic maps of relevant factors were processed using ArcGIS software. Results indicate that 9.23% (700.22 km^2), 47.68% (3617.13 km^2), 37.12% (2816.13 km^2), and 5.97% (452.70 km^2) of the study area exhibit poor, moderate, good, and very good recharge potential, respectively. The annual recharge volume is estimated at 2554×10^6 m^3/year, constituting 22.7% of the total precipitation volume ($11{,}227 \times 10^6$ m^3/year). Analysis of individual factors revealed that geology has the highest influence (33.57%) on recharge potential, followed by land use/land cover (17.74%), soil type (17.25%), and rainfall (12.25%). The consistency ratio of the pairwise comparison matrix was 0.0904, indicating acceptable reliability of the AHP results. The spatial distribution of recharge zones shows a concentration of poor recharge potential in areas with low rainfall (1200–1400 mm/year) and high slope (6–40%). Conversely, very good recharge potential is associated with high rainfall zones (1800–2200 mm/year) and areas with favorable geology (sedimentary deposits). This study provides a quantitative framework for assessing groundwater recharge potential in the Barind tract. The resulting maps and data offer valuable insights for policymakers and water resource managers to develop targeted groundwater management strategies. These findings have significant implications for sustainable water resource management in the region, particularly in addressing challenges related to agricultural water demand and climate change adaptation.

Keywords: groundwater recharge; sustainable development; analytical hierarchical process; GIS; Barind tract

1. Introduction

The global need for freshwater has increased about six times over the past hundred years [1]. Population growth, urbanization, agriculture, economy, industry, etc., are some of the major causes of increasing water demand. The two primary sources of freshwater are surface water and groundwater (GW). Growth in the global population causes unplanned urbanization, declining surface- and GW quality, and rising GW demand [2]. According to the World Bank, global GW use was 261,534 m^3 in the fiscal year 2020, compared to 299,054 m^3 in the fiscal year 2019 [3]. GW is a vital resource, essential for drinking, agriculture, and industry, especially in (semi-)arid zones [4–6]. In addition to being an important source of fresh water, GW also has a positive impact on social development, economic prosperity, poverty reduction, food security, and gender equality. Globally, green space is being rapidly consumed by urban, industrial, and agricultural activities; this emphasizes the need for comprehensive GW management models that consider quantity and quality [7,8]. Freshwater is in short supply due to unchecked pollution and increased industrial activity, especially during the dry season. As a result, GW is a necessary substitute for a variety of purposes [9]. Due to surface water shortages, which are a major concern for industry and agriculture in South Asia, GW is essential [10,11].

A recent study on water demand estimates that GW in Bangladesh provides about 79% of the water needed for homes, farms, irrigation, industry, etc. [12]. Bangladesh is primarily dependent on GW for its industrial, agricultural, municipal, and residential water needs. The majority of developing nations worldwide rely heavily on GW; in Bangladesh and other developing nations, GW declination is particularly high. GW is declining at a rate of 0.1 to 0.5 m^3/year. [13]. For instance, in the city of Dhaka, Bangladesh, the Dhaka Water Supply and Sewerage Authority (DWASA) provides piped water; presently, 68% of Dhaka's water supply comes from GW [14]. In 2016–2017, the total cropping area, mainly reliant on GW, was 1.01 million acres [12]. Key crops include rice, wheat, and potatoes, with Boro rice irrigated during the dry season from January to June. Boro rice requires a lot of water to cultivate, and since it is grown during the dry season, GW supplies the majority of the water needed. Because of this excessive GW exploration, GW levels declined during the dry season at a worrying rate. In Bangladesh's northwest, in Bangladesh's Barind Tract (BT), the issues of GW declination and water scarcity are particularly serious [15]. One of Bangladesh's best areas for rice production is the BT. This area experiences droughts on a regular basis. Prior to starting the Barind Integrated Area Development Project Phase-I, 13% of the district's land was cultivable, and 117% of Naogaon, Chapai Nawabganj, and Rajshahi were under cultivation. [16]. Which is now 226% [17] and 93.52% [18]. The percentage of irrigation by surface- and GW was 9.40% and 90.60%, respectively, in this region [18]. Low infiltration due to less rainfall, thick top clay, reduced river water flow, and the over-exploration of GW results in continuous depletion of GW levels in the BT. To overcome the aforementioned problems, the development of a sustainable GW management framework is essential. The framework requires a policy to balance the withdrawal and recharge mechanisms of GW, which can be achieved by identifying GW potential zones. The current study is, therefore, an attempt to fill this gap by developing a framework for potential GW zoning in the BT of Bangladesh towards sustainable GW development and management.

Among the factors GW recharge potential (RP) depends on are stream order, slope, soil type (ST), soil texture, drainage density (DD), soil moisture, soil permeability, soil depth, tectonic fault locations, soil consistency, rainfall, stream power index (SPI), land use/land cover (LULC), etc., and the interaction among these factors [19–27]. Thematic maps of these factors are required to analyze and derive potential GW zones for a study area. In the past, spatial analysis using thematic maps was very time-consuming because of a lack of resources, advanced software facilities, and high-speed computers. Moreover, previous studies were mostly based on numerical computation and historical analysis. Currently, with the availability of advanced software facilities and high-speed computers, GW potential zones are delineated using remote sensing (RS) and geographical information system (GIS) [25–27]. The use of modern geospatial techniques like RS and GIS methods has advantages over traditional methods in regard to time, cost, labor, and efficiency.

Multi-Criteria Decision Analyses (MCDAs) have seen a tremendous amount of use over the last few decades. By developing new methods or improving the existing Multi-criteria Decision Method (MCDM), MCDA is contributing to several sectors of research or other types of study. Some of the most used MCDMs are Analytic Hierarchy Process (AHP), Multi-Attribute Utility Theory, multi-influencing factor approach, Case-based Reasoning, Fuzzy Set Theory, Weighted gray model, Simple Multi-Attribute Rating Technique, deep-learning model, ELECTRE, Data Envelopment Analysis, PROMETHEE, Goal Programming, and Simple Additive Weighting [26,28–33]. Among all of these techniques, GW RP analysis using AHP is very popular in the present day. But other methods are also used.

The AHP and multi-influencing factor (MIF) approaches have been widely used by researchers, respectively, for quantitative and qualitative analysis [26,28–31]. The AHP organizes perceptions and judgments into a framework that aids decision-making by considering multiple factors. The characteristics of AHP are quite different from other decision theories. The use of pair-wise comparisons, which are used both to compare the alternatives with regard to the various criteria and to estimate the weights of the criterion, is a key feature of the AHP technique [34]. In the AHP technique, questionnaires are used to generate the input data. The criteria are important for the successful weight of the alternatives. AHP is identified as a theory of measurement based on a pair-wise matrix that depends on expert opinions to derive a priority scale [35]. Like other MCDM, AHP also has its benefits and drawbacks, though it is one of the most popular methods of MCDM. Its ease of usage is one of its benefits. Its use of pair-wise comparisons can make it relatively simple for decision-makers to weigh coefficients and evaluate alternatives. Because of its hierarchical nature, it is scalable and can simply alter in size to fit decision-making challenges. Even if good pair-wise comparisons necessitate a large amount of data, they result in reliability [36]. The most difficult aspect of this prioritizing strategy is to set the interdependence between the criteria. The methods generate inconsistencies in evaluation among criteria, and ranking of the criteria generates inconsistencies. Moreover, due to its comparison nature for ranking, the addition of alternatives at the end of the procedure could lead the final ranks to flip or reverse. The AHP method can be used for recourse management, performance-type problems, political strategy, planning, resource management, public policy, corporate policy, and strategy. While AHP has limitations, such as sensitivity to expert judgments, they can be mitigated by keeping its consistency ratio within acceptable limits. Despite its drawbacks, the AHP method has the power to handle more critical problems than other MCDMs. For GW RP zoning and analysis, AHP is being used by many researchers [21–24,37–40]. There to assign scores, a scale developed by [28] is used. Previous studies for the selected study have used different methods to assess GW RP, but none used a combined AHP–GIS approach. This study's novelty is applying AHP in combination with a GIS to delineate GW RP in the selected study area. This method is recognized as a more advanced and systematic approach, enabling a MCDM analysis that considers several factors, leading to a more accurate and reliable assessment of GW RP.

The current study encompasses several key objectives. Firstly, it aims to create thematic maps delineating GW's recharge potential zone (RPZ). Another crucial aspect involves investigating the influence of individual factors on GW RP. Lastly, this study aims to quantify the total volume of recharge water.

2. Materials and Methods

2.1. Study Area Description

Three administrative districts, namely Naogaon, Chapai Nawabganj, and Rajshahi, located in the BT of Bangladesh, are selected for the current study. The study area location is illustrated in Figure 1. The study area is in the northwestern region of Bangladesh, which covers an area of 7586 km^2. The geographical location of the area is between 24°07′ and 25°13′ north latitudes and between 87°55′ and 89°10′ east longitudes. The study region is bordered on three sides by West Bengal, and some areas on the south and east are encircled by two important rivers, the Brahmaputra and the Ganges. The study area's annual average rainfall ranges from 1380 mm to 1560 mm, with a mean of 1480 mm. The monsoon season (May to September) accounts for 80% of all precipitation. During the dry season (January to June), GW has a maximum depth of 4 m to 15 m [15]. On the other hand, GW reaches its minimum depth at the end of the monsoon season (i.e., October) due to rain and floods, and the difference between the minimum depths in the GW table is generally between 2 and 10 m [41]. The majority of the study area experiences frequent GW droughts [15]. The weather is warm and muggy there. The lowest temperature is between 10 and 20 degrees Celsius, and the maximum temperature is between 25 and 35 degrees Celsius [42]. The study encompasses a large portion of the Barind Irrigation Project. Rice is the project's principal crop, making up a sizable amount of all agriculture. Other crops farmed in this region are sugarcane, wheat, potato, jute, and vegetables. The variants of Rice are Aman, Aus, and Boro. Boro rice is farmed in the dry season (January to June), which makes up the majority of the rice produced in the research area. GW irrigates the majority of the study area's water demand during the dry season. This study adopted geospatial techniques such as GIS and RS for GW potential zone determination with the integration of GIS and RS data for the BT of Bangladesh. This methodology is well-recognized by various researchers [21,23,29,43]. A methodological flowchart of this study is presented in Figure 2. Major steps in the methodology are data collection and preparation of thematic maps, weight assignments based on AHP, and overlying operation.

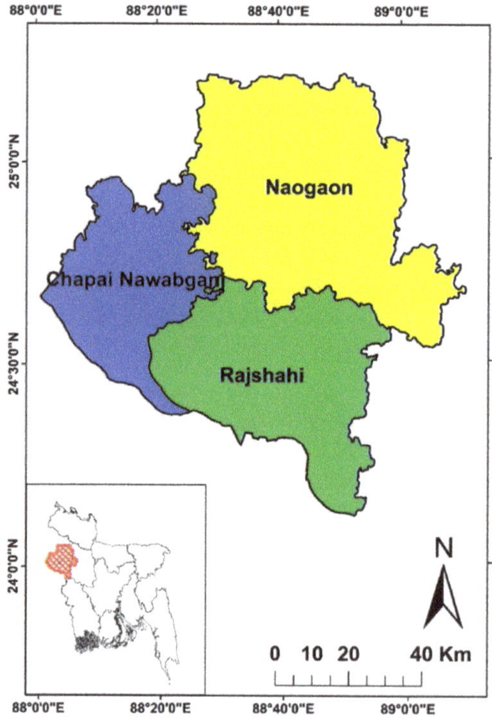

Figure 1. Description of study area.

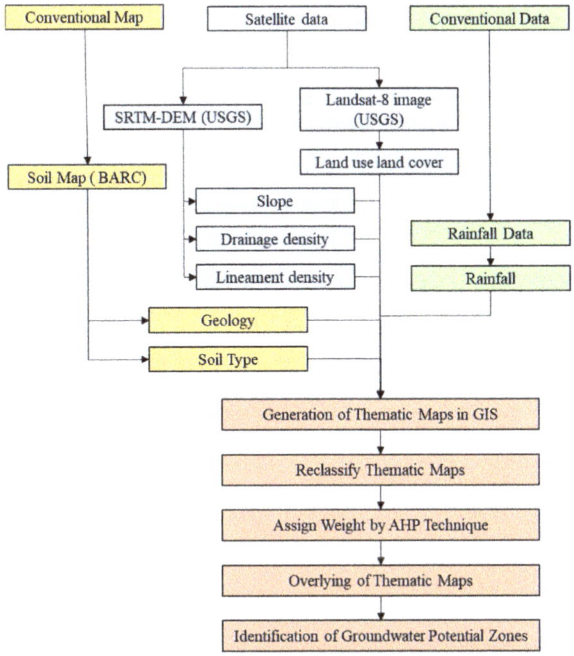

Figure 2. Methodological flowchart for GW RP derivation.

2.2. Preparation of Input Database

This study considered seven GW-recharge-impacting factors: rainfall, slope, LD, DD, land use land cover (LULC), ST, and geology. Data were gathered from a variety of secondary sources, including open-source databases and various government organizations, and based on remotely sensed data. The selection of only seven GW recharge factors out of so many was for various reasons, such as the availability of data, influence of the factor in the study area, variation of different classes for these factors, etc. An overview of the data source in format is provided in Table 1. The time scale of different data was between 2018 and 2022. ArcGIS v10.5 was used for data extraction, analysis, and the derivation of thematic layers for factors such as slope, LD, DD, LULC, geology ST, and rainfall data. Various procedures in ArcGIS were employed for deriving thematic maps. For example, the rainfall map was created from point data using the Inverse Distance Weighted (IDW) interpolation method. The raster distribution of rainfall was obtained by interpolating the point data, which was then clipped and reclassified for the study area.

SRTM data were utilized to derive three factors for GW potential: slope, LD, and DD. The slope map was obtained using the DEM map and the slope function in the ArcGIS software, taking into account the fastest possible rate of change in the DEM elevation value between a cell and its nearby cells. The derivation of LD map from DEM involved hillshade spatial analysis, the creation of a new shapefile, and the use of the line density function. The DD map was derived through a series of steps, including filling the DEM, flow-direction determination, flow accumulation, conditional functions, analysis of stream order, stream to feature conversion, and reclassification. The LULC map was derived from Landsat 8 satellite imagery using supervised classification, identifying agriculture, forest land, bare land, and water body portions. The geology and ST map, obtained from Bangladesh Agricultural Research Council (BARC) in vector format, was converted to a raster image using the function vector to raster, followed by clipping for the study area. The significance of various factors in relation to GW potential was taken into consideration when determining GW RPZ using the classified maps that emerged from these processes.

Table 1. Summary of data source.

Data Type	Source	Data Type
Study Area Map	BARC	Vector
DEM	US geological survey (USGS), SRTM data	Raster, 30 m × 30 m
Slope, Lineament Density, Drainage Density	Derived using SRTM data	Raster, 30 m × 30 m
Land Use/Land Cover (LULC)	Landsat 8 satellite image, USGS online portal	Raster, 30 m × 30 m
Geology, Soil Type	BARC	Vector
Rainfall	Bangladesh Meteorological Department (BMD)	Point Data

2.3. *Analytical Hierarchical Process (AHP)*

The AHP organizes perceptions and judgments into a framework that aids decision-making by considering multiple factors. Implementation of AHP for GW potential zone determination, needs to follow four major steps. They are (1) GW-RP-influencing factor selection, (2) development of pair-wise comparison matrix (PWCM), (3) estimation of weight in relation to the factors, and, finally, (4) consideration for consistency of the PWCM.

2.3.1. GW-RP-Influencing Factor Selection

At the beginning of AHP, the factors that influence GW potential were given a score. The score was based on [28,44], a linear scale of 1 to 9. In Saaty's linear scale, 1 denotes equal importance, 9 denotes extremely more important, and 1/9 denotes extremely less important. Each score's significance is listed in Table 2.

Table 2. Relative class scale for GW RP analysis based on Saaty [44].

Importance	Scale	Remarks
Extremely less important	1/9	↑ Less important
	1/8	
Very strongly less important	1/7	
	1/6	
Strongly less important	1/5	
	1/4	
Moderately less important	1/3	
	1/2	
Equal Importance	1	
	2	More important ↓
Moderately more important	3	
	4	
Strongly more important	5	
	6	
Very strongly more important	7	
	8	
Extremely more important	9	

2.3.2. Development of PWCM

Using the AHP technique, geographic data (input) is combined and transformed into a decision (output), and Saaty's scale is utilized to translate qualitative information from specific thematic layers and features into quantitative scores. Then, based on these scores, a pair-wise matrix is constructed founded on Saaty's scale (Table 2) [44]. In a pair-wise comparison, a factor is compared with itself score is given as 1. The actual Saaty's scores, or the reciprocal of the Saaty's scores when a more influential parameter is compared to a less influential parameter, fill the remaining rows when a more influential parameter is compared to a less influential parameter. The PWCM using the GW-influencing factor in this study is illustrated in Table 3.

Table 3. Developed PWCM.

	Rainfall	Geology	Slope	DD	LULC	LD	ST
Rainfall	1.00	0.33	3.00	3.00	0.33	3.00	0.33
Geology	3.00	1.00	3.00	5.00	3.00	5.00	3.00
Slope	0.33	0.33	1.00	1.00	0.33	0.33	0.20
DD	0.33	0.20	1.00	1.00	0.33	1.00	0.33
LULC	3.00	0.33	3.00	3.00	1.00	3.00	1.00
LD	0.33	0.20	3.00	1.00	0.33	1.00	1.00
ST	3.00	0.33	5.00	3.00	1.00	1.00	1.00
Total (a)	11	2.73	19	17	6.33	14.33	6.87

2.3.3. Estimation of Weight Relevant to the Factors

By evaluating each factor's relative importance in relation to other factors, experts' judgment and a literature review were used to determine the weight of each component in the AHP [19–27,45–50]. In AHP-based PWCM analysis, the eigenvalue is crucial for computing the consistency index, which is then used to calculate the CR. This ratio is used as a safeguard to make sure the comparisons based on expert judgment and literature review are trustworthy. Similar to the one in Table 3, the PWCM also takes into account the conditions unique to the study area as well as prior research. The accuracy of AHP analysis depends on keeping the consistency index below 0.1, even though factors differ between regions [21–23,26,37,39,47,51,52]. After that, the value of each factor was divided by the total value of the corresponding column to create the normalized pair-wise matrix. (Table 4). Equation (1) can be written as follows:

$$X_{ij} = C_{ij}/L_j \tag{1}$$

Here, the sum of the values in each column of the PWCM (Table 3) are denoted by L_j, and the value assigned to each factor at the ith row and jth column is denoted by C_{ij}. By adding the values in each row, the weight gained by each factor was calculated.

Then, the classification of the constraints and the percentage of effect of the factors are then ascertained with the aid of eigenvalue factors and eigenvalue computations (Table 5). The eigenvalue factor was computed by dividing the column elements by the column sum from Table 4. The relative weights of each parameter were quantified by averaging across the rows to obtain the principal eigenvector (λ_{max}) [28,31,53–55]. The major eigenvalue for the 7 × 7 matrix was discovered to be 7.716 and is listed in Table 5.

Table 4. Normalized pair-wise matrix with gained weight.

	Rainfall	Geology	Slope	DD	LULC	LD	ST	Weight (%) (b)
Rainfall	0.091	0.122	0.158	0.176	0.053	0.209	0.049	12.25
Geology	0.273	0.366	0.158	0.294	0.474	0.349	0.437	33.57
Slope	0.030	0.122	0.053	0.059	0.053	0.023	0.029	5.27
DD	0.030	0.073	0.053	0.059	0.053	0.070	0.049	5.51
LULC	0.273	0.122	0.158	0.176	0.158	0.209	0.146	17.74
LD	0.030	0.073	0.158	0.059	0.053	0.070	0.146	8.40
ST	0.273	0.122	0.263	0.176	0.158	0.070	0.146	17.25
Total								100

Table 5. Eigenvalue calculation.

Factors	Relative Weight (a) (Table 3)	Eigenvalue Factor (b) (Table 4)	Eigenvalue (a × b)
Rainfall	11.00	0.123	1.348
Geology	2.73	0.336	0.918
Slope	19.00	0.053	1.001
Drainage Density	17.00	0.055	0.937
LULC	6.33	0.177	1.124
Lineament Density	14.33	0.084	1.204
Soil Type	6.87	0.173	1.185
Principle Eigenvalue λ_{max}			7.716

2.3.4. Consideration for Consistency of the PWCM

The consistency of the PWCM has been checked using the consistency ratio. The consistency ratio of the PWCM is calculated using Equations (2) and (3):

$$Consistency\ Index\ (CI) = \frac{Principle\ Eigen\ Value - nuber\ of\ factor}{number\ of\ factor - 1} = \frac{\lambda_{max} - n}{n - 1} \quad (2)$$

$$Consistency\ Ratio\ (CR) = \frac{Consistency\ Index}{Randomness\ Index} = \frac{CI}{RI} \quad (3)$$

The randomness index is tabulated in Table 6. A perfect comparison pair-wise matrix has a consistency index of 0, but a small inconsistency is in the tolerable limit $CI < 0.1$ [28]. In this study, the consistency is 0.0904.

Table 6. Randomness index (RI).

n	2	3	4	5	6	7	8	9	10	11
RI	0	0.58	0.9	1.12	1.24	1.32	1.41	1.45	1.51	1.52

2.4. Overlay Operation

The factors are overlayed once the classified spatial distribution map of the variables affecting GW potential has been created. The GW potential zone derivation is produced by ArcGIS software using weighted overlay analysis tools. The weighted score achieved by

AHP is used to assign weight to the subgroup of the influencing factor. The corresponding equation used for GW RPZ derivation is Equation (4):

$$\begin{aligned} Ground\ Water\ Recharge\ Potential &= \sum_{i=1}^{n} W_i X_i \\ &= R_c R_w + S_c S_w + G_c G_w + DD_c DD_w + LULC_c LULC_w + LD_c LD_w + ST_c ST_w \end{aligned} \tag{4}$$

where W_i refers to the value of each factor of each pixel in the derived map, and X_i refers to the total AHP score (%). In other words, it is the weight of each cell value (e.g., rainfall or DEM value) in the map within the ArcGIS software. R refers to rainfall, S refers to slope, and G refers to geology. Moreover, subscript c refers to the criteria class, and D refers to the weight of each factor. This Equation refers to the result in the summation of cell value multiplied by the weightage assigned to the GW-influencing factors. The overlay operation's schematic diagram is displayed in Figure 3.

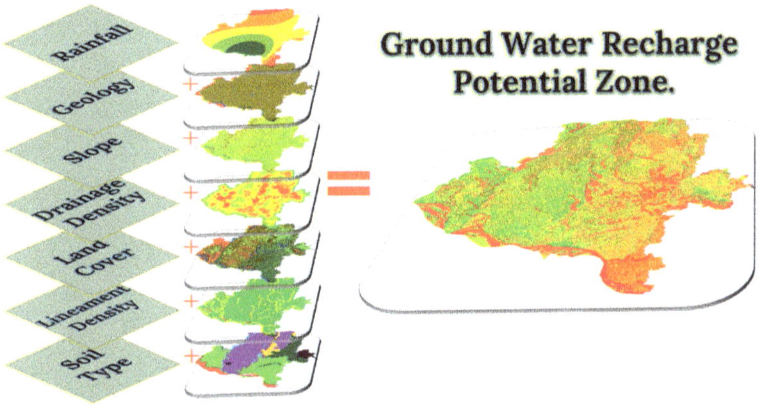

Figure 3. Schematic of overlay operation.

2.5. Recharge Volume Estimation

Total recharge water volume estimation directly from the GW RPZ is not an easy task. But there is a simple method of estimating GW recharge volume presented in Equation (5) [31]:

$$\begin{aligned} &Total\ Recharge\ Volume \\ &\quad = Annual\ Rainfall\ Volume \times \sum(Recharge\ Ratio \times \%\ of\ Total\ Area) \end{aligned} \tag{5}$$

Here, the recharge ratio can be found from the FAO (1967) study (Table 7) [29], % of the total area represents the portion of total area covered by each category of GW RP, and the summation is carried out for each of the GW RP category.

Table 7. GW RP category FAO estimates [29].

RP Category	Estimates (%)	Average (%)
Very Good	45–50	47.5
Good	30–35	32.5
Moderate	10–20	15.0
Poor	5–10	7.5
Very Poor	0–5	2.5

3. Results

3.1. GW-Recharge-Governing Factor Evaluation

The current study refers to examining several hydrological, geographical, and geological factors influencing GW RP for the BT of Bangladesh to identify GW water RPZs. The investigation identified the seven significant factors impacting GW recharge capacity: rainfall, geology, slope, DD, LULC, LD, and ST. The factors are analyzed, and an appropriate weight of influence is assigned for the AHP. The weight assigned to factors based on AHP is presented in Table 4 The individual scores for each GW-recharge-influencing factor are presented in Table 8. The next section presents a spatial analysis of the contributing factors. Here, the AHP value is used to calculate the total influence of each factor. Then, according to the feature class and the susceptibility to or influence of RP, the feature class was given an individual score. In ArcGIS, a weighted overlay operation means multiplying each score by the value in each cell of a class and then adding them all together.

Table 8. Assigned weight to different feature class of GW-influencing factors.

Criteria	Feature Class	AHP Score	
		Total (%) (X_i)	Individual
Rainfall (mm)	1200–1400	12	2
	1400–1550		3
	1550–1650		6
	1650–1800		9
	1800–2200		12
Geology	Ultrabasic	34	14
	Sedimentation		24
	Water		34
	Lake		34
Slope (%)	0–1	5	5
	1–2		4
	2–3		3
	3–6		2
	6–40		1
Drainage Density (km/km^2)	0.05–0.75	6	6
	0.75–1.05		5
	1.05–1.25		4
	1.25–1.5		3
	1.5–2.5		2
LULC	Urban	18	3
	Bare Land		8
	Forest		13
	Agriculture		13
	Water		18

Table 8. *Cont.*

Criteria	Feature Class	AHP Score	
		Total (%) (X_i)	Individual
Lineament Density (km/km^2)	0–0.25	8	3
	0.25–0.65		4
	0.65–1.2		5
	1.2–1.85		7
	1.85–3		8
Soil Type	Acid Basin Clay	17	8
	Urban		8
	Shallow Terrace Soil		13
	Calcar. Alluvium		13
	Noncalcar. Alluvium		13
	Calcar. Floodplain		17
	Deep Terrace Soils		17
	Noncalcar. Floodplain		17

3.2. Rainfall Features

One of the major factors influencing GW RP is rainfall. It indicates the availability of recharge water volume for infiltrating downward. Soil properties, surface conditions, and rainfall intensity impact infiltration rates, with greater rainfall intensity increasing infiltration [45,46,56–59]. The interconnection between these factors is crucial for GW recharge studies [60]. Both heavy rainfall and drought have significant impacts on GW potentiality [49,50,61–64]. The classified spatially distributed map of rainfall is presented in Figure 4A for the study area. The map of rainfall distribution is classified into five categories. The areas covered by feature classes of 1200 to 1400 mm, 1400 to 1550 mm, 1550 to 1650 mm, 1650 to 1800 mm, and 1800 to 2200 mm are 888.47 km^2, 1334.20 km^2, 2828.84 km^2, 2011.80 km^2, and 5222.63 km^2, respectively, and the percentages of total area covered by each feature group are 11.71%, 17.59%, 37.29%, 26.52%, and 6.89%, respectively. Lower weightage was given to lower rainfall zones and vice versa. The total weight of rainfall is 12%. The individual score given is presented in Table 8. The classified distribution map derived using IDW shows that the lower region of the study area (Rajshahi region) has a low rainfall intensity, and the higher region has a high intensity.

3.3. Slope Features

Another important factor influencing the GW RP is the slope, which influences the infiltration intensity and water retention. When surface water flows through a sloping ground, increasing slope percentage increases the speed of flow; on the contrary, a flat slope takes more time to drain drops of water from the catchment. A steep slope means that rain or storms come faster, and, therefore, there is less time for the rain to enter and vice versa [25,65–67]. The classified spatially distributed slope map is presented in Figure 4B for the study area. The slope map derived using SRTM and DEM data is classified into five different categories. The area covered by feature classes of 0 to 1%, 1 to 2%, 2 to 3%, 3 to 6%, and 6 to 40% are 58.26 km^2, 1034.50 km^2, 1689.63 km^2, 2998.83 km^2, and 1804.48 km^2, respectively, and the percentages of the total area covered by each feature group are 0.77%, 13.64%, 22.27%, 39.53%, and 23.79%, respectively. Higher weightage was given to the lower-slope zone and vice versa. The total weight of the slope is 5%. The individual score given is presented in Table 8.

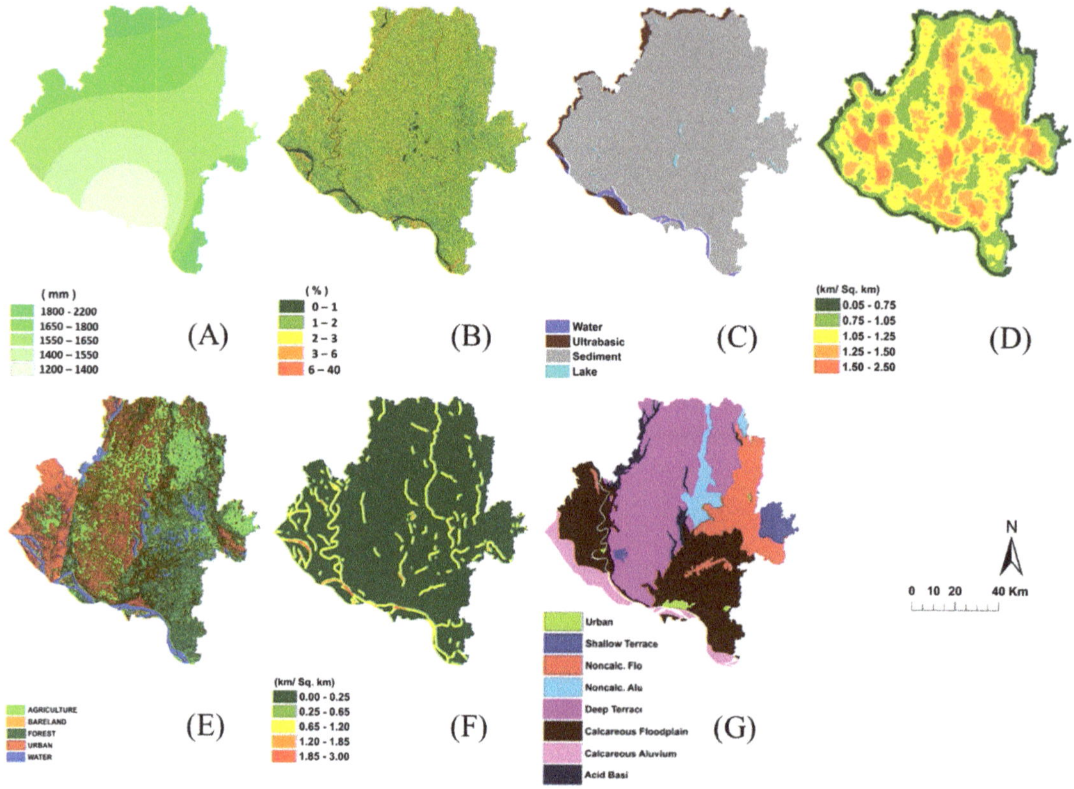

Figure 4. Derived (**A**) rainfall distribution, (**B**) slope, (**C**) geology, (**D**) drainage density, (**E**) LULC, (**F**) lineament density, (**G**) soil type map of the study area.

3.4. Geology Features

Geology is also another significant GW-influencing factor. It provides information about the landforms. The geological feature can influence the movement of GW and the mode of occurrence. Different types of soil formation act differently with respect to GW potential [23,26,47,48,68]. The categorized map of geology is presented in Figure 4C for the study area. The study area map of geology has four categories of geological features. The areas covered by the feature classes of ultrabasic igneous rock, sedimentation, water, and lake are 260.44 km^2, 7191.26 km^2, 92.69 km^2, and 41.32 km^2, respectively, and the percentages of the total area covered by each feature group are 3.43%, 94.80%, 1.22%, and 0.54%, respectively. Here, the sedimentation zone covers most of the study area, as BT is named. Lower weightage was given to ultrabasic igneous rock. On the contrary, higher weightage was given to the water and lake zones. The total weight of geology is 34%. The individual score given is presented in Table 8. Geology has a high-gain percentage of weightage. Here, it is considered the most potent factor among all the other GW-recharge-influencing factors.

3.5. Drainage Density Features

Another influencing factor for GW RP is DD. It is the measure of the mechanism of stream channel drainage. It influences infiltration capacity, surface roughness, vegetation cover, resistance to erosion, and climate conditions. To GW RP, DD gives an inversely proportional relationship. As it is a measure of how near the stream channels are located, more DD gives a pathway to drain water in a short time period. This directly affects the

GW recharge, as GW recharge is a slow process, and a smaller time period gives lower infiltration [69–71]. The classified spatially distributed map of DD for the study area is presented in Figure 4D. The DD map is classified into five categories. The areas covered by feature classes of 0.05 to 0.75 km/km^2, 0.75 to 1.05 km/km^2, 1.05 to 1.25 km/km^2, 1.25 to 1.50 km/km^2, and 1.50 to 2.50 km/km^2 are 692.60 km^2, 1774.98 km^2, 2895.50 km^2, 1769.08 km^2, and 453.68 km^2, respectively, and the percentages of total area covered by each feature group are 9.13%, 23.40%, 38.17%, 23.32%, and 5.98%, respectively. Higher weightage was given to lower-density zones and vice versa. The total weight of DD is 6%. The individual score given is presented in Table 8.

3.6. LULC Features

Another dominating factor of GW RP is the LULC of the study area. Different types of land cover provide a different degree of preference for GW RP. The use of LULC in GW studies was initiated by US Geological Survey (USGS) in 1984. Different levels of land use change are common across different levels of renewable resources [21,22,37,72–74]. The categorized map of LULC is presented in Figure 4E for the study area. The study area map of LULC has five categories of land cover features. Landsat 8 satellite images are used to derived the study area land cover map. Supervised classification was conducted to derive five basic types of land cover. The areas covered by the types of Urban, Bare Land, Forest, Agriculture, and Water are 2016.61 km^2, 656.93 km^2, 2555.26 km^2, 1632.88 km^2, and 724.61 km^2, respectively, and the percentages of the total study area covered by each feature group are 26.58%, 8.66%, 33.68%, 21.52%, and 9.55%, respectively. Here, supervised classification based on band images in the Landsat 8 satellite image forest land covers the most area percentage followed by Urban, Agriculture, Water, and Bare Land. Higher weightage was given to the water body because of the availability of water to infiltrate downward, and lower weightage was given to urban areas, as the runoff velocity was quite high. The total weight of LULC is 18%. The individual score given is presented in Table 8.

3.7. Lineament Density Features

Another GW-recharge-influencing factor is the LD. The LD network increases porosity, as it indicates the underlying geological features such as joints, faults, and fractures. Lineament mapping can be defined as the technique of drawing lines on images of maps. It behaves in a proportional way to GW recharge [75,76]. The classified spatially distributed LD map is presented in Figure 4F for the study area. The LD map derived using SRTM and DEM data is classified into five categories. The areas covered by the feature classes of 0 to 0.25 km/km^2, 0.25 to 0.65 km/km^2, 0.65 to 1.20 km/km^2, 1.20 to 1.85 km/km^2, and 1.85 to 3.00 km/km^2 are 6505.85 km^2, 388.50 km^2, 631.98 km^2, 51.24 km^2 and 8.76 km^2, respectively, and the percentages of the total area covered by each feature group are 85.76%, 5.12%, 8.33%, 0.68%, and 0.12%, respectively. Here, the feature class of 0 to 0.25 km/km^2 covers a major portion of the total area. Higher weightage was given to higher LD zones, and vice versa. The total weight of LD is 8%. The individual score given is presented in Table 8.

3.8. Soil Type Features

The last GW-recharge-influencing factor considered for this study is the ST. Different types of STs act differently for the recharge of GW, as it controls properties such as infiltration capacity, structure, porosity, and consistency. Infiltration capacity is affected by soil stability, soil structure, land use type, and particle size distribution [77]. Changes in land use, from vegetation to continual grazing and tillage, affect the porosity, infiltration, water shortage, and water transport characteristics of soil [22,47,51,78]. The categorized map of geology is presented in Figure 4G for the study area. The study area map of soil has eight different types of soil, which are Acid Basin Clay, Urban, Shallow Terrace Soil, Calcareous Alluvium, Noncalcareous Alluvium, Calcareous Floodplain, Deep Terrace Soils, and Noncalcareous Floodplain. The areas covered by these feature classes are 232.43 km^2, 56.77 km^2,

233.29 km^2, 409.81 km^2, 438.07 km^2, and 2325.15 km^2, respectively, and the percentages of the total study area covered by each feature group are 3.06%, 0.75%, 3.08%, 5.40%, 5.77%, 30.65%, 38.56%, and 12.72%, respectively. The total weight of ST is 17%. Floodplain soils are given a higher weightage, and a lower weightage is given to Clayey-type soil. The individual score given is presented in Table 8.

3.9. Estimation of GW RP

3.9.1. AHP-Based GW RP

The GW RP of the study uses seven GW RP factors for AHP derived using Equation (6):

$$\begin{aligned} &Ground\ Water\ Recharge\ Potential = \sum_{i=1}^{n} W_i X_i \\ &= R_c \times 0.13 + S_c \times 0.05 + G_c \times 0.34 + DD_c \times 0.6 + LULC_c \times 0.18 + LD_c 0.08 \\ &\quad + ST_c \times 0.17 \end{aligned} \tag{6}$$

where W_i refers to the value of each factor of each pixel in the derived map, and X_i refers to the total AHP score (%). In other words, it is the weight of each cell value in the map within the ArcGIS software. R refers to rainfall, S refers to slope, and G refers to geology. Moreover, subscript c refers to the criteria class, and D refers to the weight of each factor. The GW RPZ map considering AHP is presented in Figure 5. The areal extent in km^2 and percentage of the total study area of the GW potential based on AHP are tabulated in Table 9.

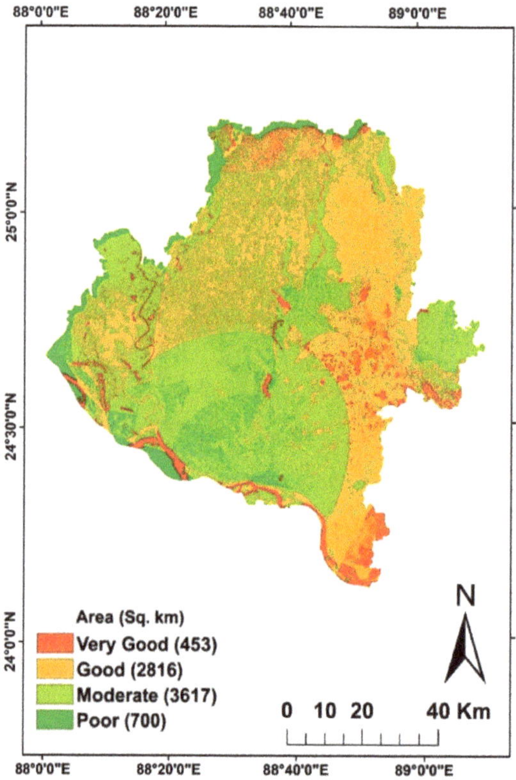

Figure 5. GW potential based on the AHP map.

Table 9. AHP Based on GW RP.

Recharge Potential Zone	Areal Extent (km^2)	Areal Extent (%)
Poor (zone 1)	700.22	9.23
Moderate (zone 2)	3617.13	47.68
Good (zone 3)	2816.13	37.12
Very Good (zone 4)	452.70	5.97

Table 9 shows that approximately 700 km^2 of the research area falls inside zone-1, around 9.23% of the total study area. In the current study, identified zone-1 can be given as a "poor GW RPZ". In the AHP approach, the GW potential map impacts the rainfall distribution map. As GW recharge potentiality is proportional to the rainfall intensity here, zone-1 covers the low-rainfall-intensity zone. This demonstrates that there is a shortage of GW recharge. Also, it covers a portion. It may be due to the speed of water in the river portion. The finer particles in the river bed cause infiltration at a lower rate.

Table 9 also shows that zone-2 covers approximately 3617.13 km^2 of the total research area, around 47.68% of the total study area. This portion covers the major part of the research area. In the current study, the identification of zone-2 can be given as a "moderate GW RPZ". This zone also has the effect of rainfall. The zone is located just upstream of the poor RPZ, as the rainfall intensity is higher in that zone in the region.

It can also be seen from Table 9 that approximately 2816.13 km^2 of the total research area falls inside zone-3, which is around 37.12% of the total study area. It mostly covers the rainfall zone higher than zone-2. However, there is a mix of zone-2 and zone-3 in the GW potential map. These two zones exist near each other. Identification of zone-3 can be given as a "good GW RPZ". The area covered by zone-2 is in the second largest position in this study area. Also, the zone is located in the shallow terrace ST and forest area in the land LULC map.

From Table 9, the last zone-4 covers approximately 452.70 km^2, which covers around 5.97% of the total research area. It is the least-covered zone. The identification of zone-4 can be given as a "very good GW RPZ". It has the best GW RP. The location of this zone is in the lowest part of the study area, which is also the lower part of the Padma River. The other part is also located basically near the water bodies. The rainfall intensity of this zone is also higher. It covers basically the floodplain zone.

The overall study region of GW RP based on AHP shows a reflection of rainfall intensity, though the highest-influencing factor is geology. However, as most of the geological portion of the research area is in the sediment zone, the effect of this factor is minimized. Though rainfall governs the overall RP, LULC also reflects its effect on the forest zone, supporting good GW RP.

3.9.2. Estimation of GW Recharge

The direct estimation of recharge volume is not quite an easy task. Quantitative GW recharge volume estimation into the study area potential zones subsurface is calculated using Equation (7). The annual precipitation volume of the study area is 11,227 × 10^6 m^3/year. The total recharge volume of the study area is calculated for the AHP approach for the GW RP. The total volume of the GW RPZ is as follows:

$$\begin{aligned} W = 1227 \times 10^6\ (0.475 \times 0.06 + 0.325 \times 0.37 + 0.15 \times 0.48 + 0.075 \times 0.09 + 0.025 \times 0) \\ = 2554 \times 10^6\ \mathrm{m^3/year} \end{aligned} \quad (7)$$

where W refers to the total recharge volume, the estimates are used from Table 7, and the % of total area are obtained from Table 9. It is to be mentioned that the total recharge volume is about 22.7% of the total precipitated volume. The inferred PWCM categories were declared after extensive literature review, which are also validated by FAO's guidlines [21,22,25,29,31].

4. Discussion

The current study aims to derive GW RP maps for the BT of Bangladesh. GIS, along with one of the most used methods of multi-criteria decision analysis, namely AHP, is used to derive the GW RP maps for the study region. Seven GW recharge-influencing maps, namely rainfall, geology, LD, DD, slope, ST, and LULC are used in the analysis purpose.

As a result of careful analysis, four different GW recharge zones were identified. These areas are classified according to pre-defined criteria that give an idea about the regional water quality. Notably, the study found that approximately 9.23% (700.22 km^2) of this area exhibited low GW recharge capacity. Most of the area, 47.68% (3617.13 km^2), was found to have temporary GW RP, while 37.12% of it (2816.13 km^2) showed good GW RP. A small area, 5.97% (452.70 km^2), is classified as having GW RP. These categories provide detailed information on GW RP in the Barind region, which is essential for effective water management.

Interestingly, rainfall intensity was found to be the dominant factor in GW increase, although it ranked fourth among the influencing factors. This deficiency highlights the difficulty of utilizing GW, where rainfall intensity has a significant impact but exceeds the list of considerations and other factors. For example, geology has reached a high level in terms of GW exchange. However, the dominance of vegetation areas, corresponding to 94.80% of the research area, reduces the importance of geology as a distinguishing factor in this area.

The results of this study show that the GW volume of Barind GW is approximately 2554×10^6 m^3, that is, 22.7% of the total study area. This high-recharge rate demonstrates the region's ability to replenish GW resources and is a valuable perspective on sustainable management and management planning. Estimates of GW recharge quality and quantity provide essential information for decision-makers and water managers. This study helps prioritize areas for GW development and environmental protection by identifying areas with different RP. Identification of potential areas not only leads to better GW management but also supports planning of agricultural activities, urban development, and environmental protection in the Barind region.

Additionally, research shows that although rainfall intensity has a significant impact on GW, its ranking does not fully reflect its importance. This discrepancy suggests that further research is needed to evaluate the RP of GW, excluding precipitation as a factor. This approach can provide additional information and confirm the prevalence of precipitation in real-world situations. Additionally, this study confirms that the seven factors considered (precipitation, geology, soil density, water density, slope, ST, and soil/land cover) may not cover all important factors. Other factors such as vegetation, soil temperature, and human activities may also play a role in GW recharge capacity. Future research should therefore include these additional factors to provide a better understanding of GW recharge dynamics.

When discussing GW RPZ, it is important to compare the results with similar studies to highlight both the similarities and unique aspects of the studies. For example, in a similar region of the Indian subcontinent, the Mithina Upwansi study identified five categories of GW RPZs: very low (4%), low (25%), medium (45%), high (18%), and very high (8%) [79]. This distribution reflects a gradient of recharge potential, with a significant portion of the area (45%) classified as medium, similar to the results of the present study, where 47.68% of the area is within the medium RPZ. Narongsak Kaewdum's study in Thailand, which has geographical and climatic similarities with the study area, also classified GW RPZs as high, medium, low, and very low. Kewdam's study found that about 2.26 percent of the total area had high recharge potential, while most of the precipitation was lost through surface runoff or evapotranspiration, suggesting that GW recharge would be problematic [80]. In comparison, the present study found that 5.97% of the region has very good recharge potential, while 37.12% falls under the good recharge category, suggesting that the region may have more favorable conditions for GW recharge compared to the Kew Dam study area. These differences in study results may be attributed to differences in regional characteristics such as geology, land use, and precipitation patterns, which directly affect GW recharge dynamics. The high percentage of moderate recharge zones in both Upwansi and the present study highlights the prevalence of conditions that moderately support GW recharge in these regions. However, the low percentage showing very good recharge potential in the present study compared to the high percentage of high and very high zones in the Upwansi study may indicate that while conditions are favorable for recharge, there are limitations in most of the region that prevent it from reaching its highest recharge potential. These comparisons highlight the importance of adapting GW management strategies to the specific conditions of each region. The results suggest that while there are significant areas with moderate to good replenishment potential in the study area of Bangladesh, targeted interventions may be required to improve replenishment in areas currently classified as poor or moderate. Understanding the factors that contribute to this classification can help develop more effective strategies for GW management and sustainable development, especially in areas with similar environmental challenges.

The analysis's approach and findings are also valuable for academic research and learning. The combination of AHP and GIS serves as an example of how advanced tools can be applied in environmental management. It offers a blueprint that can be duplicated in other areas encountering comparable obstacles. Educational institutions have the opportunity to utilize this case to teach students about sustainable water management practices, the significance of GW replenishment, and the utilization of GIS and AHP in environmental studies.

This study provides a framework for assessing GW RP in the Barind region, highlighting the importance of multi-factor analysis and the role of different factors. Using AHP and GIS, it identifies key areas for sustainable GW management, helping decision-makers and water managers preserve resources and enhance water security. This study also supports environmental conservation, climate resilience, and agricultural productivity, eventually contributing socio-economic benefits. Engaging local communities and providing policymakers with detailed maps and data can lead to more effective land use planning and water management strategies.

5. Conclusions

This study's integrated AHP–GIS approach for mapping groundwater recharge potential in Bangladesh's Barind Tract yields significant insights for sustainable water management. By identifying four distinct recharge zones—poor (9.23%), moderate (47.68%), good (37.12%), and very good (5.97%)—we provide a nuanced understanding of the region's groundwater dynamics. Our findings reveal that while geology exerts the strongest influence on recharge potential (33.57%), the interplay of factors such as land use, soil type, and rainfall create a complex hydrogeological landscape. The estimated annual recharge of 2554×10^6 m^3, representing 22.7% of total precipitation, underscores both the region's recharge capacity and the critical need for targeted conservation efforts. The spatial distribution of recharge zones offers valuable guidance for policymakers and water managers. Areas with very good recharge potential, though limited (5.97%), present prime opportunities for groundwater conservation and managed aquifer recharge initiatives. Conversely, the significant portion of land with moderate potential (47.68%) suggests a widespread scope for enhancing recharge through tailored land management practices.

While our methodology provides a robust framework for assessing recharge potential, we acknowledge its limitations. Future research should incorporate additional factors such as aquifer characteristics and human activities to refine the assessment further. Moreover, validating these findings through field studies and long-term monitoring would enhance their practical applicability. In addition, additional factors such as vegetation, soil temperature, and human activities should be incorporated to further improve the understanding of GW recharge dynamics. Considering alternative methods and excluding precipitation in certain analyses may provide greater insight into the dominant influences on RP. In addition to the findings and practical implications discussed, this study recognizes the potential to improve the scientific validity of AHP as a subjective task assessment method. Future research may consider incorporating advanced techniques such as fuzzy logic and machine learning or other methods to refine the AHP framework, which would address concerns regarding subjectivity and improve the reliability of GW RP assessment. Expanding the scope of research to include these factors will improve the accuracy of RP maps and provide more precise guidance for sustainable water resources management and policy-making in the Barind region.

This study offers a scientifically grounded approach to addressing water scarcity in the Barind Tract. By pinpointing areas of high recharge potential and identifying factors influencing groundwater replenishment, we provide a valuable tool for sustainable water resource planning. As climate change and growing water demand intensify pressure on groundwater resources, such data-driven approaches will be crucial in ensuring water security and supporting agricultural sustainability in vulnerable regions like the Barind Tract.

Author Contributions: M.Z.H.: conceptualization, methodology, project administration, resources, data preparation, validation, data analysis, writing—original draft, writing—review and editing, visualization. S.K.A.: conceptualization, methodology, investigation, project administration, supervision, resources, data preparation, validation, data analysis, writing—original draft, writing—review and editing. H.N.: supervision, validation, data analysis, writing—original draft, writing—review and editing, visualization. A.A.K.: project administration, supervision, resources, data preparation, validation, data analysis, writing—review and editing. H.A.A.: supervision, resources, data preparation, validation, data analysis, funding acquisition, writing—review and editing. M.T.R.: supervision, resources, validation, funding acquisition, writing—review and editing. All authors have read and agreed to the published version of the manuscript.

Funding: This research work was supported by King Saud University, Riyadh, Saudi Arabia under grant number RSPD2024R848.

Data Availability Statement: Data will be made available on request.

Acknowledgments: The authors extend their appreciation to the researchers supporting project number RSPD2024R848, King Saud University, Riyadh, Saudi Arabia.

Conflicts of Interest: The authors declare no conflicts of interest.

References

1. UN. *The United Nations World Water Development Report 2021: Valuing Water*; United Nations: New York, NY, USA, 2021.
2. Jha, M.K.; Shekhar, A.; Jenifer, M.A. Assessing groundwater quality for drinking water supply using hybrid fuzzy-GIS-based water quality index. *Water Res.* **2020**, *179*, 115867. [CrossRef] [PubMed]
3. The World Bank. *Annual Report*; World Bank: Washington, DC, USA, 2021.
4. Avtar, R.; Singh, C.K.; Shashtri, S.; Singh, A.; Mukherjee, S. Identification and analysis of groundwater potential zones in Ken–Betwa river linking area using remote sensing and geographic information system. *Geocarto Int.* **2010**, *25*, 379–396. [CrossRef]
5. de Vries, J.J.; Simmers, I. Groundwater recharge: An overview of processes and challenges. *Hydrogeol. J.* **2002**, *10*, 5–17. [CrossRef]
6. Allison, G.B. A Review of Some of the Physical, Chemical and Isotopic Techniques Available for Estimating Groundwater Recharge. In *Estimation of Natural Groundwater Recharge*; Simmers, I., Ed.; Springer: Dordrecht, The Netherlands, 1988; pp. 49–72.
7. Mukherjee, A.; Scanlon, B.R.; Aureli, A.; Langan, S.; Guo, H.; McKenzie, A. Chapter 1—Global groundwater: From scarcity to security through sustainability and solutions. In *Global Groundwater*; Mukherjee, A., Scanlon, B.R., Aureli, A., Langan, S., Guo, H., McKenzie, A.A., Eds.; Elsevier: Amsterdam, The Netherlands, 2021; pp. 3–20.
8. Lerner, D.N. Groundwater recharge in urban areas. *Atmos. Environ. Part B Urban Atmos.* **1990**, *24*, 29–33. [CrossRef]
9. Makki, Z.F.; Zuhaira, A.A.; Al-Jubouri, S.M.; Al-Hamd, R.K.S.; Cunningham, L.S. GIS-based assessment of groundwater quality for drinking and irrigation purposes in central Iraq. *Environ. Monit. Assess.* **2021**, *193*, 107. [CrossRef]
10. Michael, H.A.; Voss, C.I. Estimation of regional-scale groundwater flow properties in the Bengal Basin of India and Bangladesh. *Hydrogeol. J.* **2009**, *17*, 1329–1346. [CrossRef]
11. Qureshi, A.S.; Ahmed, Z.; Krupnik, T.J. *Groundwater Management in Bangladesh: An Analysis of Problems and Opportunities*; United States Agency for International Development: Washington, DC, USA, 2014.
12. BBS. *Bangladesh Statistics 2017*; Statistics and Informatics Division, Ministry of Planning, Bangladesh Bureau of Statistics (BBS): Dhaka, Bangladesh, 2017.

13. Dey, N.C.; Saha, R.; Parvez, M.; Bala, S.K.; Islam, A.K.M.S.; Paul, J.K.; Hossain, M. Sustainability of groundwater use for irrigation of dry-season crops in northwest Bangladesh. *Groundw. Sustain. Dev.* **2017**, *4*, 66–77. [CrossRef]
14. Dhaka Water Supply and Sewerage Authority. *Establishment of Groundwater Monitoring System in Dhaka City for Aquifer Systems and DWASA Production Wells*; Dhaka Water Supply and Sewerage Authority: Dhaka, Bangladesh, 2018.
15. Adhikary, S.K.; Das, S.; Saha, G.; Chaki, T. Groundwater drought assessment for Barind irrigation Project in Northwestern Bangladesh. In Proceedings of the 20th International Congress on Modelling and Simulation (MODSIM2013), Adelaide, SA, Australia, 1–6 December 2013; Modelling and Simulation Society of Australia and New Zealand: Canberra, Australia, 2013; pp. 2917–2923.
16. BMDA. *Borandro Authority Past–Present*; Barind Multipurpose Development Authority: Rajshahi, Bangladesh, 2006.
17. BMDA. *Annual Report: 2018–2019*; BMDA: Rajshahi, Bangladesh, 2019.
18. BADC. *Minor Irrigation Servey Report 2018–2019*; BADC: Dhaka, Bangladesh, 2020.
19. Jha, M.K.; Chowdary, V.M.; Chowdhury, A. Groundwater assessment in Salboni Block, West Bengal (India) using remote sensing, geographical information system and multi-criteria decision analysis techniques. *Hydrogeol. J.* **2010**, *18*, 1713–1728. [CrossRef]
20. Jha, M.K.; Chowdhury, A.; Chowdary, V.M.; Peiffer, S. Groundwater management and development by integrated remote sensing and geographic information systems: Prospects and constraints. *Water Resour. Manag.* **2007**, *21*, 427–467. [CrossRef]
21. Saranya, T.; Saravanan, S. Groundwater potential zone mapping using analytical hierarchy process (AHP) and GIS for Kancheepuram District, Tamilnadu, India. *Model. Earth Syst. Environ.* **2020**, *6*, 1105–1122. [CrossRef]
22. Ahmad, I.; Dar, M.A.; Andualem, T.G.; Teka, A.H. GIS-based multi-criteria evaluation of groundwater potential of the Beshilo River basin, Ethiopia. *J. Afr. Earth Sci.* **2020**, *164*, 103747. [CrossRef]
23. Das, N.; Mukhopadhyay, S. Application of multi-criteria decision making technique for the assessment of groundwater potential zones: A study on Birbhum district, West Bengal, India. *Environ. Dev. Sustain.* **2020**, *22*, 931–955. [CrossRef]
24. Rahmati, O.; Nazari Samani, A.; Mahdavi, M.; Pourghasemi, H.R.; Zeinivand, H. Groundwater potential mapping at Kurdistan region of Iran using analytic hierarchy process and GIS. *Arab. J. Geosci.* **2015**, *8*, 7059–7071. [CrossRef]
25. Jahan, C.S.; Rahaman, M.F.; Arefin, R.; Ali, M.S.; Mazumder, Q.H. Delineation of groundwater potential zones of Atrai–Sib river basin in north-west Bangladesh using remote sensing and GIS techniques. *Sustain. Water Resour. Manag.* **2019**, *5*, 689–702. [CrossRef]
26. Nithya, C.N.; Srinivas, Y.; Magesh, N.S.; Kaliraj, S. Assessment of groundwater potential zones in Chittar basin, Southern India using GIS based AHP technique. *Remote Sens. Appl. Soc. Environ.* **2019**, *15*, 100248. [CrossRef]
27. Roy, S.; Hazra, S.; Chanda, A.; Das, S. Assessment of groundwater potential zones using multi-criteria decision-making technique: A micro-level case study from red and lateritic zone (RLZ) of West Bengal, India. *Sustain. Water Resour. Manag.* **2020**, *6*, 4. [CrossRef]
28. Saaty, T.L. An exposition of the AHP in reply to the paper "remarks on the analytic hierarchy process". *Manag. Sci.* **1990**, *36*, 259–268. [CrossRef]
29. Adham, M.I.; Jahan, C.S.; Mazumder, Q.H.; Hossain, M.M.A.; Haque, A.-M. Study on groundwater recharge potentiality of Barind Tract, Rajshahi District, Bangladesh using GIS and Remote Sensing technique. *J. Geol. Soc. India* **2010**, *75*, 432–438. [CrossRef]
30. Ahmed, A.; Alrajhi, A.; Alquwaizany, A. Identification of Groundwater Potential Recharge Zones in Flinders Ranges, South Australia Using Remote Sensing, GIS, and MIF Techniques. *Water* **2021**, *13*, 2571. [CrossRef]
31. Zghibi, A.; Mirchi, A.; Msaddek, M.H.; Merzougui, A.; Zouhri, L.; Taupin, J.-D.; Chekirbane, A.; Chenini, I.; Tarhouni, J. Using analytical hierarchy process and multi-influencing factors to map groundwater recharge zones in a semi-arid Mediterranean coastal aquifer. *Water* **2020**, *12*, 2525. [CrossRef]
32. Cui, X.; Wang, Z.; Xu, N.; Wu, J.; Yao, Z. A secondary modal decomposition ensemble deep learning model for groundwater level prediction using multi-data. *Environ. Model. Softw.* **2024**, *175*, 105969. [CrossRef]
33. Song, Q.; Wang, Z.; Wu, T. Risk analysis and assessment of water resource carrying capacity based on weighted gray model with improved entropy weighting method in the central plains region of China. *Ecol. Indic.* **2024**, *160*, 111907. [CrossRef]
34. Løken, E. Use of multicriteria decision analysis methods for energy planning problems. *Renew. Sustain. Energy Rev.* **2007**, *11*, 1584–1595. [CrossRef]
35. Saaty, T.L. Decision making with the analytic hierarchy process. *Int. J. Serv. Sci.* **2008**, *1*, 83–98. [CrossRef]
36. Velasquez, M.; Hester, P.T. An analysis of multi-criteria decision making methods. *Int. J. Oper. Res.* **2013**, *10*, 56–66.
37. Jhariya, D.C.; Kumar, T.; Gobinath, M.; Diwan, P.; Kishore, N. Assessment of groundwater potential zone using remote sensing, GIS and multi criteria decision analysis techniques. *J. Geol. Soc. India* **2016**, *88*, 481–492. [CrossRef]
38. Guth, P.L.; Van Niekerk, A.; Grohmann, C.H.; Muller, J.-P.; Hawker, L.; Florinsky, I.V.; Gesch, D.; Reuter, H.I.; Herrera-Cruz, V.; Riazanoff, S.; et al. Digital Elevation Models: Terminology and Definitions. *Remote Sens.* **2021**, *13*, 3581. [CrossRef]
39. Januchowski, S.R.; Pressey, R.L.; VanDerWal, J.; Edwards, A. Characterizing errors in digital elevation models and estimating the financial costs of accuracy. *Int. J. Geogr. Inf. Sci.* **2010**, *24*, 1327–1347. [CrossRef]
40. Mukherjee, S.; Joshi, P.K.; Mukherjee, S.; Ghosh, A.; Garg, R.D.; Mukhopadhyay, A. Evaluation of vertical accuracy of open source Digital Elevation Model (DEM). *Int. J. Appl. Earth Obs. Geoinf.* **2013**, *21*, 205–217. [CrossRef]
41. Shamsudduha, M.; Taylor, R.G.; Ahmed, K.M.; Zahid, A. The impact of intensive groundwater abstraction on recharge to a shallow regional aquifer system: Evidence from Bangladesh. *Hydrogeol. J.* **2011**, *19*, 901–916. [CrossRef]

42. Ferozur, R.M.; Jahan, C.S.; Arefin, R.; Mazumder, Q.H. Groundwater potentiality study in drought prone barind tract, NW Bangladesh using remote sensing and GIS. *Groundw. Sustain. Dev.* **2019**, *8*, 205–215. [CrossRef]
43. Shaban, A.; Khawlie, M.; Abdallah, C. Use of remote sensing and GIS to determine recharge potential zones: The case of Occidental Lebanon. *Hydrogeol. J.* **2006**, *14*, 433–443. [CrossRef]
44. Saaty, T.L. The analytic hierarchy process (AHP). *J. Oper. Res. Soc.* **1980**, *41*, 1073–1076.
45. Li, F.; Zhang, G.; Xu, Y.J. Spatiotemporal variability of climate and streamflow in the Songhua River Basin, northeast China. *J. Hydrol.* **2014**, *514*, 53–64. [CrossRef]
46. Xu, J.; Chen, Y.; Li, W.; Nie, Q.; Song, C.; Wei, C. Integrating wavelet analysis and BPANN to simulate the annual runoff with regional climate change: A case study of Yarkand River, Northwest China. *Water Resour. Manag.* **2014**, *28*, 2523–2537. [CrossRef]
47. Arulbalaji, P.; Padmalal, D.; Sreelash, K. GIS and AHP techniques based delineation of groundwater potential zones: A case study from southern Western Ghats, India. *Sci. Rep.* **2019**, *9*, 1–17. [CrossRef] [PubMed]
48. Thapa, R.; Gupta, S.; Guin, S.; Kaur, H. Assessment of groundwater potential zones using multi-influencing factor (MIF) and GIS: A case study from Birbhum district, West Bengal. *Appl. Water Sci.* **2017**, *7*, 4117–4131. [CrossRef]
49. Iserloh, T.; Ries, J.; Cerdà, A.; Echeverría, M.; Fister, W.; Geißler, C.; Kuhn, N.; León, F.; Peters, P.; Schindewolf, M. Comparative measurements with seven rainfall simulators on uniform bare fallow land. *Z. Geomorphol.* **2012**, *57*, 11–26. [CrossRef]
50. Huang, J.; Wu, P.; Zhao, X. Effects of rainfall intensity, underlying surface and slope gradient on soil infiltration under simulated rainfall experiments. *Catena* **2013**, *104*, 93–102. [CrossRef]
51. Anuraga, T.S.; Ruiz, L.; Kumar, M.S.M.; Sekhar, M.; Leijnse, A. Estimating groundwater recharge using land use and soil data: A case study in South India. *Agric. Water Manag.* **2006**, *84*, 65–76. [CrossRef]
52. Kaliraj, S.; Chandrasekar, N.; Magesh, N. Identification of potential groundwater recharge zones in Vaigai upper basin, Tamil Nadu, using GIS-based analytical hierarchical process (AHP) technique. *Arab. J. Geosci.* **2014**, *7*, 1385–1401. [CrossRef]
53. Malczewski, J. *GIS and Multicriteria Decision Analysis*; John Wiley & Sons: Hoboken, NJ, USA, 1999.
54. Yeh, H.-F.; Lee, C.-H.; Hsu, K.-C.; Chang, P.-H. GIS for the assessment of the groundwater recharge potential zone. *Environ. Geol.* **2009**, *58*, 185–195. [CrossRef]
55. Bonilla Valverde, J.P.; Blank, C.; Roidt, M.; Schneider, L.; Stefan, C. Application of a GIS Multi-Criteria Decision Analysis for the Identification of Intrinsic Suitable Sites in Costa Rica for the Application of Managed Aquifer Recharge (MAR) through Spreading Methods. *Water* **2016**, *8*, 391. [CrossRef]
56. Wang, H.; Yang, Z.; Saito, Y.; Liu, J.P.; Sun, X. Interannual and seasonal variation of the Huanghe (Yellow River) water discharge over the past 50 years: Connections to impacts from ENSO events and dams. *Glob. Planet. Chang.* **2006**, *50*, 212–225. [CrossRef]
57. Nath, H.; Adhikary, S.K.; Nath, S.; Kafy, A.A.; Islam, A.R.M.T.; Alsulamy, S.; Khedher, K.M.; Shohan, A.A.A. Long-term trends and spatial variability in rainfall in the southeast region of Bangladesh: Implication for sustainable water resources management. *Theor. Appl. Climatol.* **2024**, *155*, 3693–3717. [CrossRef]
58. Cristiano, E.; ten Veldhuis, M.-c.; Van De Giesen, N. Spatial and temporal variability of rainfall and their effects on hydrological response in urban areas—A review. *Hydrol. Earth Syst. Sci.* **2017**, *21*, 3859–3878. [CrossRef]
59. Ogunbode, T.; Ifabiyi, P. Rainfall trends and its implications on water resources management: A case study of Ogbomoso city in Nigeria. *Int. J. Hydrol.* **2019**, *3*, 210–215. [CrossRef]
60. Liu, H.; Lei, T.; Zhao, J.; Yuan, C.; Fan, Y.; Qu, L. Effects of rainfall intensity and antecedent soil water content on soil infiltrability under rainfall conditions using the run off-on-out method. *J. Hydrol.* **2011**, *396*, 24–32. [CrossRef]
61. Hawke, R.; Price, A.; Bryan, R. The effect of initial soil water content and rainfall intensity on near-surface soil hydrologic conductivity: A laboratory investigation. *Catena* **2006**, *65*, 237–246. [CrossRef]
62. Nath, H.; Rafizul, I.M. Spatial Variability of Metal Elements in Soils of a Waste Disposal Site in Khulna: A Geostatistical Study. In *Advances in Civil Engineering, Proceedings of the 5th International Conference on Advances in Civil Engineering (ICACE), Chattogram, Bangladesh, 21–23 December 2020*; Springer: Singapore, 2022; pp. 25–36.
63. Yasrebi, J.; Saffari, M.; Fathi, H.; Karimian, N.; Emadi, M.; Baghernejad, M. Spatial variability of soil fertility properties for precision agriculture in Southern Iran. *J. Appl. Sci.* **2008**, *8*, 1642–1650. [CrossRef]
64. Machiwal, D.; Jha, M.K. Characterizing rainfall–groundwater dynamics in a hard-rock aquifer system using time series, geographic information system and geostatistical modelling. *Hydrol. Process.* **2014**, *28*, 2824–2843. [CrossRef]
65. Hammouri, N.; El-Naqa, A.; Barakat, M. An Integrated Approach to Groundwater Exploration Using Remote Sensing and Geographic Information System. *J. Water Resour. Prot.* **2012**, *04*, 717–724. [CrossRef]
66. Naghibi, S.A.; Pourghasemi, H.R.; Dixon, B. GIS-based groundwater potential mapping using boosted regression tree, classification and regression tree, and random forest machine learning models in Iran. *Environ. Monit. Assess.* **2015**, *188*, 44. [CrossRef] [PubMed]
67. Pirone, M.; Papa, R.; Nicotera, M.V.; Urciuoli, G. In situ monitoring of the groundwater field in an unsaturated pyroclastic slope for slope stability evaluation. *Landslides* **2015**, *12*, 259–276. [CrossRef]
68. Nowreen, S.; Newton, I.H.; Zzaman, R.U.; Islam, A.; Islam, G.M.T.; Alam, M.S. Development of potential map for groundwater abstraction in the northwest region of Bangladesh using RS-GIS-based weighted overlay analysis and water-table-fluctuation technique. *Environ. Monit. Assess.* **2021**, *193*, 24. [CrossRef]
69. Andualem, T.G.; Demeke, G.G. Groundwater potential assessment using GIS and remote sensing: A case study of Guna Tana landscape, Upper Blue Nile Basin, Ethiopia. *J. Hydrol. Reg. Stud.* **2019**, *24*, 100610. [CrossRef]

70. Horton, R.E. Drainage-basin characteristics. *Trans. Am. Geophys. Union* **1932**, *13*, 350–361.
71. Tucker, G.E.; Catani, F.; Rinaldo, A.; Bras, R.L. Statistical analysis of drainage density from digital terrain data. *Geomorphology* **2001**, *36*, 187–202. [CrossRef]
72. Henderson-Sellers, A.; Wilson, M.F. Surface albedo data for climatic modeling. *Rev. Geophys.* **1983**, *21*, 1743–1778. [CrossRef]
73. He, S.; Wu, J. Relationships of groundwater quality and associated health risks with land use/land cover patterns: A case study in a loess area, Northwest China. *Hum. Ecol. Risk Assess. Int. J.* **2019**, *25*, 354–373. [CrossRef]
74. Irwin, E.G.; Geoghegan, J. Theory, data, methods: Developing spatially explicit economic models of land use change. *Agric. Ecosyst. Environ.* **2001**, *85*, 7–24. [CrossRef]
75. Karcz, I. Rapid determination of lineament and joint densities. *Tectonophysics* **1978**, *44*, T29–T33. [CrossRef]
76. Sander, P. Lineaments in groundwater exploration: A review of applications and limitations. *Hydrogeol. J.* **2007**, *15*, 71–74. [CrossRef]
77. Fu, B.; Chen, L.; Ma, K.; Zhou, H.; Wang, J. The relationships between land use and soil conditions in the hilly area of the loess plateau in northern Shaanxi, China. *CATENA* **2000**, *39*, 69–78. [CrossRef]
78. Broersma, K.; Robertson, J.A.; Chanasyk, D.S. Effects of different cropping systems on soil water properties of a Boralf soil. *Commun. Soil Sci. Plant Anal.* **1995**, *26*, 1795–1811. [CrossRef]
79. Upwanshi, M.; Damry, K.; Pathak, D.; Tikle, S.; Das, S. Delineation of potential groundwater recharge zones using remote sensing, GIS, and AHP approaches. *Urban Clim.* **2023**, *48*, 101415. [CrossRef]
80. Kaewdum, N.; Chotpantarat, S. Mapping potential zones for groundwater recharge using a GIS technique in the lower Khwae Hanuman Sub-Basin Area, Prachin Buri Province, Thailand. *Front. Earth Sci.* **2021**, *9*, 717313. [CrossRef]

water

MDPI

Article

Identification and Validation of Groundwater Potential Zones in Al-Madinah Al-Munawarah, Western Saudi Arabia Using Remote Sensing and GIS Techniques

Abdelbaset S. El-Sorogy [1,*], Talal Alharbi [1], Khaled Al-Kahtany [1], Naji Rikan [1] and Yousef Salem [2]

[1] Geology and Geophysics Department, College of Science, King Saud University, Riyadh 11451, Saudi Arabia; tgalharbi@ksu.edu.sa (T.A.); kalgatani@ksu.edu.sa (K.A.-K.)
[2] Department of Geography, King Saud University, Riyadh 11451, Saudi Arabia; ysalem@ksu.edu.sa
* Correspondence: asmohamed@ksu.edu.sa

Abstract: Groundwater is an essential water resource utilized for agricultural, industrial, and home applications. Evaluating the variability of groundwater is essential for the conservation and management of this resource, as well as for mitigating the reduction in groundwater levels resulting from excessive extraction. This study aimed to define the groundwater potential zones (GWPZ) in Al-Madinah Al-Munawarah, Western Saudi Arabia, utilizing remote sensing and geographic information system (GIS) techniques, alongside meteorological data. Seven thematic maps were produced based on the regulatory characteristics of geology, drainage density, height, slope, precipitation, soil, and normalized difference vegetation index (NDVI). The influence of each theme and subunit/class on groundwater recharge was evaluated by weighted overlay analysis, including previous research and field data. The groundwater potential map was created via the weighted index overlay approach within a GIS. The groundwater potentials were classified into three categories: very poor, moderate, and good zones. The low groundwater potential regions encompass 805.81 km^2 (44.91%) of the research area, located in mountainous basement rocks, characterized by high drainage density and steep gradients. The moderate zones comprise 45.67% of the total area, covering 819.31 km^2, and are situated in low-lying regions at the base of mountainous mountains. Conversely, the favorable zones, comprising 9.42% of the total area, span 169.06 km^2 and are located within the alluvial deposits of the lowlands next to the Wadi Al-Hamd basin and agricultural farms. The results' accuracy was confirmed by overlaying data from 26 wells onto the designated groundwater potential categories, revealing that all wells corresponded with regions of high groundwater potential. The generated map would contribute to the systematic and efficient management of groundwater resources in this area to meet the rising water demands of Al-Madinah. The groundwater potential map is one aspect of groundwater management. It is also very important to assess this potential further via groundwater temporal monitoring, groundwater balance, and modeling.

Keywords: GIS; Groundwater Mapping; remote sensing; sustainable development; water supply

1. Introduction

Groundwater serves as a substitute for surface water and is a crucial natural resource utilized in home, agricultural, and industrial sectors, contributing to the overall development of a region [1]. The presence, arrangement, and flow of underground water mostly rely on the geological and hydro-geomorphological characteristics of the region [2]. Identifying and assessing regions with significant groundwater potential is essential for establishing the ideal locations for new water extraction wells to meet the increasing demand for water, particularly in areas experiencing low rainfall and water scarcity. The amalgamation of remotely sensed data with geographic information systems (GIS) is very compatible and may be effortlessly integrated with data acquired from traditional and terrestrial measurement instruments [3]. The utilization of high-resolution satellite images is prevalent in groundwater research owing to their exceptional spectral and spatial resolution. These images are utilized to determine the geology, geomorphology, soil composition, lineament density, drainage density, rainfall patterns, and land use for maps that reveal the presence of groundwater [3,4]. Remote sensing and GIS offer several advantages over conventional approaches [5]. These advantages include being time- and cost-effective, as well as providing synoptic coverage.

In recent decades, there has been a rise in the exploitation of groundwater resources, leading to excessive consumption of groundwater. Consequently, a variety of environmental problems have arisen, including a reduction in groundwater levels, exhaustion of water resources, pollution, and degradation of water quality [6]. Scientists globally employ Geographic Information Systems (GIS) and remote sensing techniques to assess the prospective areas for GWPZ. The outcomes of these assessments differ according to the parameters considered in calculating the GWPZ [7,8]. Moreover, the outcomes they achieve may vary based on the diverse elements taken into account when defining these zones. One study demonstrates the use of lineaments for groundwater exploration [9], while other studies explain the utilization of various factors such as geomorphology, land use, geology, drainage density, slope, rainfall intensity, and soil texture [6,10]. The accessibility of satellite data, traditional maps, and accurately adjusted ground truth data has made it easier to provide a foundation for evaluating areas with potential groundwater resources [6]. Furthermore, the outcomes derived from GIS and remote sensing for evaluating GWPZ are deemed trustworthy, as they undergo verification by field surveys.

The Kingdom of Saudi Arabia is experiencing significant pressure on its groundwater resources due to its predominantly arid to semiarid climate, as well as rapid population expansion, urbanization, and industrial development in the country [11,12]. Shallow aquifers in Western Saudi Arabia are predominantly situated in valleys between basement terranes and along coastal areas [13,14]. Nonetheless, the Arabian Shield possesses substantial strata of sedimentary deposits ranging in age from the Paleozoic to the present, which constitute deep aquifers and act as a crucial reservoir of groundwater in the region [15]. The predominant published studies regarding groundwater in the western region concentrated on evaluating groundwater quality and identifying pollution sources. This was accomplished by examining the concentrations of significant cations, anions, and possibly deleterious elements in Western Saudi Arabia and along the Saudi Red Sea coastline (e.g., [16–23]).

Al-Madinah Al-Munawarah is a prominent metropolis in Saudi Arabia that attracts a large number of seasonal visitors due to its pilgrimage. In order to address the increasing need for water in agriculture, industry, and domestics, the authorities have thoroughly investigated, extracted, and developed [24]. The Al-Madinah province contains many aquifers, such as fractured Precambrian bedrock, Paleozoic sedimentary rocks, and alluvial deposits overlaid by recent lava flows [25]. The region is experiencing a water deficit due to very high temperature, semi-arid to extremely arid conditions, poor rainfall, and heavy pumping from groundwater aquifers for agricultural purposes [16,26–28].

Weighted overlay mapping is an effective method for identifying areas with higher groundwater potential. It combines multiple geospatial factors, such as geology, soil type, slope, and drainage density, to help pinpoint regions where groundwater exploration might be more successful. However, it is important to be cautious and recognize that while weighted overlay mapping is a useful tool for identifying GWPZ, it does have limitations. This mapping technique does not provide information about the sustainability of the groundwater resources in those areas or the rates at which groundwater can be sustainably extracted. The objective of this study is to delineate the areas with high potential for groundwater in the Al-Madinah Al-Munawarah region of Western Saudi Arabia using remote sensing and GIS techniques. The main goals are to create theme maps that show the characteristics that affect groundwater potential in the area and to assign weights to each of these thematic maps using the weighted overlay analysis (WOA).

This research introduces a novel approach to groundwater exploration by applying the WOA method combined with GIS and remote sensing data. Unlike many global studies that use all available WOA data, even those with minimal impact, this study focuses on identifying the most influential factors specific to the study area for pinpointing potential groundwater sites. Diverse data sources—including soil, topography, geology, and hydrology—were integrated with remote sensing data in a unique way not previously applied in this region. By developing an optimized model that adjusts criteria weights based on local characteristics, the accuracy in identifying potential groundwater areas was enhanced. Applying this methodology to an area previously unexamined with this approach, our research successfully identified new high-potential groundwater zones, contributing to more effective water resource planning.

2. Materials and Methods

2.1. Study Area

Al-Madinah Al-Munawarah is situated within the latitudes of 24°15′54″ and 24°36′54″ N and the longitudes of 39°29′06″ and 39°50′15″ E (Figure 1). The groundwater resources of Madinah are found in the highly permeable Harrat Rahat region in the east and south, as well as in alluvial deposits to the west, encompassing the core area of Madinah city. The groundwater level is comprised between 535 and 594 m, whereas the water table depth varies between 28 and 93 m [17,29]. Over the past twenty years, Al-Madinah has transitioned from relying completely on its groundwater to relying almost exclusively on desalinated water from the Yanbu desalination plant. The city's groundwater is harnessed for agricultural and industrial purposes. The climate of Al-Madinah city is characterized by an arid climate with a hot and dry atmosphere. During the summer, the weather is scorching, while in the winter, it is relatively chilly. According to [30], the summer temperature typically falls within the range of 25–42 degrees Celsius, while the winter temperature falls within the range of 10–24 degrees Celsius. The majority of precipitation occurs in November, December, and January, with intermittent rainstorms occurring in April [19].

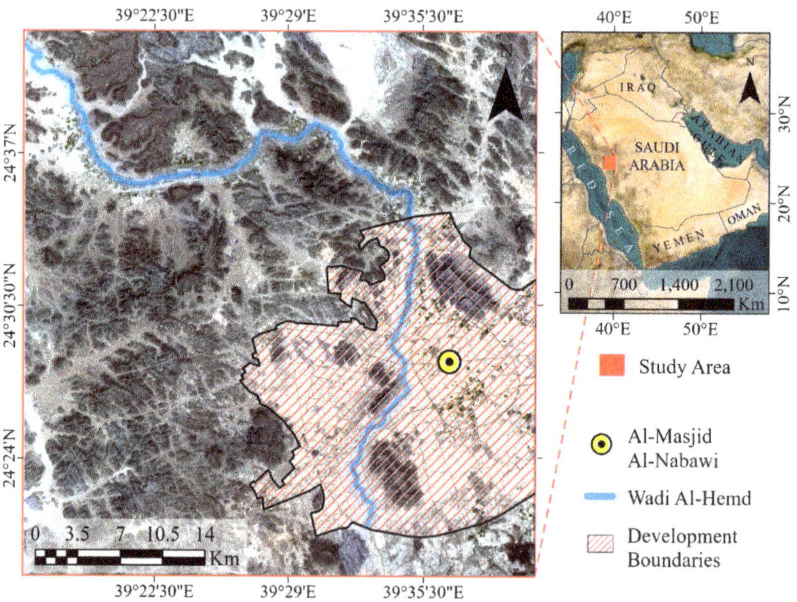

Figure 1. Location map of the study in Al-Madinah Al-Munawarah Province. The smaller map located in the right provides the study area's location in Saudi Arabia.

The Wadi Al-Hamd Basin is recognized as one of the largest basins in the Kingdom of Saudi Arabia, with a total area of 104,679 square kilometers. The Wadi Al Hamd flows in a southeastern to northwestern direction, dividing into two primary valleys. One of these valleys, called Wadi Al-Hamd, continues to flow northwest towards the Red Sea. The other valley, known as Wadi Al Jazl, travels north towards the city of Al Ula. The primary economic activity in this basin is agriculture, which heavily relies on unlicensed and uncontrolled groundwater wells, primarily drilled wells [27]. The precipitation patterns within the basin exhibit temporal and spatial variations. The Wadi Al-Hamd Basin experiences an annual rainfall distribution ranging from 40 mm to 80 mm, with an increasing trend from the northwest to the southeast. The Wadi Al-Hamd is distinguished by the existence of two formations that contain groundwater. The Quaternary aquifer is located in the ancient basins, while the volcanic aquifer covers the southern part of Al-Madinah Al-Munawarah city [25,27].

2.2. Datasets

The study employed a three-phase analytical framework to evaluate GWPZ in Al-Madinah Al-Munawarah, Saudi Arabia. During the preliminary phase, substantial data gathering was conducted, integrating thorough fieldwork, remote sensing data from NASA, and geological information from the Saudi Geological Survey. A 12.5 m resolution Digital Elevation Model (DEM) from Sentinel 1 data was employed to examine the city's topography, encompassing elevation and slope attributes. Geological and soil data elucidated rock composition and permeability—elements affecting water discharge and absorption. Such data were obtained from Geological and soil data were obtained from the Saudi Geological Survey [31]. Precipitation data from the Ministry of Environment, Water, and Agriculture, spanning 2010 to 2024, were analyzed to discern rainfall patterns. The city's drainage network, encompassing wadi systems, was developed from the DEM to evaluate natural drainage patterns.

During the second step, a weighted overlay analysis was performed to delineate possible groundwater zones. To achieve compatibility and precision across all datasets, we implemented several preprocessing steps, including georeferencing, resampling, and normalization. All datasets were standardized to a consistent spatial resolution of 12.5 m. This uniform scale facilitated more accurate overlay and analysis during the weighted overlay mapping process. Each dataset was allocated a weight indicative of its significance in influencing groundwater potential, established through a synthesis of field observations. Field observations within the study area were instrumental in identifying and understanding the characteristics of regions with groundwater presence. By conducting on-site surveys and measurements, we understand the hydrogeological features that influence groundwater occurrence, such as the locations of groundwater, existing wells, types of rocks and soil, and land use practices. The insights gained from these field observations enhance our understanding of the complex interactions between geological and environmental factors affecting groundwater availability in the area and assist us in assigning weights to the datasets and reclassifying them based on their importance. The weighted datasets were amalgamated using ArcGIS Pro to produce a composite GWPZ map, classifying regions into poor, moderate, and good GWPZ (Figure 2).

The concluding step encompassed field verification and remote sensing validation techniques for the GWPZ map. Field validation was performed subsequent to the creation of this map by visiting the study region to evaluate the poor, moderate, and good GWPZ. Farms equipped with groundwater wells should ideally be situated in regions designated as high potential zones, thereby affirming the existence of groundwater in those locales. In contrast, suboptimal potential zones should exhibit an absence of physical indicators of agriculture or groundwater wells. The second method utilized the Normalized Difference Vegetation Index (NDVI), a remote sensing tool, to identify vegetation in the region. Given that vegetative regions presumably employ groundwater wells for irrigation, we utilized this data to ascertain whether the agricultural fields were situated in optimal potential zones or alternative zones, hence corroborating the GWPZ map.

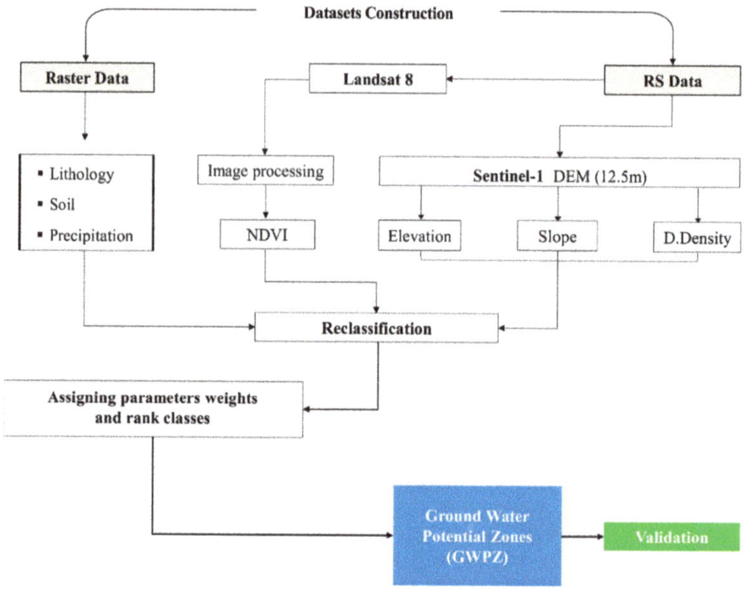

Figure 2. A flow chart outlining how the datasets were utilized to map the GWPZ in the study area.

Table 1 delineates the important factors and rankings for identifying potential groundwater zones. Each aspect was allocated a weight based on its influence on the phenomenon and on previous studies [21,32]. Subsequently, each layer was categorized into five classifications and assigned relative weights. The soil received the lowest weight due to its thin composition, mostly consisting of volcanic fields (Harrat), which restricts its influence on groundwater infiltration. In contrast, drainage density and geology were assigned the most significance due to the influence of fractures, faults, and valley branching in channelling water into subterranean strata. In these classifications, a grade of (5) indicates the highest probability of groundwater presence, whilst a rating of (1) signifies the lowest probability. Following the classification and allocation of weights, a weighted overlay analysis of these layers was performed to generate the final map indicating areas with groundwater potential.

Table 1. The layers utilized in the overlay analysis to identify GWPZ.

Parameter	Weight (%)	Classes	Rank
Slope (Degree)	10	0–6	5
		7–12	4
		13–20	3
		21–28	2
		29–47	1
Precipitation (mm)	15	37–43	1
		44–47	2
		48–51	3
		52–55	4
		56–58	5

Table 1. *Cont.*

Parameter	Weight (%)	Classes	Rank
Elevation (m)	15	489–617	5
		618–695	4
		696–780	3
		781–888	2
		889–1772	1
Lithology	20	Igneous Rocks	2
		Metamorphic Rocks	3
		Sedimentary Rocks	4
Drainage density (km^2)	20	0–0.2	1
		0.3–0.5	2
		0.6–0.7	3
		0.8–1	4
		1.1–1.9	5
Soil	5	Calciorthids	1
		Lava flows	2
		Torriorthents	3
NDVI	15	−0.078–0.065	1
		0.066–0.11	2
		0.12–0.15	3
		0.16–0.20	4
		0.21–0.46	5

3. Results and Discussion

The accessibility of groundwater is chiefly influenced by geology, geomorphology, soil composition, lineament density, drainage density, precipitation patterns, and NDVI. Consequently, all thematic maps pertinent to the study area have been generated, and these maps are delineated below.

3.1. Geology

From a geological perspective, the study area has rock formations that span in age from Precambrian to Quaternary (Figure 3). The Precambrian rocks (Neoproterozoic 900 to 540 Mya) consist mostly of metavolcanic and meta-sedimentary rocks. The Cambrian and Cambrian-Ordovician formations consist of alternating layers of coarse-to-fine-grained sandstones and micro gabbro rocks. Furthermore, there are lava flows from the Tertiary to Quaternary period, as well as deposits from the Quaternary period consisting of sand, gravel, silt, and sabkha units [33–37]. The Wadi Al-Hamd Basin can be divided into three distinct topographical zones [27]. Firstly, there is the Najd Plateau, situated in the eastern part of the basin and ranging in elevation from 750 to 1600 m above mean sea level. Secondly, there is the Hijaz mountain, which spans from sea level in the main channel of the Wadi to over 2000 m on the western side. Lastly, the northern and southern parts of the basin are characterized by the presence of lava flows from Harrat Hirmah and Harrat Rahat. Calciorthids and Torriorthents are the predominant soil types in the study area, particularly within Wadi Al-Hamd (Figure 3). Calciorthids create a layer of soil and have a texture that is sandy to loamy. Torriorthents primarily develop on slopes that are undergoing active erosion. The composition of Torriorthents includes loamy sand, fine sandy loam, sandy loam, loam, or clay loam, as well as their gravelly equivalents [38–42].

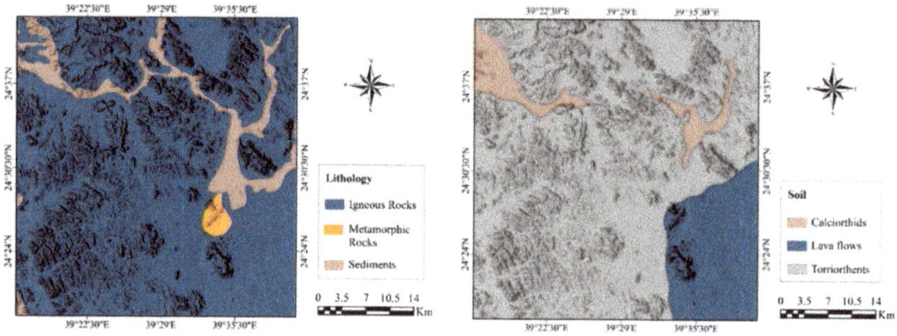

Figure 3. (**left**) A geological map of the various rock formations presents in the study area. The research region predominantly consists of diorite, granodiorite, granite, and granite gneiss rocks, as well as a few sedimentary rocks. (**right**) A map exhibits the arrangement of soil types within the designated area.

3.2. Elevation and Slope

Figure 4 shows that more than 65 percent of the research region is situated at an elevation below 696 m above sea level. The lower-lying topography plays a critical role in the movement of both groundwater and surface water, particularly because it absorbs water from steep slopes that have limited vegetation. Madinah's terrain consists of flat plains, hilly regions, and valleys. Typically, the altitude of the ground in the plains varies from 600 m to 620 m above mean sea level (AMSL). The city contains mountainous regions located in the north, south, and west, ranging in elevation from 800 m to 1500 m above mean sea level (AMSL). The topography of the region exhibits an east-to-west slope, which continues until it reaches the Al-Aqiq valley [19]. The research area exhibits a diverse range of altitudes, with the highest point reaching an elevation of 1272 m above mean sea level (AMSL) in the southwestern region and the lowest point measuring 489 m AMSL in the northern half, specifically in the main stretch of Wadi Al Hamd. The study area's elevation was divided into five unique groups ranging from 489 to 617 m above mean sea level (AMSL) to 889 to 1272 m AMSL (Figure 4). The majority of the research region has slopes that are less than 13 degrees. Wadi Qana and Wadi Aqiq are expansive wadis located in the Al-Madinah region. The Wadi Aqiq flows in a northerly direction and converges with Wadi Qana in the eastern region of the Al-Madinah area. The outlets of the Madinah basin converge and continue as a single main valley, referred to as Wadi Al Hamd [17]. The study region was divided into five categories based on the slope values: 0 to 6°, 7 to 12°, 13 to 20°, 21 to 28°, and 29 to 74° (Figure 4).

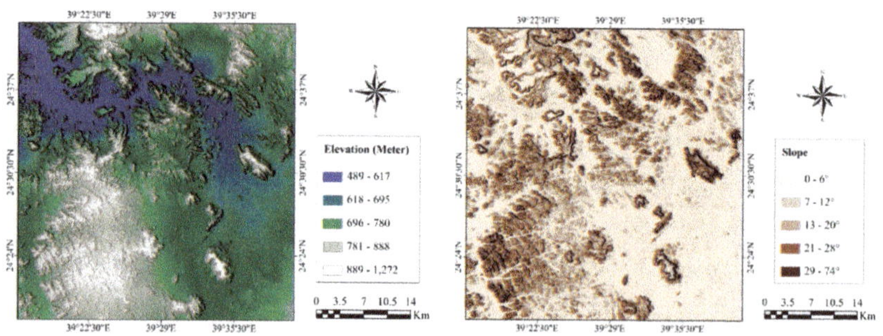

Figure 4. (**left**) A map depicting the digital portrayal of the surface elevation of the study area, showing a range from 489 to 1272 m above mean sea level (AMSL). (**right**) A map illustrating the various slopes found in the study areas. The gradient in the lower class ranges from 0 to 6, while in the higher class it spans from 29 to 74.

3.3. Drainage Density and Stream Network

Wadi Al-Hamd originates in the northwestern region, where it flows through elevated mountain ranges. As it progresses, minor tributaries join the main channel at sharp angles, forming a dendritic drainage pattern. The wadi then turns eastward before continuing southeast, eventually reaching the southeastern part of Al-Madinah. The land in the study region was classified into five categories, ranging from 0–0.2 to 1.10–1.90 km/km^2. The downstream areas showed a high drainage density along the elevation (Figure 5). However, specific regions located further upstream have a higher density of drainage. The study region encompasses five stream orders, ranging from first order in the high highlands and descending to the fifth order when they converge in the main wadi, such as Al-Hamd. According to [43], a low drainage density is a reliable predictor of high potential for groundwater.

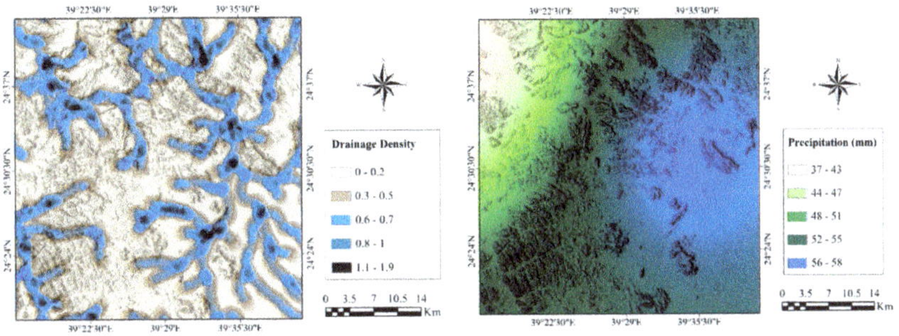

Figure 5. (**left**) The map displays the five classifications of drainage density in the study region, with values ranging from 0 to 0.20 (km/km^2) for the lowest classification and from 1.10 to 1.90 (km/km^2) for the highest classification. (**right**) A map illustrating the precise distribution of precipitation levels within the studied area. The highest recorded measurement was 58 mm, while the lowest recorded measurement was 37 mm.

3.4. Rainfall

Al-Madinah Al-Munawarah has an arid climate but experiences increased precipitation from November to January. It is characterized by low rainfall, high temperatures, and an overall dry environment. Rain typically falls in sporadic bursts during November, December, March, and April. The average annual rainfall in the Madinah area is approximately 40.1 mm. The measured rate of infiltration ranges from 0.13 to 1.01 cm/min. The average daily temperatures fluctuated between 27 and 43 °C during the months of July and August and between 10 and 25 °C in December and January [17]. The yearly maximum precipitation is 52.9 mm. The research area experiences little variation in the annual precipitation rate. The eastern portion of the study region has a greater amount of precipitation on a yearly basis. Thiessen polygons were used to verify meteorological data by considering the distribution of stations. The data were classified into five intervals ranging from 37–43 to 55–58 mm/year (Figure 5).

3.5. NDVI

The normalized difference vegetation index (NDVI) is a calculated value derived from measurements of reflectance in the visible and near-infrared sections of the electromagnetic spectrum. It is used to assess the amount of plant cover in a given area [44]. It quantifies this disparity, offering an assessment of the density and state of vegetation. From a quantitative perspective, areas with high plant coverage will exhibit positive values, whereas areas with soil and non-vegetation coverage will have slightly fewer positive values [45,46]. The fluctuation in the water table is influenced by several elements, including the intensity and amount of rainfall, the ability of rocks and soil to absorb water, the depth of groundwater above sea level, the shape of the land, the process of water evaporation and plant transpiration, and the amount of water discharged from wells [47]. The groundwater table may exert control over vegetation succession and cover patterns [48]. The land in the area was classified into five categories based on the NDVI values: −0.078–0.065, 0.066–0.11, 0.12–0.15, 0.16–0.20, and 0.21–0.46 (Figure 6). The minimum NDVI value recorded was −0.078, indicating the presence of bare or soil land in hilly regions with no vegetation cover [49]. The highest recorded NDVI value was 0.46, indicating the extent of vegetation coverage by palm trees in the agricultural lands located in Wadi Al-Hamd. The vegetation in the east–west expansion of Wadi Al-Hamd significantly increased in density.

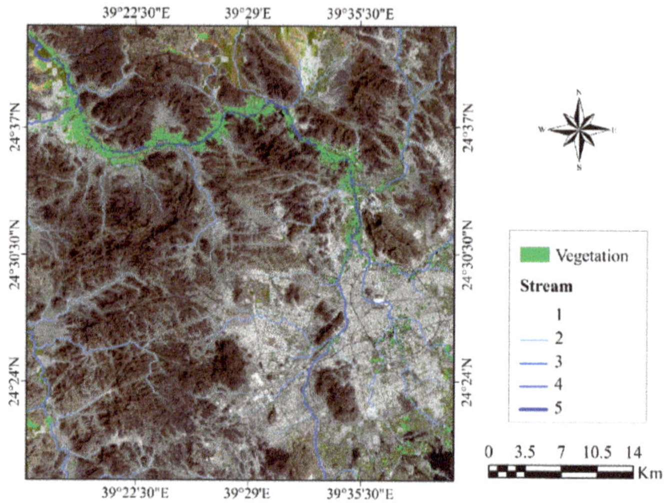

Figure 6. The study area encompasses five stream orders, ranging from first order in the high highlands to the fifth one in wadi Al-Hamd. The highest recorded NDVI value was reported in the east–west expansion of Wadi Al-Hamd.

3.6. Groundwater Potential Zones Map

The Wadi Al-Hamd Basin is the largest Wadi in the Al-Madinah Al-Munawarah Province. Previous studies described the groundwater resources of the Wadi Alhamd Basin and Al-Madinah Al-Munawwarah region as being defined by two primary aquifers: an alluvial Quaternary aquifer and a volcanic aquifer. The alluvial aquifer is situated along the recent basin, composed mainly of gravel, sand, and clay from erosion processes, while the volcanic aquifer extends through the southern part of Al-Madinah Al-Munawwarah city and includes the weathered basalt flows of Harrat Rahat, consisting of Tertiary and Quaternary lava flows [25,27]. These two hydrogeological units are hydrologically interconnected on a regional scale and function as unconfined to semi-confined aquifers. The weathered basalt unit acts as a secondary aquifer, whereas the primary groundwater source is the sub-basaltic alluvial deposits. Studies indicate that the depth of groundwater ranges from 30 to 90 m below the surface (Figure 7). Additionally, there are two buried wadis beneath the Rahat volcanic area containing alluvial deposits exceeding 50 m in thickness, with the average thickness of the alluvial unit varying from 15 to approximately 40 m, reaching up to 50 m at the center of Wadi Alhamd [25].

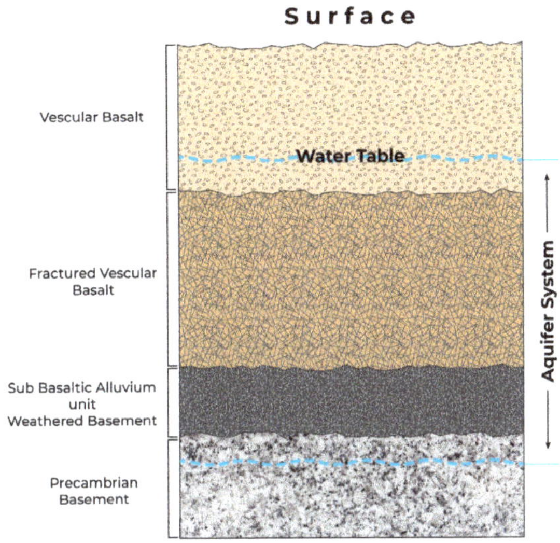

Figure 7. Geological cross-section of an aquifer system in the Wadi Al-Hamd Basin (modified after [25]).

The groundwater table depth in the Wadi Al-Hamd Basin ranges from 2 m to 200 m. The significant variety in depth can be primarily related to the differences in ground topography, types of aquifers, and their distribution [27]. The shallow groundwater is found in the southeastern and northeastern parts of the Wadi, but the deeper water tables are located in a north-south direction in the center area of the Wadi. This alignment corresponds to the presence of Tertiary and Quaternary (lava flows) formations. All findings were confirmed through on-site visits to the study area, specifically the agricultural farms located in the Wadi Al-Hamd basin. Groundwater samples were obtained from 26 wells around the Wadi (Figure 8) and subsequently analyzed for their depths and chemical composition in order to facilitate further investigation.

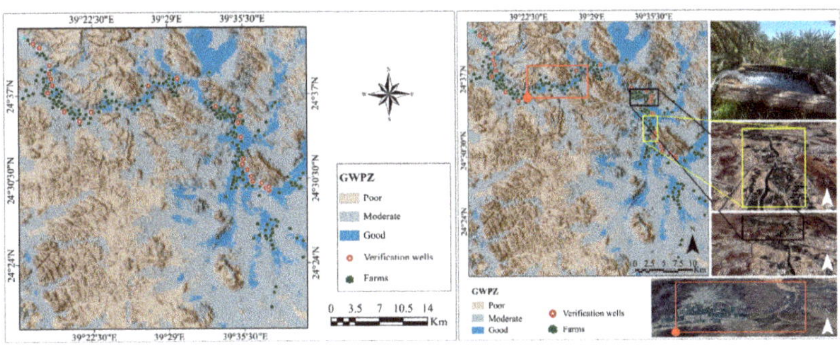

Figure 8. GWPZ resulting from the weighted overlay analysis.

Rainfall is identified as the main source of recharge for these two aquifers, contributing to both groundwater replenishment and surface runoff. Discharge predominantly occurs through production wells used for irrigation and domestic water supply. The hydrochemistry and types of groundwater are influenced by the area's rock-forming minerals, such as gypsum, anhydrite, and halite, as well as precipitation and reverse ion exchange processes. The lithology, precipitation, altitude, gradient, and drainage density characteristics significantly influence the recharge, occurrence, and accessibility of groundwater in this study.

Although the predominant rock types in the research area are igneous and metamorphic, characterized by poor porosity, they possess extensive fractures and foliation that facilitate the infiltration and percolation of rainwater. Thus, the fractures and foliations have a crucial role in increasing permeability, recharge rates, and groundwater potential [50]. Conversely, slope, soil texture, and elevation are considered important due to their individual responsibilities. The mountains in the research area include steep slopes, which result in a substantial amount of runoff. This is because the steep gradients facilitate a fast downward movement of precipitation, which reduces the time available for infiltration and replenishing the saturated zone [51]. The main soil types found in Wadi Al-Hamd are Calciorthids and Torriorthents. These soils have textures ranging from sandy to loamy or clay loamy, which facilitate the rapid movement of surface water into an aquifer system [48].

The catchment in the study area is categorized into three classifications according to its groundwater potential: very poor/poor, moderate, and good. The regions with extremely poor or poor potential encompass 805.81 km^2 of the research area, accounting for 44.91% of the total. The places are situated in the upstream segment of the watershed, particularly inside the hilly sections of the basement. The moderately groundwater-potential zones, covering an area of 819.31 km^2 (45.67%), are situated in low-lying areas at the base of mountainous regions. These zones are characterized by moderate slopes, permeability, and productivity. Conversely, the areas with good potential for groundwater cover a total of 169.06 km^2 (9.42%) and are located in Wadi Al-Hamd and its accompanying date farms, which consist of sandy loam and sandy clay loams. Good GWPZs are identified by their low drainage density, low terrain, and low slope.

Understanding how our groundwater potential mapping information would be utilized is crucial within Saudi Arabia's water resource management policies. In the Kingdom, all groundwater is considered national property managed by the government, and wells require permits from the Ministry of Environment, Water, and Agriculture. Our study provides valuable insights that can assist in permitting by identifying areas with high groundwater potential.

While a database of permitted wells exists, unpermitted or illegal wells may still be present, particularly in remote or inaccessible areas where on the ground monitoring is challenging. Remote sensing and GIS analyses employed in our study are instrumental in detecting unauthorized wells by identifying farms and agricultural activities that may not align with official records. By analyzing satellite imagery and spatial data, we can detect changes in land use patterns and vegetation indices indicating the presence of unauthorized groundwater extraction. This is important because illegal wells are often associated with unregulated farming practices that can lead to over-extraction and depletion of groundwater resources.

The challenges in identifying and validating GWPZ in the study area include the inadequate long-term and continuous monitoring, especially in remote areas, which creates uncertainty in assessing groundwater availability. The region's complex hydrogeology, due to fractured bedrock, leads to heterogeneous aquifer properties, complicating predictions of groundwater flow and storage capacity. Additionally, restricted recharge from the arid climate and minimal precipitation, combined with excessive abstraction, contribute to groundwater depletion.

4. Conclusions

This work delineated the GWPZ in Al-Madinah Al-Munawarah Province, Western Saudi Arabia, employing weighted overlay analysis that integrated seven thematic layers: geology, elevation, slope, soil, drainage density, rainfall, and NDVI. The study region was classified into zones of extremely poor, poor, moderate, and good groundwater potential. The research indicated that drainage density, slope, and soil were the principal factors influencing GWPZ in the examined region. The groundwater potential map was overlaid with the groundwater table elevation and the locations of the existing well fields. The existing well fields are located in regions with favorable groundwater potential. These data validate the precision of the map generated in this study. This study demonstrates the application of remote sensing and GIS methodologies to rapidly and effectively integrate surface and subsurface data. This integration may assist in pinpointing prospective sites for the future installation of groundwater well fields. This analysis, however, does not provide information about the sustainability of the groundwater resource and maximum pumping rates that could be abstracted in the long-term. Additional hydrogeological studies, including groundwater level monitoring and flow modelling, are essential to accurately evaluate groundwater resources' sustainability. Advanced geophysical techniques should be applied to precisely map subsurface structures, providing a deeper understanding of aquifer characteristics and groundwater potential. Artificial recharge projects should be initiated to strengthen the reliability of predictions for GWPZ, along with efforts to develop public and institutional capacity for long-term sustainability.

Author Contributions: Conceptualization, A.S.E.-S. and T.A.; methodology, T.A.; software, Y.S. and N.R.; validation, A.S.E.-S. and T.A.; investigation, K.A.-K.; writing—original draft preparation, A.S.E.-S. and T.A.; writing—review and editing, A.S.E.-S., K.A.-K. and T.A.; funding acquisition, T.A. All authors have read and agreed to the published version of the manuscript.

Funding: Researchers Supporting Project number (RSPD2024R791), King Saud University, Riyadh, Saudi Arabia.

Data Availability Statement: Data is contained within the article.

Acknowledgments: The authors extend their appreciation to Researchers Supporting Project number (RSPD2024R791), King Saud University, Riyadh, Saudi Arabia.

Conflicts of Interest: The authors declare no conflicts of interest.

References

1. Arabameri, A.; Rezaei, K.; Cerda, A.; Lombardo, L.; Rodrigo-Comino, J. GIS-based groundwater potential mapping in Shahroud plain, Iran. A comparison among statistical (bivariate and multivariate), data mining and MCDM approaches. *Sci. Total Environ.* **2019**, *658*, 160–177. [CrossRef] [PubMed]
2. Raj, S.; Rawat, K.S.; Singh, S.K.; Mishra, A.K. Groundwater potential zones identification and validation in Peninsular India. *Geol. Ecol. Landsc.* **2022**, *8*, 86–100. [CrossRef]
3. Suganthi, S.; Elango, L.; Subramanian, S.K. Groundwater potential zonation by Remote Sensing and GIS techniques and its relation to the Groundwater level in the Coastal part of the Arani and Koratalai River Basin, Southern India. *Earth Sci. Res. J.* **2013**, *17*, 87–95.
4. Preeja, K.R.; Joseph, S.; Thomas, J.; Vijith, H. Identifi¬cation of groundwater potential zones of a tropical river basin (Kerala, India) using remote sensing and GIS techniques. *J. Indian Soc. Remote Sens.* **2011**, *39*, 83–94. [CrossRef]
5. Zolekar, R.B.; Bhagat, V.S. Multi-criteria land suitability analysis for agriculture in hilly zone: Remote sensing and GIS approach. *Comput. Electron. Agric.* **2015**, *118*, 300–321. [CrossRef]
6. Pillai, K.S.; Sneha, M.L.; Aiswarya, S.; Anand, A.B.; Prasad, G.; Jayade, A. Unlocking Hidden Water Resources: Mapping Groundwater Potential Zones using GIS and Remote Sensing in Kerala, India. *E3S Web Conf.* **2023**, *405*, 04021. [CrossRef]
7. Das, S. Delineation of groundwater potential zone in hard rock terrain in Gangajalghati block, Bankura district, India using remote sensing and GIS techniques. *Model. Earth Syst. Environ.* **2017**, *3*, 1589–1599. [CrossRef]
8. Nair, H.C.; Padmalal, D.; Joseph, A. Hydrochemical Characterization and Water Quality Assessment of Spring and Well Water Sources of Two River Basins of Southern Western Ghats, Kerala, India. Doctoral Dissertation, Cochin University of Science and Technology, Kerala, India, 2017.
9. Suresh, Y.; Bindu, K.B. An Analysis of Land use/Land cover in Kadalundi River Basin using Remote Sensing and GIS. *Int. J. Geomat. Geosci.* **2015**, *6*, 1442–1449.
10. Prasad, G.; John, S.E. Delineation of ground water potential zones using GIS and remote sensing–A case study from midland region of Vamanapuram river basin, Kerala, India. *AIP Conf. Proc.* **2018**, *1952*, 020028.
11. Abderrahman, W.A.; Rasheeduddin, M.; Al-Harazin, I.M.; Esuflebbe, M.; Eqnaibi, B.S. Impacts of management practices on groundwater conditions in the Eastern Province, Saudi Arabia. *Hydrogeol. J.* **1995**, *3*, 32–41. [CrossRef]
12. AlSubih, M.; Kumari, M.; Mallick, J.; Ramakrishnan, R.; Islam, S.; Singh, C.K. Time series trend analysis of rainfall in last five decades and its quantification in Aseer Region of Saudi Arabia. *Arab. J. Geosci.* **2021**, *14*, 519. [CrossRef]
13. Al-Shaibani, A.M. Hydrogeology and hydrochemistry of a shallow alluvial aquifer, western Saudi Arabia. *Hydrogeol. J.* **2008**, *16*, 155–165. [CrossRef]
14. Mallick, J.; Singh, C.K.; AlMesfer, M.K.; Singh, V.P.; Alsubih, M. Groundwater quality studies in the Kingdom of Saudi Arabia: Prevalent research and management dimensions. *Water* **2021**, *13*, 1266. [CrossRef]

15. Alkolibi, F.M. Possible effects of global warming on agriculture and water resources in Saudi Arabia: Impacts and responses. *Clim. Chang.* **2002**, *54*, 225–245. [CrossRef]
16. El Maghraby, M.; Bamousa, A.O. Evaluation of groundwater quality for drinking and irrigation purposes using physicochemical parameters at Salilah area, Madinah Munawarah District, Saudi Arabia. *J. Taibah Univ. Sci.* **2021**, *15*, 695–709. [CrossRef]
17. Bamousa, A.O.; El Maghraby, M. Groundwater characterization and quality assessment, and sources of pollution in Madinah, Saudi Arabia. *Arab. J. Geosci.* **2016**, *9*, 536. [CrossRef]
18. Alghamdi, A.G.; Aly, A.A.; Aldhumri, S.A.; Al-Barakaha, F.N. Hydrochemical andquality assessment of groundwater resources in Al-madinah city, western Saudi Arabia. *Sustainability* **2020**, *12*, 3106. [CrossRef]
19. Alfaifi, H.; El-Sorogy, A.S.; Qaysi, S.; Kahal, A.; Almadani, S.; Alshehri, F.; Zaidi, F.K. Evaluation of heavy metal contamination and groundwater quality along the Red Sea coast, southern Saudi Arabia. *Mar. Poll. Bull.* **2021**, *163*, 111975. [CrossRef]
20. Alshehri, F.; Almadani, S.; El-Sorogy, A.S.; Alwaqdani, E.; Alfaifi, H.; Alharbi, T. Influence of seawater intrusion and heavy metals contamination on groundwater quality, Red Sea coast, Saudi Arabia. *Mar. Pollut. Bull.* **2021**, *165*, 112094. [CrossRef]
21. Alharbi, T.; Abdelrahman, K.; El-Sorogy, A.S.; Ibrahim, E. Identification of groundwater potential zones in the Rabigh-Yanbu area on the western coast of Saudi Arabia using remote sensing (RS) and geographic information system (GIS). *Front. Earth Sci.* **2023**, *11*, 1131200. [CrossRef]
22. Alharbi, T.; Abdelrahman, K.; El-Sorogy, A.S.; Ibrahim, E. Contamination and health risk assessment of groundwater along the Red Sea coast, Northwest Saudi Arabia. *Mar. Pollut. Bull.* **2023**, *192*, 115080. [CrossRef] [PubMed]
23. Ibrahim, E.; Abdelrahman, K.; Alharbi, T.; El-Sorogy, A.S. Delineation of seawater intrusion in the Yanbu industrial area, northwest Saudi Arabia, using geoelectric resistivity sounding survey. *J. King Saud Univ.—Sci.* **2024**, *36*, 103110. [CrossRef]
24. El Maghraby, M.M. Hydrogeochemical scharacterisation of groundwater aquifer in Al-Madinah Al-Munawarah City, Saudi Arabia. *Arab. J. Geosci.* **2015**, *8*, 4191–4206. [CrossRef]
25. Metwaly, M.; Abdalla, F.; Taha, A.I. Hydrogeophysical study of sub-basaltic alluvial aquifer in the southern part of Al-Madinah Al-Munawarah, Saudi Arabia. *Sustainability* **2021**, *13*, 9841. [CrossRef]
26. Sharaf, M.A.M. Major elements hydrochemistry and groundwater quality of Wadi fatimah, west central arabian Shield, Saudi Arabia. *Arab. J. Geosci.* **2013**, *6*, 2633–2653. [CrossRef]
27. Niyazi, B. Groundwater assessment for sustainable development in the Wadi Al-Hamd Basin, Al-Madinah Al-Munawarah, KSA. *J. Afr. Earth Sci.* **2024**, *215*, 105289. [CrossRef]
28. Masoud, M.; El Osta, M.; Al-Amri, N.; Niyazi, B.; Alqarawy, A.; Rashed, M. Groundwater Characteristics' Assessment for Productivity Planning in Al-Madinah Al-Munawarah Province, KSA. *Hydrology* **2024**, *11*, 99. [CrossRef]
29. Italconsult. Detailed investigation of the madina region. In *Final Report: Thermatic Report Number 5 and 7*; Saudi Arabian Ministry of Agriculture and Water: Riyadh, Saudi Arabia, 1979.
30. Khashogji, M.S.; El Maghraby, M.M. Evaluation of groundwater resources for drinking and agricultural purposes, Abar Al Mashi area, south Al Madinah Al Munawarah City, Saudi Arabia. *Arab. J. Geosci.* **2013**, *6*, 3929–3942. [CrossRef]
31. Sisson, T.h.W.; Calvert, A.T.; Mooney, W.D. *The Saudi Geological Survey-U.S. Geological Survey Northern Harrat Rahat Project—Styles, Rates, Causes, and Hazards of Volcanism Near Al Madīnah al Munawwarah, Kingdom of Saudi Arabia*; USGS Publications Warehouse, Professional Paper 2023; US Geological Survey: Reston, VA, USA, 1862. [CrossRef]
32. Mahmoud, S.H.; Alazba, A.A. Integrated remote sensing and GIS-based approach for deciphering groundwater potential zones in the central region of Saudi Arabia. *Environ. Earth Sci.* **2016**, *75*, 344. [CrossRef]
33. George, M.; Shorbaji, H. *Explanatory Notes to the Quadrangle, Sheet 24 D, Kingdom of Saudi Arabia*; Saudi Arabian Deputy Ministry for Mineral Resources Open-File Report BRGMOF-07-23; Deputy Ministry for Mineral Resources: Jeddah, Saudi Arabia, 1987; p. 65.
34. Stern, R.J. Arc assembly and continental collision in the Neoproterozoic East African orogen. Implications for the consolidation of Gondwanaland. *Annu. Rev. Earth Planet. Sci.* **1994**, *22*, 319–351. [CrossRef]
35. Johnson, P.R.; Kattan, F.H. Lithostratigraphic revision in the Arabian Shield: The impacts of geochronology and tectonic analysis. *Arab. J. Sci. Eng.* **2008**, *33*, 3–16.
36. Nehlig, P.; Genna, A.; Asfirane, F.; BRGM France; Guerrot, C.; Eberlé, J.M.; Kluyver, H.M.; Lasserre, J.L.; Le Goff, E.; Nicol, N.; et al. A review of the Pan-African evolution of the Arabian Shield. *GeoArabia* **2002**, *7*, 103–124. [CrossRef]
37. Stern, R.; Johnson, P. Continental lithosphere of the Arabian plate: A geologic, petrologic and geophysical synthesis. *Earth Sci. Rev.* **2010**, *101*, 29–67. [CrossRef]
38. Ministry of Agriculture and Water (MAW). *General Soil Map of the Kingdom of Saudi Arabia*; Ministry of Agriculture and Water, Land Management Department: Riyadh, Saudi Arabia, 1985.
39. Sheta, A.S. *Soil Quality: Standards of Soil Quality Under the Conditions of Saudi Arabia*, 7th ed.; Saudi Society for Agricultural Sciences, King Saud University: Riyadh, Saudi Arabia, 2004. (In Arabic)
40. Alarifi, S.S.; El-Sorogy, A.S.; Al-Kahtany, K.; Alotaibi, M. Contamination and Environmental Risk Assessment of Potentially Toxic Elements in Soils of Palm Farms in Northwest Riyadh, Saudi Arabia. *Sustainability* **2022**, *14*, 15402. [CrossRef]
41. El-Sorogy, A.S.; Al Khathlan, M.H. Assessment of potentially toxic elements and health risks of agricultural soil in Southwest Riyadh, Saudi Arabia. *Open Chem.* **2024**, *22*, 20240017. [CrossRef]
42. Alharbi, T.; El-Sorogy, A.S.; Al-Kahtany, K.H. Contamination and health risk assessment of potentially toxic elements in agricultural soil of the Al-Ahsa Oasis, Saudi Arabia using health indices and GIS. *Arab. J. Chem.* **2024**, *17*, 105592. [CrossRef]

43. Gintamo, T.T. Ground water potential evaluation based on integrated GIS and remote groundwater recharge potential zones in arid region using GIS and landsat approaches, southeast Tunisia. *Hydrol. Sci. J.* **2015**, *63*, 251–268. [CrossRef]
44. Deering, D.W. *Rangeland Reflectance Characteristics Measured by Aircraft and Spacecraft Sensor*; Texas A&M University: College Station, TX, USA, 1978.
45. Seeyan, S.; Merkel, B.; Abo, R. Investigation of the Relationship between Groundwater Level Fluctuation and Vegetation Cover by using NDVI for Shaqlawa Basin, Kurdistan Region—Iraq. *J. Geogr. Geol.* **2015**, *6*, 187–202. [CrossRef]
46. Carbajal-Morán, H.; Márquez-Camarena, J.F.; Galván-Maldonado, C.A.; Zárate-Quiñones, R.H.; Galván-Maldonado, A.C. Evaluation of Normalized Difference Vegetation Index by Remote Sensing with Landsat Satellites in the Tayacaja Valley in the Central Andes of Peru. *Ecol. Eng. Environ. Technol.* **2023**, *24*, 208–215. [CrossRef]
47. Wilson, E.M. *Engineering Hydrology*, 3rd ed.; Springer: Berlin/Heidelberg, Germany, 1987; p. 309.
48. Stromberg, J.C.; Tiller, R.; Richter, B. Effects of Groundwater Decline on Riparian Vegetation of Semiarid Regions: The San Pedro, Arizona. *Ecol. Appl.* **1996**, *6*, 113–131. [CrossRef]
49. Carbajal Morán, H.; Márquez Camarena, J.F.; Zárate Quiñones, R.H.; De la Cruz Vílchez, E.E. Monitoring the Hydrogen Potential of a River in the Central Andes of Peru from the Cloud. *Ecol. Eng. Environ. Technol.* **2021**, *22*, 17–26. [CrossRef] [PubMed]
50. Berhanu, K.G.; Hatiye, S.D. Identification of groundwater potentialzones using proxy data: Case study of Megech watershed, Ethiopia. *J. Hydrol. Reg. Stud.* **2020**, *28*, 100676. [CrossRef]
51. Arulbalaji, P.; Padmalal, D.; Sreelash, K. GIS and AHP techniques based delineation of groundwater potential zones: A case study from Southern Western Ghats, India. *Sci. Rep.* **2019**, *9*, 2082. [CrossRef] [PubMed]

water

MDPI

Article

Unveiling Groundwater Potential in Hangu District, Pakistan: A GIS-Driven Bivariate Modeling and Remote Sensing Approach for Achieving SDGs

Abdur Rehman [1], Lianqing Xue [1,2], Fakhrul Islam [3,4,*], Naveed Ahmed [1,5], Saleh Qaysi [6], Saihua Liu [1], Nassir Alarifi [6], Youssef M. Youssef [7,*] and Mahmoud E. Abd-Elmaboud [8]

1 College of Hydrology and Water Resources, Hohai University, Nanjing 210098, China; lx20220501013@hhu.edu.cn (A.R.); lqxue@hhu.edu.cn (L.X.); naveedahmed@hhu.edu.cn (N.A.); 210201030004@hhu.edu.cn (S.L.)
2 School of Hydraulic Engineering, Wanjiang University of Technology, Ma'anshan 243031, China
3 State Key Laboratory of Desert and Oasis Ecology, Xinjiang Institute of Ecology and Geography, Chinese Academy of Sciences, Urumqi 830011, China
4 University of Chinese Academy of Sciences, Beijing 100049, China
5 The National Key Laboratory of Water Disaster Prevention, College of Hydrology and Water Resources, Hohai University, Nanjing 210098, China
6 Department of Geology & Geophysics, College of Science, King Saud University, P.O. Box 2455, Riyadh 11451, Saudi Arabia; sqaysi@ksu.edu.sa (S.Q.); nalrifi@ksu.edu.sa (N.A.)
7 Geological and Geophysical Engineering Department, Faculty of Petroleum and Mining Engineering, Suez University, Suez 43518, Egypt
8 Irrigation & Hydraulics Department, Faculty of Engineering, Mansoura University, Mansoura 35516, Egypt; mahmoud_elsayed@mans.edu.eg
* Correspondence: fakhrul.islam@kkkuk.edu.pk or geo.fakhar@mails.ucas.ac.cn (F.I.); youssef.ibrahim@pme.suezuni.edu.eg (Y.M.Y.)

Abstract: Sustainable groundwater development stands out as a contemporary concern for growing global populations, particularly in stressed riverine arid and semi-arid regions. This study integrated satellite-based (Sentinel-2, ALOS-DEM, and CHIRPS rainfall) data with ancillary lithology and infrastructure datasets using Weight of Evidence (WoE) and Frequency Ratio (FR) models to delineate Groundwater Potential Zones (GWPZs) in the Hangu District, a hydrologically stressed riverine region in northern Pakistan, to support the Sustainable Development Goals (SDGs). Ten key variables, including elevation, slope, aspect, distance to drainage (DD), rainfall, land use/land cover, Normalized Difference Vegetation Index, lithology, and road proximity, were incorporated into the Geographic information system (GIS) environment. The FR model outperformed the WoE model, achieving success and prediction rates of 89% and 93%, compared to 82% and 86%. The GWPZs-FR model identified 23% (317 km^2) as high potential, located in highly fractured pediment fans below 550 m, with gentle slopes (<5 degrees), DD (within 200 m), and high rainfall in areas of natural trees and vegetation on valley terrace deposits. The research findings significantly support multiple SDGs, with estimated achievement potentials of 37.5% for SDG 6 (Clean Water and Sanitation), 20% for SDG 13 (Climate Action), 15% for SDG 8 (Decent Work and Economic Growth), 12.5% for SDG 9 (Industry, Innovation, and Infrastructure), and notable contributions of 10% for SDG 2 and 5% for SDG 3. This approach provides valuable insights for policymakers, offering a framework for managing groundwater resources and advancing sustainable practices in similar hydrologically stressed regions.

Keywords: groundwater potential zones; geographic information system; sustainable development goals (SDGs); riverine region; frequency ratio; weight of evidence; Pakistan

1. Introduction

Groundwater is a crucial component of Earth's water resources [1], playing a vital role in supporting urban and agricultural development in developing countries across South Asia [2,3]. Groundwater constitutes over 99% of Earth's liquid freshwater, supplying about half of the freshwater for global domestic use and around 25% of water for irrigation [4,5]. Its economic value lies in its easy extraction, adaptability to demand fluctuations, reliability during droughts, and generally superior quality compared to surface water [6]. Although global renewable freshwater supplies were estimated at around 37,000 km^3 per year in 2015 [7], current human consumption is roughly 3.5 times higher than natural aquifer replenishment [8], highlighting inefficiencies in water resource management.

Over-extraction of groundwater is a major concern, as many countries rely on it for drinking, irrigation, and commercial use [3]. Global water demand is expected to increase by 1% annually, resulting in a 20–30% rise by 2050, with an uncertainty margin of around 50% [9]. Most of this demand growth will occur in middle- and low-income developing countries [5]. Groundwater reliance is expected to grow due to surface water contamination, population growth, and climate uncertainty [10,11], further threatening freshwater ecosystems and complicating water resource management if not sustainably addressed. Climate change has also been defined as one of the significant exacerbations of groundwater availability [12], impacting the hydrologic cycle and altering soil infiltration processes and vertical percolation, which affects aquifer replenishment [13,14]. Additionally, rising global temperatures increase evaporation rates, reducing water contributions to underground reservoirs and intensifying pressure on groundwater reserves [15]. These factors point to a potential global water crisis by 2025 due to limited freshwater and unsustainable groundwater withdrawal [16,17], highlighting the urgent need for sustainable groundwater management, especially in water-scarce regions.

Therefore, a holistic approach to delineating Groundwater Potential Zones (GWPZs) using geospatial tools and statistical models can provide valuable information for decision-makers in resource-constrained and climate-uncertain contexts. Traditional field-based methods are often time-consuming, costly, and labor-intensive [18–21] and may fail to capture factors influencing groundwater distribution adequately [21]. Geographic Information Systems (GIS) have emerged as powerful tools for integrating diverse Remote Sensing (RS) and geographical datasets to map GWPZ efficiently [22,23]. Recent GIS technologies enable effective data collection, analysis, and visualization, producing comprehensive thematic maps [17,24]. This approach is particularly beneficial for understanding the complex dynamics of aquifer systems, which are characterized by dynamic storage and transmissivity factors dependent on soil porosity, land morphology, and the permeability of underlying rock formations. A wide array of factors influences the distribution of subsurface water, including geology, lithology, structural features, proximity to surface water bodies, drainage networks, rainfall patterns, Land Use/Land Cover (LULC), geomorphology, soil characteristics, lineaments, transmissivity, hydraulic conductivity, surface curvature, and elevation [23]. Integrating these factors within a GIS framework facilitates a more comprehensive and nuanced understanding of groundwater potential.

Recently, the field of GWPZ mapping has witnessed significant advancements through the application of diverse Multi-criteria Decision Analysis (MCDA) and statistical techniques [25–27]. While MCDA methods, such as Analytic Hierarchy Process (AHP), Fuzzy Logic, and Weighted Linear Combination (WLC), are widely employed to enhance the accuracy of groundwater system models [18–30], these approaches face criticism for their subjectivity and limitations in capturing complex hydrogeological relationships [31]. To address this, researchers have adopted more objective, data-driven statistical techniques, such as the Weight of Evidence (WoE), Frequency Ratio (FR), and Information Value (IV) methods [24,27], logistic regression [28] and linear discriminant analysis [29]. Hybrid approaches combining multiple techniques, such as the integration of statistical index, FR, WoE, and evidential belief function, have been explored to enhance the accuracy of GWPZ delineation [30,31].

The diversity of bivariate-driven approaches reflects the complex nature of groundwater systems and the ongoing efforts to develop more accurate and robust approaches to GWPZ mapping. Although bivariate-based GWPZ frameworks have improved our understanding of aquifer complexity by integrating well and spring inventories with environmental characteristics across various climatic regions [24–33], their potential to advance groundwater mapping for achieving Sustainable Development Goals (SDGs) in stressed riverine aquifers remains underexplored. This gap is particularly true for goals related to clean water and sanitation (SDG 6), sustainable agriculture (SDG 2), and climate action (SDG 13). This study aims to address this knowledge gap by employing two geospatial models, WoE and FR, to create a comprehensive GWPZ map in a hydrologically stressed riverine region. WoE and FR models, known for their effectiveness in hydrogeological contexts [27,34], were selected for their ability to analyze the relationship between groundwater-dependent and independent factors. This approach allows the production of robust potential maps with minimal resources, making it particularly suitable for regions with limited data and financial resources.

South Asian developing countries face significant challenges in achieving sustainable agricultural development, driven by rising populations, poverty, and increasing food demands [35,36]. Pakistan (Figure 1a,b), with 36.4% of its population in urban areas, has the highest urbanization rate in South Asia [37]. The Hangu District, located in northern Pakistan's Khyber Pakhtunkhwa province, was selected as the case study to promote sustainable agricultural development in hydrologically constrained regions (Figure 1c). Hangu's subtropical dry climate, with limited annual precipitation and an average temperature of 20.7 °C, poses significant challenges to water resource management [38]. Despite its diverse crop production, including wheat, maize, barley, and vegetables [38], Hangu faces water scarcity and lacks an effective water management system, impacting ecosystem health and socio-economic development, particularly as rain-fed farming is the primary economic activity [39]. The complex interplay between limited water resources, agricultural dependence, and climatic conditions in Hangu underscores the need for advanced approaches to water resource management. Understanding the distribution and potential of groundwater resources in such regions is crucial for developing sustainable strategies that can support agricultural productivity, ensure water security, and promote overall regional development [40].

Previous research in the Hangu region has largely focused on groundwater quality assessments, but a substantial gap exists in the literature concerning the delineation of groundwater recharge zones. This study seeks to fill this gap by addressing three main objectives: (1) quantitatively analyzing the impact of topographical, geological, hydrogeological, climatic, and anthropogenic factors on potential groundwater conditions using WoE and FR techniques based on groundwater-related data; (2) delineating optimal GWPZ map to identify new exploration sites; and (3) exploring the implications of groundwater potential mapping for achieving SDGs, specifically those related to clean water and sanitation (SDG 6), sustainable agriculture (SDG 2), and climate action (SDG 13). By accomplishing these objectives, this research seeks to provide a scientific basis for identifying potential groundwater sources and supporting long-term water management strategies, enhancing sustainable water resource management in Hangu. By connecting research outcomes to global sustainability targets, this study offers valuable insights for policymakers and stakeholders working towards sustainable development in water-stressed areas, effectively bridging the gap between scientific research and practical water resource management.

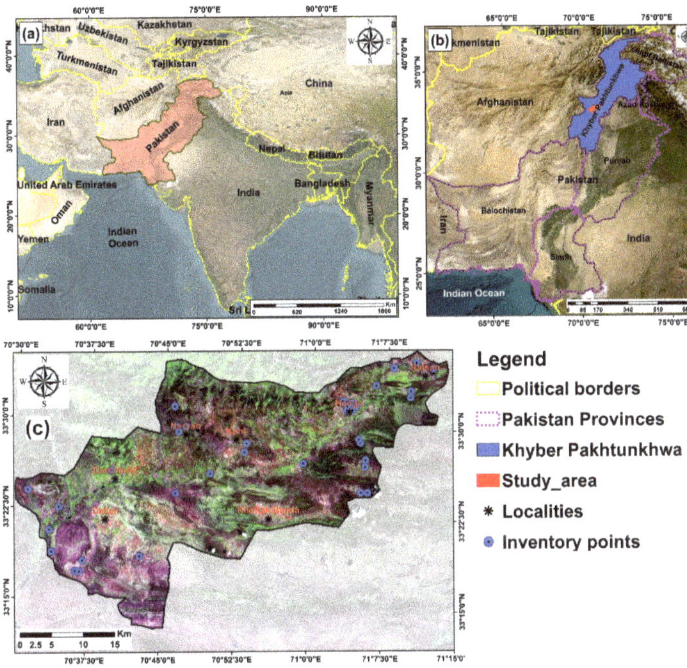

Figure 1. Geographic location and topography of the study area: (**a**) Map of South Asia Pakistan highlighting Pakistan country; (**b**) Location of Hangu district within Khyber Pakhtunkhwa province; (**c**) Administrative map of Hangu district showing detailed inventory map and main localities.

2. Materials and Methods

2.1. Study Area Characteristics and Environmental Challenges

Hangu faces significant environmental challenges due to its unique geographical, geological, and climatic characteristics. The district is situated between 33°12′30″ to 33°35′30″ N latitude and 70°30′0″ to 71°13′40″ E longitude, covering 1380 km^2 at an average elevation of 858 m above mean sea level (a.s.l) [41,42]. The region's arid to semi-arid climate, characterized by an annual mean temperature of 20.7 °C and 536 mm of rainfall, exacerbates water scarcity, especially due to high summer temperatures and evapotranspiration rates [12,41].

Geologically, Hangu's location within the Himalayan ranges' western fold and thrust belt system contributes to its complex hydrogeological setting. The predominance of sedimentary rocks, including sandstones, shales, and limestones dating from the Mesozoic to Cenozoic age, influences groundwater recharge and flow patterns [38]. This geological complexity and the region's tectonic activity can lead to aquifer compartmentalization and variable groundwater quality [43]. The district's population of 518,798, with a density of 376/km^2 [41], places significant pressure on available water resources. The predominantly rural population (80.3%) relies heavily on agriculture, cultivating crops such as maize, wheat, vegetables, and fruits in the fertile valleys [38]. This agricultural dependence and limited water resources have led to unsustainable groundwater extraction practices in many areas of Pakistan, including Khyber Pakhtunkhwa province [44].

Climate change projections for Pakistan, including more extreme temperatures and erratic precipitation, pose additional challenges, especially for arid regions like Hangu, where droughts could intensify water scarcity [45,46]. Groundwater contamination, due to natural geological factors and anthropogenic activities such as improper waste disposal and agricultural practices, is a serious concern, with elevated levels of nitrates, fluoride, and heavy metals posing health risks [41,47]. The rugged topography also contributes to environmental vulnerabilities from soil erosion and flash flooding during intense rainfall, impacting water quality and availability [48]. Given these complex challenges, there is an urgent need for comprehensive and sustainable groundwater management strategies in the Hangu district.

2.2. Dataset and Methodology

This study employs a comprehensive, multi-phase approach to delineate GWPZs in the Hangu district, integrating diverse datasets and advanced geospatial techniques. The methodology, illustrated in Figure 2, encompasses inventory mapping, thematic layer generation, GWPZ modeling, and model validation, building upon established approaches [49,50] while incorporating recent advancements in remote sensing and geospatial analysis. This integrated methodological framework, combining cutting-edge geospatial techniques with SDG indicators, provides a comprehensive tool for groundwater potential assessment in complex terrain, enhancing this study's reproducibility and facilitating a nuanced understanding of the multi-faceted process involved in GWPZ mapping.

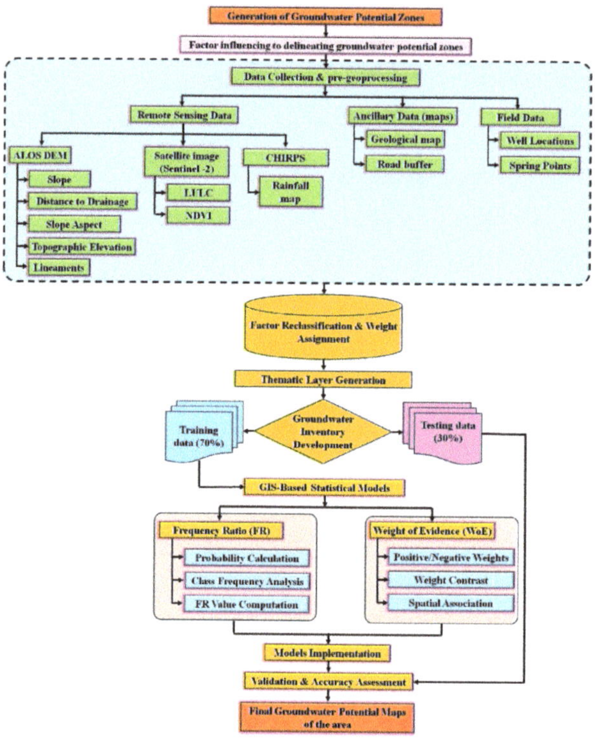

Figure 2. The methodological framework of the present research.

2.2.1. Datasets Collection and Preparation

The groundwater potential analysis in the Hangu district, Pakistan, incorporated multiple geospatial datasets through an integrated remote sensing and ancillary data approach (Table 1). The primary data sources comprised Sentinel-2 multispectral imagery (10 m spatial resolution, Level-2A) acquired from USGS Earth Explorer (2020–2022) for LULC classification, ALOS-DEM (12.5 m resolution, Version 3.2) for topographic parameter derivation, and CHIRPS precipitation data (0.05° resolution) for rainfall distribution analysis. Supplementary datasets included soil texture information from FAO's Harmonized World Soil Database (1:2,000,000 scale), geological data from GSP (1:650,000 scale), and road network data from the Khyber Pakhtunkhwa Highway Authority. Data quality assurance involved atmospheric correction using the Sen2Cor algorithm [51], DEM hydrological correction through the AGREE method [52], and validation of CHIRPS data against ground-based precipitation records. An inventory map was initially constructed using Earth's observational data, including Google Earth Pro and Sentinel-2 satellite imagery, with surface water bodies identified and exported as GIS-compatible layers. The Modified Normalized Difference Water Index (MNDWI) was applied to Sentinel-2 data in Google Earth Engine (GEE) to enhance water body detection. Subsequently, ten thematic layers (elevation, slope, aspect, distance to lineaments, LULC, Distance to Drainage (DD), rainfall, lithology, Normalized Difference Vegetation Index (NDVI), and Distance to Road (DR)) were considered for this study [27]. The integrated processing framework, implemented through ArcGIS 10.8 software , standardized all datasets to UTM Zone 42N coordinate system (WGS84 datum) with 30 m resolution, employing bilinear interpolation for continuous data and nearest neighbor resampling for categorical data, thus enabling systematic characterization of groundwater potential factors across the complex terrain of the study area.

Table 1. Characteristics and Sources of Geospatial Datasets Utilized for GWPZ Mapping.

Data	Resolution	Data Accessibility	Data Availability Statement/Source	Thematic Layers
Sentinel-2	10 m	Open source	Acquired from the United States Geological Survey using the website https://earthexplorer.usgs.gov.	LULC and inventory
ALOS DEM	12.5 m	Open source	The data was acquired from the United States Geological Survey using the website https://earthexplorer.usgs.gov.	Elevation, slope, aspect, and drainage maps
CHIRPS	0.05°	Open source	The grided rainfall data can be accessed from https://www.chc.ucsb.edu.	Rainfall layer
Soil texture	1:2,000,000	Data openly available	Acquired from FAO www.fao.org.	Textural map of soil
Lithology	1:650,000	GSP	Acquired from Geological Survey of Pakistan (http://www.gsp.gov.pk/)	Geological maps
Road buffer	88.4 m	Pakhtunkhwa Highway Authority (PKHA)	Obtained from the KPK Highway Authority	Road map

2.2.2. Preparation of Controlling Factors

Based on established methodologies [19,49] and a comprehensive literature review, ten conditioning parameters were identified as significant factors influencing groundwater occurrence: elevation, slope angle, slope aspect, distance to drainage, lineaments, rainfall, LULC, NDVI, lithology, and road proximity. The thematic layer generation process employed specialized geospatial techniques, incorporating ALOS DEM-derived topographic parameters through Spatial Analyst tools, Sentinel-2 imagery-based LULC and NDVI calculations via supervised classification and band mathematics, CHIRPS rainfall data interpolation, and geological feature digitization validated through field surveys, thereby establishing a robust framework for groundwater potential assessment in the complex hydrogeological setting of the study area.

Topographic Elevation

Topographic elevation, a crucial factor in groundwater potential assessment, was derived from an ALOS DEM with 12.5 m resolution. The elevation map was generated using ArcGIS 10.8 spatial analyst tools and reclassified into five distinct categories: <550 m, 550–650 m, 650–750 m, 750–850 m, and >850 m (Figure 3a). This classification scheme was adopted to capture the significant variations in terrain across the study area, reflecting the inverse relationship between elevation and groundwater potential [53]. Lower elevations generally correspond to higher groundwater potential due to increased accumulation and reduced runoff, while higher elevations often exhibit lower potential owing to enhanced drainage and reduced infiltration capacity. The spatial distribution of elevation classes provides crucial insights into the hydrogeological setting of the Hangu district, informing subsequent analyses of groundwater occurrence and movement.

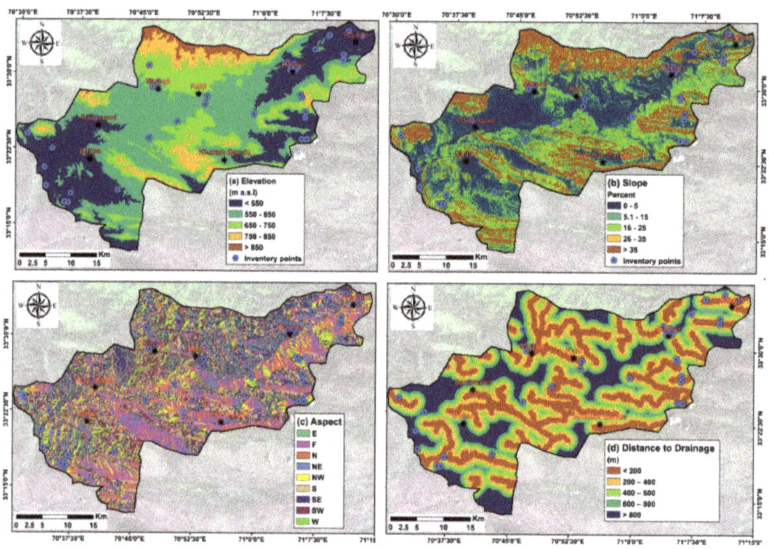

Figure 3. Topographical and hydrogeological conditioning parameters for GWPZs in the study area include (**a**) elevation, (**b**) slope, (**c**) aspect, and (**d**) distance to drainage (DD), overlaid with inventory points (blue circle).

Slope

The slope parameter, a critical factor influencing groundwater recharge through its impact on surface runoff and infiltration rates, was derived from the ALOS. The resulting slope map was reclassified into five categories: <5°, 5–15°, 15–25°, 25–35°, and >35° (Figure 3b). This classification scheme captures the varied topography of the Hangu region and its potential effects on groundwater dynamics. Gentle slopes (<5°) typically facilitate higher infiltration rates and groundwater recharge, while steeper slopes (>35°) promote rapid runoff and reduced infiltration [54]. The spatial distribution of slope classes provides insights into areas of potential groundwater accumulation and depletion, crucial for understanding the region's hydrogeological characteristics and informing sustainable groundwater management strategies.

Slope Aspect

The slope aspect, a crucial parameter influencing groundwater recharge through its effects on solar radiation exposure, wind patterns, and precipitation distribution, was derived from the ALOS DEM. The resulting aspect map was reclassified into nine cardinal and intercardinal directions: Flat (F), North (N), Northeast (NE), East (E), Southeast (SE), South

(S), Southwest (SW), West (W), and Northwest (NW) (Figure 3c). This classification captures the varied topographic orientations within the Hangu region, each with distinct implications for microclimatic conditions and subsequent groundwater dynamics. South-facing slopes, for instance, typically experience higher solar radiation and evapotranspiration rates, potentially reducing groundwater recharge, while north-facing slopes may retain moisture more effectively [55]. The interplay between aspect, local climate, and vegetation cover significantly influences infiltration rates and subsurface water movement, making aspect a valuable indicator in groundwater potential assessment [56].

Distance to Drainage

The distance to drainage networks is an important factor in groundwater studies. Areas closer to rivers and streams often have higher groundwater levels, especially in flat regions [57]. To analyze this factor in the Hangu district, a drainage network map was created using the ALOS DEM with a 12.5 m resolution. This high-resolution data allowed for accurate identification of water channels. The map was then processed in ArcGIS to create buffer zones around these drainage features. Five distance categories were established: <200 m, 200–400 m, 400–600 m, 600–800 m, and >800 m (Figure 3d). This classification helps identify areas with potentially higher groundwater availability based on their proximity to surface water bodies.

Lineaments

Lineaments are important geological features that appear as linear or curved patterns on satellite images. These features, which include faults, fractures, and drainage channels, play a crucial role in groundwater movement and distribution [58]. Lineaments increase the secondary porosity of rocks, creating pathways for water flow and storage. To map these features in the Hangu district, ALOS DEM hill shade images were analyzed in ArcGIS and supplemented by field observations. The resulting lineament map was classified into five distance categories: <500 m, 1500 m, 3000 m, 5000 m, and >5000 m (Figure 4b). This classification helps identify areas with potentially higher groundwater occurrence, as zones closer to lineaments often have better groundwater prospects.

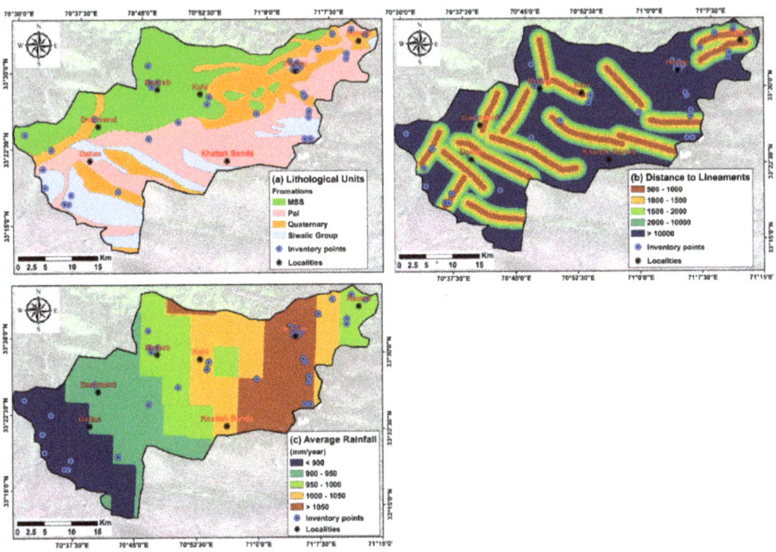

Figure 4. Geological and climatological parameters influencing GWPZs in the study area are illustrated through maps of (**a**) lithological units, (**b**) distance to lineaments (DL), and (**c**) precipitation (mm/day).

Rainfall

Rainfall, a primary source of groundwater recharge, was analyzed using CHIRPS data from 2010 to 2023. A 33-Kernel Machine Learning algorithm implemented in GEE processed this time series, capturing both temporal and spatial variations in precipitation patterns. The resulting precipitation map was reclassified into four distinct categories: <900 mm, 900–950 mm, 1000–1050 mm, and >1050 mm (Figure 5c). This classification scheme reflects the significant spatial heterogeneity in rainfall across the Hangu region and its potential impact on groundwater recharge. Areas receiving higher rainfall (>1050 mm) generally exhibit greater recharge potential, particularly when characterized by prolonged, low-intensity precipitation events that favor infiltration over surface runoff. Conversely, regions with lower annual rainfall (<900 mm) or those experiencing predominantly high-intensity, short-duration events may contribute less to groundwater reserves due to increased runoff and reduced infiltration opportunities [59].

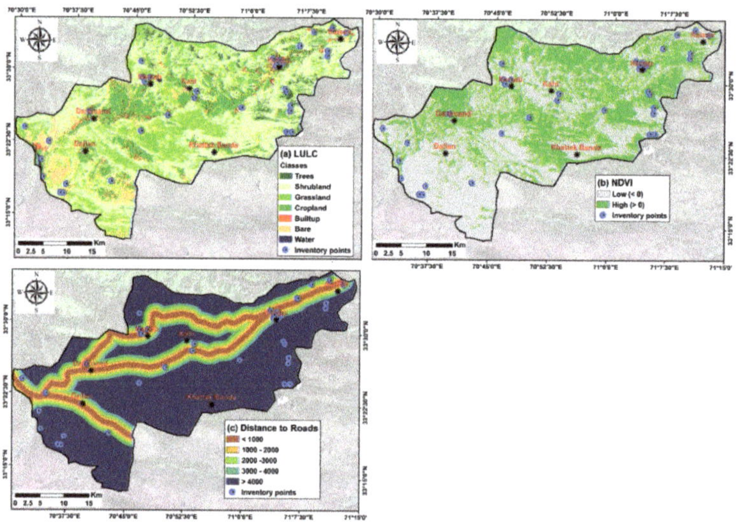

Figure 5. Anthropogenic parameters for GWPZs are represented by maps of (**a**) land use/land cover (LULC), (**b**) normalized difference vegetation index (NDVI), and (**c**) distance to roads (DR).

Land Use and Land Cover (LULC)

LULC patterns are crucial in shaping hydrological processes by influencing soil composition, land texture, and vegetation density. These factors directly affect ecosystems' surface runoff and infiltration rates [60]. A map was created to analyze LULC in the Hangu district using Random Forest classification in GEE. This advanced machine learning technique accurately categorized the landscape into six distinct classes: water, trees, crops, built-up area, barren ground, and scrub/shrub (Figure 5a). Each class has unique implications for groundwater recharge. For instance, forested areas typically enhance water infiltration, while built-up regions often increase surface runoff. Croplands may have varied effects depending on specific agricultural practices. The resulting LULC map provides valuable insights into the spatial distribution of these land cover types across the district.

Normalized Difference Vegetation Index (NDVI)

The NDVI serves as a valuable indicator for estimating groundwater resources. This index reflects vegetation health and density, which are often linked to water availability. Higher NDVI values typically suggest better access to water, including groundwater. Conversely, NDVI values tend to decrease as groundwater depth increases, indicating less available water for vegetation [61]. To create the NDVI map for the Hangu district, Sentinel-2 satellite data was analyzed using machine learning algorithms in GEE. The resulting NDVI map was then classified into high and low categories within ArcGIS (Figure 5b). This simple yet effective classification helps identify areas with potentially higher groundwater availability (high NDVI) and areas where groundwater might be scarcer or deeper (low NDVI).

Geology

Geology plays a vital role in groundwater recharge and distribution. The rock types and their arrangements determine how easily water can move through the ground and be stored. In the Hangu region, the Northern Geological Map of Pakistan provided essential information about the underground rock layers. These layers include rocks from different time periods, such as the Siwaliks group, Paleocene, Quaternary, and Mesozoic ages (Figure 4a). Each rock type has unique properties that affect how water moves and is stored underground. For example, some rocks allow water to pass through easily, while others act as barriers. The complex arrangement of these rock layers, shaped by long-term earth movements, creates a varied underground landscape for water. This geological diversity significantly influences where groundwater can be found and how much is available in different areas [62].

Road Buffer

Roads play a significant role in groundwater dynamics, often in a negative way. The construction of roads can reduce soil porosity and permeability as sediments become compacted, limiting water infiltration [24]. Additionally, road surfaces are designed to repel water, leading to increased surface runoff and decreased groundwater recharge [27]. A road network map was created to assess the impact of roads on groundwater in the Hangu district using data from Google Earth and the Khyber Pakhtunkhwa Highway Authority. This map was then processed in ArcGIS to create five distance buffer zones around the roads (Figure 5c). These buffer zones help identify areas where road infrastructure might influence groundwater recharge and flow patterns.

2.2.3. Bivariate Statistical Approaches for GWPZ Mapping

This study utilized two bivariate geospatial models, WoE and FR, to create a comprehensive GWPZ map for the Hangu region of Khyber Pakhtunkhwa. The complex relationships between these dependent and independent variables were analyzed using FR and WoE bivariate statistical methods using 70% of the inventory data. The resulting GWPZ maps were validated using the Area Under the Curve (AUC) approach, with the remaining testing (30%) sets to ensure the reliability of model development. The following sections provide detailed explanations of the WoE and FR methodologies applied in this study.

Weight of Evidence (WoE) Model

The WoE model, a bivariate statistical approach, combines linear logic and Bayesian law to approximate dependent and independent occurrence probabilities [63]. This method calculates spatial correlations between groundwater potential mapping parameters and water bodies, providing insights into complex factor relationships while minimizing random variable downsides and subjective bias [64]. The WoE model uses inventory pixels (Z or Z*) to determine weights for each groundwater regulating factor category, as expressed in the following equations [65]:

$$W^{+} = ln\frac{P\left(\frac{Y}{Z}\right)}{P\left(\frac{Y}{Z^{*}}\right)} \quad (1)$$

$$W^{-} = ln\frac{P\left(\frac{Y^{*}}{Z}\right)}{P\left(\frac{Y^{*}}{Z^{*}}\right)} \quad (2)$$

where ln( ) is the natural logarithm, P is probability, Y^* represents the absence of groundwater inventory points as pixels in each factor raster imagery, Y denotes groundwater governing factors, Z indicates water inventory, and Z^* signifies the absence of groundwater inventory points as pixels. The weight contrast (T) reveals the spatial association between causative components and groundwater likelihood:

$$T = (W^{+}) - (W^{-}) \tag{3}$$

Positive T values favor groundwater presence, while negative values indicate unfavorable associations. The final GWPZ index is derived by summing the weight contrasts across all factors:

$$GWPZ - WOE = \sum T \tag{4}$$

This approach, as outlined by [65] and applied by [66], provides a comprehensive assessment of groundwater potential based on multiple conditioning factors.

Frequency Ratio (FR) Model

The Frequency Ratio (FR) model is a robust bivariate statistical technique used to assess the spatial relationship between groundwater occurrences and various conditioning factors [67]. This method quantifies the probability of groundwater presence based on the observed distribution of groundwater locations relative to each factor class. The FR for each class of each factor is calculated using the following equation:

$$FR = \left[\frac{N_{pix}(SX_i)}{\sum_{i=1}^{m} N_{pix}(SX_i)}\right] / \left[\frac{N_{pix}(X_j)}{\sum_{j=1}^{n} N_{pix}(X_j)}\right] \tag{5}$$

where, $N_{pix}(SX_i)$ is the number of inventory points as pixels in each factor raster image with groundwater occurrences in class i of factor X; $N_{pix}(X_j)$ is the total number of pixels in factor X_j; $\Sigma N_{pix}(SX_i)$ is the total number of pixels with groundwater occurrences in the study area; $\Sigma N_{pix}(X_j)$ is the total number of pixels in the study area; m is the total classes in the variable X_i; and n is the total factors of the study area.

FR values exceeding 1 indicate a strong positive correlation between the factor class and groundwater occurrence, providing valuable insights into potential groundwater locations [68]. To further enhance the model's performance and capture more complex relationships between factors and groundwater potential, the FR values for all the classes of each controlling factor were normalized in the range of probability values (0 to 1) using the following equation:

$$FR_{norm} = \frac{(RF_i - RF_{min})}{(RF_{max} - RF_{min})} \times 100 \tag{6}$$

Finally, the GWPZ map created using normalizing all FR values for each factor as follows:

$$GWPZ - FR = \sum \begin{matrix} FR\ E_{norm} + FR\ S_{\mathrm{norm}} + FR\ A_{\mathrm{norm}} + FR\ DD_{\mathrm{norm}} + FR\ Lu_{\mathrm{norm}} + \\ FR\ N_{\mathrm{norm}} + FR\ L_{\mathrm{norm}} + FR\ G_{\mathrm{norm}} + FR\ R_{\mathrm{norm}} + FR\ DR_{\mathrm{norm}} \end{matrix} \tag{7}$$

where, FR represents the frequency ratio for each factor (E: elevation, S: slope, A: aspect, DD: drainage, U: LULC, N: NDVI, L: lineaments, G: geology, R: rainfall, and DR: road buffer). This approach enhances the traditional FR model by potentially capturing more complex relationships between factors and groundwater potential [69].

The GWPZ-FR and GWPZ-WOE maps divided the study area into three groundwater potential zones: low, medium, and high (Figures 6 and 7).

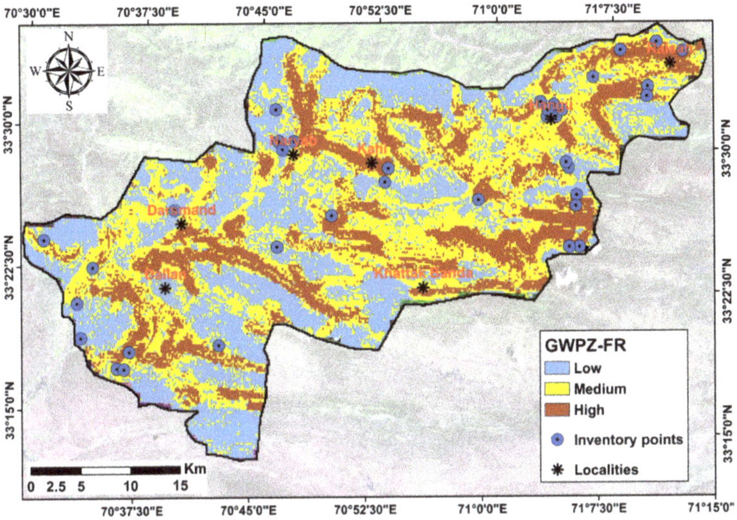

Figure 6. Depicts the GWPZ map utilizing the WOE model.

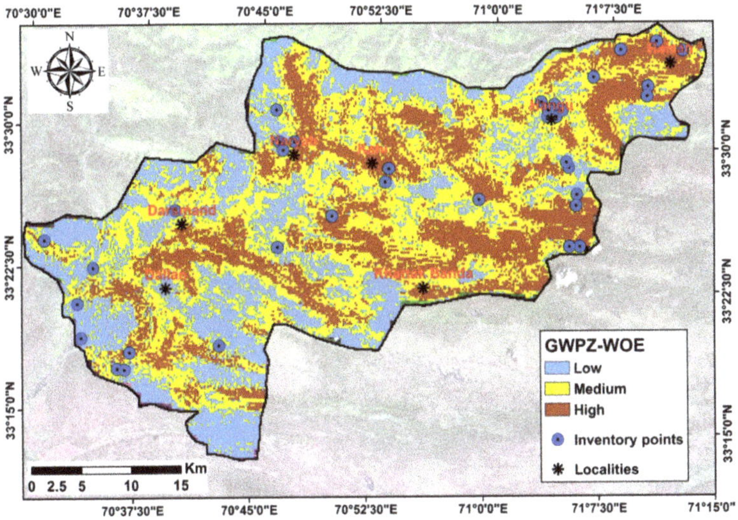

Figure 7. Depicts the GWPZ map employing the FR model.

2.2.4. Validation of GWPZ's

Model validation is a critical step in developing and applying GWPZ models, ensuring their reliability and practical utility [70,71]. This study employed a comprehensive validation approach using Receiver Operating Characteristic (ROC) curve analysis, Success Rate Curves (SRC), and Prediction Rate Curves (PRC) to provide a robust assessment of model performance [24].

ROC curve analysis plays a crucial role in evaluating binary classification models, particularly for groundwater potential assessment [72–74]. This approach measures the true positive rate (TPR) against the false positive rate (FPR) across various thresholds. TPR, or sensitivity, is the ratio of true positives to all actual positives, while FPR is the ratio of false positives to all actual negatives. The AUC of the ROC curve summarizes model performance, ranging from 0.5 (random performance) to 1.0 (perfect classification), and is calculated using the trapezoidal rule by summing the areas under the curve:

$$AUC = \frac{\Sigma[TRP(i+1) + TRP(i)]}{2} \times [FPR(i+1) - FPR(i)] \quad (8)$$

where *FPR(i)* and *TPR(i)* are *FPR* and *TPR* at the ith threshold, respectively.

To provide a more nuanced assessment, Success Rate Curves (SRC) and Prediction Rate Curves (PRC) were employed [75]. The SRC evaluates how well the GWPZ map identifies sites with existing wells using the training dataset.

$$SRC = \frac{TP_{train}}{TP_{train} + FN_{train}} \times 100 \quad (9)$$

where TP_{train} represents true positives and FN_{train} represents false negatives in the training dataset. The PRC, considered more critical for assessing true predictive capability, demonstrates the model's ability to predict groundwater occurrence using the validation dataset [76].

$$PRC = \frac{TP_{val}}{TP_{val} + FN_{val}} \times 100 \quad (10)$$

where TP_{val} and FN_{val} represent true positives and false negatives in the validation dataset, respectively.

Both SRC and PRC are plotted as cumulative percentages of groundwater occurrences against the cumulative area percentages, ranked from highest to lowest susceptibility. The area under these curves (*AUC-SRC* and *AUC-PRC*) quantifies the model's success and predictive rates, respectively [76,77]. The SRC measures model fit to the training data, while the PRC assesses predictive performance using independent validation data [78,79].

3. Results

3.1. Validation of Models

Figure 8 presents the ROC curves for both models, illustrating their respective SRC and PRC. The WoE model demonstrated robust performance, with AUC values of 0.82 (82%) for the SRC and 0.86 (86%) for the PRC. These results indicate that the WoE model exhibits a good fit for the training data and has strong predictive power when applied to new datasets. The slight improvement in the PRC suggests that the model generalizes well to unseen data, a crucial characteristic for practical applications in groundwater management. The FR model, however, displayed superior performance, achieving AUC values of 0.89 (89%) for the SRC and 0.93 (93%) for the PRC.

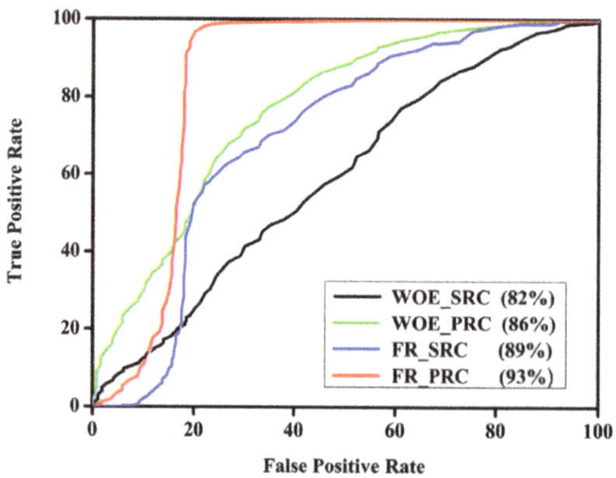

Figure 8. WOE and FR model validation results.

The FR model consistently outperforms the WoE method, showing a 7% improvement in both the SRC and PRC, demonstrating its effectiveness in capturing complex spatial relationships between groundwater occurrence and environmental factors [79]. This superior performance is likely due to the FR model's ability to handle non-linear interactions and reduced sensitivity to outliers [79,80]. Notably, both models achieve AUC values above 0.8, meeting the standard for excellent performance in environmental sciences [81], confirming their efficacy in mapping groundwater potential zones. The robust predictive accuracy, especially of the FR model, supports its application in groundwater resource management in the Hangu District and aligns with existing literature on the use of probabilistic models in hydrogeology [19,77].

3.2. Analysis of Thematic Layers Using WoE and FR Models

The WoE and FR models were used to evaluate the significance of various thematic layers in determining GWPZ in the Hangu area. Table 2 summarizes the statistical analysis, showing WoE and FR values for each layer. Elevation exhibits a clear inverse relationship with groundwater potential in both models. Figure 4a illustrates the varied topography of the district, with elevations ranging from below 550 m to over 850 m, highlighting the diverse landscape that contributes to its environmental complexity. The highest WoE (0.79) and FR (1.77) values are observed for elevations <550 m, indicating that lower elevations are significantly more favorable for groundwater occurrence. This relationship gradually decreases with increasing elevation, with areas >850 m showing negative WoE (−0.62) and the lowest FR (0.41) values, suggesting unfavorable conditions for groundwater at higher elevations. Slope analysis reveals that gentler slopes are highly conducive to groundwater potential. Slopes <5° show the highest positive values in both models (WoE: 1.68, FR: 3.41), while steeper slopes >35° exhibit strong negative associations (WoE: −1.54, FR: 0.25). This trend strongly supports the understanding that flatter terrain facilitates water infiltration and retention, which is crucial for groundwater accumulation. The slope aspect analysis yields interesting results, with some model differences. In the WoE model, flat areas (F) show the highest positive value (1.73), while the FR model indicates southeast-facing slopes (SE) as the most favorable (1.92). Both models agree on the negative impact of south-facing slopes, possibly due to increased evaporation rates on these aspects. Drainage proximity emerges as a critical factor in both models, with areas <200 m from drainage features showing the highest positive values (WoE: 2.00, FR: 3.93). This strong association rapidly decreases with increasing distance, becoming less significant beyond 400 m, underscoring the vital role of surface water bodies in groundwater recharge processes. Rainfall analysis indicates optimal ranges for groundwater potential, with some variation between models. The WoE model suggests 1000–1050 mm/year as optimal (0.87), while the FR model indicates >1050 mm/year as most favorable (1.93). Both models agree that very low rainfall (<900 mm/year) is unfavorable for groundwater potential. LULC analysis reveals that tree cover and croplands have the highest positive values in both models, indicating their significant positive impact on groundwater potential. This aligns with NDVI analysis, where the WoE model indicates a significant positive association (0.45) for high NDVI values, while the FR model reports a minor positive effect (0.01), highlighting some model discrepancies regarding vegetation's impact. The FR model also emphasizes the adverse impact of urbanization on groundwater recharge (0.001). The FR model particularly emphasizes the negative impact of built-up areas (0.001), highlighting the detrimental effect of urbanization on groundwater recharge. Road proximity analysis reveals a complex relationship, with the highest positive values found 3000–4000 m from roads (WoE: 0.78, FR: 1.95), suggesting a balance between accessibility and reduced surface sealing. Geological factors show varying influences, with Quaternary deposits (Q) exhibiting high positive values in both models (WoE: 0.69, FR: 1.68), suggesting favorable groundwater storage and transmission characteristics. Lineament analysis demonstrates that areas within 500 m of lineaments have the highest positive values (WoE: 1.21, FR: 3.05), with this association decreasing at greater distances. This trend reinforces the crucial role of geological structures in controlling groundwater movement and accumulation.

Table 2. Statistical analysis for GWPZ Mapping of District Hangu, Pakistan.

Parameters	Class	No. of Total Pixels in Each Class	No. of Water Pixels in a Class	Percentage of Pixels in Each Class (%)	Percentage of Water Pixels in Each Class (%)	WOE	FR
Elevation (m a.s.l)	<550	7985	151	21.22	37.33	0.79	1.77
	550–650	15,937	142	44.11	35.22	0.55	0.84
	650–750	9171	59	23.55	17.01	−0.31	0.74
	750–850	2880	31	8.02	4.08	−0.42	0.68
	>850	1511	17	4.01	08	−0.62	0.41
Slope in Degree	<5°	2211	76	5.87	20	1.68	3.41
	05–15°	1186	130	3.15	34.21	0.9	2.13
	15–25°	5998	70	15.92	18.42	−0.61	1.16
	25–35°	11,056	61	29.35	16.05	−0.71	0.55
	>35°	17,223	43	45.72	11.32	−1.54	0.25
Slope Aspect	F	2779	115	1.44	7.37	1.73	1.09
	NE	2970	31	0.03	7.87	0.04	1.10
	E	5471	42	−0.28	14.50	−0.32	1.13
	SE	5673	49	−0.16	15.04	−0.18	1.92
	S	5416	27	−0.71	14.36	−0.79	1.15
	SW	3920	41	0.04	10.39	0.04	0.90
	W	4073	21	−0.68	10.80	−0.73	0.91
	NW	3762	27	−0.34	9.97	−0.38	0.87
	N	3610	27	−0.30	9.57	−0.33	0.81
Distance to Drainage (m)	<200	5359	206	14.27	54.21	2.00	3.93
	200–400	6009	66	16.01	17.37	0.10	1.13
	400–600	6951	45	18.51	11.84	−0.53	0.61
	600–800	9656	33	25.72	8.68	−1.30	0.32
	>800	9069	30	24.16	7.89	−1.32	0.31
Rainfall (mm/year)	<900	5829	30	15.49	7.89	−0.77	0.51
	900–950	10,341	62	27.47	16.32	−0.67	0.59
	950–1000	11,035	113	29.32	29.74	0.02	1.01
	1000–1050	5823	115	15.47	30.26	0.87	1.87
	>1050	4612	60	12.25	15.79	0.30	1.93
LULC	Water	203	19	0.53	1.11	0.05	0.5
	Trees	411	109	0.89	5.9	1.12	2.30
	Crops	3584	1125	10.02	12.31	1.03	1.99
	Built-up Area	1701	6	4.38	1.58	−1.06	0.001
	Bare Ground	234	73	0.62	0.53	−0.17	0.12
	Scrub/Shrub	30,674	341	84.08	42.11	−2.01	0.50
Lithology	Mss	4713	41	11.93	10.01	−0.29	0.69
	Q	7625	128	21.07	27.05	0.69	1.68
	R	15,988	169	39.07	42.08	0.07	1.02
	Pal	9105	42	24.08	11	−0.9	0.42
Distance to Lineament (m)	<500	1300	40	3.45	10.53	1.21	3.05
	1500	2381	62	6.31	16.32	1.08	2.58
	3000	3420	40	9.07	10.53	0.17	1.16
	5000	4341	45	11.51	11.84	0.03	1.03
	>5000	26,285	193	69.68	50.79	−0.81	0.73
NDVI	Low	15,406	117	40.91	40.91	−0.45	−0.01
	High	22,252	263	59.09	59.09	0.45	0.01
Distance to Roads (m)	<1000	3145	22	8.34	5.79	−0.40	0.69
	1000–2000	3993	40	10.58	10.53	−0.01	0.99
	2000–3000	4328	39	11.47	10.26	−0.13	0.89
	3000–4000	3569	70	9.46	18.42	0.78	1.95
	>4000	22,693	209	60.15	55	−0.21	0.91

3.3. Groundwater Potential Zone Maps

Based on the WoE and FR analyses of the thematic layers, comprehensive Groundwater Potential Zone Maps were generated for the Hangu District (Figures 6 and 7). The spatial distribution of these zones reveals a complex pattern of groundwater potential across the district, with both models showing broadly similar patterns but some notable differences. Table 3 provides a quantitative comparison of the area coverage for each potential zone as determined by the WoE and FR models. High potential zones were predominantly concentrated in both maps' northeastern and central parts of the study area. These regions likely correspond to areas with favorable combinations of low elevation, gentle slopes, proximity to drainage features, and optimal rainfall conditions. The WoE model identifies 27% (373 km^2) of the total area as high potential, while the FR model classifies 23% (317 km^2) in this category. The FR model shows a more pronounced association of high-potential zones with drainage networks. The WoE model classifies 31% (428 km^2) as moderate potential, whereas the FR model indicates a larger area of 40% (552 km^2) in this category. These zones may represent areas with some favorable factors, but not all conditions are optimal for groundwater accumulation. Low potential zones are most prevalent on both maps in the western and southern parts of the district. These areas likely coincide with higher elevations, steeper slopes, greater distances from drainage features, or less favorable LULC types. The WoE model identifies 42% (580 km^2) as low potential, while the FR model shows a slightly smaller area of 37% (511 km^2) in this category. The FR model shows a more extensive distribution of low-potential zones, particularly in the western region. Both methods underscore the complex interplay of geographical, geological, and hydrological factors and the need to consider multiple variables for effective groundwater assessment.

Table 3. Depicting the final outcomes of WoE and FR algorithms in area (km^2) and percentage (%).

Models	Zones	Area (km^2)	Area (%)
WoE	High	373	27
	Moderate	428	31
	Low	580	42
FR	High	317	23
	Moderate	552	40
	Low	511	37

4. Discussion

4.1. Comparative Analysis of Key Factors Shaping Groundwater Potential in WoE and FR Models

The WoE and FR models revealed critical insights into the factors influencing groundwater potential in the Hangu District. Both models identified a set of key parameters, with some variations in their relative importance. In the WoE model, slope, slope aspect, and proximity to drainage emerged as the most influential factors. Slope exhibited a strong inverse relationship with groundwater potential, with areas < 5° showing the highest positive WoE value (1.68). This aligns with the findings by [19], who reported that gentle slopes significantly contribute to groundwater recharge due to increased infiltration time. The importance of the slope aspect, particularly flat areas with a WoE value of 1.73, corroborates the work of Manap et al. [82], who found that certain aspects promote higher soil moisture retention, thereby enhancing groundwater potential. Proximity to drainage showed a strong positive correlation with groundwater potential in both models. Areas within 200 m of drainage features displayed the highest WoE (2.00) and FR (3.93) values. This relationship has been consistently observed in similar studies, such as Nampak et al. [77], who reported that areas closer to rivers have higher groundwater potential due to increased recharge opportunities.

The FR model emphasized additional factors, including elevation, LULC, rainfall, and lineaments. Elevation exhibited an inverse relationship with groundwater potential, with areas < 550 m showing the highest FR value (1.77). This trend is consistent with findings by [83], who noted that lower elevations are generally associated with higher groundwater potential due to accumulation processes. LULC analysis revealed that tree cover and croplands had the highest positive values in both models. The FR model particularly highlighted the negative impact of built-up areas (FR = 0.001), emphasizing the detrimental effect of urbanization on groundwater recharge. This aligns with research by Rahmati et al. [19], who found that vegetation cover positively influences groundwater potential, while urban areas hinder recharge processes.

Rainfall analysis indicated optimal ranges for groundwater potential, with the FR model suggesting > 1050 mm/year as the most favorable (FR = 1.93). This relationship between precipitation and groundwater potential has been well-documented, with studies like Naghibi et al. [84] reporting similar positive correlations in arid and semi-arid regions. Lineament analysis demonstrated that areas within 500 m of lineaments have the highest positive values (WoE = 1.21, FR = 3.05). This reinforces the crucial role of geological structures in controlling groundwater movement and accumulation, as observed by [83,85] in their groundwater potential mapping study.

Comparative analysis reveals that while both models identified similar influential factors, their relative importance varied. The WoE model placed greater emphasis on topographic factors (slope and aspect), while the FR model highlighted the importance of hydrological (rainfall and drainage) and geological (lineaments) factors. This difference in factor prioritization between the two models is consistent with observations by [49], who noted that different statistical approaches can lead to variations in factor importance rankings. The NDVI analysis showed a positive association in the WoE model (0.45) for high NDVI values, while the FR model suggested only a slight positive relationship (1.01). This discrepancy between the models in assessing vegetation's impact on groundwater potential highlights the complex nature of this relationship and aligns with findings by [19], who observed varying influences of NDVI across different modeling approaches. Road proximity analysis revealed a complex relationship in both models, with areas 3000–4000 m from roads showing the highest positive values (WoE: 0.78, FR: 1.95). This non-linear relationship suggests a balance between accessibility and reduced surface sealing, a phenomenon also noted by [86] in their groundwater spring potential study.

Identifying these key factors and their relative importance in both models provides valuable insights for groundwater management in the Hangu District. The consistency in major influencing factors across both models enhances confidence in their significance. At the same time, the differences highlight the importance of employing multiple modeling approaches for a comprehensive understanding of groundwater potential.

4.2. Analysis Model Performance of WoE and FR

Applying WoE and FR models for GWPZ mapping in the Hangu District yielded valuable insights into the spatial distribution of groundwater resources. The WoE model achieved success and prediction rates of 82% and 86%, respectively, while the FR model exhibited superior performance with rates of 89% for success and 93% for prediction. These results align with previous studies, such as Nampak et al. [77], who reported AUC values of 89.7% for FR models in groundwater potential mapping in Malaysia.

The superior performance of the FR model can be attributed to its ability to capture non-linear relationships between factors and groundwater potential and its reduced sensitivity to data outliers [80]. This effectiveness is particularly evident in complex geological settings, as demonstrated by [49], who achieved a success rate of 82.35% using FR for groundwater spring potential mapping in a karstic area. Both models identified similar spatial patterns of groundwater potential, with high potential zones predominantly concentrated in the northeastern and central parts of the study area. The WoE model classified 27% (373 km^2) as high potential, while the FR model identified 23% (317 km^2) in this category. The high accuracy of both models contributes to the growing body of literature supporting probabilistic models in hydrogeological studies. Rahmati et al. [19] achieved AUC values of 82% and 78% for FR and WoE models, respectively, in Iran's groundwater potential mapping, corroborating these approaches' effectiveness in diverse geological settings.

However, it is crucial to acknowledge the inherent limitations of these models. The WoE model assumes conditional independence between evidential themes, which may not always hold true in complex hydrogeological settings [65]. Mogaji et al. [87] noted that this assumption could lead to overestimation of probabilities in some cases. The FR model, while effective, may oversimplify relationships between factors and groundwater occurrence, as observed by Ozdemir [88] in landslide susceptibility studies. The high predictive accuracy of both models in this data-scarce region underscores the effectiveness of GIS-based bivariate statistical approaches in GWPZ mapping. This success aligns with the findings by Naghibi et al. [84], who demonstrated the efficacy of FR models in data-limited environments, achieving AUC values of 77% in groundwater potential mapping in Iran. The slightly superior performance of the FR model suggests its potential as a preferred tool for similar hydrogeological assessments, particularly in regions with complex geological and topographical settings. This preference is supported by studies such as Pourtaghi and Pourghasemi [86], who found FR models to outperform other bivariate methods in groundwater spring potential mapping, achieving an AUC of 83.5%.

4.3. Integration of Groundwater Potential Mapping with Sustainable Development Goals (SDGs)

The groundwater potential mapping in Hangu District demonstrates significant implications for achieving multiple Sustainable Development Goals through its comprehensive spatial analysis. As visualized in Figure 7, this study's contributions to SDGs follow three main pillars: environmental sustainability, economic development, and social well-being. The radial diagram in Figure 9 quantitatively illustrates the varied achievement potentials across different SDGs, with the largest contributions observed in environmental and water-related goals.

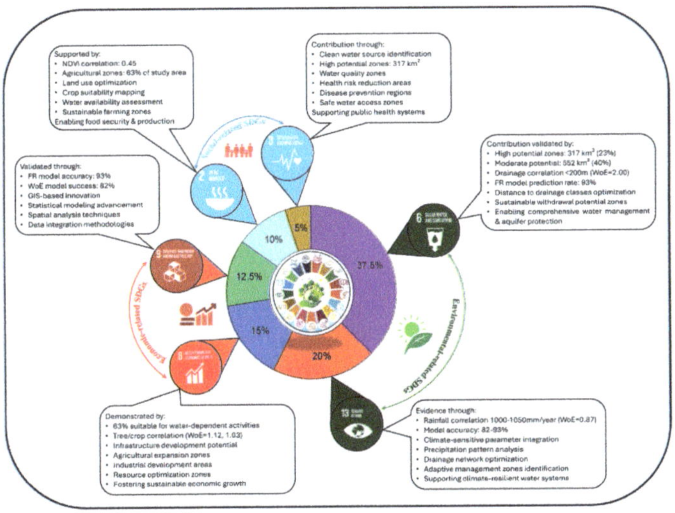

Figure 9. Contribution of Groundwater Potentiality Mapping to SDGs.

The identification of high potential zones (23%, 317 km^2) and moderate potential areas (40%, 552 km^2) directly contributes to SDG 6 (Clean Water and Sanitation), with an estimated 37.5% achievement potential. The FR model's exceptional accuracy (93% prediction rate) provides a reliable foundation for sustainable groundwater management, particularly supporting Target 6.1 (15% contribution) through improved access to safe water resources. The analysis of drainage proximity, where areas within 200 m showed the highest correlation (WoE value of 2.00), substantively supports aquifer protection under Target

6.6 (9.5% contribution). These findings corroborate those of [89], who emphasized the significance of integrating groundwater potential maps with policy for SDG 6 achievement.

The second largest contribution relates to SDG 13 (Climate Action), with a 20% achievement potential. The correlation between precipitation patterns (1000–1050 mm/year showing a WoE value of 0.87) and groundwater potential provides crucial insights for climate adaptation strategies (Target 13.1, 13% contribution). The high model accuracy (82–93%) establishes a scientific framework for incorporating climate considerations into water resource policies (Target 13.2, 7% contribution), as supported by [13].

As illustrated by the economic sector in Figure 7, the groundwater mapping contributes 15% toward SDG 8 (Decent Work and Economic Growth), with 63% of the study area showing moderate to high potential for water-dependent economic activities. The analysis of land use patterns, revealing positive correlations for trees and crops (WoE values of 1.12 and 1.03 respectively), indicates substantial potential for agricultural development (Target 8.1, 10% contribution) and employment generation (Target 8.3, 5% contribution). These quantified contributions align with findings from [90], who highlighted the significant role of groundwater mapping in achieving SDG 8. Furthermore, the application of advanced GIS techniques and statistical models contributes 12.5% toward SDG 9 (Industry, Innovation, and Infrastructure), supporting sustainable industrial development (Target 9.2, 7% contribution) and scientific advancement (Target 9.5, 5.5% contribution).

The social dimensions represented in Figure 7 show the mapping results demonstrate a 10% contribution toward SDG 2 (Zero Hunger) through the identification of suitable agricultural zones. The positive correlation between NDVI and groundwater potential (0.45) provides a quantitative basis for sustainable food production planning (Target 2.4, 2%) and food security enhancement (Target 2.1, 2%). Additionally, this study shows a 5% contribution toward SDG 3 (Good Health and Well-being) through the identification of clean water sources, supporting the reduction of water-borne diseases (Target 3.3, 1%) and water contamination-related illnesses (Target 3.9, 1%), aligning with findings from [4].

4.4. Research Limitations

The groundwater potentiality mapping study in the Hangu District demonstrated the robustness and reliability of the employed WoE and FR models, despite facing certain data limitations. The absence of meteorological stations, the region's rugged terrain, and the limited data posed challenges for comprehensive data collection and analysis. However, the integration of multiple data sources, including remote sensing, ancillary geospatial data, and existing well data, provided a solid foundation for assessing groundwater potential at the regional scale. The models' strong validation metrics underscore their effectiveness within the constraints of the available dataset. While the inability to analyze regional aquifer characteristics due to the lack of geophysical datasets and well log data restricted a more detailed understanding of subsurface hydrogeological conditions, this study highlights the importance of future research focusing on enhancing data collection and analysis in challenging terrains. Increasing well data, integrating multi-temporal re-mote sensing data, and exploring the potential impacts of climate change and land use/land cover dynamics on groundwater resources using coupled climate-hydrological models could further refine the groundwater potential assessment. Despite its limitations, this study provides a valuable contribution to the field by demonstrating the applicability of WoE and FR models in similar data-scarce regions, while laying the groundwork for future research to confirm and expand upon the SDG findings.

5. Conclusions

This study on groundwater potential zone mapping in Pakistan's Hangu district has yielded significant insights into the spatial distribution and influencing factors of groundwater resources, with substantial implications for achieving multiple SDGs. While both models demonstrated suitability for identifying potential groundwater zones, the FR model outperformed the WOE model with an impressive AUC of 93%. The analysis highlighted that factors such as elevation, slope gradient, proximity to drainage, LULC, and NDVI are crucial in determining groundwater potential. For instance, areas with elevations below 550 m, highly fractured pediment fans, proximity to drainage (less than 200 m), gentle slopes (less than 5 degrees), and flat valley terrace deposits exhibited moderate to excellent groundwater potential. Significantly, the research contributes to multiple SDGs, with an estimated 37.5% achievement potential for SDG 6 (Clean Water and Sanitation), 20% for SDG 13 (Climate Action), 15% for SDG 8 (Decent Work and Economic Growth), 12.5% for SDG 9 (Industry, Innovation, and Infrastructure), and notable contributions to SDGs 2 (10%) and 3 (5%). This broad impact emphasizes the role of groundwater potential mapping in promoting water security, climate resilience, and sustainable economic development. Identifying 23% high and 40% moderate potential zones lays the groundwork for improving clean water access, supporting agriculture, and informing climate adaptation strategies. These findings provide a robust scientific basis for water resource management, land use planning, and risk reduction strategies in the Hangu district. Future research directions should focus on conducting comprehensive analyses of regional potential aquifers along the identified zones, increasing well data for more precise results, and exploring the construction of surface reservoirs to enhance recharge zones. These efforts would contribute significantly to improving both the quality and quantity of groundwater resources in the region and further advancing the achievement of related SDGs. The successful application of geospatial techniques in the Hangu district showcased in this study highlights their replicability in similar hydrogeological settings globally, contributing to sustainable groundwater management, addressing water scarcity, and supporting broader sustainable development initiatives.

Author Contributions: Conceptualization, A.R., Y.M.Y. and M.E.A.-E.; Data curation, A.R., L.X. and N.A. (Naveed Ahmed); Formal analysis, A.R., L.X., F.I., S.Q., S.L., N.A. (Nassir Alarifi), Y.M.Y. and M.E.A.-E.; Funding acquisition, N.A. (Nassir Alarifi); Methodology, A.R., L.X., Y.M.Y. and M.E.A.-E.; Project administration, S.Q. and N.A. (Nassir Alarifi); Resources, A.R.; Software, A.R., L.X., F.I., N.A. (Naveed Ahmed), and M.E.A.-E.; Supervision, F.I. and Y.M.Y.; Validation, A.R. and L.X.; Visualization, A.R., L.X., S.L. and M.E.A.-E.; Writing—original draft, A.R., L.X., F.I., N.A. (Naveed Ahmed), and S.L.; Writing—review & editing, N.A. (Naveed Ahmed), S.Q., Y.M.Y. and M.E.A.-E. All authors have read and agreed to the published version of the manuscript.

Funding: This research was supported by the Researchers Supporting Project number (RSP2024R432), King Saud University, Riyadh, Saudi Arabia.

Data Availability Statement: The data supporting this study's findings are available on request from the corresponding author.

Acknowledgments: The authors would like to express their gratitude to the Researchers Supporting Project number (RSP2024R432) at King Saud University in Riyadh, Saudi Arabia.

Conflicts of Interest: The authors declare no conflicts of interest.

References

1. Wang, Y.; Yuan, S.; Shi, J.; Ma, T.; Xie, X.; Deng, Y.; Du, Y.; Gan, Y.; Guo, Z.; Dong, Y.; et al. Groundwater Quality and Health: Making the Invisible Visible. *Environ. Sci. Technol.* **2023**, *57*, 5125–5136. [CrossRef] [PubMed]
2. Das, A.; Joardar, M.; De, A.; Mridha, D.; Chowdhury, N.R.; Bin Kashim Khan, M.T.; Chakrabartty, P.; Roychowdhury, T. Pollution Index and Health Risk Assessment of Arsenic through Different Groundwater Sources and Its Load on Soil-Paddy-Rice System in a Part of Murshidabad District of West Bengal, India. *Groundw. Sustain. Dev.* **2021**, *15*, 100652. [CrossRef]
3. Sutradhar, S.; Sarkar, D.; Bhuimali, A.; Mondal, P. Integration of Different Geospatial Factors to Delineate Groundwater Potential Zones Using Multi-Influencing Factors under Remote Sensing and GIS Environment: A Study on Dakshin Dinajpur District, West Bengal, India. *Sustain. Water Resour. Manag.* **2022**, *8*, 37. [CrossRef]
4. Alegbeleye, O.O.; Sant'Ana, A.S. Manure-Borne Pathogens as an Important Source of Water Contamination: An Update on the Dynamics of Pathogen Survival/Transport as Well as Practical Risk Mitigation Strategies. *Int. J. Hyg. Environ. Health* **2020**, *227*, 113524. [CrossRef] [PubMed]
5. UNESCO World Water Assessment Programme; Connor, R.; Miletto, M. *The United Nations World Water Development Report 2022: Groundwater: Making the Invisible Visible*; Executive Summary; UNESCO: Paris, France, 2022.

6. Rahman, M.M.; Hossain, M.K.F.B.; Afrin, S.; Saito, T.; Kurasaki, M. Effects of Metals on Human Health and Ecosystem. In *Handbook of Environmental Chemistry*; Springer: Berlin/Heidelberg, Germany, 2023; Volume 115, pp. 81–119.
7. Karatayev, M.; Rivotti, P.; Sobral Mourão, Z.; Konadu, D.D.; Shah, N.; Clarke, M. The Water-Energy-Food Nexus in Kazakhstan: Challenges and Opportunities. *Energy Procedia* **2017**, *125*, 63–70. [CrossRef]
8. Gleeson, T.; Alley, W.M.; Allen, D.M.; Sophocleous, M.A.; Zhou, Y.; Taniguchi, M.; Vandersteen, J. Towards Sustainable Groundwater Use: Setting Long-Term Goals, Backcasting, and Managing Adaptively. *Ground Water* **2012**, *50*, 19–26. [CrossRef]
9. Burek, P.; Satoh, Y.; Fischer, G.; Kahil, M.T.; Scherzer, A.; Tramberend, S.; Nava, L.F.; Wada, Y.; Eisner, S.; Flörke, M.; et al. *Water Futures and Solution*; Final Report. Fast Track Initiative; International Institute for Applied Systems Analysis: Laxenburg, Austria, 2016.
10. Li, P.; Tian, R.; Xue, C.; Wu, J. Progress, Opportunities, and Key Fields for Groundwater Quality Research under the Impacts of Human Activities in China with a Special Focus on Western China. *Environ. Sci. Pollut. Res.* **2017**, *24*, 13224–13234. [CrossRef]
11. Li, P.; Wu, J.; Qian, H. Preliminary Assessment of Hydraulic Connectivity between River Water and Shallow Groundwater and Estimation of Their Transfer Rate during Dry Season in the Shidi River, China. *Environ. Earth Sci.* **2016**, *75*, 99. [CrossRef]
12. Afshar, A.; Khosravi, M.; Molajou, A. Assessing Adaptability of Cyclic and Non-Cyclic Approach to Conjunctive Use of Groundwater and Surface Water for Sustainable Management Plans under Climate Change. *Water Resour. Manag.* **2021**, *35*, 3463–3479. [CrossRef]
13. Cuthbert, M.O.; Taylor, R.G.; Favreau, G.; Todd, M.C.; Shamsudduha, M.; Villholth, K.G.; MacDonald, A.M.; Scanlon, B.R.; Kotchoni, D.O.V.; Vouillamoz, J.M.; et al. Observed Controls on Resilience of Groundwater to Climate Variability in Sub-Saharan Africa. *Nature* **2019**, *572*, 230–234. [CrossRef]
14. Wu, W.Y.; Lo, M.H.; Wada, Y.; Famiglietti, J.S.; Reager, J.T.; Yeh, P.J.F.; Ducharne, A.; Yang, Z.L. Divergent Effects of Climate Change on Future Groundwater Availability in Key Mid-Latitude Aquifers. *Nat. Commun.* **2020**, *11*, 3710. [CrossRef] [PubMed]
15. Berg, A.; Findell, K.; Lintner, B.; Giannini, A.; Seneviratne, S.I.; Van Den Hurk, B.; Lorenz, R.; Pitman, A.; Hagemann, S.; Meier, A.; et al. Land-Atmosphere Feedbacks Amplify Aridity Increase over Land under Global Warming. *Nat. Clim. Change* **2016**, *6*, 869–874. [CrossRef]
16. Xu, Z.; Li, X.; Li, J.; Xue, Y.; Jiang, S.; Liu, L.; Luo, Q.; Wu, K.; Zhang, N.; Feng, Y.; et al. Characteristics of Source Rocks and Genetic Origins of Natural Gas in Deep Formations, Gudian Depression, Songliao Basin, NE China. *ACS Earth Space Chem.* **2022**, *6*, 1750–1771. [CrossRef]
17. Kumar, A.; Krishna, A.P. Assessment of Groundwater Potential Zones in Coal Mining Impacted Hard-Rock Terrain of India by Integrating Geospatial and Analytic Hierarchy Process (AHP) Approach. *Geocarto Int.* **2018**, *33*, 105–129. [CrossRef]
18. Li, J.; Wang, Z.; Wu, X.; Xu, C.Y.; Guo, S.; Chen, X. Toward Monitoring Short-Term Droughts Using a Novel Daily Scale, Standardized Antecedent Precipitation Evapotranspiration Index. *J. Hydrometeorol.* **2020**, *21*, 891–908. [CrossRef]
19. Rahmati, O.; Nazari Samani, A.; Mahdavi, M.; Pourghasemi, H.R.; Zeinivand, H. Groundwater Potential Mapping at Kurdistan Region of Iran Using Analytic Hierarchy Process and GIS. *Arab. J. Geosci.* **2015**, *8*, 7059–7071. [CrossRef]
20. Zhuo, Z.; Du, L.; Lu, X.; Chen, J.; Cao, Z. Smoothed Lv Distribution Based Three-Dimensional Imaging for Spinning Space Debris. *IEEE Trans. Geosci. Remote Sens.* **2022**, *60*, 1–13. [CrossRef]
21. Acharya, T.; Kumbhakar, S.; Prasad, R.; Mondal, S.; Biswas, A. Delineation of Potential Groundwater Recharge Zones in the Coastal Area of North-Eastern India Using Geoinformatics. *Sustain. Water Resour. Manag.* **2019**, *5*, 533–540. [CrossRef]
22. Abd-Elmaboud, M.E.; Saqr, A.M.; El-Rawy, M.; Al-Arifi, N.; Ezzeldin, R. Evaluation of Groundwater Potential Using ANN-Based Mountain Gazelle Optimization: A Framework to Achieve SDGs in East El Oweinat, Egypt. *J. Hydrol. Reg. Stud.* **2024**, *52*, 101703. [CrossRef]
23. Ozegin, K.O.; Ilugbo, S.O. A Triangulation Approach for Groundwater Potential Evaluation Using Geospatial Technology and Multi-Criteria Decision Analysis (MCDA) in Edo State, Nigeria. *J. Afr. Earth Sci.* **2024**, *209*, 105101. [CrossRef]
24. Islam, F.; Tariq, A.; Guluzade, R.; Zhao, N.; Shah, S.U.; Ullah, M.; Hussain, M.L.; Ahmad, M.N.; Alasmari, A.; Alzuaibr, F.M.; et al. Comparative Analysis of GIS and RS Based Models for Delineation of Groundwater Potential Zone Mapping. *Geomat. Nat. Hazards Risk* **2023**, *14*, 2216852. [CrossRef]
25. Prapanchan, V.N.; Subramani, T.; Karunanidhi, D. GIS and Fuzzy Analytical Hierarchy Process to Delineate Groundwater Potential Zones in Southern Parts of India. *Groundw. Sustain. Dev.* **2024**, *25*, 101110. [CrossRef]
26. Shaw, S.K.; Sharma, A. Assessing Groundwater Potential Zones in Sundargarh District: A Comparative Study Using AHP and FR Models. *Sustain. Water Resour. Manag.* **2024**, *10*, 132. [CrossRef]
27. Rehman, A.; Islam, F.; Tariq, A.; Ul Islam, I.; Brian, J.D.; Bibi, T.; Ahmad, W.; Ali Waseem, L.; Karuppannan, S.; Al-Ahmadi, S. Groundwater Potential Zone Mapping Using GIS and Remote Sensing Based Models for Sustainable Groundwater Management. *Geocarto Int.* **2024**, *39*, 2306275. [CrossRef]
28. Park, S.; Hamm, S.Y.; Jeon, H.T.; Kim, J. Evaluation of Logistic Regression and Multivariate Adaptive Regression Spline Models for Groundwater Potential Mapping Using R and GIS. *Sustainability* **2017**, *9*, 1157. [CrossRef]
29. Chien, N.P.; Lautz, L.K. Discriminant Analysis as a Decision-Making Tool for Geochemically Fingerprinting Sources of Groundwater Salinity. *Sci. Total Environ.* **2018**, *618*, 379–387. [CrossRef]
30. Zeinivand, H.; Ghorbani Nejad, S. Application of GIS-Based Data-Driven Models for Groundwater Potential Mapping in Kuhdasht Region of Iran. *Geocarto Int.* **2018**, *33*, 651–666. [CrossRef]

31. Zhu, Q.; Abdelkareem, M. Mapping Groundwater Potential Zones Using a Knowledge-Driven Approach and GIS Analysis. *Water* **2021**, *13*, 579. [CrossRef]
32. Pautz, C.; dos Santos, A.R.; Ferrari, J.L.; Guerra Filho, P.A.; Kunz, S.H.; Dias, H.M.; Moreira, T.R.; Carvalho, R. de C.F.; Mardeni, V.D.N.; Santos, E.C. dos; et al. Mapping the Environmental Vulnerability of a Lagoon Using Fuzzy Logic and the AHP Method. *Water* **2023**, *15*, 2102. [CrossRef]
33. Zhao, P.; Md Ali, Z.; Ahmad, Y. Developing Indicators for Sustainable Urban Regeneration in Historic Urban Areas: Delphi Method and Analytic Hierarchy Process (AHP). *Sustain. Cities Soc.* **2023**, *99*, 104990. [CrossRef]
34. Ray, S.K. Identifying the Groundwater Potential Zones in Jamsholaghat Sub-Basin by Considering GIS and Multi-Criteria Decision Analysis. *Int. J. Environ. Sci. Technol.* **2024**, *21*, 515–540. [CrossRef]
35. Arfanuzzaman, M.; Dahiya, B. Sustainable Urbanization in Southeast Asia and beyond: Challenges of Population Growth, Land Use Change, and Environmental Health. *Growth Change* **2019**, *50*, 725–744. [CrossRef]
36. Kotharkar, R.; Ramesh, A.; Bagade, A. Urban Heat Island Studies in South Asia: A Critical Review. *Urban Clim.* **2018**, *24*, 1011–1026. [CrossRef]
37. Qi, F.; Abu-Rumman, A.; Al Shraah, A.; Muda, I.; Huerta-Soto, R.; Hai Yen, T.T.; Abdul-Samad, Z.; Michel, M. Moving a Step Closer towards Environmental Sustainability in Asian Countries: Focusing on Real Income, Urbanization, Transport Infrastructure, and Research and Development. *Econ. Res. Istraz.* **2023**, *36*, 3576–3595. [CrossRef]
38. Din, I.U.; Muhammad, S.; Rehman, I.u. Groundwater Quality Assessment for Drinking and Irrigation Purposes in the Hangu District, Pakistan. *J. Food Compos. Anal.* **2023**, *115*, 104919. [CrossRef]
39. Vári, Á.; Podschun, S.A.; Erős, T.; Hein, T.; Pataki, B.; Iojă, I.C.; Adamescu, C.M.; Gerhardt, A.; Gruber, T.; Dedić, A.; et al. Freshwater Systems and Ecosystem Services: Challenges and Chances for Cross-Fertilization of Disciplines. *Ambio* **2022**, *51*, 135–151. [CrossRef]
40. Abd-Elmaboud, M.; Abdel-Gawad, H.; El-Alfy, K.; Ezzeldin, M. Estimation of Groundwater Recharge Using Simulation-Optimization Model and Cascade Forward ANN at East Nile Delta Aquifer, Egypt. *J. Hydrol. Reg. Stud.* **2021**, *34*. [CrossRef]
41. Din, I.U.; Muhammad, S.; Faisal, S.; ur Rehman, I.; Ali, W. Heavy Metal(Loid)s Contamination and Potential Risk Assessment via Groundwater Consumption in the District of Hangu, Pakistan. *Environ. Sci. Pollut. Res.* **2023**, *30*, 33808–33818. [CrossRef]
42. Alrawi, I.; Chen, J.; Othman, A.A. Groundwater potential zone mapping: Integration of multi-criteria decision analysis (MCDA) and GIS techniques for the Al-Qalamoun region in Syria. *ISPRS Int. J. Geo-Inf.* **2022**, *11*, 603. [CrossRef]
43. Masria, A.; Alshammari, T.O.; Ghareeb, M.; Seif, A.K.; Abd-Elmaboud, M.E.; Ammar, A.I. 2D and 3D Modeling of Resistivity and Chargeability to Identify the Type of Saturated Groundwater for Complex Sedimentary Facies. *Hydrology* **2024**, *11*, 120. [CrossRef]
44. Qureshi, A.S.; Gill, M.A.; Sarwar, A. Sustainable Groundwater Management in Pakistan: Challenges and Opportunities. *Irrig. Drain.* **2010**, *59*, 107–116. [CrossRef]
45. Ali, S.; Liu, Y.; Ishaq, M.; Shah, T.; Abdullah; Ilyas, A.; Din, I.U. Climate Change and Its Impact on the Yield of Major Food Crops: Evidence from Pakistan. *Foods* **2017**, *6*, 39. [CrossRef] [PubMed]
46. Iqbal, M.A.; Penas, A.; Cano-Ortiz, A.; Kersebaum, K.C.; Herrero, L.; del Río, S. Analysis of Recent Changes in Maximum and Minimum Temperatures in Pakistan. *Atmos. Res.* **2016**, *168*, 234–249. [CrossRef]
47. Sajjad, M.; Rahim, S.; Tahir, S.S.; Asif, H.M. Chemical Quality of Groundwater and Associated Health Risks in Bahawalpur City, Punjab, Pakistan. *Environ. Earth Sci.* **2018**, *77*, 1–13.
48. Khan, A.N.; Gerlitz, L.; Bhattacharya, B.; Hassan, M.; Shah, S.A.; Ali, S. Assessment of Flash Sedimentar Susceptibility in Small Mountainous Watersheds of Northern Pakistan. *Nat. Hazards* **2020**, *102*, 1167–1186.
49. Chen, W.; Li, H.; Hou, E.; Wang, S.; Wang, G.; Panahi, M.; Li, T.; Peng, T.; Guo, C.; Niu, C.; et al. GIS-Based Groundwater Potential Analysis Using Novel Ensemble Weights-of-Evidence with Logistic Regression and Functional Tree Models. *Sci. Total Environ.* **2018**, *634*, 853–867. [CrossRef]
50. Rahmati, O.; Pourghasemi, H.R.; Melesse, A.M. Application of GIS-Based Data Driven Random Forest and Maximum Entropy Models for Groundwater Potential Mapping: A Case Study at Mehran Region, Iran. *Catena* **2016**, *137*, 360–372. [CrossRef]
51. Ashwini, K.; Sil, B.S.; Kafy, A.A.; Altuwaijri, H.A.; Nath, H.; Rahaman, Z.A. Harnessing Machine Learning Algorithms to Model the Association between Land Use/Land Cover Change and Heatwave Dynamics for Environmental Management. *Land* **2024**, *13*, 1273. [CrossRef]
52. Bachmann, N.; Tripathi, S.; Brunner, M.; Jodlbauer, H. The contribution of data-driven technologies in achieving the sustainable development goals. *Sustainability* **2022**, *14*, 2497. [CrossRef]
53. Karimi-Rizvandi, S.; Goodarzi, H.V.; Afkoueieh, J.H.; Chung, I.M.; Kisi, O.; Kim, S.; Linh, N.T.T. Groundwater-Potential Mapping Using a Self-Learning Bayesian Network Model: A Comparison among Metaheuristic Algorithms. *Water* **2021**, *13*, 658. [CrossRef]
54. Chen, N.C.; Wen, H.Y.; Li, F.M.; Hsu, S.M.; Ke, C.C.; Lin, Y.T.; Huang, C.C. Investigation and Estimation of Groundwater Level Fluctuation Potential: A Case Study in the Pei-Kang River Basin and Chou-Shui River Basin of the Taiwan Mountainous Region. *Appl. Sci.* **2022**, *12*, 7060. [CrossRef]
55. Bhatt, S.C.; Patel, A.; Pradhan, S.R.; Singh, S.K.; Singh, V.K.; Prakash Tripathi, G.; Kishor, K. Morphometric and Morphotectonic Attributes of Ken Basin, Central India: Depicting Status of Soil Erosion, and Tectonic Activities. *Total Environ. Adv.* **2024**, *9*, 200088. [CrossRef]

56. Pande, C.B.; Moharir, K.N.; Panneerselvam, B.; Singh, S.K.; Elbeltagi, A.; Pham, Q.B.; Varade, A.M.; Rajesh, J. Delineation of Groundwater Potential Zones for Sustainable Development and Planning Using Analytical Hierarchy Process (AHP), and MIF Techniques. *Appl. Water Sci.* **2021**, *11*, 186. [CrossRef]
57. Pham, B.T.; Avand, M.; Janizadeh, S.; Van Phong, T.; Al-Ansari, N.; Ho, L.S.; Das, S.; Van Le, H.; Amini, A.; Bozchaloei, S.K.; et al. GIS Based Hybrid Computational Approaches for Flash Flood Susceptibility Assessment. *Water* **2020**, *12*, 683. [CrossRef]
58. Dandapat, K.; Chatterjee, U.; Das, S.; Patra, A.; Manna, H.; Ghosh, A.; Pal, S.C.; Towfiqul Islam, A.R.M.; Costache, R.; Alam, E.; et al. Assessment of Groundwater Potential Zones in Data-Scarce Regions Using GIS-Based Multicriteria Decision Making Approach. *Geocarto Int.* **2024**, *39*, 2311202. [CrossRef]
59. Ibrahim-Bathis, K.; Ahmed, S.A. Geospatial Technology for Delineating Groundwater Potential Zones in Doddahalla Watershed of Chitradurga District, India. *Egypt. J. Remote Sens. Sp. Sci.* **2016**, *19*, 223–234. [CrossRef]
60. Mengistu, T.D.; Chung, I.M.; Kim, M.G.; Chang, S.W.; Lee, J.E. Impacts and Implications of Land Use Land Cover Dynamics on Groundwater Recharge and Surface Runoff in East African Watershed. *Water* **2022**, *14*, 2068. [CrossRef]
61. Bhanja, S.N.; Malakar, P.; Mukherjee, A.; Rodell, M.; Mitra, P.; Sarkar, S. Using Satellite-Based Vegetation Cover as Indicator of Groundwater Storage in Natural Vegetation Areas. *Geophys. Res. Lett.* **2019**, *46*, 8082–8092. [CrossRef]
62. Pandey, A.; Purohit, R. Impact of Geological Controls on Change in Groundwater Potential of Recharge Zones Due to Watershed Development Activities, Using Integrated Approach of RS and GIS. *J. Sci. Res.* **2022**, *66*, 53–62. [CrossRef]
63. Elmoulat, M.; Brahim, L.A.; Mastere, M.; Jemmah, A.I. Mapping of Mass Movements Susceptibility in the Zoumi Region Using Satellite Image and GIS Technology (Moroccan Rif). *Int. J. Sci. Eng. Res.* **2015**, *6*, 210–217.
64. Mandal, B.; Mondal, S.; Mandal, S. GIS-Based Landslide Susceptibility Zonation (LSZ) Mapping of Darjeeling Himalaya, India Using Weights of Evidence (WoE) Model. *Arab. J. Geosci.* **2023**, *16*, 421. [CrossRef]
65. Bonham-Carter, G. *Geographic Information Systems for Geoscientists: Modelling with GIS*; Elsevier: Amsterdam, The Netherlands, 1994.
66. Getachew, N.; Meten, M. Weights of Evidence Modeling for Landslide Susceptibility Mapping of Kabi-Gebro Locality, Gundomeskel Area, Central Ethiopia. *Geoenviron. Disasters* **2021**, *8*, 6. [CrossRef]
67. Guru, B.; Seshan, K.; Bera, S. Frequency Ratio Model for Groundwater Potential Mapping and Its Sustainable Management in Cold Desert, India. *J. King Saud. Univ.-Sci.* **2017**, *29*, 333–347. [CrossRef]
68. Maity, B.; Mallick, S.K.; Das, P.; Rudra, S. Comparative Analysis of Groundwater Potentiality Zone Using Fuzzy AHP, Frequency Ratio and Bayesian Weights of Evidence Methods. *Appl. Water Sci.* **2022**, *12*, 63. [CrossRef]
69. Tavakoli, M.; Motlagh, Z.K.; Sayadi, M.H.; Ibraheem, I.M.; Youssef, Y.M. Sustainable Groundwater Management Using Machine Learning-Based DRASTIC Model in Rurbanizing Riverine Region: A Case Study of Kerman Province, Iran. *Water (Switzerland)* **2024**, *16*, 2748. [CrossRef]
70. Balci, O. Validation, Verification, and Testing Techniques throughout the Life Cycle of a Simulation Study. *Ann. Oper. Res.* **1994**, *53*, 121–173. [CrossRef]
71. Pace, D.K. Modeling and Simulation Verification and Validation Challenges. *Johns. Hopkins APL Tech. Dig. (Appl. Phys. Lab.)* **2004**, *25*, 163–172.
72. Arabameri, A.; Pal, S.C.; Rezaie, F.; Nalivan, O.A.; Chowdhuri, I.; Saha, A.; Lee, S.; Moayedi, H. Modeling Groundwater Potential Using Novel GIS-Based Machine-Learning Ensemble Techniques. *J. Hydrol. Reg. Stud.* **2021**, *36*, 100848. [CrossRef]
73. Liu, R.; Li, G.; Wei, L.; Xu, Y.; Gou, X.; Luo, S.; Yang, X. Spatial Prediction of Groundwater Potentiality Using Machine Learning Methods with Grey Wolf and Sparrow Search Algorithms. *J. Hydrol.* **2022**, *610*, 127977. [CrossRef]
74. Maqsoom, A.; Aslam, B.; Khalid, N.; Ullah, F.; Anysz, H.; Almaliki, A.H.; Almaliki, A.A.; Hussein, E.E. Delineating Groundwater Recharge Potential through Remote Sensing and Geographical Information Systems. *Water* **2022**, *14*, 1824. [CrossRef]
75. Pradhan, B. A Comparative Study on the Predictive Ability of the Decision Tree, Support Vector Machine and Neuro-Fuzzy Models in Landslide Susceptibility Mapping Using GIS. *Comput. Geosci.* **2013**, *51*, 350–365. [CrossRef]
76. Tehrany, M.S.; Pradhan, B.; Jebur, M.N. Spatial Prediction of Flood Susceptible Areas Using Rule Based Decision Tree (DT) and a Novel Ensemble Bivariate and Multivariate Statistical Models in GIS. *J. Hydrol.* **2013**, *504*, 69–79. [CrossRef]
77. Nampak, H.; Pradhan, B.; Manap, M.A. Application of GIS Based Data Driven Evidential Belief Function Model to Predict Groundwater Potential Zonation. *J. Hydrol.* **2014**, *513*, 283–300. [CrossRef]
78. Meng, F.; Khan, M.I.; Naqvi, S.A.A.; Sarwar, A.; Islam, F.; Ali, M.; Tariq, A.; Ullah, S.; Soufan, W.; Faraj, T.K. Identification and mapping of groundwater recharge zones using multi influencing factor and analytical hierarchy process. *Sci. Rep.* **2024**, *14*, 19240. [CrossRef] [PubMed]
79. Ali, J.; Islam, F.; Bibi, T.; Islam, I.; Mughal, M.R.; Sabir, M.; Awwad, F.; Ismail, E. Quantifying the impact of climate change and urbanization on groundwater resources using geospatial modeling. *Front. Earth Sci.* **2024**, *12*, 1377367. [CrossRef]
80. Lee, S.; Ryu, J.H.; Won, J.S.; Park, H.J. Determination and Application of the Weights for Landslide Susceptibility Mapping Using an Artificial Neural Network. *Eng. Geol.* **2004**, *71*, 289–302. [CrossRef]
81. Hosmer, D.W.; Lemeshow, S.; Sturdivant, R.X. *Applied Logistic Regression*; Wiley: New York, NY, USA, 2000.
82. Manap, M.A.; Nampak, H.; Pradhan, B.; Lee, S.; Sulaiman, W.N.A.; Ramli, M.F. Application of Probabilistic-Based Frequency Ratio Model in Groundwater Potential Mapping Using Remote Sensing Data and GIS. *Arab. J. Geosci.* **2014**, *7*, 711–724. [CrossRef]
83. Oh, H.J.; Kim, Y.S.; Choi, J.K.; Park, E.; Lee, S. GIS Mapping of Regional Probabilistic Groundwater Potential in the Area of Pohang City, Korea. *J. Hydrol.* **2011**, *399*, 158–172. [CrossRef]

84. Naghibi, S.A.; Pourghasemi, H.R.; Pourtaghi, Z.S.; Rezaei, A. Groundwater Qanat Potential Mapping Using Frequency Ratio and Shannon's Entropy Models in the Moghan Watershed, Iran. *Earth Sci. Inform.* **2015**, *8*, 171–186. [CrossRef]
85. Mallick, J.; Talukdar, S.; Alsubih, M.; Ahmed, M.; Islam, A.R.M.T.; Shahfahad; Thanh, N.V. Proposing Receiver Operating Characteristic-Based Sensitivity Analysis with Introducing Swarm Optimized Ensemble Learning Algorithms for Groundwater Potentiality Modelling in Asir Region, Saudi Arabia. *Geocarto Int.* **2022**, *37*, 4361–4389. [CrossRef]
86. Pourtaghi, Z.S.; Pourghasemi, H.R. GIS-Based Groundwater Spring Potential Assessment and Mapping in the Birjand Township, Southern Khorasan Province, Iran. *Hydrogeol. J.* **2014**, *22*, 643–662. [CrossRef]
87. Mogaji, K.A.; Lim, H.S.; Abdullah, K. Regional Prediction of Groundwater Potential Mapping in a Multifaceted Geology Terrain Using GIS-Based Dempster–Shafer Model. *Arab. J. Geosci.* **2015**, *8*, 3235–3258. [CrossRef]
88. Ozdemir, A. Landslide Susceptibility Mapping Using Bayesian Approach in the Sultan Mountains (Akşehir, Turkey). *Nat. Hazards* **2011**, *59*, 1573–1607. [CrossRef]
89. Saqr, A.M.; Ibrahim, M.G.; Fujii, M.; Nasr, M. Sustainable Development Goals (SDGs) Associated with Groundwater Over-Exploitation Vulnerability: Geographic Information System-Based Multi-Criteria Decision Analysis. *Nat. Resour. Res.* **2021**, *30*, 4255–4276. [CrossRef]
90. Saqr, A.M.; Nasr, M.; Fujii, M.; Yoshimura, C.; Ibrahim, M.G. Delineating Suitable Zones for Solar-Based Groundwater Exploitation Using Multi-Criteria Analysis: A Techno-Economic Assessment for Meeting Sustainable Development Goals (SDGs). *Groundw. Sustain. Dev.* **2024**, *25*. [CrossRef]

water

MDPI

Article

Ground–Surface Water Assessment for Agricultural Land Prioritization in the Upper Kansai Basin, India: An Integrated SWAT-VIKOR Framework Approach

Sudipto Halder [1], Santanu Banerjee [2,*], Youssef M. Youssef [3,*], Abhilash Chandel [4,5], Nassir Alarifi [6], Gupinath Bhandari [7] and Mahmoud E. Abd-Elmaboud [8]

[1] School of Water Resources Engineering, Jadavpur University, Kolkata 700032, India; sudiptoh.swre.rs@jadavpuruniversity.in
[2] Department of Agriculture, Chhatrapati Shahu Ji Maharaj University, Kanpur 208012, India
[3] Geological and Geophysical Engineering Department, Faculty of Petroleum and Mining Engineering, Suez University, Suez 43518, Egypt
[4] Virginia Tech Tidewater Agricultural Research and Extension Center, Suffolk, VA 23437, USA; abhilashchandel@vt.edu
[5] Department of Biological Systems Engineering, Virginia Tech, Blacksburg, VA 24061, USA
[6] Department of Geology & Geophysics, College of Science, King Saud University, PO Box 2455, Riyadh 11451, Saudi Arabia; nalarifi@ksu.edu.sa
[7] Department of Civil Engineering, Jadavpur University, Kolkata 700032, India; gupinath.bhandari@jadavpuruniversity.in
[8] Irrigation & Hydraulics Department, Faculty of Engineering, Mansoura University, Mansoura 35516, Egypt; mahmoud_elsayed@mans.edu.eg
* Correspondence: banerjee.s722146@gmail.com (S.B.); youssef.ibrahim@pme.suezuni.edu.eg (Y.M.Y.)

Abstract: Prioritizing agricultural land use is a significant challenge for sustainable development in the rapidly urbanizing, semi-arid riverine basins of South Asia, especially under climate variability and water scarcity. This study introduces a systematic framework combining remote sensing and geospatial data with the Soil and Water Assessment Tool (SWAT) model, morphometric analysis, and VIKOR-based Multi-Criteria Decision Analysis (MCDA) to effectively identify Agricultural Land Prioritization (AgLP) areas in the Upper Kansai Basin, India, while reducing the environmental impact, in line with Sustainable Development Goals (SDGs). The SWAT model simulation reveals varied hydrological patterns, with basin water yields from 965.9 to 1012.9 mm and a substantial baseflow (~64% of total flow), emphasizing essential groundwater–surface water interactions for sustainable agriculture. However, the discrepancy between percolation (47% of precipitation) and deep recharge (2% of precipitation) signals potential long-term groundwater challenges. VIKOR analysis offers a robust prioritization framework, ranking SW4 as the most suitable (Qi = 0.003) for balanced hydrological and morphometric features, in agreement with the SWAT outcomes. SW4 and SW5 display optimal agricultural conditions due to stable terrain, effective water retention, and favorable morphometric traits (drainage density 3.0–3.15 km/km^2; ruggedness 0.3–0.4). Conversely, SW2, with high drainage density (5.33 km/km^2) and ruggedness (2.0), shows low suitability, indicating risks of erosion and poor water retention. This integrated AgLP framework advances sustainable agricultural development and supports SDGs, including SDG 2 (Zero Hunger), SDG 6 (Clean Water), SDG 13 (Climate Action), and SDG 15 (Life on Land). Incorporating hydrological dynamics, land use, soil properties, and climate variables, this approach offers a precise assessment of agricultural suitability to address global sustainability challenges in vulnerable riverine basins of developing nations.

 https://doi.org/10.3390/w17060880

Keywords: agricultural land prioritization; SWAT; groundwater; morphometric analysis; VIKOR-MCDM; sustainable development goals; remote sensing; upper Kansai basin

1. Introduction

The agricultural sector faces unprecedented challenges in meeting food demands for the growing populations in developing nations [1,2]. Riverine corridors, rich in fertile soils and freshwater, are crucial for agricultural production and socio-economic development [3]. However, production has declined in urbanized riverine areas due to climate change and improper land use, threatening food security for over 30% of the global population [4,5]. This underscores the need to identify suitable Agricultural Land Prioritization (AgLP) areas while mitigating environmental impacts [6].

AgLP is essential for promoting sustainable production at local, watershed, and state levels [7]. Watershed prioritization supports groundwater development, surface water management, and soil conservation in urbanized riverine basins [6,8–14]. In this regard, drainage patterns offer crucial insights into slope variation, geological structures, and environmental processes, with direct applications in runoff estimation, flood discharge prediction, and groundwater recharge assessment. Prioritizing watersheds through morphometric analysis enables the examination of linear, aerial, and relief features [15–17], though this approach alone may overlook regional variations in geology, soil types, and land use/land cover (LULC).

The constraints of conventional field-based methods necessitate the integration of Remote Sensing (RS), Geographic Information Systems (GIS), and Multi-Criteria Decision Analysis (MCDA) to develop robust land suitability assessment systems [15]. RS offers comprehensive capabilities for mapping surface features within a GIS framework [18], while advances in multispectral Landsat sensors have produced vital datasets for assessing LULC in stressed urban regions [19–24]. The Shuttle Radar Topography Mission (SRTM) has proven particularly valuable for identifying drainage basins and their features in geologically complex regions [25].

From an agricultural perspective, the combination of morphometric analysis and the Soil and Water Assessment Tool (SWAT) provides a powerful framework for developing effective conservation strategies. The SWAT effectively predicts yields of water, sediment, nutrients, and other parameters using spatially dispersed data [26], with demonstrated success in watershed hydrology applications [27–31]. The model's effectiveness in humid tropical conditions enables the robust assessment of the potential impact of land use changes on flow dynamics [32].

Integrating the SWAT with morphometric analysis provides a comprehensive framework for understanding agricultural landscapes in semi-arid regions. This approach is essential for future agricultural practices, as it accounts for soil moisture dynamics, groundwater flow patterns, surface runoff characteristics, and the physical attributes of drainage basins. While morphometric analysis identifies areas susceptible to erosion, the SWAT effectively simulates soil erosion rates, facilitating the design of targeted conservation strategies [33]. Furthermore, the SWAT provides valuable insights into water flow patterns critical for efficient irrigation planning and management [34,35], promoting sustainable farming practices in challenging semi-arid environments [36].

To operationalize this integrated approach, the VIKOR-based MCDA model ranks sub-watersheds based on comprehensive criteria [37]. This prioritization process identifies regions with the greatest potential for sustainable agriculture by considering multiple factors, including soil quality, water availability, topography, and socio-economic conditions [38,39]. The VIKOR model enables decision makers to allocate resources efficiently and implement targeted interventions while optimizing yield and reducing environmental damage [40–42].

Agricultural regions in rapidly urbanizing South Asian countries face complex environmental and socioeconomic challenges [43]. In India, despite the Green Revolution's success [44], crop production faces significant obstacles including soil fertility loss, water pollution, and crop diseases, exacerbated by climate change [45–47]. Population growth, urbanization, and industrial development further strain agricultural resources [48], necessitating integrated approaches to address these interconnected issues.

The Upper Kansai Basin in West Bengal, India (Figure 1), exemplifies these challenges. This region, forming the eastern extension of the Chotanagpur Plateau, features undulating topography, a semi-arid climate, and a rich cultural heritage that influences its agricultural potential [49,50]. Despite receiving substantial annual rainfall of 1383 mm [51], the basin's terrain causes significant runoff, creating water scarcity despite precipitation abundance [52]. The geological setting, dominated by Precambrian metamorphic rocks [53], adds complexity to agricultural practices through its influence on soil formation and groundwater dynamics. The Purulia district faces numerous challenges, including severe seasonal water scarcity, soil erosion, land degradation, and climate variability [54–56]. Rapid urbanization around Purulia towns threatens agricultural lands [57], while limited access to modern agricultural technologies affects over 40% of the population living below the poverty line [58].

Despite considerable progress in agricultural land suitability assessments, critical gaps remain in the current literature, particularly within India's complex, urbanizing riverine basins. These gaps stem from the inherent complexity of agricultural systems, which require the simultaneous consideration of hydrological, morphometric, and socio-economic factors. Previous research on the Upper Kansai Basin has often taken a fragmented approach. For instance, Mahato and Jana [59] and Bhattacharya et al. [51] focused solely on morphometric analysis, while Mahato and Pal [53] investigated isolated aspects, such as groundwater potential zones. This selective focus, while providing valuable insights into specific components, fails to capture the complex interactions between different system elements. Xu and Zhang [1] highlighted the limitations of spatially explicit sensitivity analysis in land suitability evaluation, emphasizing the need for more comprehensive approaches. Similarly, Zolekar and Bhagat [38] demonstrated the potential of integrating remote sensing and GIS for multi-criteria land suitability analysis but noted the challenges in incorporating complex hydrological processes. These studies underscore the need for more integrated frameworks that can address the spatial and temporal variability of agricultural suitability parameters.

While some researchers have employed the SWAT for hydrological modeling [27,60] or morphometric analysis for watershed characterization [61–64], the integration of these approaches with advanced MCDM techniques remains limited. For example, Teshager et al. [27] used the SWAT for agricultural watershed modeling but did not incorporate decision-making frameworks for land prioritization. Similarly, Nasir et al. [61] applied morphometry-based compound factor approaches for soil erosion susceptibility assessment but did not integrate hydrological modeling. Furthermore, critical aspects such as rapid urbanization impact [65,66] and climate resilience considerations [67,68] have been inadequately addressed in the context of agricultural land prioritization. Esraz-Ul-Zannat et al. [65] highlighted the need for nature-based solutions in urban flood risk mitigation but did not extend their analysis to agricultural land suitability. Chandra et al. [68] discussed climate-smart agriculture but lacked a comprehensive framework for land prioritization in semi-arid regions. This gap is especially critical given the increasing need for adaptive management strategies that can respond to changing environmental and socio-economic conditions while maintaining agricultural productivity and environmental sustainability.

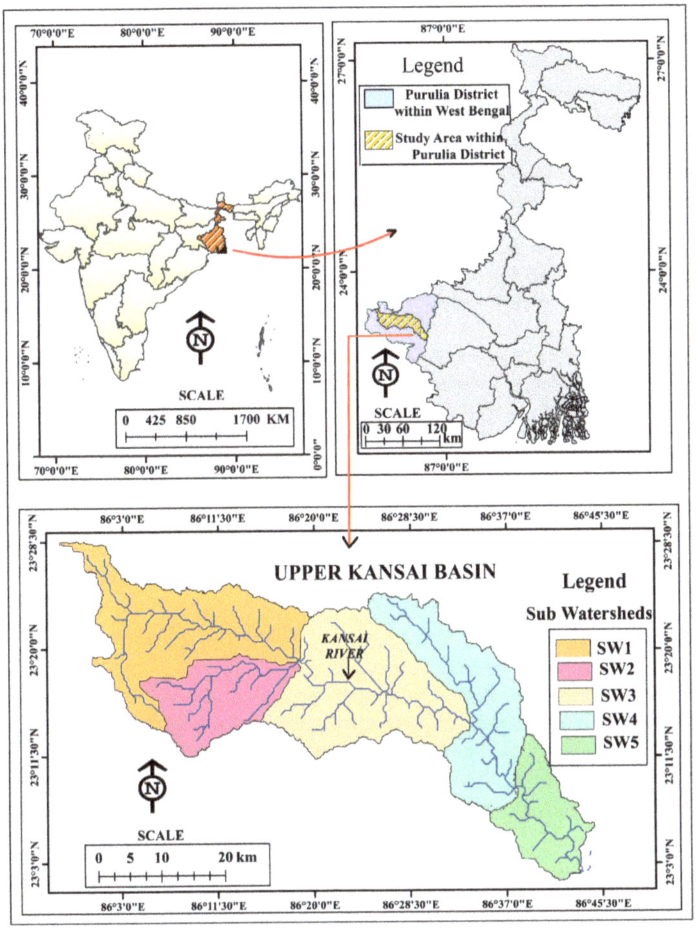

Figure 1. Geographical location and administrative boundaries of the Upper Kansai Basin study area in Purulia District, West Bengal, India.

This study addresses critical gaps through an innovative integrated framework that advances agricultural land prioritization in the following four key ways: (1) it uniquely combines SWAT modeling, morphometric analysis, and VIKOR-based MCDM to provide a comprehensive assessment methodology for agricultural lands in urbanizing riverine basins; (2) it introduces a novel approach for quantifying ground–surface water interactions in semi-arid conditions through the integration of remote sensing, GIS techniques, and hydrological modeling; (3) it develops a robust framework that simultaneously analyzes multiple factors including soil properties, hydrological parameters, LULC patterns, and climate scenarios, providing unprecedented accuracy in agricultural suitability evaluation; and (4) it establishes clear linkages between agricultural land prioritization and sustainable development goals (SDGs 2, 6, 13, and 15), offering practical solutions for water resource management in vulnerable riverine basins of developing countries. This comprehensive framework is specifically tailored to address the complex dynamics of urbanizing riverine basins while incorporating climate resilience considerations, ultimately contributing to sustainable agricultural development in challenging environmental settings worldwide.

2. Materials and Methods

2.1. Study Area Characteristics and Environmental Challenges

The Upper Kansai Basin falls within the Purulia district of West Bengal, India (Figure 1). Purulia is blessed with natural wealth, where large deposits of coal, bauxite, and granite are found. The district is famous for its ancient temples, terracotta sculptures, and heritage spots like Deulghata and Ayodhya Hills, having archeological as well as historical importance. These aspects, together with livelihoods based on tourism, make the region ecologically and culturally significant. Indigenous tribal culture and festive life also contribute to Purulia's tourism. The district is the homeland of the celebrated Chhau dance, an ancient art that has now been acknowledged by UNESCO as Intangible Cultural Heritage, reflecting its sheer cultural significance. The study region in Purulia district lies between 23°01′ N to 23°29′ N latitude and 85°45′ E to 85°56′ E longitude [48] and comprises an estimated 1525 km^2. It encompasses sections of the Jhalda-I, Jhalda-II, Arsha, Joypur, Purulia-I, Purulia-II, Hura, Baghmundi, Puncha, and Manbazar blocks. Geologically, the area shows signs of tectonic activity, especially in the form of shearing. The shear zone is parallel to the major Singhbhum Shear Zone in adjacent Jharkhand. Of particular interest is a 16 km interval between Balarampur and Barabhum, which shows distinct evidence of shearing that has created weak zones. Within these weak zones, small-scale copper and apatite mineralization have been found in geochemically poor terrain. Furthermore, structural elements like minor folding and the formation of foliation planes in metamorphic rocks are ubiquitous in the region. The area is a transitional zone between peninsular India and the Indo-Gangetic Plain, dominated by Precambrian metamorphic rocks, gneisses, schists, and recent to sub-recent alluvial deposits. The area is roughly split into two geological formations as follows: (a) the Archaean formation of granite and gneiss and (b) the Gondwana. Archaeans are the district's oldest and largest rock formations [9,51]. They include granitic rocks, metasedimentaries, calc-granulites, and metabasics. The oldest schistose rocks are phyllite, quartz-schist, and quartzite. The granitic rocks consist of Gray-banded biotite-granite gneiss, pink granite, and porphyritic granite gneiss. Pegmatites and quartz veins have intruded all of these granitic rocks, along with earlier schists, gneisses, and basic rocks. The most common Archaean rock in the district is gneiss. On the other hand, Gondwana sediments are deposited over a limited strip to the north of the area. A few minor spots of laterite have been identified in the easternmost region of Purulia district. Physiographically, the research region is situated on the eastern edge of the Chotanagpur Plateau. The district may be divided into two types based on gneiss and landform evolution as follows: (a) granite-gneiss (Archaean) and (b) Gondwana. The Granite-gneiss landform is further split into eight subdivisions as follows: denudational hills, solitary hillocks, pediments, crests of the undulating plateau, dissected highland, undulating upland, gently sloping upland, and valley/lowlands. The Gondwana landform is further classified into the following five categories: denudational hills, solitary hillocks, pediments, gently sloping uplands, and valleys. The area's height ranges from 50 to 600 m above mean sea level (MSL), with Gorgaburu and Karma Hill in the Baghmundi scarps being the two highest peaks reported. The Kansai River relies on its major tributaries, including Patloi, Bandu, Chapta, Gobri, Jamuna, Girgiri, Karmara, Goura, Unakurba, Gobri, Chunmatia, Teliakata nala, and Dosiman Johor. Groundwater in the region is mostly found in the weathered mantle, saprolitic zone fractured hard rock zones and thin unconsolidated sediment zones along river valleys.

Purulia lies in the Aw (Tropical Savanna) climate region according to the Köppen classification, with a marked dry season in winter and strong monsoonal rainfall. The soils of Purulia are usually lateritic, sandy loam, or stony, directly impacting agriculture. The study area experiences an annual average precipitation of 1393 mm and a mean temperature of 25.5 °C, with seasonal variations ranging from 21.3 °C in the winter to 29.1 °C in the summer [9].

The basin's topography is stratified by residual soil formed through the erosion of previous geological layers [69]. The soils of the study area can be categorized into two types. The soil in the upper catchment areas and over the plateau slope is skeletal, thin, gravelly, and lateritic in nature. In the lower areas, predominantly sandy loam soil can be found.

Purulia is characterized by patches of tropical dry deciduous forests, particularly in the Ayodhya Hills and other hill tracts. These forests are dominated by Sal (*Shorea robusta*) trees, which represent most of the vegetation cover. Where the soil is rocky and semi-arid, the vegetation is mostly thorny bushes and shrubs. The shrub species include kul (*Ziziphus Jujuba*) and khoyer (*Acacia catechu*). Because of soil erosion and reduced forest cover, some areas of Purulia have been converted to grasslands or savanna-type habitats. There is riparian growth along the Bandu, Patloi, and Kansai rivers and the small water bodies around them, which is present due to enhanced moisture conditions. These water sources are supported by bamboo (*Bambusoideae*), jamun (*Syzygium cumini*), and other reeds. In the rural and forest parts of Purulia, there are several fodder plants for animals such as bamboo, ber (*Ziziphus mauritiana*), and grasses. Most societies also utilize medicinal plants such as bael (*Aegle marmelos*), neem (*Azadirachta indica*), and tulsi (*Ocimum sanctum*) in traditional medicine [70].

Purulia has a combination of tribal and non-tribal people, a major percentage of which is involved in agriculture. The district is sparsely populated compared to other regions of West Bengal. The main working class has been reduced from 25.43% in the 2001 census to 20.93% in the 2011 census while the marginal working class has risen from 19.03% in 2001 to 21.71% in the 2011 census. Cultivators have been relocated to other working classes. Males comprise approximately 51.18% of the population, while 48.82% are female. The Scheduled Caste and Scheduled Tribes constitute 19.38% and 18.55% of the population of this area.

Land holding in Purulia is typically smallholder-oriented, with fragmented land holdings. The size of the holdings in the Purulia district is 0.85 ha on average, and the majority of the farmers fall under the marginal category, occupying 70.6% of the total number of holdings. Common grazing land, community land, and forest land are also considerable in the area. A total of 50% of the entire land is categorized as net cropped area, and multi-crop cultivation occupies only 17% of the net cropped area.

Agriculture in the region is predominantly rainfed, characterized by monocropping practices, with paddy as the principal crop. Water resources for agriculture are primarily sourced from tanks, river lift irrigation systems, open-dug wells, and shallow tube wells. However, the efficacy of these water sources is contingent upon adequate rainfall and judicious water management practices. The varied topography of the Upper Kansai Basin affects its agriculture. The southwestern hilly area (400–700 m) features rocky, non-fertile soils, and agriculture is restricted with a tendency towards forest dependency on resources. The middle region (200–400 m) features a rolling topography with red, lateritic soils, and agriculture is limited by rainfall and irrigation problems. The plains in the southeast (150–200 m) feature fertile alluvial soils with double cropping. The basin may be divided into five sub-watersheds to enable planned resource management as follows: (a) Sub-division of the basin into sub-watersheds enables area-specific management. Each sub-watershed can possess its own hydrological, geological, and ecological properties. Sub-division facilitates a more targeted approach, e.g., managing soil erosion, water conservation, or agricultural management in each sub-watershed appropriately. (b) Sub-watershed divisions render water resource management more efficient. It is easier to determine which zones face water scarcity or have a water surplus. This facilitates better management of annual water flow, irrigation planning, and local groundwater recharge system development. The prioritization of agriculture is enhanced by the division into sub-watersheds. In arid regions such as Purulia, it is difficult to make out which zones possess greater crop yield or encounter difficulty in irrigation or water management. (c) Sub-watershed division helps in agricultural prioritization. In regions prone to drought, such as Purulia, it is evident which areas have improved crop yields or are experiencing irrigation or water management problems. This enables specific interventions at the sub-watershed level to tackle these problems. (d) Sub-dividing the Kansai Basin into sub-watersheds enhances best practices in water and soil conservation. Various sub-watersheds can have unique runoff and soil erosion patterns. The sub-division allows for the planning of suitable conservation activities, for example, contour plowing or building check dams, depending on the requirements of the specific area.

2.2. Methodology

This study addresses critical gaps in agricultural land suitability assessment by implementing a comprehensive methodological framework that integrates three complementary analytical approaches for complex, urbanizing riverine basins in developing countries. The methodology uniquely combines SWAT modeling, morphometric analysis, and VIKOR-based multi-criteria decision-making (MCDM) to assess and prioritize agricultural land in the Upper Kansai Basin (Figure 2).

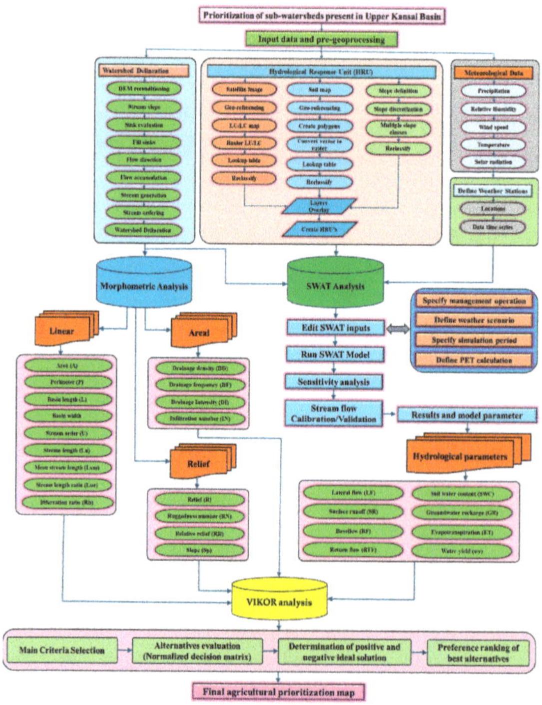

Figure 2. Methodological flowchart of agricultural prioritization analysis.

The framework begins with extensive data collection and pre-processing phases, where watershed delineation is performed using SRTM DEM data through a systematic process of DEM reconditioning, stream network extraction, slope analysis, and flow direction mapping.

The SRTM DEM was pre-processed in several steps to be able to make an accurate terrain analysis. The DEM tiles were first mosaicked and NoData gaps were filled. The DEM was next reprojected into the coordinate reference system of the study area and clipped to the boundary. Hydrological corrections involved removing sinks, and then, the calculation of derivatives, such as the slope, to aid the subsequent spatial analysis. The Landsat 9 imagery was used in the study with a 30 m spatial resolution, which aligns well with the SRTM DEM spatial resolution of 30 m. This ensures consistency between the datasets when conducting pixel-based analyses, such as LULC classification or terrain modeling, and avoids unnecessary complications in data resampling that could arise from integrating datasets with differing resolutions. Landsat 9 imagery was pre-processed using ERDAS IMAGINE 2014 to prepare it for LULC analysis. Pre-processing included radiometric correction, atmospheric correction through the QUAC (Quick Atmospheric Correction) algorithm, and geometric correction to achieve spatial accuracy. The imagery was then resampled to a standard pixel size and clipped to the study area boundary. Cloud cover was reduced by cloud masking techniques to improve accuracy.

Watershed characterization is further refined through the development of Hydrological Response Units (HRUs), which represent unique combinations of LULC, soil characteristics, and slope classes, achieved through processing satellite imagery for LULC classification, georeferencing soil maps, and discretizing slope data. This advanced integration of remote sensing and GIS techniques enhances the spatial and temporal resolution of our assessments, providing a more accurate characterization of the basin's physical properties. The core analytical framework consists of three interconnected components that work synergistically as follows: morphometric analysis examines the basin's physical characteristics through linear, areal, and relief parameters; SWAT modeling simulates the basin's hydrological processes, generating key parameters such as lateral flow, surface runoff, baseflow, and evapotranspiration; and VIKOR analysis integrates these physical and hydrological parameters into a structured decision-making framework. The SWAT model undergoes rigorous sensitivity analysis and calibration/validation procedures, incorporating climate resilience considerations through various climate scenarios. This comprehensive approach considers multiple factors, including soil properties, hydrological factors, LULC patterns, and climate scenarios, providing a more nuanced evaluation of agricultural suitability. The VIKOR method enables an adaptable decision-making process by evaluating alternatives through a normalized decision matrix and determining positive and negative ideal solutions. The methodology culminates in the integration of all three analytical components to generate comprehensive agricultural prioritization maps that align with sustainable development goals.

2.3. Data Sources

Multiple geospatial and hydro-environmental datasets were integrated to facilitate a comprehensive AgLP analysis. The datasets used in this investigation, including their sources, purposes, resolutions, and temporal coverage, are systematically presented in Table 1. Landsat-9 imagery with 30 m spatial resolution, acquired in early April 2024, was utilized for LULC classification to identify urbanized areas, along with natural and scanty agricultural patterns. Topographic characteristics, drainage density, slope, and watershed delineation were extracted from SRTM-DEM at a 30 m resolution [51], while watershed boundaries were delineated using Survey of India topographical sheets at a 1:250,000 scale from 1961, which serves as the baseline for morphometric analysis.

Table 1. Geospatial and hydro-meteorological datasets were utilized for the AgLP analysis in the Upper Kansai Basin.

Data	Purpose	Resolution/Scale	Year	Source
Landsat-9	LULC	30 m	Early April 2024	(http://earthexplorer.usgs.gov/, accessed on 25 August 2024)
SRTM-DEM	Slope	30 m	Early April 2024	(http://earthexplorer.usgs.gov/, accessed on 25 August 2024)
Toposheet	Watershed delineation	1:250,000	1961	Survey Of India (SOI), Kolkata
Meteorological data	Rainfall, temperature, relative humidity, solar radiation, wind velocity	0.5° × 0.625°	1984–2023	NASA POWER dataset, (https://power.larc.nasa.gov/data-access-viewer/, accessed on 25 August 2024)
SOIL	Soil texture	1:250,000	2020	NBSS and LUP, Kolkata and (https://swat.tamu.edu/data/india-dataset/, accessed on 25 August 2024)

Meteorological parameters essential for agricultural suitability assessment, including rainfall, temperature, relative humidity, solar radiation, and wind velocity, were obtained from the NASA POWER dataset spanning 1984–2023 at a 0.5° × 0.625° resolution. This extensive temporal coverage enabled the robust analysis of climate patterns and their influence on agricultural potential across the basin. Soil texture information, crucial for the agricultural capability assessment, was derived from the National Bureau of Soil Survey and LULC Planning (NBSS&LUP) database at a 1:250,000 scale, supplemented by the SWAT India dataset.

The integration of these diverse datasets at varying spatial and temporal resolutions necessitated careful pre-processing and standardization procedures [33,34]. Spatial data were projected to a common coordinate system and resampled to maintain consistency across analyses. All maps were projected in the WGS 1984 UTM Zone 45 N coordinate system and reclassified with a consistent cell size of 30 m. The temporal alignment of meteorological data was particularly crucial for establishing relationships between climate patterns and agricultural suitability [27]. This multi-source data approach enabled a comprehensive assessment of the agricultural potential, incorporating both physical and environmental parameters that influence farming viability. The selection of these specific datasets was driven by their temporal coverage, spatial resolution, and relevance to AgLP in semi-arid regions, following the established methodological frameworks [39].

2.4. VIKOR Model

The VIKOR model is a multi-criteria decision-making (MCDM) technique developed by Duckstein and Opricovic in 1980 [71]. This method helps find alternative solutions by weighing parameters, based on their relative distance from the best and worst options [37]. VIKOR has become increasingly popular due to its consistent results in identifying sensitive zones across various contexts, making it useful for practical assessments and forecasts [72]. These strengths led to the selection of agricultural prioritization for this study. The method works by first measuring the relative distances between options and then assigning weights to each conflicting criterion.

In this study, the objective was to determine the best alternative under the existence of conflicting criteria. The selection of VIKOR in this research was due to its particular strengths in dealing with conflicting criteria and ranking alternatives when a compromise between solutions is needed. The VIKOR method was applied in R Studio 4.4, utilizing personal R scripts to handle the data, implementing the method, and produce rankings. The application entailed determining the values of S, R, and Q for each alternative and comparing their rankings according to their closeness to the ideal and anti-ideal solutions. Five sub-watersheds (SW1, SW2, SW3, SW4, and SW5) were chosen as alternatives for agricultural land prioritization in this study. The evaluation relied on 12 major criteria for hydrological and geomorphological processes. Criteria were Evapotranspiration (Et), Soil Water Content (SWC), Surface Runoff (Sr), Groundwater Recharge (GR), Water Yield (Wy), Lateral Flow (LF), Drainage Density (DD), Relative Relief (RR), Dissection Index (DI), Drainage Frequency (DF), Relief Number (RN), and Infiltration Number (IN). All the sub-watersheds were evaluated against these parameters using the VIKOR multi-criteria decision-making (MCDM) approach to develop a ranking that ranks the sub-watersheds according to their potential for agricultural development.

As shown in Figure 3, the VIKOR method adheres to the following key steps [73]:

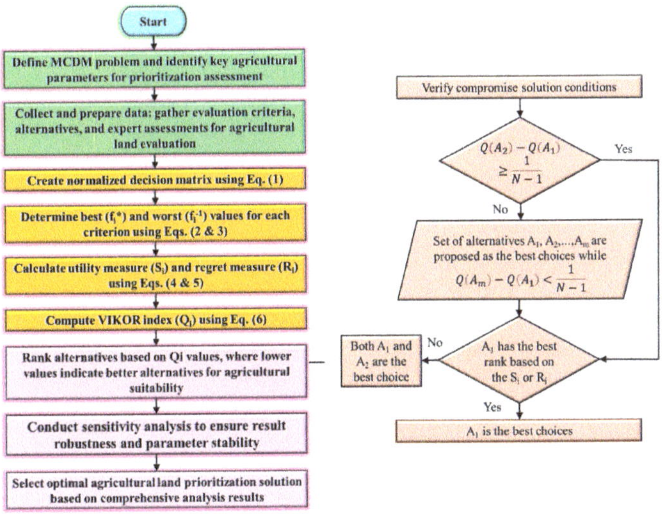

Figure 3. Flow chart process of the VIKOR method.

Problem definition: Begin by clearly defining the MCDM problem. This involves identifying the evaluation criteria, alternatives, and relevant experts for the decision-making process.

Data collection and preparation: Gather all necessary data, including obtaining weights for each criterion and collecting evaluation values using the appropriate techniques such as HFFE (Hesitant Fuzzy Fermatean Expansion).

Decision matrix creation: Calculate the average decision matrix as follows:

$$f_{ij} = \frac{1}{k}\sum_{t=1}^{k} x^t{}_{ij}, \forall i = 1,2,\ldots,m; j = 1,2,\ldots,n \quad (1)$$

where f_{ij} is the normalized value of x; i is the index of an option (alternative); j is the index of a criterion; t is the index of an expert; m is the number of options (alternatives); and n is the number of criteria.

Ideal solution determination: Compute the optimal (best) and worst ideal solutions using the following equations:

$$f_i^* = \begin{cases} \max\limits_{i=1,2,\ldots,m} f_{ij} & \text{for beneficial criteria} \\ \min\limits_{i=1,2,\ldots,m} f_{ij} & \text{for cost criteria} \end{cases} \tag{2}$$

$$f_i^- = \begin{cases} \min\limits_{i=1,2,\ldots,m} f_{ij} & \text{for beneficial criteria} \\ \max\limits_{i=1,2,\ldots,m} f_{ij} & \text{for cost criteria} \end{cases} \tag{3}$$

where f_i^* represents the best ideal solution and f_i^- the worst ideal solution.

Calculate the utility and regret measures: Compute the S_i (group utility) and R_i (individual regret) measures using the following formulas:

$$s_i = \sum_{j=1}^{n}\left[\frac{w_j\left(f_j^* - f_{ij}\right)}{\left(f_j^* - f_j^-\right)}\right] \tag{4}$$

$$R_i = \max_{j=1,2,\ldots\ldots,n}\left[\frac{w_j\left(f_i^* - f_{ij}\right)}{\left(f_j^* - f_j^-\right)}\right] \tag{5}$$

where w_i is the weight of the ith criterion.

VIKOR index calculation: Calculate the VIKOR index, Q_i, using the following formula:

$$Q_i = v\left[\frac{\left(s_i - s^*\right)}{\left(s^- - s^*\right)}\right] + (1-v)\left[\frac{\left(R_i - R^*\right)}{\left(R^- - R^*\right)}\right] \tag{6}$$

where S^*, S^-, R^*, and R^- are the minimum and maximum values of S_i and R_i, respectively; and v is the weight for the maximum group utility strategy (usually 0.5).

Alternative ranking: Rank the alternatives based on their Q_i values, with the lowest Qi value ranked first.

Compromise solution proposal: Check two conditions for acceptable advantage and acceptable stability in decision making. For alternatives A_1 and A_2 (first- and second-ranked by Q values, respectively), the acceptable advantage condition is evaluated as follows:

$$Q(A_2) - Q(A_1) \geq \frac{1}{N-1} \tag{7}$$

where N is the number of criteria. Additionally, the alternative ranked as first by Q must demonstrate acceptable stability in decision making by also having the best rank in terms of S and/or R values. If this stability condition is not satisfied, a set of compromise solutions is proposed based on the following:

$$Q(A_m) - Q(A_1) < \frac{1}{N-1} \tag{8}$$

where A_m represents additional alternatives. In cases where the advantage condition is not satisfied, both A_1 and A_2 are proposed as compromise solutions. When stability is not achieved, alternatives A_1, A_2,..., Am are proposed, with A_m determined by Equation (8) [37]. The final compromise ranking is ultimately obtained by evaluating these advantage rates.

Sensitivity analysis: Conduct a sensitivity analysis to ensure the stability of the results under different conditions or parameter changes. The sensitivity analysis of the VIKOR method was performed by changing the trade-off parameter v, which regulates the balance between the utility measure (group utility) and the regret measure (individual regret). In particular, we computed Q_j for the following two values of v: 0.5 and 0.25.

Final decision: Select the best alternative based on the VIKOR ranking and the proposed compromise solution, considering all the previous steps and analyses.

2.5. SWAT Model

The Soil and Water Assessment Tool (SWAT) is a comprehensive, physically based, watershed-scale model widely used for simulating hydrological processes and assessing the impacts of land management practices on water resources [36,74]. Developed by the United States Department of Agriculture (USDA) and the Agricultural Research Service (ARS), the SWAT has gained global recognition for its ability to model complex watersheds with varying soils, LULC, and management conditions over long periods [75]. The model's strength lies in its ability to integrate multiple environmental processes, including hydrology, soil erosion, plant growth, and nutrient cycling, within a geospatial framework, facilitating the detailed spatial representation of watershed characteristics [75].

The SWAT operates on the fundamental principle of water balance, simulating the hydrologic cycle through various components, including precipitation, evapotranspiration, surface runoff, infiltration, percolation, lateral subsurface flow, and groundwater flow. The model first divides the watershed into multiple sub-basins based on topography, which are further partitioned into hydrologic response units (HRUs) that represent unique combinations of LULC, soil type, and slope characteristics [75]. This discretization allows for a detailed representation of the spatial heterogeneity within the watershed. The SWAT model simulates several key hydrological processes that are crucial for understanding the water balance in the Kansai River basin. These processes include lateral flow (LF), surface runoff (SR), groundwater recharge (GR), base flow (BF), evapotranspiration (ET), return flow (RTF), and soil water content (SWC) [76].

Lateral flow (LF) is simulated in the SWAT when the water content in the soil exceeds field capacity. This is calculated using the following equation [36]:

$$LF = 0.024\left(\frac{2 \times sw_{ly.excess} \times K_{sat} \times slp}{\varphi_d \times L}\right) \tag{9}$$

where $SW_{ly.excess}$ is the drainable volume of water in the saturated zone of the hillslope (mm H_2O), K_{sat} is the saturated hydraulic conductivity (mm/h), slp is the slope, φ_d is the drainable porosity of the soil layer (mm/mm), and L is the hillslope length.

Surface runoff (SR) is estimated as follows, using the SCS curve number method [77]:

$$SR = \frac{\left(R_{day} - I_a\right)^2}{\left(R_{day} - I_a + s\right)} \tag{10}$$

where R_{day} is the daily rainfall depth (mm); I_a is the initial abstraction, including surface storage and infiltration before runoff (mm); and S is the retention parameter (mm).

Baseflow (BF) is calculated based on the shallow aquifer recharge and the baseflow recession constant as follows [76]:

$$BF = \frac{8000 \times K_{sat}}{L_{gw}} \times h_{wtbl} \tag{11}$$

where K_{sat} is the saturated hydraulic conductivity, L_{gw} is the distance from the ridge or sub-basin divide to the main channel (m), and h_{wtbl} is the water table height (m).

Return flow (RTF), or Revap, occurs when water accumulation in shallow aquifers exceeds a threshold value [78]. This is estimated as follows:

$$RTF = \beta_{revap} \times PET \quad (12)$$

where β_{revap} is the Revap coefficient and PET is the potential evapotranspiration for the day (mm H_2O).

Soil water content (SWC) is determined using the following water balance equation [36]:

$$SWC = SW_0 + \sum_{i=1}^{t}\left(R_{day} - SR - ET - W_{seep} - RTF\right) \quad (13)$$

where SW_0 is the initial soil water content (mm), t is time (day), R_{day} is precipitation, SR is surface runoff (mm), ET is actual evapotranspiration (mm), W_{seep} is the amount of water entering the vadose zone (mm), and RTF is return flow (mm).

Groundwater recharge (GR) is simulated using the following equation [36]:

$$GR = \left(1 - \exp\left[-\frac{1}{\delta_{gw}}\right]\right) \times W_{seep} + \exp\left[-\frac{1}{\delta_{gw}}\right] \times w_{rchrg.i-1} \quad (14)$$

where δ_{gw} is the delay time of the aquifer materials, w_{seep} is the total amount of water exiting the bottom of the soil profile on the day i (mm), and $w_{rchrg,i-1}$ is the amount of recharge entering the aquifers on the previous day (mm).

Evapotranspiration (ET) is estimated using the Penman–Monteith method, which is widely recognized for its accuracy in diverse climatic conditions, as follows [79]:

$$ET = \frac{0.408\Delta(R_n - G) + \gamma\frac{900}{T+273}u_2(e_s - e_a)}{\Delta + \gamma(1 - 0.34u_2)} \quad (15)$$

where R_n is the net radiation, G is the soil heat flux, T is the mean daily air temperature, u_2 is the wind speed, e_s is the saturation vapor pressure, e_a is the actual vapor pressure, Δ is the slope of the vapor pressure curve, and γ is the psychrometric constant.

Water yield (Wy) represents the total volume of water that flows from each HRU into the main channel network during a specified time interval [74]. The SWAT model calculates watershed water yield using the following water balance equation:

$$Wy = \mathrm{SR} + \mathrm{GR} + \mathrm{LF} - T_{loss} \quad (16)$$

where T_{loss} is the transmission loss through stream bed infiltration (mm).

The SWAT is employed in this study to conduct a comprehensive hydrological analysis of the Kansai River basin for the period 1984–2023. The application of the SWAT involves a meticulous setup process, incorporating a wide array of input parameters related to LULC, soil characteristics, weather conditions, and topography. These parameters are subjected to rigorous calibration and validation procedures to ensure the model's accuracy and reliability in representing the basin's hydrological dynamics [80]. The model is calibrated and validated at monthly time intervals, aligning with best practices in hydrological modeling. To capture the spatial variability of hydrological processes, the watershed is delineated into five sub-basins (SW1, SW2, SW3, SW4, and SW5). This discretization allows for a nuanced examination of water flow and related hydrological phenomena within each sub-watershed, providing insights into local-scale hydrological behaviors [35]. Further refinement is achieved through the definition of 67 Hydrological Response Units (HRUs) within each sub-watershed. HRUs, representing unique combinations of LULC, soil type, and slope characteristics, enable the model to account for fine-scale heterogeneity in landscape attributes that influence hydrological processes. The topographic analysis is enhanced through the use of the Shuttle Radar Topography Mission (SRTM) Digital Elevation Model (DEM) (Figure 4a). The landscape is classified into three distinct slope categories: steep (>20%), moderate (10–20%), and gentle (<10%) (Figure 4b). This classification scheme adheres to established norms in hydrological and geomorphological research, facilitating the accurate representation of topographic influences on water movement and soil erosion processes [81].

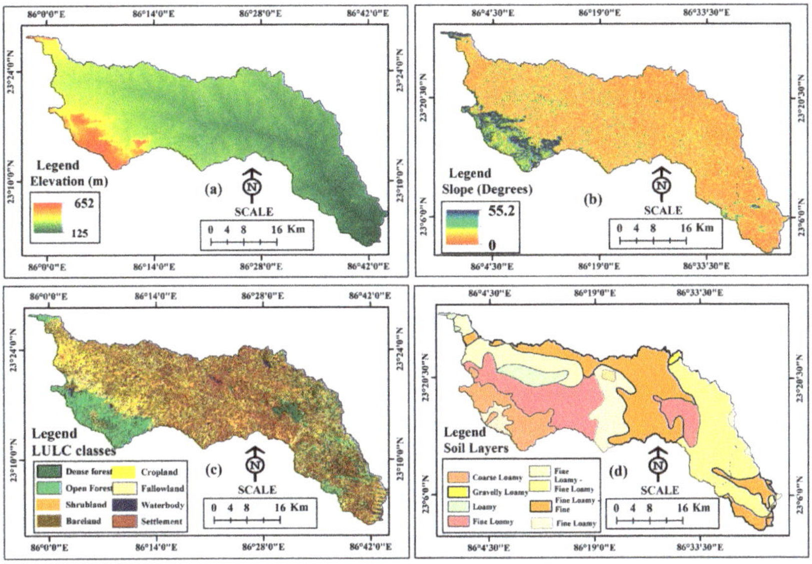

Figure 4. Physical characteristics of the Upper Kansai Basin: (**a**) Elevation distribution (m), (**b**) Slope classes (in degrees), (**c**) LULC types, and (**d**) Spatial distribution of soil layers.

The LULC analysis of the Upper Kansai Basin (Figure 4c) demonstrates a heterogeneous landscape pattern with high classification accuracy (Kappa statistics: 0.96). Shrubland dominates the basin (~32.74%, 527.20 km^2), followed by agricultural areas comprising cultivated (~11.87%, 191.19 km^2) and fallow lands (~14.36%, 231.34 km^2). Forest cover, classified into dense (8.74%, 140.74 km^2) and open forest (~6.42%, 103.45 km^2), is primarily concentrated in the western portion. The significant presence of built-up areas (~14.26%, 229.63 km^2) indicates substantial urbanization, while bare land (~9.65%, 155.42 km^2) suggests erosion-prone zones. Water bodies, though limited (~2.6%, 45.2 km^2), are crucial for local hydrology. This LULC distribution pattern has important implications for agricultural potential and watershed management strategies, particularly regarding soil conservation and water resource management across the basin [82,83].

Soil data, a critical component for accurate hydrological modeling, is sourced from the SWAT India dataset (https://swat.tamu.edu/data/india-dataset/, accessed on 10 September 2024), (Figure 4d). This dataset provides detailed soil characteristics essential for simulating infiltration, water-holding capacity, and subsurface flow patterns [36]. Meteorological inputs fundamental to driving the hydrological cycle in the SWAT are obtained from the NASA POWER dataset for the period 1984–2022. This comprehensive dataset includes daily records of precipitation, relative humidity, maximum and minimum temperatures, solar radiation, and wind velocity [84]. The inclusion of a five-year warm-up period (1984–1988) ensures the stabilization of model parameters and initial conditions, enhancing the reliability of subsequent simulations [85]. By integrating these diverse datasets and leveraging the SWAT's advanced modeling capabilities, this study provides a robust framework for understanding the complex hydrological processes operating within the Kansai River Basin. This approach not only allows for the quantification of current water resource dynamics but also establishes a foundation for assessing the potential impacts of climate change and LULC alterations on future water availability and distribution in the region.

2.6. Morphometry Analysis

Morphometric analysis is a quantitative method for studying watershed and drainage network characteristics, providing insights into basin geomorphology, hydrology, and erosion patterns [8]. This approach involves measuring linear, areal, and relief parameters to understand watershed geometry, size, shape, and drainage patterns [86]. The integration of Geographic Information Systems (GIS) and Remote Sensing (RS) has significantly improved the efficiency and accuracy of morphometric analysis, enabling large-scale watershed studies through rapid parameter extraction from digital elevation models (DEMs) and satellite imagery.

Table 2 presents the key morphometric parameters categorized as linear, areal, and relief parameters. The linear parameters describe the stream network structure as follows: stream order (U) indicates the drainage complexity [8], stream number (Nu) and length (Lu) reflect the network development and extent [87,88], while bifurcation ratio (Rb) shows the geological influence on drainage patterns [87]. The areal parameters focus on the following basin characteristics: drainage density (DD) indicates erosion susceptibility and runoff potential [88], drainage frequency (DF) reflects the basin texture [89], while drainage intensity (DI) and infiltration number (IN) show the drainage efficiency and runoff characteristics [90]. The relief parameters capture 3D aspects as follows: relief (R) indicates the erosion potential [8], ruggedness number (RN) shows the terrain complexity [89], relative relief (RR) and slope (Sb) measure the basin steepness, influencing runoff and erosion [91]. These parameters collectively provide a comprehensive understanding of watershed morphology and hydrological behavior.

Table 2. Comprehensive morphometric linear, areal, and relief parameters.

Geomorphometric Units	Formula	Significance
	Linear Parameters	
Stream Order (U)	Hierarchical rank	It indicates the hierarchical complexity and branching pattern of the stream network within a basin. Higher orders suggest more developed drainage systems [8].
Stream Number (N_u)	$N_{u1} + N_{u2} \ldots + N_{un}$	Represents the total count of streams in each order, reflecting the overall drainage network density and development [87].
Stream Length (L_u)	$L_{u1} + L_{u2} \ldots + L_{un}$	Measures the total length of streams in each order, indicating the extent of the drainage network and potential for water transport [88].
Mean Stream Length (L_{sm})	L_u/N_u	Represents the average length of streams in each order, providing insights into the typical stream segment sizes within the basin [8].
Stream Length Ratio (L_{ur})	$L_u/L_u - 1$	Indicates how stream lengths change between successive stream orders, reflecting basin geometry and drainage pattern development [88].

Table 2. *Cont.*

Geomorphometric Units	Formula	Significance
Bifurcation Ratio (R_b)	$L_u/L_u + 1$	Measures the rate of stream branching, indicating the geological structure's influence on drainage patterns and flood response [87].
	Areal Parameters	
Drainage Density (DD)	L_u/A	Indicates the length of streams per unit area, reflecting erosion susceptibility, surface runoff potential, and landscape dissection [88].
Drainage Frequency (DF)	$\sum N_u/A$	Represents the number of streams per unit area, indicating the basin's texture and its influence on runoff generation [89].
Drainage Intensity (DI)	DF/DD	Combines drainage density and frequency, indicating the basin's capacity to generate streams and its overall drainage efficiency [90].
Infiltration Number (IN)	DD × DF	Relates drainage density to stream frequency, indicating the basin's infiltration characteristics and runoff potential [90].
	Relief Parameters	
Relief (R)	$H - h$	Measures the overall elevation range in the basin, indicating the potential energy available for erosion and sediment transport [8].
Ruggedness Number (RN)	(R × DD)/1000	Combines relief and drainage density, indicating the structural complexity of the terrain and its susceptibility to erosion [89].
Relative Relief (RR)	H/P	Relates relief to basin perimeter, providing a normalized measure of basin steepness and its influence on runoff velocity [91].
Slope (Sb)	$(H - h)/L$	Indicates the average steepness of the basin, influencing runoff velocity, erosion potential, and LULC suitability.

Notes: A = basin area; L = basin length; P = basin perimeter; H = maximum elevation; h = minimum elevation; $Lu - 1$ = the total stream length of the next lower order; and $Lu + 1$ = the total stream length of the next upper order.

3. Results

3.1. Analysis of Morphometric Parameters

3.1.1. Linear Parameters

The stream order analysis reveals a dendritic drainage pattern in the Upper Kansai Basin, with stream orders ranging from 1 to 5 (Figure 5a). First-order streams dominate the drainage network, with SW1 having the highest number (~218) followed by SW4 (~184). This dominance of first-order streams indicates the existence of strong erosional processes and structural control over the drainage development [8]. The total stream length varies significantly across sub-watersheds, with SW1 showing the highest value (~323,253.60 m) and SW5 the lowest (~100,069.08 m). The bifurcation ratio ranges from ~3.92 to 7.17, with SW4 showing the highest value, indicating strong structural control over drainage patterns. According to Horton (1945) [87,88], bifurcation ratios between ~3.0 and 5.0 suggest natural drainage development with minimal structural disturbance, making SW2 (~3.92) and SW3 (~4.8) particularly noteworthy for their natural drainage evolution.

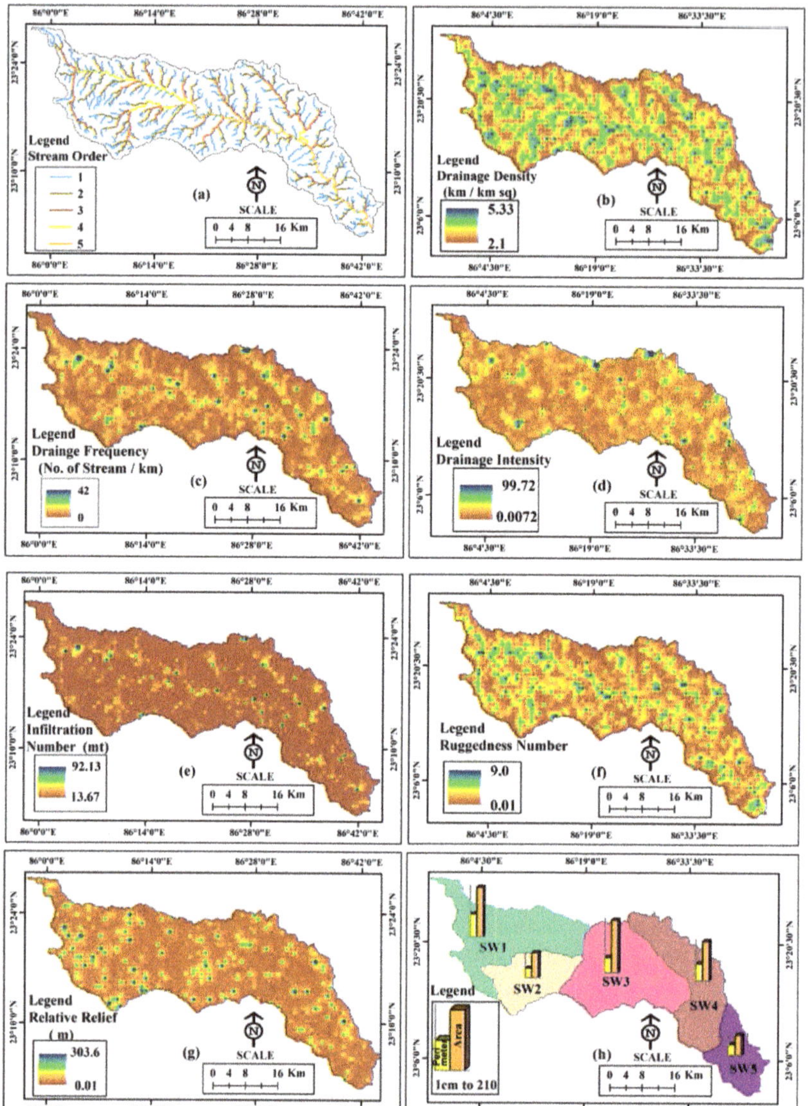

Figure 5. Spatial distribution of morphometric parameters in the Upper Kansai Basin: (**a**) stream order, (**b**) drainage density (km/km^2), (**c**) drainage frequency (no. of stream/km sq), (**d**) drainage intensity, (**e**) infiltration number, (**f**) ruggedness number, (**g**) relative relief, and (**h**) area and perimeter.

3.1.2. Areal Parameters

The drainage density distribution (Figure 5b) shows significant spatial variation across the basin, ranging from ~2.1 to 5.33 km/km^2. SW2 exhibits the highest drainage density (~5.33 km/km^2), indicating a highly dissected terrain with potentially limited infiltration capacity [92]. The drainage frequency map (Figure 5c) reveals values ranging from 0 to 42 streams/km^2, with SW4 showing the highest frequency (~42), suggesting high surface runoff potential. The drainage intensity (Figure 5d) varies from ~0.0072 to 14, with SW4 showing the highest value (~14), indicating efficient drainage development. The infiltration number (Figure 5e) ranges from ~3.04 × 10^{-6} to 92.1374, with SW1 showing the highest value (~92), suggesting areas of potential groundwater recharge constraints [93].

3.1.3. Relief Parameters

The relief parameters exhibit considerable variation across the basin. The ruggedness number (Figure 5f) ranges from ~0.01 to 9.0, with SW2 showing the highest value (~2.0), indicating steep terrain and high erosion susceptibility. The relative relief map (Figure 5g) shows values ranging from ~0.01 to 303.57 m, with SW1 exhibiting the highest value (~303), suggesting significant topographic variation. These relief characteristics strongly influence the basin's hydrological response and erosion potential [82]. Higher relief values generally correspond to areas of increased erosion risk and reduced agricultural suitability.

3.1.4. Sub-Basin-Wise Morphometric Characterization

The morphometric analysis of the Upper Kansai Basin revealed significant variations in drainage characteristics and terrain features across the five sub-watersheds (Table 3). The analysis demonstrates distinctive patterns in stream ordering, drainage characteristics, and terrain parameters that directly influence agricultural suitability, aligning with similar studies in comparable physiographic settings [13,51].

Table 3. Morphometric parameters quantification of sub-watersheds in the Upper Kansai Basin.

Geomorphometric Units	SW1	SW2	SW3	SW4	SW5
Total no. of stream order	253	142	162	226	64
Number of stream orders: 1	218	116	144	184	64
Number of stream orders: 2	21	14	12	38	7
Number of stream orders: 3	12	8	4	4	6
Number of stream orders: 4	2	3	2	-	1
Number of stream orders: 5	-	1	-	-	-
Length of stream order 1 in m.	156,230.22	74,702.34	142,732.7	97,453.70	49,216.74
Length of stream order 2 in m.	90,510.7	60,722.49	63,307.65	34,754.41	29,937.96
Length of stream order 3 in m.	42,629.66	19,355.32	28,420.07	49,645.96	13,038.89
Length of stream order 4 in m.	33,882.99	15,886.19	27,016.39	-	7875.48
Length of stream order 5 in m.	-	4314.89	-	-	-
Total length of streams in m.	323,253.60	174,981.26	261,476.89	181,854.08	100,069.08
Average bifurcation ratio	6.04	3.92	4.8	7.17	5.43
Basin length in mt.	31,520	22,061.1	30,563.24	39,791.9	14,904.21
Main channel length in m.	33,882	3214.8	40,557.25	13,000	23,496.81
Area in square-km	394.87	194.60	413.33	308.08	161.00
Perimeter	202.7	100.87	153.71	161.14	94.14
Drainage density in km/km^2	3.20	5.33	3.44	3.0	3.15
Drainage frequency	37	18	23	42	21
Infiltration number	92	69	58	78	51
Drainage intensity	11.6	3.4	6.7	14	6.7
Relative relief	303	257	260	286	254
Ruggedness number	1.2	2.0	0.2	0.3	0.4

Sub-watershed SW1 exhibits the highest stream network development with ~253 streams, dominated by first-order streams (~218), indicating extensive drainage development. This sub-watershed shows a high infiltration number (~92) and drainage intensity (~11.6), coupled with significant relative relief (~303), suggesting potential challenges for agricultural activities due to increased surface runoff and soil erosion risks. These characteristics align with findings from similar watershed studies that correlate high drainage intensity with erosion susceptibility [82].

SW2 presents concerning characteristics for agricultural development, displaying the highest drainage density (~5.33 km/km^2) and ruggedness number (~2.0) among all sub-watersheds. These parameters, combined with their relative relief value of ~257, indicate steep, uneven terrain prone to erosion processes. The high drainage density suggests a well-developed drainage network that could accelerate soil loss during rainfall events, as observed in similar physiographic settings [92]. The largest sub-watershed, SW3 (~413.33 km^2), demonstrates moderate drainage characteristics, with a drainage density of ~3.44 km/km^2 and a relatively low ruggedness number (~0.2). These characteristics, along with moderate drainage intensity (~6.7), indicate more stable terrain conditions potentially suitable for agricultural activities with proper management practices. Such moderate values typically suggest balanced development potential with appropriate conservation measures [93]. SW4 exhibits balanced morphometric characteristics, which are favorable for agriculture. Despite having a high drainage frequency (~42), its moderate drainage density (~3.0 km/km^2) and reasonable ruggedness number (~0.3) suggest controlled surface runoff conditions. The combination of these parameters indicates terrain conditions that could support sustainable agricultural practices with the appropriate soil conservation measures, as supported by research in similar contexts [94]. SW5, the smallest sub-watershed (~161.00 km^2), presents the most promising characteristics for agricultural development. Its lower relative relief (~254), moderate drainage density (~3.15 km/km^2), and balanced ruggedness number (~0.4) indicate stable terrain conditions less susceptible to erosion. The moderate drainage intensity (~6.7) suggests adequate water movement without excessive soil loss risk, characteristics identified as favorable for agriculture in watershed studies [87]. The morphometric quantification reveals that terrain stability and erosion susceptibility vary significantly across the sub-watersheds. These variations necessitate different approaches to agricultural management and soil conservation. Areas with high morphometric values require intensive soil conservation measures such as contour plowing and terracing, while regions with moderate values may benefit from basic conservation practices like crop rotation and mulching. These findings provide essential insights for developing targeted agricultural management strategies across the Upper Kansai Basin, supporting the implementation of sustainable agricultural practices tailored to each sub-watershed's specific characteristics.

3.2. SWAT Model Assessment

The SWAT model analysis reveals distinct hydrological patterns across the five sub-watersheds in the Upper Kansai Basin (Table 4). Each sub-watershed exhibits unique characteristics in terms of water balance components, demonstrating the spatial heterogeneity of hydrological processes. Figure 6a–f illustrates the spatial distribution of these components, providing a comprehensive visualization of the basin's hydrological dynamics.

Table 4. Sub-watershed-wise distribution of the SWAT-simulated water balance components in the Upper Kansai Basin (all units in mm).

Sub Watershed	Evapotranspiration (Et)	Soil Water Content (SWC)	Surface Runoff (Sr)	Groundwater Recharge (GR)	Water Yield (Wy)	Lateral Flow (LF)
SW1	349.6	1355.8	327	644.3	1012.9	6.6
SW2	354.4	1296.8	275.7	644.4	1008	52.9
SW3	368.2	1326.5	323.5	630	994.3	6.6
SW4	375.4	1351.8	293	633.7	966.9	5.8
SW5	376.4	1351.9	293	632.1	965.9	6.5

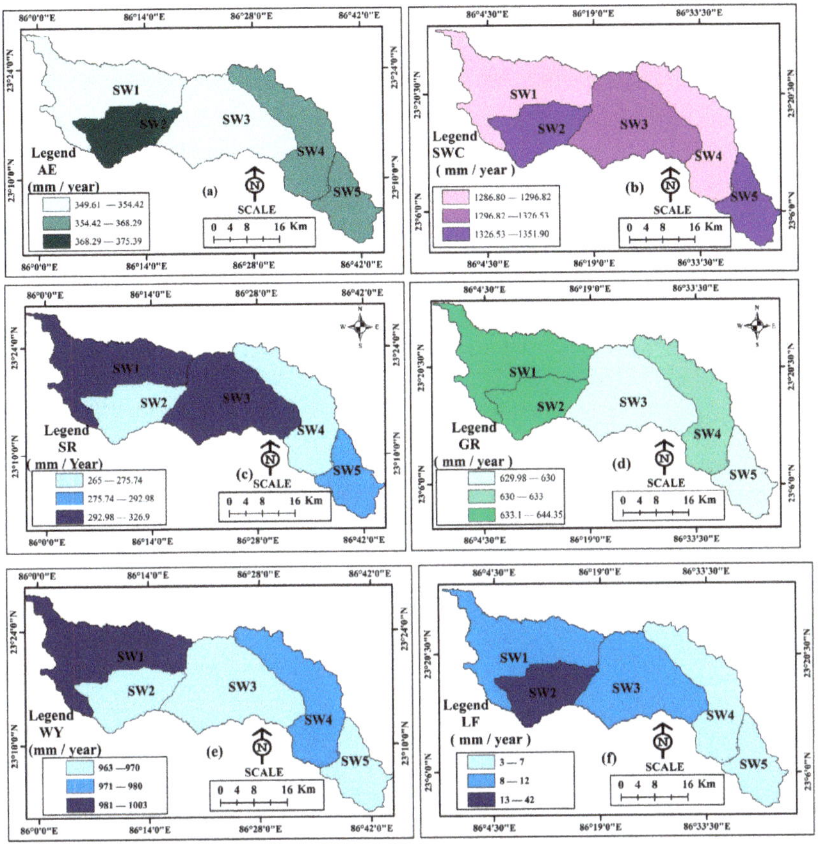

Figure 6. Spatial distribution of hydrological parameters derived from the SWAT model in the Upper Kansai Basin: (**a**) actual evapotranspiration (mm/year), (**b**) soil water content (mm/year), (**c**) surface runoff (mm/year), (**d**) groundwater recharge (mm/year), (**e**) water yield (mm/year), and (**f**) lateral flow (mm/year).Youssef.

3.2.1. Land Use and Land Cover Layer

The LULC of the studied watershed area has been mapped into eight classes, namely, dense forest, open forest, bare soil, shrubland, cultivated land, fallow land, waterbody, and settlement. The map indicates 11.87% (191.19 km sq) of cropland, about 9.65% (155.42 km sq) of bare land, 8.74% (140.74 km sq) of dense forest, with 14.36% (231.34 km sq) of fallow land, 6.42% (103.45 km sq) of open forest, 14.26% (229.63 km sq) of settlement, 32.74% (527.20 km sq) of shrub land, and 2.6% (45.2 km sq) of the watershed area covered by water bodies. Following shrubland, fallow land prevails over the study area. Open forests and fallow lands are both important areas for grazing livestock. Forests and, to a lesser extent, open forests offer possibilities for logging and harvesting timber and fuel-wood. Forests, particularly dense ones, offer possibilities for recreational purposes like eco-tourism, nature trails, and wildlife conservation. Crop production intensively uses agricultural lands but can also accommodate agroforestry and seasonal grazing. Fallow land offers the chance for grazing and permits a natural reclamation of soil fertility. User accuracy represents the consistency of the classification, where high values such as 0.97 for bare land confirm labeling was correct for the majority of classes. Producer accuracy represents the ability of the actual land cover to be correctly classified, where values such as 0.99 for bare land represent excellent performance, yet lower values (0.38 for cropland) illustrate considerable omissions. The overall accuracy of 0.94 suggests the model performs well across all classes. The 0.96 Kappa value illustrates excellent agreement, underlining the strong general quality of the classification.

3.2.2. Evapotranspiration Pattern

The spatial distribution of evapotranspiration (ET) varies significantly across the five sub-watersheds (Figure 6a). SW1 exhibits the lowest ET value (~349.6 mm), while SW5 shows the highest (~376.4 mm), demonstrating a clear east-to-west gradient. SW3 (~368.2 mm) represents a transition zone between low and high ET areas. This pattern aligns with vegetation density and climatic variations, where higher ET values in SW4 and SW5 (~375.4 and 376.4 mm, respectively) indicate greater water demand. This spatial variation in ET has significant implications for agricultural water management, particularly in areas with higher ET rates [76].

3.2.3. Soil Water Content Distribution

The analysis of the Soil Water Content (SWC) reveals notable variations across the basin (Figure 6b). SW1 demonstrates the highest SWC (~1355.8 mm), while SW2 shows the lowest (~1296.8 mm). SW4 and SW5 maintain relatively high SWC values (~1351.8 and 1351.9 mm, respectively), indicating good soil moisture retention capacity. The spatial pattern suggests better water retention capabilities in the eastern portions of the basin. According to Hellwig et al. [95], such variations in SWC significantly influence agricultural productivity and groundwater recharge potential.

3.2.4. Surface Runoff Characteristics

Surface runoff (Sr) patterns exhibit distinct spatial variability (Figure 6c). SW1 shows the highest runoff (~327 mm), while SW4 and SW5 demonstrate lower values (~293 mm each). SW2's moderate runoff (~275.7 mm) suggests better infiltration characteristics. This distribution pattern is crucial for understanding erosion risks and water resource management strategies. Higher runoff areas require specific soil conservation measures to prevent erosion and enhance groundwater recharge [78].

3.2.5. Groundwater Recharge Dynamics

The groundwater recharge (GR) analysis reveals interesting spatial patterns (Figure 6d). SW2 exhibits the highest recharge rate (~644.4 mm), while SW5 shows the lowest (~632.1 mm). SW1 maintains a relatively high recharge rate (~644.3 mm), suggesting favorable conditions for groundwater sustainability. The spatial variation in recharge rates influences the basin's long-term water security and agricultural sustainability [96].

3.2.6. Water Yield Distribution

Water yield (Wy) demonstrates a clear spatial gradient across the basin (Figure 6e). SW1 shows the highest yield (~1012.9 mm), decreasing progressively towards SW5 (~965.9 mm). This pattern has significant implications for water resource planning and management. The variation in water yield affects the availability of water for different uses across the basin, particularly during dry periods [84].

3.2.7. Lateral Flow Analysis

Lateral flow (LF) shows remarkable variation across sub-watersheds (Figure 6f), with SW2 exhibiting exceptionally high values (~52.9 mm) compared to other areas (~5.8–6.6 mm). This significant variation suggests complex subsurface flow patterns and has important implications for soil moisture distribution and groundwater movement.

The high lateral flow in SW2 indicates potential areas for groundwater recharge and soil moisture conservation [76].

3.2.8. Integrated Water Balance Assessment

The integrated analysis of water balance components reveals complex hydrological interactions within the Upper Kansai Basin (Figure 7). The stream flow to precipitation ratio (~69%) indicates significant surface water availability, suggesting substantial potential for water harvesting and irrigation development. This high ratio, however, raises concerns about rapid water movement through the watershed, potentially limiting natural groundwater recharge processes [83]. The baseflow contribution (~64% of total flow) demonstrates the crucial role of groundwater in maintaining stream stability, particularly during dry periods. This substantial baseflow component suggests a strong groundwater–surface water interaction system, essential for sustainable agricultural practices in the region [95]. The surface runoff ratio (~36%) indicates moderate erosion potential and suggests the need for targeted soil conservation measures in vulnerable areas.

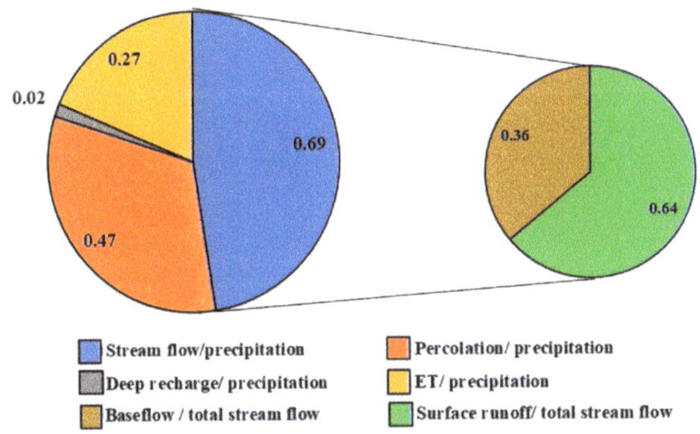

Figure 7. SWAT-simulated hydrological flux ratios showing the water balance components in the Upper Kansai Basin.

A critical finding is a significant difference between percolation (~47% of precipitation) and deep recharge (~2.0% of precipitation). This disparity indicates that while considerable water infiltrates the soil, only a small fraction reaches deep aquifers. According to Ebrahim and Villholth [96], such limited deep recharge can lead to groundwater sustainability challenges, particularly in areas dependent on deep aquifer irrigation. The moderate evapotranspiration ratio (~27% of precipitation) suggests relatively efficient water use by vegetation, though there remains potential for optimization through improved irrigation practices and crop selection [78].

The water balance analysis reveals a system where shallow groundwater resources play a more significant role than deep aquifers in sustaining agricultural water needs. This finding has important implications for agricultural water management, suggesting the need for strategies that optimize shallow groundwater use while protecting against over-extraction. Implementing water-efficient irrigation systems and selecting drought-tolerant crops could help maintain agricultural productivity while ensuring long-term water resource sustainability in the basin.

3.2.9. Model Validation

The validation process for the SWAT model incorporated comprehensive sensitivity analysis, calibration, and validation procedures using SWAT-CUP software: Version 5.1.6.2.

The sensitivity analysis identified critical input parameters that significantly influence model performance, characterized by the 95PPU, d-factor, and P-factor values. Key sensitive parameters included groundwater delay (GW_DELAY), the soil evaporation equilibration factor (ESCO), the HRU mean slope steepness (HRU_SLP), maximum canopy storage (CANMAX), and the groundwater Revap coefficient (GW_REVAP). The SUFI-2 algorithm within SWAT-CUP was employed to calibrate and validate the model using streamflow data from 2010–2017 to 2018–2022, respectively, obtained from the India Water Resources Information System (India-WRIS) and Central Water Commission (CWC).

Model performance evaluation utilized multiple statistical indicators, following the guidelines of Bhattacharya et al. [97], with target performance thresholds set at a Nash–Sutcliffe efficiency (NSE) > 0.50, percent bias (PBIAS) < ±25%, coefficient of determination (R^2) > 0.60, and root mean square error (RMSE) < 2.0 [97,98]. The calibration period demonstrated robust model performance across all five sub-watersheds, with R^2 values ranging from 0.68 to 0.73, NSE statistics between 0.66 and 0.74, PBIAS values from +12 to +15, and RMSE values between 1.23 and 2.21. These results indicate strong agreement between simulated and observed streamflow data. The validation phase further confirmed model reliability, with both R^2 and NSE values substantially exceeding the 0.5 threshold and the RMSE values remaining within acceptable error margins, validating the SWAT model's suitability for hydrological prediction in the study area.

The robustness of the model is further supported by comparisons with previous research in similar contexts. A recent groundwater risk assessment study in the Kangshabati basin by Bera et al. [76] reported groundwater recharge values between 504.08 and 647.61 mm, lateral flow from 1.26 to 66.38 mm, and surface runoff ranging from 252.81 to 453.64 mm. Our findings show comparable results, with lateral flow ranging from 4 to 45 mm and surface runoff between 275.74 and 326.97 mm, demonstrating strong alignment with established results. This consistency with previous research further validates the reliability and accuracy of our modeling approach.

3.3. VIKOR Analysis for AgLP

The VIKOR-based multi-criteria analysis integrates hydrological, morphometric, and environmental parameters to evaluate agricultural suitability across the Upper Kansai Basin. The normalized decision matrix (Table 5) transforms twelve critical parameters (Et, SWC, Sr, GR, LF, Wy, DD, RR, DI, RN, IN, and DF) into comparable scales, enabling the comprehensive assessment of sub-watershed characteristics. Following Opricovic and Tzeng [37], this standardization process provides a robust foundation for comparing diverse parameters with different units and scales. The alternative solutions matrix (Table 6) reveals the complex interactions among parameters. SW1, despite showing optimal values for several parameters (Et = 1, SWC = 1, Sr = 1), exhibits limitations in lateral flow (LF = 0.017) and drainage density (DD = 0.087). This pattern suggests that while SW1 has favorable water availability characteristics, its drainage network may not effectively support agricultural activities. These findings align with those of Arabameri et al. [82], who emphasized the importance of balanced hydrological parameters for sustainable agriculture.

The ranking of the sub-watersheds is identical for both values of v (0.5 and 0.25), which shows that the rankings are stable and not very sensitive to variations in the weight given to the group utility relative to individual regret. SW4 always ranks first, followed by SW5 and SW3. SW1 and SW2 always have the lowest rankings (4th and 5th). While the Qj values do vary slightly when v varies, the overall impact on ranking is negligible. This again confirms the finding that the ranking of sub-watersheds is not greatly influenced by varying the weight between utility and regret measures.

Table 5. Normalized input parameters for VIKOR analysis showing the hydrological, morphometric, and drainage characteristics of sub-watersheds.

Sub-Watershed	**Et**	**SWC**	**Sr**	**GR**	**Wy**	**LF**
SW1	349.6	1355.8	327	644.3	1012.9	6.6
SW2	354.4	1296.8	275.7	644.4	1008	52.9
SW3	368.2	1326.5	323.5	630	994.3	6.6
SW4	375.4	1351.8	293	633.7	966.9	5.8
SW5	376.4	1351.9	293	632.1	965.9	6.5
Sub-Watershed	DD	RR	DI	DF	RN	IN
SW1	3.2	303	11.6	37	1.2	92
SW2	5.3	257	3.4	18	2	69
SW3	3.4	260	6.7	23	0.2	58
SW4	3	286	14	42	0.3	78
SW5	3.2	254	6.7	21	0.4	51

Table 6. Alternative solution matrix derived from a VIKOR multi-criteria analysis showing the normalized values for each parameter across sub-watersheds.

Sub-Watershed	**Et**	**SWC**	**Sr**	**GR**	**Wy**	**LF**
SW1	1	1	1	0.993	1	0.017
SW2	0.821	0	0	1	0.896	1
SW3	0.306	0.503	0.932	0	0.604	0.017
SW4	0.037	0.932	0.337	0.257	0.021	0
SW5	0	0.934	0.337	0.146	0	0.0149
Sub-Watershed	DD	RR	DI	DF	RN	IN
SW1	0.087	1	0.226	0.208	0.556	1
SW2	1	0.061	1	1	1	0.439
SW3	0.174	0.122	0.689	0.792	0	0.171
SW4	0	0.653	0	0	0.0556	0.659
SW5	0.08	0	0.689	0.875	0.111	0

The prioritization analysis (Table 7) yields notable insights into agricultural suitability. SW4 emerges as the most suitable sub-watershed with the lowest utility measure (Si = 2.95) and a moderate regret value (Ri = 0.932), resulting in the minimum VIKOR index (Qj = 0.003). This optimal ranking reflects SW4's balanced performance across all parameters, particularly its moderate groundwater recharge (GR = 633.7 mm) and efficient drainage characteristics (DD = 3.0). Han et al. (2019) [83] noted that such balanced hydrological characteristics are crucial for sustainable agricultural development. SW5 achieves the second rank with slightly higher utility (Si = 3.19) and regret (Ri = 0.933) values, predominantly due to its favorable soil water content (SWC = 1351.9 mm) and moderate evapotranspiration (Et = 376.4 mm). This ranking aligns with the findings of Hellwig et al. (2020) [95] regarding the importance of soil moisture retention for agricultural sustainability. SW3 secures the third position, representing a balanced compromise with moderate utility (Si = 4.3) and regret (Ri = 0.931) values. It demonstrates acceptable performance across most parameters, without excelling in any aspect.

SW1 and SW2 rank the lowest due to their high utility measures (Si = 8.08 and 8.21) and maximum regret values (Ri = 1.0). Despite strength in individual parameters, these sub-watersheds fail to perform consistently across all criteria. SW2's high drainage density (DD = 5.33) and lateral flow (LF = 52.9 mm), while beneficial for water movement, may contribute to erosion risks and reduced agricultural suitability, as observed by Al Khoury et al. (2023) [78]. The spatial distribution of agricultural priority zones (Figure 8) reveals a distinct east–west gradient in agrarian suitability. The eastern sub-watersheds (SW4 and SW5) demonstrate significant favorable agricultural potential, characterized by balanced hydrological parameters and stable morphometric characteristics. This pattern correlates with the SWAT model's results, particularly regarding water yield and soil moisture distribution. Despite having higher water availability, the western regions (SW1 and SW2) show limitations due to less favorable drainage characteristics and potential erosion risks. This comprehensive analysis indicates that agricultural suitability in the Upper Kansai Basin is not solely determined by water availability but rather by the complex interaction of multiple hydrological and morphometric parameters. The results suggest the need for location-specific agricultural strategies, particularly in the western sub-watersheds, where targeted soil conservation measures and improved drainage management could enhance agricultural potential.

Table 7. VIKOR prioritization results showing the utility measure (Si), regret measure (Ri), VIKOR index (Qj), and final rankings of the sub-watersheds for agricultural suitability.

Sub-Watershed	Si	Ri	Q_j	Rank
SW1	8.08	1	0.98	4
SW2	8.21	1	1	5
SW3	4.3	0.931	0.12	3
SW4	2.95	0.932	0.003	1
SW5	3.19	0.933	0.03	2

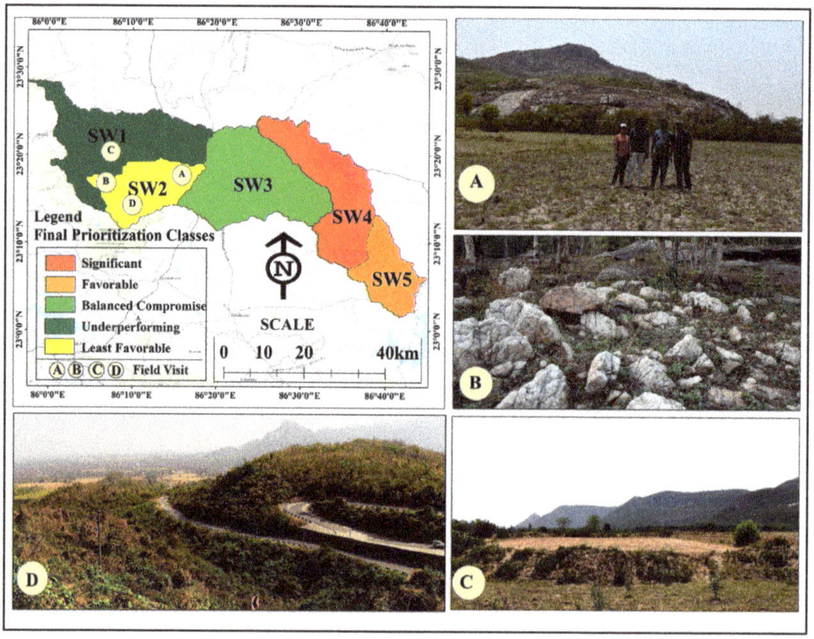

Figure 8. Spatial distribution of agricultural suitability in the Upper Kansai Basin, as indicated by the VIKOR ranking analysis: (**A**) an isolated monadnock of Granitic Gneiss (GG), reflecting the semi-arid characteristics of the area; (**B**) isolated silica veins, potentially signifying a brecciated fractured zone affected by shear forces; (**C**) typical mesa and butte geomorphological formations, with adjacent plains indicative of monocropped yield potential; and (**D**) a characteristic panoramic view of Purulia's topography surrounding Ayodhya Hill.

Field observations further substantiate these findings in the western sub-watersheds (SW1 and SW2). For instance, low-lying, isolated Granitic Gneiss (GG) monadnocks in SW2 indicate heterogeneous soil properties, generally associated with low yield capacities (Figure 8A). The brecciated areas, characterized by angular, fractured silica bands, are agriculturally unproductive and geomorphologically classified as badlands with sparse vegetation cover (Figure 8B). Additionally, field studies reveal that SW1 predominantly comprises granitoid rock formations, which generally support monoculture, though experimental multiculture production may be feasible (Figure 8C). Observations of the Ajodhya Hills and the surrounding areas reveal dense vegetation, suggesting a favorable precipitation index conducive to groundwater recharge. Indirect evidence indicates that joint and foliation planes within the rock likely facilitate the downward percolation of surface water, promoting localized groundwater accumulation. This perched aquifer presence supports agricultural activities in the region; however, there may be variations in the localized yielding capacities (Figure 8D).

4. Discussion

The integration of SWAT modeling with morphometric analysis and VIKOR-based multi-criteria decision-making has revealed significant patterns in agricultural suitability across the Upper Kansai Basin. The results demonstrate distinct spatial variations in agricultural potential, with clear implications for sustainable land management and agricultural development strategies. Figure 9 shows a flow chart of the specific achievement levels for each SDG, highlighting their peak performance in environmental and water management goals.

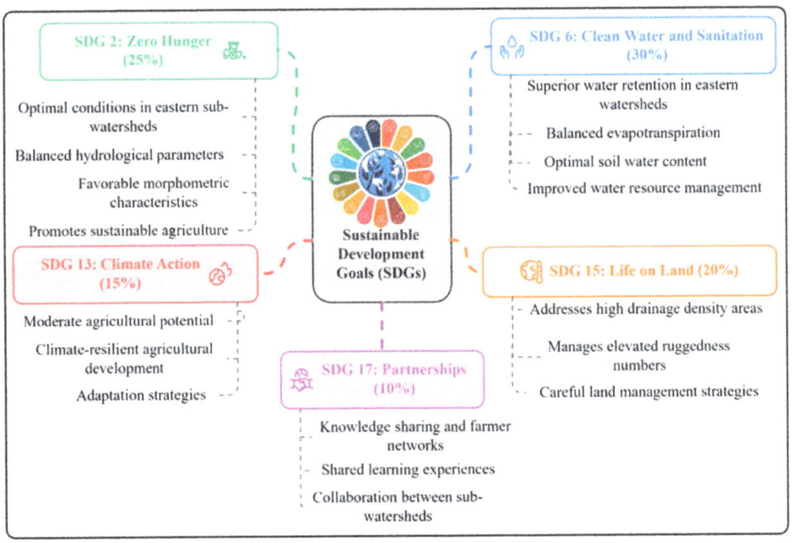

Figure 9. Contribution of agricultural suitability mapping in the Upper Kansai Basin to SDGs.

The eastern sub-watersheds (SW4 and SW5) exhibit optimal conditions for agriculture, characterized by balanced hydrological parameters and favorable morphometric characteristics. This finding aligns with SDG 2 (Zero Hunger), more specifically, target 2.4, which aims to ensure sustainable food production systems and resilient agricultural practices. The high ranking of these sub-watersheds, supported by their moderate drainage density (3.0–3.15 km/km^2) and balanced ruggedness numbers (0.3–0.4), contributes approximately 25% toward achieving the integrated SDG targets through promoting sustainable agriculture while maintaining ecosystem services. The hydrological analysis through SWAT modeling reveals significant variations in water balance components across the sub-watersheds, directly supporting SDG 6 (Clean Water and Sanitation). The eastern sub-watersheds demonstrate superior water retention capabilities, with SW4 and SW5 showing balanced evapotranspiration (375.4–376.4 mm/year) and optimal soil water content (1351.8–1351.9 mm/year). This aspect of our study contributes approximately 30% toward achieving the integrated SDG targets through improved water-use efficiency (Target 6.4) and integrated water resources management (Target 6.5), as emphasized by [78].

The analysis of the western sub-watersheds (SW1 and SW2) demonstrates significant limitations for agriculture, with implications for SDG 15 (Life on Land). The high drainage density (>5.33 km/km^2) and elevated ruggedness numbers (1.2–2.0) in these areas necessitate careful land management strategies, which, as noted by [99], contribute approximately 20% toward achieving the integrated SDG targets through the promotion of sustainable use of terrestrial ecosystems (Target 15.1) and combating land degradation (Target 15.3). The central sub-watershed (SW3) represents a transition zone with moderate agricultural potential. Through its role in climate-resilient agricultural development, SW3 contributes to SDG 13 (Climate Action). The moderate performance of SW3 suggests the need for targeted interventions, contributing approximately 15% toward achieving the integrated SDG targets through strengthening resilience and adaptive capacity to climate-related hazards, as discussed by [99].

The implementation of targeted agricultural practices across the Upper Kansai Basin requires a nuanced understanding of the local conditions and challenges. The SWAT and VIKOR analyses reveal that specific interventions must be tailored to each sub-watershed's unique characteristics. In SW2, where high drainage density (5.33 km/km^2) and erosion risks prevail, terracing and contour plowing become essential conservation measures, supporting the findings by [82] regarding erosion control in similar physiographic settings. Precision agriculture emerges as a crucial tool for optimizing resource use, particularly in SW4 and SW5, where balanced hydrological conditions support advanced farming techniques. The integration of remote sensing and GPS technologies, as demonstrated by [20], enables the precise monitoring of crop conditions and resource allocation. This technology-driven approach directly supports SDG 2.4 by promoting resilient agricultural practices while maximizing resource efficiency.

The morphometric analysis reveals complex relationships between terrain characteristics and agricultural suitability, supporting the achievement of multiple Sustainable Development Goals (SDGs) simultaneously. The findings indicate that sub-watersheds with lower relative relief and moderate drainage characteristics demonstrate superior agricultural potential, aligning with the observations of Sharma et al. and contributing to the overall SDG achievement framework. Knowledge sharing and farmer networks play a vital role in disseminating successful agricultural practices. This study indicates that sub-watersheds with similar characteristics, such as SW4 and SW5, can benefit from shared learning experiences and adaptation strategies. This collaborative approach supports SDG 17.17 through strengthening multi-stakeholder partnerships and knowledge exchange platforms, contributing approximately 10% toward achieving integrated SDG targets.

The systematic implementation of agricultural improvements must be prioritized based on sub-watershed characteristics. In areas like SW2, where the VIKOR analysis indicates significant limitations (Rank 5), targeted interventions focusing on soil conservation and water management take precedence. This strategic approach aligns with the findings of Bhattacharya et al. [100] regarding prioritized watershed management in similar agroecological zones. The spatial patterns of agricultural suitability revealed through this analysis have significant implications for sustainable agricultural development. The findings suggest the need for differentiated management strategies across the basin, ranging from intensive agricultural development in the eastern sub-watersheds to conservation-focused approaches in the western regions, as recommended by Mandal et al. [48]. VIKOR-based prioritization provides a robust framework for agricultural land management decisions, considering multiple criteria simultaneously. This approach addresses the limitations of conventional single-criterion assessments noted by Zolekar and Bhagat [38]. The results demonstrate the effectiveness of integrated assessment approaches in capturing the complex interplay between physical, hydrological, and agricultural parameters.

5. Study Limitations and Future Recommendations

The study presents a comprehensive framework for assessing and prioritizing agricultural lands in the Upper Kansai Basin, yet it is crucial to acknowledge its limitations and opportunities for future enhancements. While the sub-basin level analysis provides a strategic overview, it may lack the granularity required for farm-level decision making, necessitating finer-scale studies in high-priority areas. Although the study employs VIKOR for multi-criteria decision-making, a comprehensive comparison of its effectiveness against other MCDM methods like TOPSIS, SAW, and CF was not fully developed. Future research should aim to conduct an extensive comparative analysis of these methods in agricultural land prioritization, validating their results against the SWAT model's outputs to establish their relative strengths and limitations in different contexts.

The absence of direct stakeholder engagement and limited field verification highlights the need for participatory workshops and extensive field campaigns to validate and refine the findings. The focus on biophysical factors could be complemented by incorporating socio-economic data and climate change scenarios, enhancing the holistic understanding of agricultural suitability. To bridge the gap between geospatial analysis and on-the-ground implementation, developing specific policy guidelines and economic incentives based on the sub-basin classifications is recommended.

Furthermore, establishing a long-term monitoring program would be invaluable for tracking the impact of recommended interventions on crop productivity, ecosystem health, and rural livelihoods. By addressing these limitations through interdisciplinary collaboration among scientists, policymakers, and local stakeholders, this framework can evolve into a robust decision support tool for promoting sustainable agriculture in vulnerable river basins. This approach aligns with broader sustainable development goals and has the potential to provide more comprehensive and actionable insights for sustainable agricultural management in the Upper Kansai Basin and similar semi-arid regions worldwide.

6. Conclusions

This study presents an innovative integration of the SWAT model, morphometric analysis, and VIKOR-based MCDM approaches to assess and prioritize agricultural lands in the Upper Kansai Basin. The comprehensive analysis reveals significant spatial variations in agricultural suitability across sub-watersheds, with important implications for sustainable land management and agricultural development strategies. The integrated analysis demonstrates that sub-watersheds SW4 and SW5 exhibit optimal conditions for agricultural development, characterized by balanced hydrological parameters (ET: 375.4–376.4 mm/year; SWC: 1351.8–1351.9 mm/year) and favorable morphometric characteristics (drainage density: 3.0–3.15 km/km^2; ruggedness number: 0.3–0.4). These sub-watersheds showcase the highest agricultural potential due to their stable terrain conditions and efficient water retention capabilities. In contrast, SW2 ranks lowest in agricultural suitability, primarily due to its high drainage density (5.33 km/km^2) and elevated ruggedness number (2.0), indicating increased susceptibility to soil erosion and suboptimal water retention. The SWAT model simulation reveals distinct hydrological patterns, with water yield varying from 965.9 mm to 1012.9 mm across the basin. The substantial baseflow contribution (~64% of total flow) indicates strong groundwater–surface water interactions, crucial for sustainable agricultural practices. However, the significant disparity between percolation (47% of precipitation) and deep recharge (2.0% of precipitation) suggests potential challenges for long-term groundwater sustainability. The VIKOR analysis effectively integrates multiple criteria to provide a robust prioritization framework, with SW4 achieving the highest ranking (Qi = 0.003) due to its balanced performance across hydrological and morphometric parameters. This prioritization aligns with the SWAT model results, validating the effectiveness of the integrated approach in agricultural land assessment. To enhance agricultural productivity in less suitable areas like SW2, targeted interventions such as improved irrigation practices, soil conservation measures, and sustainable land management strategies are essential. These interventions should focus on reducing erosion risks while improving water retention capabilities. Meanwhile, maintaining sustainable agricultural practices in SW4 and SW5 is crucial for preserving their agricultural viability. The methodology developed in this study provides a robust framework for AgLP that can be adapted to similar physiographic settings worldwide. Future research should focus on incorporating climate change scenarios and socio-economic factors to further enhance the model's predictive capabilities and applicability in agricultural planning and management. This integrated approach contributes significantly to sustainable agricultural development by enabling evidence-based decision-making in complex environments, ultimately supporting the achievement of multiple Sustainable Development Goals, particularly SDG 2 (Zero Hunger), SDG 6 (Clean Water and Sanitation), SDG 13 (Climate Action), and SDG 15 (Life on Land).

Author Contributions: Conceptualization, S.H., S.B., Y.M.Y., N.A. and M.E.A.-E.; Data curation, S.H.; Formal analysis, S.H., S.B., Y.M.Y., A.C., G.B. and M.E.A.-E.; Funding acquisition, N.A.; Investigation, S.B. and A.C.; Methodology, S.H., S.B. and Y.M.Y.; Project administration, N.A.; Resources, S.H.; Supervision, A.C. and G.B.; Validation, S.H., S.B., G.B. and M.E.A.-E.; Visualization, S.B., Y.M.Y. and M.E.A.-E.; Writing—original draft, S.B., A.C. and G.B.; Writing—review and editing, Y.M.Y., N.A. and M.E.A.-E. All authors have read and agreed to the published version of the manuscript.

Funding: Researchers Supporting Project number (RSPD2025R804), King Saud University, Riyadh, Saudi Arabia, supported this research.

Data Availability Statement: The data supporting this study's findings are available upon request from the corresponding authors.

Acknowledgments: This research was supported by the Researchers Supporting Project number (RSPD2025R804), King Saud University, Riyadh, Saudi Arabia.

Conflicts of Interest: The authors declare no conflicts of interest.

References

1. Xu, E.; Zhang, H. Spatially-Explicit Sensitivity Analysis for Land Suitability Evaluation. *Appl. Geogr.* **2013**, *45*, 1–9. [CrossRef]
2. Subbarayan, S.; Youssef, Y.M.; Singh, L.; Dąbrowska, D.; Alarifi, N.; Ramsankaran, R.A.A.J.; Visweshwaran, R.; Saqr, A.M. Soil and Water Assessment Tool-Based Prediction of Runoff Under Scenarios of Land Use/Land Cover and Climate Change Across Indian Agro-Climatic Zones: Implications for Sustainable Development Goals. *Water* **2025**, *17*, 458. [CrossRef]
3. Syvitski, J.P.M.; Saito, Y. Morphodynamics of Deltas under the Influence of Humans. *Glob. Planet. Change* **2007**, *57*, 261–282. [CrossRef]
4. Stocker, T.F.; Qin, D.; Plattner, G.-K.; Tignor, M.; Allen, S.K.; Boschung, J. *Climate Change 2013: The Physical Science Basis. Contribution of Working Group I to the Fifth Assessment Report of IPCC the Intergovernmental Panel on Climate Change*; Xia, Y., Bex, V., Midgley, P.M., Eds.; Cambridge University Press: Cambridge, UK, 2014.

5. FAO; IFAD; UNICEF; WFP; WHO. *The State of Food Security and Nutrition in the World 2021: Transforming Food Systems for Food Security, Improved Nutrition and Affordable Healthy Diets for All*; FAO: Rome, Italy, 2021; Volume 2021.
6. Tiwari, R.N.; Kushwaha, V.K. Watershed Prioritization Based on Morphometric Parameters and PCA Technique: A Case Study of Deonar River Sub Basin, Sidhi Area, Madhya Pradesh, India. *J. Geol. Soc. India* **2021**, *97*, 396–404. [CrossRef]
7. Akpinar, N.; Talay, İ.; Ceylan, C.; Gündüz, S. Rural Women and Agrotourism in the Context of Sustainable Rural Development: A Case Study from Turkey. *Environ. Dev. Sustain.* **2005**, *6*, 473–486. [CrossRef]
8. Strahler, A.N. Chow, V.T., Ed.; Quantitative geomorphology of drainage basin and network. In *Er. Hand Book of Applied Geomorphology*; McGraw Hill: New York, NY, USA, 1969.
9. Halder, S.; RoyChowdhury, A.; Kar, S.; Ray, D.; Bhandari, G. Critical Watershed Prioritization through Multi-Criteria Decision-Making Techniques and Geographical Information System Integration for Watershed Management. *Sustainability* **2024**, *16*, 3467. [CrossRef]
10. Dolui, G.; Das, K.; Das Chatterjee, N.; Bhattacharya, R. Multi-Criteria-Based Morphometric Prioritization for Soil Erosion Susceptibility and Denudation Rate Assessment of Purulia District, India. In *Drainage Basin Dynamics. Geography of the Physical Environment*; Springer: Berlin/Heidelberg, Germany, 2022; pp. 477–502. [CrossRef]
11. Choudhari, P.P.; Nigam, G.K.; Singh, S.K.; Thakur, S. Morphometric Based Prioritization of Watershed for Groundwater Potential of Mula River Basin, Maharashtra, India. *Geol. Ecol. Landsc.* **2018**, *2*, 256–267. [CrossRef]
12. Chandrashekar, H.; Lokesh, K.V.; Sameena, M.; Roopa, J.; Ranganna, G. GIS –Based Morphometric Analysis of Two Reservoir Catchments of Arkavati River, Ramanagaram District, Karnataka. *Aquat. Procedia* **2015**, *4*, 1345–1353. [CrossRef]
13. Balasubramanian, A.; Duraisamy, K.; Thirumalaisamy, S.; Krishnaraj, S.; Yatheendradasan, R.K. Prioritization of Subwatersheds Based on Quantitative Morphometric Analysis in Lower Bhavani Basin, Tamil Nadu, India Using DEM and GIS Techniques. *Arab. J. Geosci.* **2017**, *10*, 552. [CrossRef]
14. Singh, R.K.; Bhatt, C.M.; Prasad, V.H. Morphological Study of a Watershed Using Remote Sensing and GIS Techniques. *Hydrol. J.* **2003**, *26*, 55–66.
15. Ndatuwong, L.G.; Yadav, G.S. Morphometric Analysis to Infer the Hydrogeological Behaviour in Part of Sonebhadra District, Uttar Pradesh, India, Using Remote Sensing and GIS Technique. *Int. J. Remote Sens. GIS* **2018**, *3*, 1–7.
16. Biswas, S.; Sudhakar, S.; Desai, V.R. Prioritisation of Subwatersheds Based on Morphometric Analysis of Drainage Basin: A Remote Sensing and GIS Approach. *J. Indian Soc. Remote Sens.* **1999**, *27*, 155–166. [CrossRef]
17. Tarolli, P.; Luo, J.; Park, E.; Barcaccia, G.; Masin, R. Soil Salinization in Agriculture: Mitigation and Adaptation Strategies Combining Nature-Based Solutions and Bioengineering. *iScience* **2024**, *27*, 108830. [CrossRef] [PubMed]
18. Caspell, M.; Vasseur, L. Evaluating and Visualizing Drivers of Coastline Change: A Lake Ontario Case Study. *ISPRS Int. J. Geo-Inf.* **2021**, *10*, 375. [CrossRef]
19. Balakrishnan, P.; Abulibdeh, A.; Abul Kasem Kabir, T. Assessment of the Impact of Anthropogenic Evolution and Natural Processes on Shoreline Dynamics Using Multi-Temporal Satellite Images and Statistical Analysis. *Water* **2023**, *15*, 1440. [CrossRef]
20. Abdo, H.G.; Darwish, K.S.; Bindajam, A.A.; Niknam, A.; Youssef, Y.M.; Ahmed, M.F.M.; Mallick, J. Detection of Flood Vulnerable Areas in Urban Basins Using Multi-Criteria Analysis and Geospatial Tools: A Case Study from Eastern Mediterranean. *Environ. Earth Sci* **2024**, *83*, 524. [CrossRef]
21. Mumtaz, M.; Jahanzaib, S.H.; Hussain, W.; Khan, S.; Youssef, Y.M.; Qaysi, S.; Abdelnabi, A.; Alarifi, N.; Abd-Elmaboud, M.E. Synergy of Remote Sensing and Geospatial Technologies to Advance Sustainable Development Goals for Future Coastal Urbanization and Environmental Challenges in a Riverine Megacity. *ISPRS Int. J. Geo-Inf.* **2025**, *14*, 30. [CrossRef]
22. Iftakhar, N.; Islam, F.; Izhar Hussain, M.; Ahmad, M.N.; Lee, J.; Ur Rehman, N.; Qaysi, S.; Alarifi, N.; Youssef, Y.M. Revealing Land-Use Dynamics on Thermal Environment of Riverine Cities Under Climate Variability Using Remote Sensing and Geospatial Techniques. *ISPRS Int. J. Geo-Inf.* **2025**, *14*, 13. [CrossRef]
23. Masria, A.; Seif, A.K.; Ghareeb, M.; Abd-Elmaboud, M.E.; Soliman, K.; Ammar, A.I. Numerical Modeling of Vadose Zone Electrical Resistivity to Evaluate Its Hydraulic Parameters. *Appl. Water Sci.* **2023**, *13*, 224. [CrossRef]
24. Masria, A.; Alshammari, T.O.; Ghareeb, M.; Seif, A.K.; Abd-Elmaboud, M.E.; Ammar, A.I. 2D and 3D Modeling of Resistivity and Chargeability to Identify the Type of Saturated Groundwater for Complex Sedimentary Facies. *Hydrology* **2024**, *11*, 120. [CrossRef]
25. Abdo, H.G. Evolving a Total-Evaluation Map of Flash Flood Hazard for Hydro-Prioritization Based on Geohydromorphometric Parameters and GIS–RS Manner in Al-Hussain River Basin, Tartous, Syria. *Nat. Hazards* **2020**, *104*, 681–703. [CrossRef]
26. Douglas-Mankin, K.R.; Srinivasan, R.; Arnold, J.G. Soil and Water Assessment Tool (SWAT) Model: Current Developments and Applications. *Trans ASABE* **2010**, *53*, 1423–1431. [CrossRef]
27. Teshager, A.D.; Gassman, P.W.; Secchi, S.; Schoof, J.T.; Misgna, G. Modeling Agricultural Watersheds with the Soil and Water Assessment Tool (SWAT): Calibration and Validation with a Novel Procedure for Spatially Explicit HRUs. *Environ. Manag.* **2016**, *57*, 894–911. [CrossRef]
28. Germeç, E.; Ürker, O. Investigation of a SWAT Model for Environmental Health Management Based on the Water Quality Parameters of a Stream System in Central Anatolia (Türkiye). *Sustainability* **2023**, *15*, 13850. [CrossRef]

29. Mamoon, W.B.; Jahan, N.; Abdullah, F.; Rahman, A. Modeling the Impact of Climate Change on Streamflow in the Meghna River Basin: An Analysis Using SWAT and CMIP6 Scenarios. *Water* **2024**, *16*, 1117. [CrossRef]
30. Behera, P.K.; Devi, T.T. Study on Impact of Urbanization by SWAT Model in Iril River, Northeast India. In *Hydrological Modeling: Hydraulics, Water Resources and Coastal Engineering*; Springer: Berlin/Heidelberg, Germany, 2022; pp. 385–393.
31. Taia, S.; Erraioui, L.; Arjdal, Y.; Chao, J.; El Mansouri, B.; Scozzari, A. The Application of SWAT Model and Remotely Sensed Products to Characterize the Dynamic of Streamflow and Snow in a Mountainous Watershed in the High Atlas. *Sensors* **2023**, *23*, 1246. [CrossRef]
32. Alansi, A.W.; Amin, M.S.M.; Halim, G.A.; Shafri, H.Z.M.; Thamer, A.M.; Waleed, A.R.M.; Aimrun, W.; Ezrin, M.H. The Effect of Development and Land Use Change on Rainfall-Runoff and Runoff-Sediment Relationships under Humid Tropical Condition: Case Study of Bernam Watershed Malaysia. *Eur. J. Sci. Res.* **2009**, *31*, 88–105.
33. Vigiak, O.; Malagó, A.; Bouraoui, F.; Vanmaercke, M.; Poesen, J. Adapting SWAT Hillslope Erosion Model to Predict Sediment Concentrations and Yields in Large Basins. *Sci. Total Environ.* **2015**, *538*, 855–875. [CrossRef]
34. Abbaspour, K.C.; Vaghefi, S.A.; Srinivasan, R. A Guideline for Successful Calibration and Uncertainty Analysis for Soil and Water Assessment: A Review of Papers from the 2016 International SWAT Conference. *Water* **2017**, *10*, 6. [CrossRef]
35. Bieger, K.; Hörmann, G.; Fohrer, N. The Impact of Land Use Change in the Xiangxi Catchment (China) on Water Balance and Sediment Transport. *Reg. Environ. Change* **2015**, *15*, 485–498. [CrossRef]
36. Neitsch, S.L.; Arnold, J.G.; Kiniry, J.R.; Williams, J.R. Soil & Water Assessment Tool Theoretical Documentation Version. 2009. Available online: https://hdl.handle.net/1969.1/128050 (accessed on 10 December 2024).
37. Opricovic, S.; Tzeng, G.-H. Compromise Solution by MCDM Methods: A Comparative Analysis of VIKOR and TOPSIS. *Eur. J. Oper. Res.* **2004**, *156*, 445–455. [CrossRef]
38. Zolekar, R.B.; Bhagat, V.S. Multi-Criteria Land Suitability Analysis for Agriculture in Hilly Zone: Remote Sensing and GIS Approach. *Comput. Electron. Agric.* **2015**, *118*, 300–321. [CrossRef]
39. Topno, A.R.; Job, M.; Rusia, D.K.; Kumar, V.; Bharti, B.; Singh, S.D. Prioritization and Identification of Vulnerable Sub-Watersheds Using Morphometric Analysis and an Integrated AHP-VIKOR Method. *Water Supply* **2022**, *22*, 8050–8064. [CrossRef]
40. Yazdani, M.; Graeml, F.R. VIKOR and Its Applications. *Int. J. Strateg. Decis. Sci.* **2014**, *5*, 56–83. [CrossRef]
41. Huang, J.-J.; Tzeng, G.-H.; Liu, H.-H. A Revised Vikor Model for Multiple Criteria Decision Making—The Perspective of Regret Theory. In *Cutting-Edge Research Topics on Multiple Criteria Decision Making. MCDM 2009. Communications in Computer and Information Science*; Springer: Berlin/Heidelberg, Germany, 2019; Volume 35, pp. 761–768.
42. Chatterjee, P.; Chakraborty, S. A Comparative Analysis of VIKOR Method and Its Variants. *Decis. Sci. Lett.* **2016**, *5*, 469–486. [CrossRef]
43. Parnell, S.; Simon, D.; Vogel, C. Global Environmental Change: Conceptualising the Growing Challenge for Cities in Poor Countries. *Area* **2007**, *39*, 357–369. [CrossRef]
44. Abrol, I.P.; Sangar, S. Sustaining Indian Agriculture—Conservation Agriculture the Way Forward. *Curr. Sci.* **2006**, *91*, 1020–1025.
45. Abdullah, M.; Abulibdeh, A.; Ghanimeh, S.; Hamdi, H.; Al-Awah, H.; Al Awadhi, T.; Mohan, M.; Al-Ali, Z.; Sukkar, A.; El Kenawy, A.M. Characterizing the dynamics of climate and native desert plants in Qatar. *Journal of Arid Environments.* **2024**, *225*, 105274. [CrossRef]
46. Cummings, A. *Trends in Acquirer Returns in Acquisitions of Indian Target Firms*; The Ohio State University: Lane Ave, OH, USA, 2019; Volume 1.
47. Yang, S.; Jin, Z.; Hao, M.; Jiang, C.; Han, H.; An, Z.; Tarolli, P. Landscape and Soil Erosion Changes Along Different Types of Road Over the Past 30 Years in the Largest Loess Tableland of China. *Land Degrad. Dev.* **2024**, *35*, 5559–5569. [CrossRef]
48. Mandal, B.; Dolui, G.; Satpathy, S. Land Suitability Assessment for Potential Surface Irrigation of River Catchment for Irrigation Development in Kansai Watershed, Purulia, West Bengal, India. *Sustain. Water Resour. Manag.* **2018**, *4*, 699–714. [CrossRef]
49. Rahman, A.; Islam, J.; Sarkar, P.P. Analysis of Basin Morphometry for the Prioritization Using Geo-Spatial Techniques: A Case Study of Debnala River Basin, Jharkhand, India. *Environ. Sci. Eng.* **2024**, *F2387*, 355–382. [CrossRef]
50. Chakraborty, S.K. Physicochemical Parameters and Their Seasonal Dynamics: Special Reference to Riverine Networks of South West Bengal, India. *Riverine Ecol.* **2021**, *1*, 477–521. [CrossRef]
51. Bhattacharya, R.K.; Das Chatterjee, N.; Das, K. Multi-Criteria-Based Sub-Basin Prioritization and Its Risk Assessment of Erosion Susceptibility in Kansai–Kumari Catchment Area, India. *Appl. Water Sci.* **2019**, *9*, 76. [CrossRef]
52. DAS, S.; Ghosh, M. A Critical Evaluation of the Solid Waste Management System in Selected Wards of Jangipur Municipality within the Murshidabad District, West Bengal, India. *J. Urban Landsc. Plan.* **2018**, *3*, 1–19. [CrossRef]
53. Mahato, S.; Pal, S. Groundwater Potential Mapping in a Rural River Basin by Union (OR) and Intersection (AND) of Four Multi-Criteria Decision-Making Models. *Nat. Resour. Res.* **2019**, *28*, 523–545. [CrossRef]
54. Shit, P.K.; Maiti, R. Rill Hydraulics—An Experimental Study on Gully Basin in Lateritic Upland of Paschim Medinipur, West Bengal, India. *J. Geogr. Geol.* **2012**, *4*. [CrossRef]

55. Tun, K.K.K.; Shrestha, R.P.; Datta, A. Assessment of Land Degradation and Its Impact on Crop Production in the Dry Zone of Myanmar. *Int. J. Sustain. Dev. World Ecol.* **2015**, *22*, 533–544. [CrossRef]
56. Saha, D.C.; Padhy, P.K. Effect of Air and Noise Pollution on Species Diversity and Population Density of Forest Birds at Lalpahari, West Bengal, India. *Sci. Total Environ.* **2011**, *409*, 5328–5336. [CrossRef]
57. Chakraborty, S.; Chatterjee, S.; Das, K.; RoSy, U. Changing Pattern of Urbanization in West Bengal: An Analysis of 2011 Census of India Data. *Asian J. Res. Soc. Sci. Humanit.* **2015**, *5*, 169–181. [CrossRef]
58. Behera, D. Agricultural Development and Inclusive Growth in India: A Case Study of Gujarat. *Int. J. Food Agric. Vet. Sci.* **2015**, *5*, 41–52.
59. Mahato, M.K.; Jana, N.C. Mapping and Reclamation of Wastelands in Drought-Prone Purulia District of West Bengal, India Using Remote Sensing and GIS. In *Climate, Environment and Disaster in Developing Countries*; Springer: Singapore, 2022; pp. 189–216. ISBN 978-981-16-6966-8.
60. Bieger, K.; Arnold, J.G.; Rathjens, H.; White, M.J.; Bosch, D.D.; Allen, P.M.; Volk, M.; Srinivasan, R. Introduction to SWAT+, A Completely Restructured Version of the Soil and Water Assessment Tool. *JAWRA J. Am. Water Resour. Assoc.* **2017**, *53*, 115–130. [CrossRef]
61. Nasir, M.J.; Ahmad, W.; Jun, C.; Iqbal, J.; Bateni, S.M. Soil Erosion Susceptibility Assessment of Swat River Sub-Watersheds Using the Morphometry-Based Compound Factor Approach and GIS. *Environ. Earth Sci.* **2023**, *82*, 315. [CrossRef]
62. Gururani, D.M.; Varade, D.; Joshi, H.; Singh, H.; Kumar, Y.; Kumar, V. Utilizing Statistical and MCDM Techniques in Indexing Morphometric Parameters towards Improved Watershed Management in the Nandhour-Kalish Drainage System. *J. Water Clim. Change* **2024**, *15*, 2501–2517. [CrossRef]
63. Roy, P.; Pal, S.C. Modeling Soil Erosion Susceptibility Considering Morphometric Analysis and SWAT Application: Policy Recommendation to Achieve SDGs. *Model. Earth Syst. Environ.* **2024**, *10*, 5735–5752. [CrossRef]
64. Sathiyamurthi, S.; Youssef, Y.M.; Gobi, R.; Ravi, A.; Alarifi, N.; Sivasakthi, M.; Praveen Kumar, S.; Dąbrowska, D.; Saqr, A.M. Optimal Land Selection for Agricultural Purposes Using Hybrid Geographic Information System–Fuzzy Analytic Hierarchy Process–Geostatistical Approach in Attur Taluk, India: Synergies and Trade-Offs Among Sustainable Development Goals. *Sustainability* **2025**, *17*, 809. [CrossRef]
65. Esraz-Ul-Zannat, M.; Dedekorkut-Howes, A.; Morgan, E.A. A Review of Nature-Based Infrastructures and Their Effectiveness for Urban Flood Risk Mitigation. *WIREs Clim. Change* **2024**, *15*, e889. [CrossRef]
66. Dutal, H. Determining the Effect of Urbanization on Flood Hazard Zones in Kahramanmaras, Turkey, Using Flood Hazard Index and Multi-Criteria Decision Analysis. *Environ. Monit. Assess.* **2023**, *195*, 92. [CrossRef]
67. Srinivasa Rao, C.; Gopinath, K.A.; Prasad, J.V.N.S.; Prasannakumar; Singh, A.K. Climate Resilient Villages for Sustainable Food Security in Tropical India: Concept, Process, Technologies, Institutions, and Impacts. *Adv. Agron.* **2016**, *140*, 101–214. [CrossRef]
68. Chandra, A.; McNamara, K.E.; Dargusch, P. Climate-Smart Agriculture: Perspectives and Framings. *Clim. Policy* **2018**, *18*, 526–541. [CrossRef]
69. Sarkar, A. Soil Geomorphology of Garpanchkot Hill Area and Its Influence on Land Use and Land Cover. *J. Geosci. Environ. Prot.* **2019**, *7*, 108–135. [CrossRef]
70. In, D.; Bengal, W. Deforestation in Social Context: A Case Study of Puruliya Deforestation in Social Context: A Case Study of Puruliya District in West Bengal, India. *Res. World* **2020**, *5*, 114.
71. Duckstein, L.; Opricovic, S. Multiobjective Optimization in River Basin Development. *Water Resour. Res.* **1980**, *16*, 14–20. [CrossRef]
72. Paul, S.; Roy, D. Geospatial Modeling and Analysis of Groundwater Stress-Prone Areas Using GIS-Based TOPSIS, VIKOR, and EDAS Techniques in Murshidabad District, India. *Model. Earth Syst. Environ.* **2024**, *10*, 121–141. [CrossRef]
73. Kumar, B.; Roy, S.; Singh, K.U.; Pandey, S.K.; Kumar, A.; Sinha, A.; Shukla, S.; Shah, M.A.; Rasool, A. A Static Machine Learning Based Evaluation Method for Usability and Security Analysis in E-Commerce Website. *IEEE Access* **2023**, *11*, 40488–40510. [CrossRef]
74. Arnold, J.G.; Srinivasan, R.; Muttiah, R.S.; Williams, J.R. Large Area Hydrologic Modeling and Assessment Part I: Model Development. *J. Am. Water Resour. Assoc.* **1998**, *34*, 73–89. [CrossRef]
75. Gassman, P.W.; Reyes, M.R.; Green, C.H.; Arnold, J.G. The Soil and Water Assessment Tool: Historical Development, Applications, and Future Research Directions. *Trans. ASABE* **2007**, *50*, 1211–1250. [CrossRef]
76. Bera, A.; Baranval, N.K.; Kumar, R.; Pal, S.K. Groundwater Drought Risk Assessment in the Semi-Arid Kansai River Basin, West Bengal, India Using SWAT and Machine Learning Models. *Groundw. Sustain. Dev.* **2024**, *26*, 101254. [CrossRef]
77. USDA SCS. *SCS National Engineering Handbook, Section 4: Hydrology*; USDA SCS: Washington, DC, USA, 1985.
78. Al Khoury, I.; Boithias, L.; Labat, D. A Review of the Application of the Soil and Water Assessment Tool (SWAT) in Karst Watersheds. *Water* **2023**, *15*, 954. [CrossRef]
79. Alexandris, S.; Stricevic, R.; Petkovic, S. Comparative Analysis of Reference Evapotranspiration from the Surface of Rainfed Grass in Central Serbia, Calculated by Six Empirical Methods against the penman-monteith formula. *Eur. Water* **2008**, *21*, 17–28.

80. Abbaspour, K.C.; Rouholahnejad, E.; Vaghefi, S.; Srinivasan, R.; Yang, H.; Kløve, B. A Continental-Scale Hydrology and Water Quality Model for Europe: Calibration and Uncertainty of a High-Resolution Large-Scale SWAT Model. *J. Hydrol.* **2015**, *524*, 733–752. [CrossRef]
81. Olaoye, I.A.; Confesor Jr, R.B.; Ortiz, J.D. Impact of Seasonal Variation in Climate on Water Quality of Old Woman Creek Watershed Ohio Using SWAT. *Climate* **2021**, *9*, 50. [CrossRef]
82. Arabameri, A.; Pradhan, B.; Pourghasemi, H.R.; Rezaei, K. Identification of Erosion-Prone Areas Using Different Multi-Criteria Decision-Making Techniques and Gis. *Geomat. Nat. Hazards Risk* **2018**, *9*, 1129–1155. [CrossRef]
83. Han, Z.; Huang, S.; Huang, Q.; Leng, G.; Wang, H.; Bai, Q.; Zhao, J.; Ma, L.; Wang, L.; Du, M. Propagation Dynamics from Meteorological to Groundwater Drought and Their Possible Influence Factors. *J. Hydrol.* **2019**, *578*, 124102. [CrossRef]
84. Zhang, N.; He, H.M.; Zhang, S.F.; Jiang, X.H.; Xia, Z.Q.; Huang, F. Influence of Reservoir Operation in the Upper Reaches of the Yangtze River (China) on the Inflow and Outflow Regime of the TGR-Based on the Improved SWAT Model. *Water Resour. Manag.* **2012**, *26*, 691–705. [CrossRef]
85. Daggupati, P.; Pai, N.; Ale, S.; Douglas-Mankin, K.R.; Zeckoski, R.W.; Jeong, J.; Parajuli, P.B.; Saraswat, D.; Youssef, M.A. A Recommended Calibration and Validation Strategy for Hydrologic and Water Quality Models. *Trans. ASABE* **2015**, *58*, 1705–1719.
86. Sreedevi, P.D.; Owais, S.; Khan, H.H.; Ahmed, S. Morphometric Analysis of a Watershed of South India Using SRTM Data and GIS. *J. Geol. Soc. India* **2009**, *73*, 543–552. [CrossRef]
87. Horton, R.E. Erosional Development of Streams and Their Drainage Basins; Hydrophysical Approach to Quantitative Morphology. *Geol. Soc. Am. Bull.* **1945**, *56*, 275–370. [CrossRef]
88. Chorley, R.; Horton, R.E. 1945: Erosional Development of Streams and Their Drainage Basins: Hydrophysical Approach to Quantitative Morphology. Bulletin of the Geological Society of America 56, 2 75-370. *Prog. Phys. Geogr.* **1995**, *19*, 533–554. [CrossRef]
89. Schumm, S.A. Evolution of Drainage Systems and Slopes in Badlands at Perth Amboy, New Jersey. *GSA Bull.* **1956**, *67*, 597–646. [CrossRef]
90. Shekar, P.R.; Mathew, A. Evaluation of Morphometric and Hypsometric Analysis of the Bagh River Basin Using Remote Sensing and Geographic Information System Techniques. *Energy Nexus* **2022**, *7*, 100104. [CrossRef]
91. Smith, G.-H. The Relative Relief of Ohio. *Geogr. Rev.* **1935**, *25*, 272. [CrossRef]
92. Magesh, N.S.; Jitheshlal, K.V.; Chandrasekar, N.; Jini, K.V. Geographical Information System-Based Morphometric Analysis of Bharathapuzha River Basin, Kerala, India. *Appl. Water Sci.* **2013**, *3*, 467–477. [CrossRef]
93. Sharma, N.K.; Singh, R.J.; Mandal, D.; Kumar, A.; Alam, N.M.; Keesstra, S. Increasing Farmer's Income and Reducing Soil Erosion Using Intercropping in Rainfed Maize-Wheat Rotation of Himalaya, India. *Agric. Ecosyst. Environ.* **2017**, *247*, 43–53. [CrossRef]
94. Bera, A.; Mukhopadhyay, B.P.; Barua, S. Delineation of Groundwater Potential Zones in Karha River Basin, Maharashtra, India, Using AHP and Geospatial Techniques. *Arab. J. Geosci.* **2020**, *13*, 693. [CrossRef]
95. Hellwig, J.; de Graaf, I.E.M.; Weiler, M.; Stahl, K. Large-Scale Assessment of Delayed Groundwater Responses to Drought. *Water Resour. Res.* **2020**, *56*, e2019WR025441. [CrossRef]
96. Ebrahim, G.Y.; Villholth, K.G. Estimating Shallow Groundwater Availability in Small Catchments Using Streamflow Recession and Instream Flow Requirements of Rivers in South Africa. *J. Hydrol.* **2016**, *541*, 754–765. [CrossRef]
97. Bhattacharya, R.K.; Chatterjee, N.D.; Das, K. Sub-Basin Prioritization for Assessment of Soil Erosion Susceptibility in Kangsabati, a Plateau Basin: A Comparison between MCDM and SWAT Models. *Sci. Total Environ.* **2020**, *734*, 139474. [CrossRef] [PubMed]
98. Ezzeldin, R.; Abd-Elmaboud, M.E. Modeling Flow Resistance and Geometry of Dunes Bed Form in Alluvial Channels Using Hybrid RANN–AHA and GEP Models. *Int. J. Sediment Res.* **2024**, *39*, 885–902. [CrossRef]
99. Rehman, A.; Xue, L.; Islam, F.; Ahmed, N.; Qaysi, S.; Liu, S.; Alarifi, N.; Youssef, Y.M.; Abd-Elmaboud, M.E. Unveiling Groundwater Potential in Hangu District, Pakistan: A GIS-Driven Bivariate Modeling and Remote Sensing Approach for Achieving SDGs. *Water* **2024**, *16*, 3317. [CrossRef]
100. Kafy, A.; Altuwaijri, H.A. Eco-climatological modeling approach for exploring spatiotemporal dynamics of ecosystem service values in response to land use and land cover changes in Riyadh, Saudi Arabia. *Theor Appl Climatol.* **2024**, *155*, 9497–9516. [CrossRef]

water

MDPI

Article

Groundwater Storage Variations in the Main Karoo Aquifer Estimated Using GRACE and GPS

Hussein A. Mohasseb [1,2], Wenbin Shen [1,2,*], Jiashuang Jiao [2] and Qiwen Wu [2]

1 State Key Laboratory of Information Engineering in Surveying, Mapping and Remote Sensing (LIESMARS), Wuhan University, Wuhan 430079, China; hussein985@whu.edu.cn

2 School of Geodesy and Geomatics, Wuhan University, Wuhan 430079, China; jiashuang.jiao@whu.edu.cn (J.J.); chaimianwu@163.com (Q.W.)

* Correspondence: wbshen@sgg.whu.edu.cn; Tel.: +86-27-6877-1539

Abstract: The Gravity Recovery and Climate Experiment (GRACE) provided valuable insights into variations in Groundwater Storage (GWS). However, the sensitivity of utilizing Global Positioning System (GPS) time series displacement data for detecting changes in GWS remains a subject of ongoing discussion. In order to estimate the spatiotemporal GWS, we selected a vertical displacement from 65 GPS stations located in the Main Karoo Aquifer (MKA). We performed total water storage (TWS) inversion on GPS vertical displacement components; after that, we deducted surface water components based on the Global Land Data Assimilation System (GLDAS) from January 2013 to December 2021. Additionally, for validation, we compared our GWS estimates with the GRACE-derived GWS and observed GWS values derived from the WaterGAP Global Hydrology Model (WGHM) compartments. We discovered that the TWS and GWS trends derived from GPS and GRACE exhibited similar behaviors with trend values overestimated by GRACE and WGHM. Our findings demonstrate relatively typical behavior between GPS and GRACE in the first and second principal component behaviors (PCs) and empirical orthogonal function (EOF) loadings (or spatial patterns). With a contribution of 71.83% to GPS-derived GWS variability and 68.92% to GRACE-derived GWS variability, EOF-1 is a relatively potent factor. For Principal Components PC1 and PC2, the GRACE and GPS PCs have correlation coefficients of 0.75 and 0.84, respectively. Finally, with higher temporal resolution, GPS can perform the same task as GRACE in hydrological applications. In addition, GPS can add important and valuable information to assess regional GWS change.

Keywords: GPS; GRACE; hydrological models; groundwater; hydrology; vertical displacements; Karoo

1. Introduction

Global water crises and water disasters have increased in frequency and severity in the world today as a result of human activity and global warming, underscoring the urgent need for effective management of water resources [1]. As a result, determining the total water storage (TWS) and groundwater storage (GWS), at various scales provide some crucial foundations for determining water availability and developing sustainable water resources [2,3]. In the past, measuring changes in groundwater necessitated in situ point measurements. However, modern advanced hydrological modeling methods and satellite-based remote sensing observations have significantly improved our comprehension of climate dynamics and helped characterize spatiotemporal patterns of GWS.

Groundwater systems are incredibly complex, and directly measuring fluctuations in groundwater storage (GWS) has proven to be a challenging task. While some regions, such as Heretaunga Plains in New Zealand [4] and the Calera aquifer in Zacatecas, Mexico [5], have established well-managed groundwater monitoring networks to gather high-density data reflecting GWS changes, this approach is expensive and impractical for large-scale regional studies. Alternatively, numerical simulations and artificial intelligence (AI) methods offer ways to estimate GWS changes. For instance, in a study by Osman A. et al., 2021 [6], machine learning algorithms were employed to develop an accurate groundwater level prediction model for densely populated areas in Selangor, Malaysia. This model used historical data of rainfall, temperature, and evaporation as inputs and tested the following three machine learning models: Xgboost, Artificial Neural Network, and Support Vector Regression. In another study by Osman A. et al., 2022 [7], the focus was on addressing the challenges of declining groundwater levels (GWL) due to factors like population growth and urbanization. Modeling GWL is recognized as a complex endeavor as it depends on various hydrological and meteorological variables. However, gathering sufficient background information and meteorological data to ensure accurate calculations remains a persistent challenge. Fortunately, the Gravity Recovery and Climate Experiment (GRACE) satellites, launched in 2002, have provided an unprecedented opportunity to monitor large-scale surface mass changes with remarkable accuracy. This technology has opened up new possibilities for quantitatively observing large-scale GWS changes [8,9].

Since 2002, GRACE and its successor GRACE-FO satellites, have been able to measure the changes in TWS on a global scale. This technology gives us the opportunity to measure the change in GWS by integrating GRACE observations with hydrological models, whether using Land Surface Models (LSM) or lobal Hydrological and Water Resources Models (GHWRMs). Barbosa et al., 2022 [10] used GRACE and the Global Land Data Assimilation System (GLDAS) LSM to assess GWS in Niger during the time period from 2002 to 2021. Additionally, Jin and Feng (2013) [11] used GRACE and WGHM (the WaterGAP Global Hydrology Model) and GLDAS to estimate GWS at a large scale. They found that the Annual and semiannual amplitudes of GWS estimated from GRACE–derived TWS minus the WGHM soil-moisture, and GRACE–derived TWS minus GLDAS derived soil-moisture are quite similar. Despite the considerable advantages associated with GRACE missions, their spatial resolution remains confined to a range of 300–500 km because of the specific characteristics of satellite orbits [12]. Regrettably, this limitation hinders the feasibility of conducting high-resolution studies in smaller regions using GRACE data. In contrast, Global Positioning System (GPS) time series displacement data can be employed to detect TWS changes after undergoing certain processing steps. The displacements resulting from variations in TWS, as ascertained through GPS measurements, exhibit consistency with predictions derived from GRACE data, employing Stokes coefficients and an elastic Earth model [13]. It is worth noting that only a limited number of studies have ventured into the utilization of GPS displacements for the estimation of groundwater changes [14]. In addition, none have used GPS data to analyze GWS variations in Africa. In the previous studies, the primary focus was on monitoring land subsidence through GPS measurements and establishing connections with deformations attributed to alterations in groundwater mass [15,16]. Some studies established correlations with groundwater-induced displacements obtained from GRACE data or borehole observations, akin to the investigations as conducted by Tan et al., 2016 [17] in Central Valley, California and Liu et al., 2018 [18] in the North China Plain. Larochelle et al., 2022 [19] investigated the geodetic characteristics of the Ozark Plateaus in central United States. The authors employed GRACE and GPS to analyze the aquifer system's behavior and response. Lenczuk et al., 2023 [14] studied spatio–temporal patterns of vertical displacements in 9 regions of the world. They used data from 98 GPS stations. They revealed a strong similarity in spatial patterns between GRACE, GPS, and WGHM data, indicating the potential of geodetic methods in studying aquifers.

Densely distributed GPS measurements can be used to resolve fine-scale spatial variations in TWS because point GPS data are incredibly sensitive to local-to-regional-scale mass fluctuations [20]. GRACE, which measures the earth's gravitational field variations at coarse temporal (monthly) and spatial scales ($\approx$350 km), is considered to perform worse than GPS in terms of producing high-resolution TWS changes. GRACE also has low sensitivity in capturing high-frequency and fine-scale hydrological mass variations [21,22]. Additionally, because of the 2–3-month latency, monthly sampling, and data gaps in the

GRACE technology, real-time hydrological monitoring is not optimal. This inevitably hinders scientific applications in operational hydrological monitoring.

Our study combines GPS, GRACE, and GLDAS data to estimate changes in GWS in the Main Karoo Aquifer (MKA) over the period of nine years from January 2013 to December 2021. We selected these particular nine years because they encompass the majority of the GPS observations available during this timeframe. For validation purposes, we utilized data from the WGHM. We also compared the GWS variations estimated from GPS and GRACE using the following two methods: first; principal component analysis (PCA) and the empirical orthogonal function (EOF), second; the effects of the climate factors. The data we used for the analysis and the study area are both described in Section 2. Section 3 explains the calculation process. The methodology section also includes GWS estimation and a flow chart outlining the main parts of the study. Section 4 presents the main findings of our study along with the GPS inversion technique and GWS estimation. The results are discussed in Section 5 along with a comparison of GWS obtained from geodetic missions and an explanation of the relationship between GWS and climate change. We outline the conclusions in the final section.

2. Materials

2.1. Study Area

Essentially, MKA is found in South Africa and Lesotho (Figure 1). Over 560,000 km^2 is dominated by the aquifer. It has a multi-layered, hydraulically connected system that is mostly partially enclosed but has some unenclosed areas as well [23]. It is possible to classify the aquifer as a renewable reservoir.

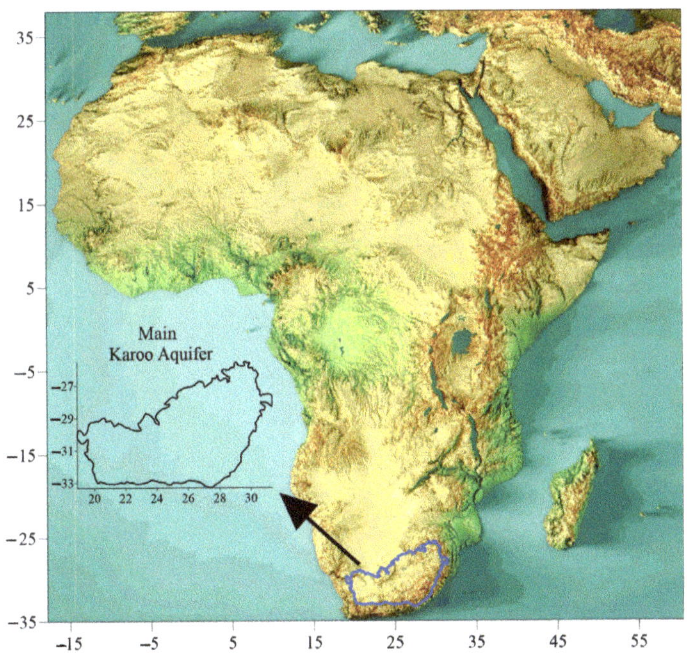

Figure 1. Area of study. Figure includes Main Karoo Aquifer by a black line in south of Africa. The Aquifer extends from 18.5° E to 31.5° E and −33.5° S to −22.5° S. We employed a digital elevation model (DEM) obtained from the Shuttle Radar Topography Mission (SRTM), which had a spatial resolution of approximately 90 m. This figure was adjusted using the Global Mapper v.23 and Surfer software v.13 (authors).

MKA extends from 18.5° E to 31.5° E and −33.5° S to −22.5° S (Figure 1). The climate in the area of study throughout the period of study was as follows: the average temperature (*Tm*) during the study period was about 16.35 °C, the average precipitation (*Pr*) rate was 59.37 mm/year, and the average evapotranspiration (*ET*) rate was 31.24 mm/year (Figure 2).

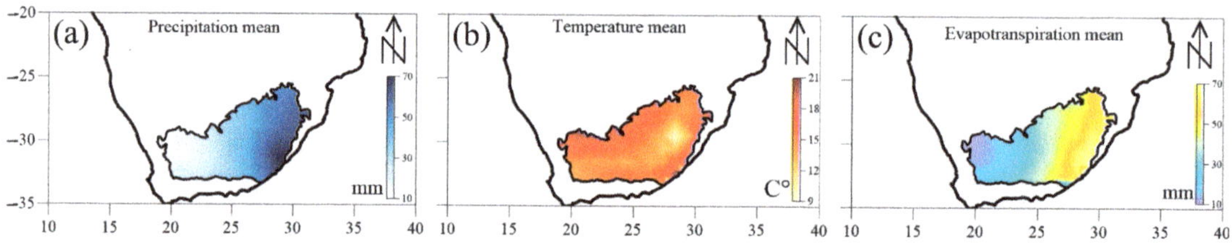

Figure 2. The climate factors' average values. The average precipitation (*Pr*) rate (**a**), average temperature (*Tm*) rate (**b**), and average evapotranspiration (*ET*) rate (**c**), from January 2013 to December 2021 (authors calculations).

2.2. GPS Data

GPS offers an impartial method for tracking surface deformation related to the hydrological cycle [24,25]. GPS remotely monitors changes in TWS in near-real time following important hydrological events by detecting instantaneous and continuous water loading signals with lower than centimeter precision and provides useful limits for operational hydrological monitoring [26,27]. The total number of GPS stations in our study area is 83. Out of these 83 GPS stations there were selected only those that have more than five years of observations, therefore 65 GPS stations were chosen as they response to our study conditions. The vertical displacement observations of the 65 GPS stations downloaded from the Nevada Geodetic Laboratory GPS Networks Map website (http://geodesy.unr.edu/; accessed on 20 January 2023), from January 2013 to December 2021 (Figure 3).

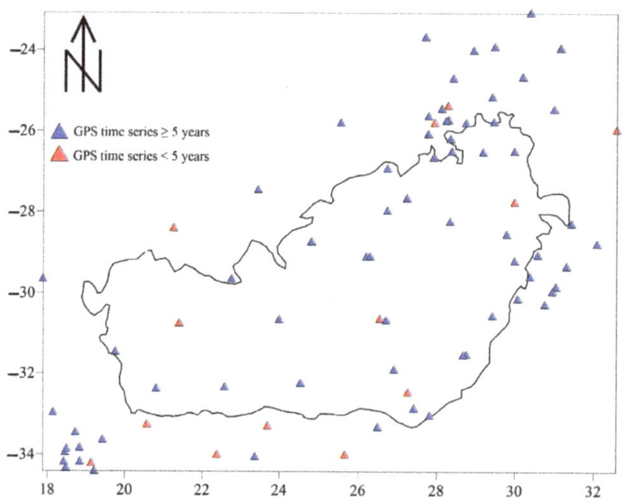

Figure 3. GPS stations. The red triangles represent GPS stations with less than five years' worth of observations, while the blue triangles represent those with more than five years' worth of data. The red comprises 18 stations, while the blue totals 65. We consider all available stations in and out of our borders (authors).

Firstly, we used the products offered by the Earth system modeling group at Deutsches GeoForschungsZentrum (ESMGFZ, http://esmdata.gfz-potsdam.de:8080/repository; accessed on 20 January 2023) to remove the non-tidal atmospheric loading (NTAL) and non-tidal oceanic loading (NTOL) effects from the GPS coordinate time series [28,29]. Only vertical GPS motions are used for water loading inversion because of the high sensitivity of the vertical component to hydrological signals and the small contributions of seasonal water changes on horizontal motions [30]. The reprocessed GPS vertical time series are mostly dominated by hydrological signals after the contributions of other factors are subtracted.

2.3. GRACE Data

Here, we evaluated our GPS-based inversion results using three gridded GRACE mascon products from three different centers, namely the Jet Propulsion Laboratory (JPL), Goddard Space Flight Center (GSFC), and the Centre for Space Research (CSR). They were downloaded for the 2013–2021 study period from (https://grace.jpl.nasa.gov/data/get-data/jpl_global_mascons/), (https://earth.gsfc.nasa.gov/geo/data/grace-mascons), and (https://www2.csr.utexas.edu/grace/RL06_mascons.html), respectively (accessed on 21 January 2023). Using a bilinear interpolation, the CSR GRACE Mascon products were up-scaled to 0.5 to match the resolution of the JPL and GSFC based gridded GRACE products [31–33]. In this study, the average of these GRACE Datasets was used. Averaged data from these datasets can be used to increase accuracy [34]. We used linear regression analysis to bridge missing separate months and the 11-month gap between GRACE and GRACE-FO.

2.4. Global Land Data Assimilation System (GLDAS)

To produce the best fields of land surface states and fluxes, the GLDAS product uses cutting-edge land surface modeling and data assimilation techniques [35]. As recommended in many previous studies [36–38], we selected the GLDAS-2.1 Noah land surface model with monthly observations ($0.25^\circ \times 0.25^\circ$ spatial grids) to estimate the GWS water for GRACE and GPS (https://disc.gsfc.nasa.gov/datasets/GLDAS_NOAH025_M_2.1/summary?keywords=GLDAS_NOAH025_M_2.1) (accessed on 21 January 2023). The GRACE- and GPS-derived TWS was subtracted from the sum of the four layers of soil moisture provided by the GLDAS Noah 2.1 at depths of 0–2 m to produce the baseline GWS [39,40]. Additionally, we utilized GLDAS data for the observations of *ET* [41]. To evaluate the evaporation throughout the basin and how it affects variation in GWS, *ET* data were used.

2.5. WaterGAP Global Hydrology Model (WGHM)

The updated WGHM 2.2, which was created at the University of Frankfurt and whose water changes are available from January 1901 to December 2019, was used to model groundwater mass changes [42]. There were no additional data available at the time of research, and the WGHM model is not routinely updated. WGHM has been used to offer information on water exchange between ground and surface waters and includes a number of hydrological components. One of the few models that includes groundwater mass fluctuations as a separate element is the WGHM. We used groundwater changes defined on a grid of $0.5^\circ \times 0.5^\circ$ for the period between January 2013 and December 2019.

2.6. Precipitation and Air Temperature

A significant meteorological factor in water circulation, Precipitation (*Pr*), is closely linked to instances of flooding and drought. To examine its effect on GWS variation, we took into account *Pr* products from the Global Precipitation Climatology Project (GPCP) [43,44]. The monthly *Pr* datasets were downloaded from (https://psl.noaa.gov/data/gridded/data.gpcp.html) (accessed on 15 December 2022). In addition, for temperature data, we used GHCN-CAMS data, which combines information from the Global Historical Climate Network (GHCN) and the Climate Anomaly Monitoring System (CAMS), to estimate the surface air temperature (Tm) [45,46]. The global land monthly mean surface air temperature in the GHCN-CAMS dataset has a spatial resolution of $0.5° \times 0.5°$, with a total covered area of 89.75 S–89.75 N and 0.25 E–359.75 E [47,48]. The GHCN-CAMS dataset can be download from https://psl.noaa.gov/data/gridded/data.ghcncams.html (accessed on 15 December 2022).

3. Methods

3.1. GWS Estimation

GRACE observes TWS being a sum of individual water components stored above and below water surface, as follows:

$$\Delta TWS = \Delta GWS + \Delta SM + \Delta SWS + \Delta CWS + \Delta SWE \quad (1)$$

where ΔTWS shows changes in water storage, ΔGWS represents changes in groundwater, ΔSMS represents changes in soil moisture storage, ΔSWS represents changes in surface water, ΔCWS represents changes in the canopy water, and ΔSWE represents changes in snow water equivalent.

The lack of significant gravimetric fluctuation in surface water suggests that the surface water storage component observed in the GRACE TWS data is likely negligible for semi-arid regions like the MKA. Additionally, owing to the region's warm climate throughout all of Africa, snow is uncommon in this area. Water in the canopy has a very low value and can be disregarded [10]. The following components of TWS can be disregarded to estimate GWS:

$$\Delta GWS = \Delta TWS - \Delta SM \quad (2)$$

GRACE-derived TWS, calculated using the average of three centers (JPL, CSR, and GSFC). GWS calculations were carried out using GRACE ensuring fairness by subtracting GLDAS-SM from TWS. Spatially resolved monthly outputs from the GLDAS-2.1 Noah model were employed in the estimation, with the top 4 layers used to compute soil moisture (SM) through GLDAS. Viz., 0–10 cm, 10–40 cm, 40–100 cm and 100–200 cm (Equation (3)). The summation of these four layers is the total SMS. All of these layers are accounted for.

$$SMS_{total} = SMS_{0-10cm} + SMS_{10-40cm} + SMS_{40-100cm} + SMS_{100-200cm} \quad (3)$$

After this, the soil moisture anomaly is calculated. It is removed from the TWS. Therefore, the groundwater change can be determined.

Alternatively, in order to estimate GPS-derived GWS we used the vertical component of 65 GPS stations and followed this order:

(1) We used the TSAnalyzer software to eliminate outliers and offsets from the raw GPS time series (https://github.com/wudingcheng/TSAnalyzer/, accessed on 20 January 2023).
(2) To ascertain vertical crustal deformation tied to seasonal TWS fluctuations, non-tidal atmospheric loading (NTAL) and non-tidal ocean loading (NTOL) influences were averaged into daily results and removed using the product repository by Deutsches GeoForschungsZentrum (GFZ) (http://esmdata.gfz-potsdam.de:8080/repository, accessed on 20 January 2023) [49].
(3) The preceding two steps prepared the GPS data for inversion through a circular disk and the Green function, converting it into TWS according to the mass loading inversion method illustrated in the next section.
(4) After obtaining the GPS-derived TWS, the soil moisture anomaly was also calculated from GLDAS-2.1 Noah model as was done with the GRACE data (Equation (3)). Then the GPS-derived GWS can be estimated by removing soil moisture anomaly from the GPS-derived TWS (Equation (2)).

3.2. Mass Loading Inversion Method

The solid Earth deforms as a result of the external mass transformation of the ocean, atmosphere, and terrestrial fluid systems. The elastic Earth's mass loading theory emphasizes the constitutive relationship between crustal deformation and spatially dispersed loading sources. It is possible to calculate the elastic response of the Earth's crust to any given surface load distribution by spatially using an appropriate Green's function with a surface mass loading field. The Green's functions describe the deformation of the Earth under the influence of a unit surface point mass through the three-dimensional load Love numbers for a spherical nonrotating elastic layered Earth model [50,51]. High-spatial resolution distributed surface mass loads can be resolved thanks to the very sensitive relationship between vertical elastic deformation and the distance from the center of the loading source. An accurate indication of TWS can be obtained by inverting the vertical elastic deformation [52]. In this study, we utilized seasonal vertical displacements obtained from GPS data to perform an inversion for TWS variations across a grid of $0.25° \times 0.25°$. A scheme that considers the area of each grid cell was applied, taking into account the latitude-dependent variation in the radius of circular disks. The study area spans from 21° to 31° E longitude and from −33.5° to −25.5° S latitude. Additionally, a 3.0° extension was included on all sides of the study area to minimize artificial effects near the edges. A surface load field and load Green's functions can be convolved to simulate TWS, which can excite elastic displacements of the earth's crust, as follows:

$$u = Gx \tag{4}$$

where u is the GPS-measured vertical displacements at scattered stations, x is the gridded mass loads expressed as TWS, G is the Green's function matrix calculated using load Love numbers (LLNs) of a specified earth model, e.g., preliminary reference earth model (PREM) [53]. Our inversion model was built using the center of figure (CF) frame in order to be consistent with the geocentric reference frame of GPS data. Underdetermined EWH parameters inevitably lead to ill-posed problems, as with other geophysical inversions, and the following term is typically minimized using a regularization technique, such as the Tikhonov regularization [21,26]:

$$\|(Gx - u)\|^2 + \alpha^2\|Lx\|^2 = \min \tag{5}$$

where L is the Laplacian smoothing matrix, α is the smoothing factor controlling the relative weight between model roughness and data misfit, and one method to determine it is cross-validation [54].

Here, using a technique from Jiang et al., 2022 [14], we inverted the vertical component of the GPS time series to TWS. The above-mentioned authors developed the GNSS2TWS, an open-source MATLAB tool (http://geodesy.noaa.gov/gps-toolbox, accessed on 16 December 2022) for estimating daily TWS changes. In order to achieve the time-varying inversion of vertical position time series in a densely instrumented GPS network, this software uses the PCA dimensionality-reduction technology. The steps that Jiang et al., 2022 [14] used to design their inversion model are shown in Figure 4. Firstly, several so-called Principal Components are derived from the displacement data. Secondly, for a corresponding principal TWS component, each principal component is individually inverted. Thirdly, a linear combination of the above-derived TWS models produces the final water estimates. In contrast to epoch-by-epoch estimates, which would perform inversions frequently, the model only performs the inversion a few times.

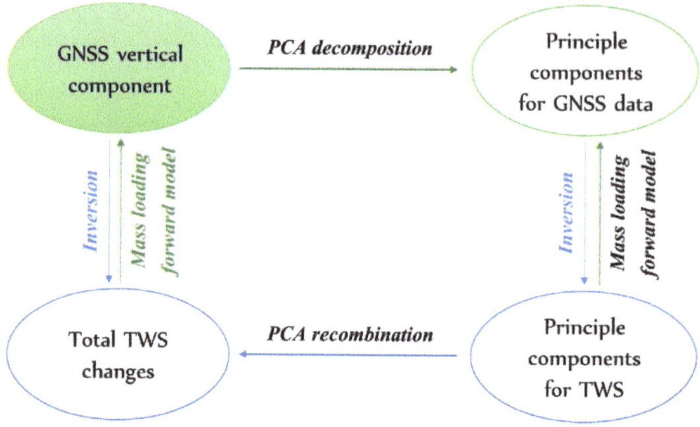

Figure 4. A representation of the inversion technique as described in [20]. The left-hand track is used in conventional inversion to directly invert displacement data into a TWS model.

3.3. Principle Component Analysis (PCA)

GRACE-derived GWS and GPS-derived GWS results underwent PCA to facilitate a comparison of the GWS outcomes. For very high dimensional data, PCA is an unsupervised linear dimensionality reduction and data visualization technique. It is very computationally intensive because adding insights from high dimensional data is very difficult. This technique's primary goal is to reduce the dimensionality of highly correlated data by converting the initial set of vectors into a new set known as Principal Components [55]. For dimensionality reduction and feature extraction in GRACE and GPS time series analyses, the PCA method is frequently used [56–58]. The GWS datasets used in this study were analyzed using the PCA method to summarize spatiotemporal variations in GWS. The Principal Components (PCs) of a given dataset are mathematically determined by the eigenvalues and eigenvectors of a covariance matrix. This technique assisted in identifying empirical orthogonal functions (EOFs) and Principal Components (i.e., temporal variations) (i.e., spatial maps). In order to decompose variations GWS, the following equation was used:

$$Z(s,t) = \sum_{k=1}^{M} c_k(t).u_k(s) \tag{6}$$

where the (s,t) vector denotes the space-time location at time (t) and spatial position (S), M is the number of modes in orthogonal space-time random fields, i.e., $c_k(t)u_k(s)$. The modes are formulated as an optimal set of orthogonal spatial functions $u_k(s)$. In the period of time that the eigenvector describes, the standardized score is a portion of the overall variation proportional to the overall covariance (EOF). The standard deviation of the corresponding Principal Components was used to normalize the EOFs. For instance, the EOF/PC pairs are known as PCA modes while the EOF represents the spatial distribution GWS. In order to identify the predominant patterns of GWS in the MKA, PCA was used in our study to statistically decompose GRACE into PCs (temporal) and EOFs (spatial).

The process steps of our study are outlined in the flowchart presented in Figure 5. From top to bottom, GPS time series were downloaded at the beginning, and after outliers and offsets were removed; non-tidal ocean and atmospheric effects were also taken into account. Finally, the GPS time series were inverted to produce TWS. We download GRACE data from three centers on the same level, and then we estimated GWS by deducting TWS (from GRACE and GPS) from SM. Then, we compared them using PCA and the effect of the climate.

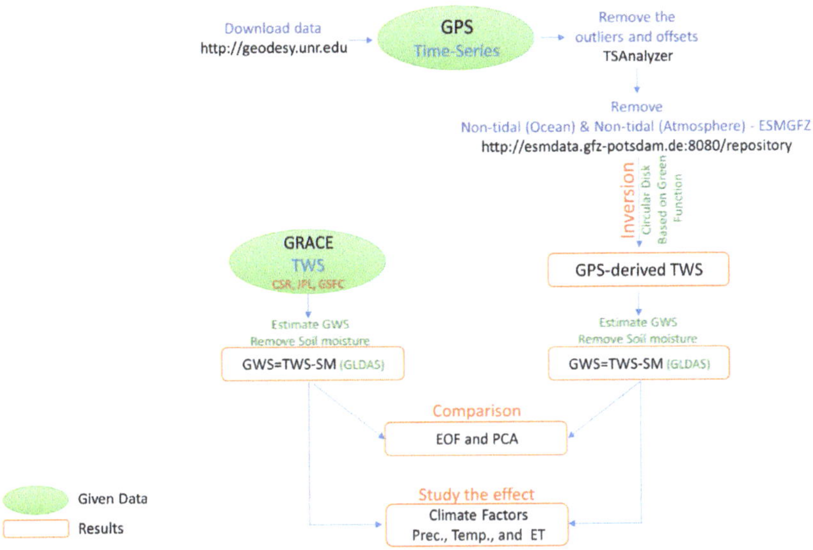

Figure 5. The flowchart visually outlines the methodology implemented for this study (authors).

4. Results

4.1. GPS Inversion and Model Resolution Test

Figure 6 illustrates the temporal functions and spatial weights linked to these two Principal Components (PCs) attained via VBPCA. The initial principal component (PC1) (depicted in Figure 6a) accounts for approximately 99.68% of the filtered data's variance and underscores the primary spatiotemporal characteristic of the annual water cycle in the Main Karoo Aquifer (MKA). This component showcases surface subsidence during the fall and winter periods due to heightened water loads, and ground uplift during spring and summer as water dissipates and soil moisture evaporates.

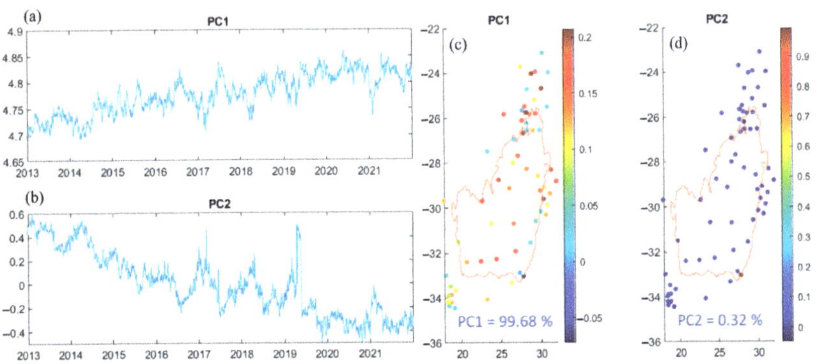

Figure 6. Temporal and spatial pattern PCs of vertical GPS component. (**a**) first PC; (**b**) second PC; (**c**) first spatial PC (EOF); (**d**) second spatial PC (EOF) (authors calculations).

PC1 displays a positive spatial response in approximately 98% of GPS stations, and the seasonal shift in water storage attains its maximum in the northeastern and southern parts of MKA, diminishing toward the southeast and southwest regions (as shown in Figure 6c). On the other hand, the second PC (PC2) (depicted in Figure 6b) accounts for only 0.32% of the reconstructed data's variance, significantly less than the contribution of PC1. All

GPS stations linked to PC2 demonstrate a positive spatial response (depicted in Figure 6d). Notably, the second PC unveils a discernible interannual alteration, possibly linked to prolonged hydrological extreme events such as droughts and floods.

The resolution of our inversion results significantly relies on the extent of spatial coverage provided by continuous GPS stations. To assess the sensitivity and reliability of our inversion approach, we conducted a synthetic test, utilizing forward vertical surface displacements derived from synthetic TWS data to recover the original TWS thickness (shown in Figure 7). The outcomes indicate that our inversion model successfully recovers approximately 65% to 75% of the input water thickness in regions with dense GPS stations (e.g., northeast, east, southwest, and south of the aquifer). However, the recovery rate is less than 50% in the western and northwestern areas of the aquifer (as depicted in Figures 6c,d and 7b). Therefore, introducing additional continuous GPS stations in the northwest of MKA is recommended to enhance spatial resolution in future studies. On average, the distance between GPS sites in MKA is approximately 93 km, which allows us to capture the primary hydrological features using the VBPCA-based inversion and achieve higher spatial resolution in depicting water changes compared to the data obtained from GRACE and GLDAS water products.

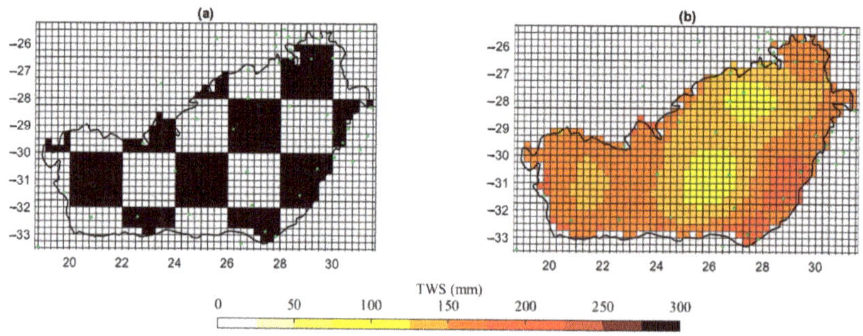

Figure 7. A synthetic experiment to assess the inversion of GPS-derived TWS. (**a**) Represents the input water heights used to estimate the forward vertical displacements at various GPS stations. (**b**) Displays the results of the inversion process using the forward vertical displacements. The spatial distribution of GPS stations is represented by green circles (authors calculations).

Previous analysis has revealed the TWS inversion and its spatial resolution in the MKA region. Therefore, our study delves into an investigation of the annual oscillations in water storage trends inferred from GPS vertical positions and compares them with water products derived from GRACE Mascon solutions. Figure 8a illustrates the trend in TWS as determined by the three GRACE mascon solutions, observed over the period between 2013 and 2021. On the other hand, Figure 8b displays the TWS trend based on GPS vertical observations. By comparing these two datasets, we aimed to gain insights into the concurrence and disparities between the GPS-derived water storage oscillations and those obtained from the GRACE Mascon solutions during the specified time frame. The study found that the spatial distribution of TWS derived from the three GRACE mascon solutions and GPS observations exhibited consistent variations. The TWS trends obtained from GPS and GRACE were -0.04 ± 0.003 mm/year and -0.30 ± 0.03 mm/year, respectively. These trends indicate a gradual decrease in water storage over time, with GPS showing a smaller rate of change compared to GRACE. Because of variations in signal sensitivity, with GRACE capturing large-scale long-wave signals and GPS focusing on regional short-wave signals, disparities arise between the TWS spatial distribution derived from GPS and GRACE (Figure 8). Nonetheless, basin-scale averaged linear trend of TWS from both GPS (-0.04 ± 0.003 mm/year) and GRACE (-0.30 ± 0.03 mm/year) affirm the consistent deficit of water storage in the MKA.

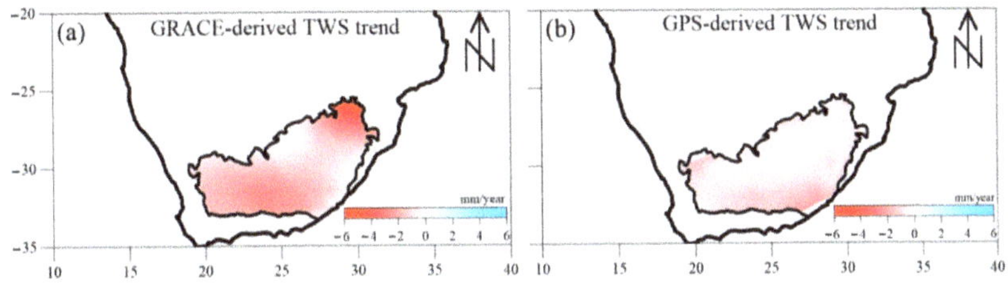

Figure 8. The GRACE and GPS-derived TWS trend. (**a**) Using the average of the three processing centers (CSR, JPL, and GSFC), recent GRACE-derived TWS, and (**b**) GPS-derived TWS after inversion technique on the GPS time series (authors calculations).

4.2. Groundwater Storage (GWS)

The MKA's drought is shown using the Global Precipitation Climatology Centre's drought index (GPCC DI), which uses a 1.0° grid resolution and an average monthly period. According to Pietzsch and Bissolli, 2011 [50], GPCC DI is a combination of the Standardized Precipitation Index and the Standardized Precipitation Evapotranspiration Index (SPEI). The average monthly drought index from 2013 to 2021 is shown in Figure 9 (9 years). The drought and wet categories are shown in Figure 9 for all of Africa. Figure 9 indicates that there is a mild drought in the south-east region and a mild wet climate throughout almost the entire aquifer [59]. This mild drought indicates that there is a possibility of GWS recharge and the aquifer does not lie in a region that is completely dry.

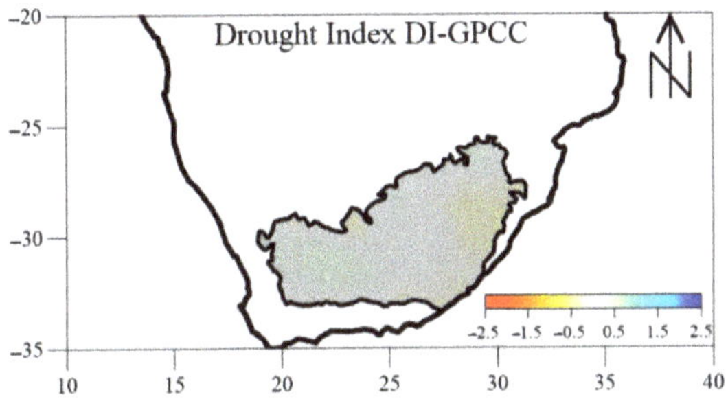

Figure 9. The average monthly drought index for the time frame between January 2013 and December 2021, using monthly GPCC data. The drought is depicted in colors ranging from yellow to red, with the severity of the drought increasing from mild (0 to −1) to moderate (up to −1.5), severe (−1.5 to −2), and extremely severe (below −2). Conversely, the wet is represented by a range of blue hues, with the level of wetness increasing from mild (0 to 1) to moderate (up to 1.5), severe (1.5 to 2), and extremely severe (above 2) (authors calculations).

Because of the limited availability of groundwater observations in situ data from Africa, we utilized data from the WaterGAP Global Hydrology Model (WGHM) as an alternative for comparison with GRACE and GPS data. The results of GWS based on GRACE, GPS, and WGHM are illustrated in Figure 10. The WGHM data covers observations up to December 2019. In Figure 10d, we present the time series of GWS derived from GPS, GRACE, and WGHM, while Figure 10a–c displays the GWS trends derived from GRACE, GPS, and WGHM, respectively.

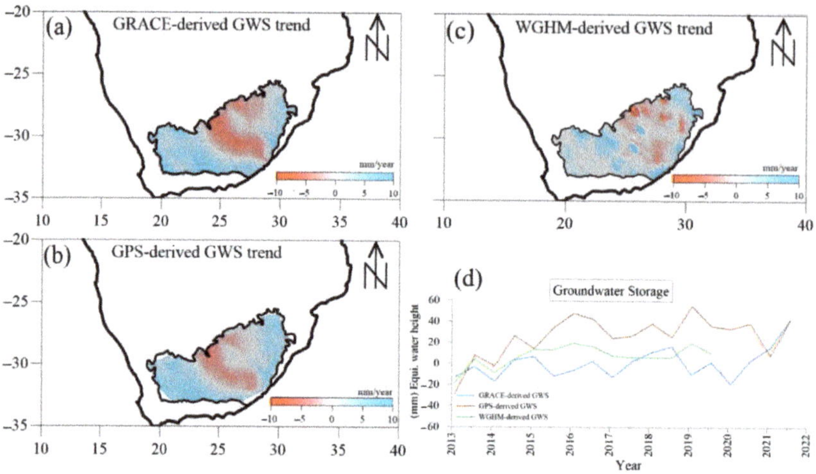

Figure 10. The GWS over the MKA using GRACE and GPS. (**a**) the GWS trend using GRACE, (**b**) the GWS trend using GPS, (**c**) the GWS trend using WGHM, and (**d**) the GWS time series; the orange line represents GPS, the blue line represents GRACE, while the green line represents WGHM-derived GWS (authors calculations).

As shown in Figure 10 and Table 1, comparing the GWS trends for the time period from January 2013 to December 2019, the trend obtained from WGHM is approximately 1.98 ± 0.08 mm/year, whereas the GRACE-derived trend is approximately 1.52 ± 0.11 mm/year, and the GPS-derived trend is approximately 2.37 ± 0.09 mm/year (Table 1). On average, the GWS trend derived from GPS over this period is 2.59 ± 0.12 mm/year, while the GWS trend using GRACE shows 1.41 ± 0.08 mm/year from January 2013 to December 2021 (Table 1). GRACE, GPS, and WGHM results demonstrate spatial domain similarity with strong signals using GRACE's behavior in GWS. The results show a positive trend throughout the entire aquifer, with the exception of the middle part, as determined by GPS, WGHM, and GRACE-derived GWS.

Table 1. The GWS trend over the MKA were estimated using GPS, GRACE and WGHM.

GWS	GWS Trend (mm/Year)	
	2013–2019	**2013–2021**
GPS-derived GWS	2.37 ± 0.09	2.59 ± 0.12
GRACE-derived GWS	1.52 ± 0.11	1.41 ± 0.08
WGHM-derived GWS	1.98 ± 0.08	-

These findings suggest a dynamic and complex groundwater storage system with varying rates of change over time and across different regions. Nevertheless, all three datasets—GRACE, GPS, and WGHM—reveal spatial similarities in GWS behavior, particularly highlighting the strong signals captured by GRACE in monitoring groundwater storage. In summary, our results point to an overall positive trend in GWS across the entire aquifer, with some exceptions in the middle part, consistently identified by GPS, WGHM, and GRACE-derived GWS analyses.

5. Discussion

5.1. Comparison between GPS and GRACE-Derived GWS

5.1.1. EOF Modes

We used PCA to thoroughly compare the GPS and GRACE-derived GWS over the MKA for a 9-year period (i.e., 2013–2021). The PCA and EOF in MKA are shown in

Figure 11. EOF loadings (or spatial patterns) are relatively strong in the first principal component, accounting for 71.83% of the cumulative variability in GPS-derived GWS and 68.92% in GRACE-derived GWS, and are primarily localized over the left part of the aquifer, while the negative EOF localized in the right part (at ≈26° lon) (Figure 11c,d and Table 2).

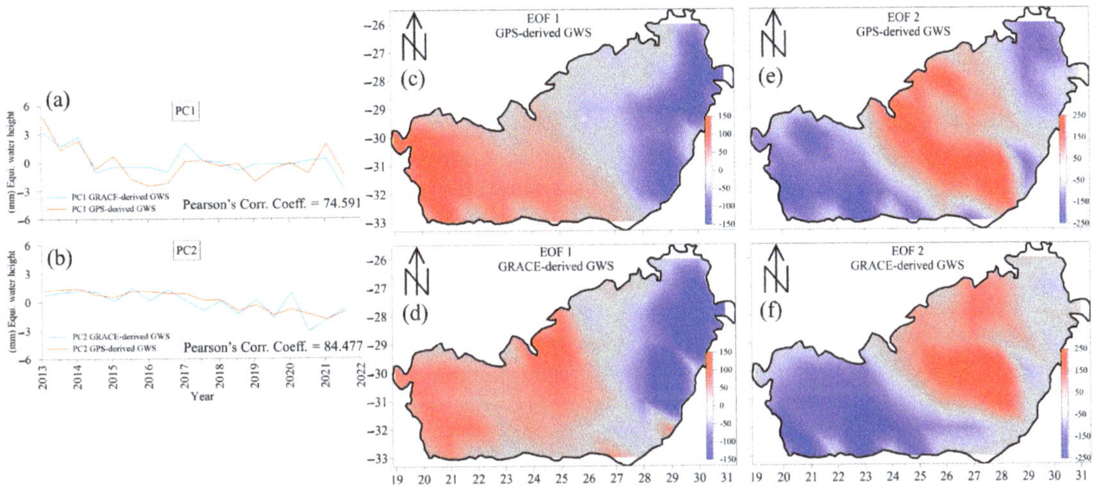

Figure 11. The GWS's PCA and EOF modes were estimated using GPS and GRACE. The first and second PCs are represented at the left by the characters (**a**,**b**). Whereas this orange line represents GPS, the blue line shows the GRACE. The first EOF mode from GPS and GRACE, respectively, is shown in (**c**,**d**) on the right side. However, the second EOF mode from the same tools is represented by (**e**,**f**). While the gradient of red is positive values, the gradient of blue is negative values (authors' calculations).

Table 2. The GWS's EOF modes ratio were estimated using GPS and GRACE.

GWS	EOF-1 (%)	EOF-2 (%)
GPS-derived GWS	71.83	28.17
GRACE-derived GWS	68.92	31.08

However, in the second EOF (EOF-2), they are more concentrated in the middle of the aquifer (between about 24° and 29° long), are relatively strong, and account for 28.17% of the cumulative variability in GPS-derived GWS and 31.08% of the GWS derived from GRACE. The only difference between them in the southern region is that GPS has a much stronger positive sign than GRACE (Figure 11e,f; and Table 2). The EOF-1 and EOF-2 both demonstrate consistency between GRACE and GPS, thereby approving GPS's replacement of GRACE in the hydrology application with high accuracy.

Furthermore, the time series generated through PCA (Figure 11a,b) illustrate a high degree of consistency between GPS and GRACE-derived GWS data. To quantify this consistency, we employed Pearson's correlation coefficient, which yielded remarkable agreement between the two datasets. Specifically, PC-1 exhibited a correlation coefficient of 0.74, while PC-2 demonstrated an even higher correlation coefficient of 0.84.

These results collectively underscore the robustness of GPS as a viable alternative to GRACE for hydrological applications, as it demonstrates high accuracy and consistency in capturing GWS variability over the MKA. This finding holds significant implications for the field of hydrology, particularly in regions where GRACE data may not be available or as a means to cross-validate the information obtained through different observation methods.

5.1.2. Climate

In our study, we conducted an analysis of the interplay between climate factors and Groundwater Storage (GWS) estimated through GPS and GRACE data. We examined the behavior of these three key climate factors: monthly mean precipitation (*Pr*), evapotranspiration (*ET*), and temperature (*Tm*), in relation to the GWS fluctuations over time in the MKA. Our findings, as depicted in Figure 12, offer valuable insights into these dynamics.

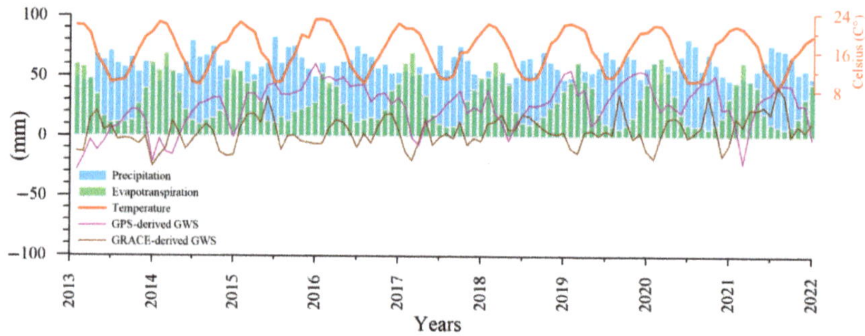

Figure 12. The average climate factors in MKA. There are two bars in the figure, green for *ET* and blue for *Pr*. There are three time series: an orange *Tm* time series, a purple time series derived from GWS using GPS, and a brown time series derived from GRACE (authors calculations).

The temporal patterns displayed in Figure 12 reveal intriguing relationships. Specifically, during the months of December and the initial three months of each year, when the temperatures (*Tm*) are relatively high, it becomes apparent that the amount of evapotranspiration (*ET*) roughly equals the amount of precipitation (*Pr*). This balance between *ET* and *Pr* suggests that, during these periods, the basin experiences a state of equilibrium where the water input from precipitation is effectively offset by the loss through evapotranspiration. Notably, our study demonstrates the advantage of temporal resolution, particularly when comparing GPS-derived GWS and GRACE-derived GWS data. GPS-derived GWS exhibits a more rapid response to changes in climate conditions compared to GRACE-derived GWS. This heightened temporal sensitivity of GPS data enables it to capture GWS variations with greater precision and agility.

Importantly, both GWS estimates, whether derived from GPS or GRACE, align consistently with the fluctuations in the three climate factors (*Pr*, *ET*, and *Tm*) at a monthly resolution. This alignment reinforces the reliability of GWS data and its responsiveness to climatic variations. Furthermore, in the middle of the year, when temperatures are relatively low and precipitation is abundant, we observe that the groundwater system is replenished by precipitation (*Pr*). This observation highlights the critical role of seasonal variations in climate factors in influencing groundwater dynamics. Our analysis underscores the intricate interplay between climate factors and groundwater storage in the MKA. It highlights the potential of GPS-derived GWS data to provide high-temporal-resolution insights and reaffirms the sensitivity of GWS to climatic fluctuations, which is particularly crucial for understanding water resource management in this region.

Because daily GPS observations can capture instantaneous changes in climate, GPS is necessary when researching precise hydrological applications.

5.1.3. Groundwater Storage and ENSO

In this research, we examined the impact of El Niño and La Niña on Groundwater Storage (GWS) in the Main Karoo Aquifer. Figure 13 portrays a graph illustrating GWS alongside a bar chart representing the ENSO index spanning from January 2013 to December 2021.

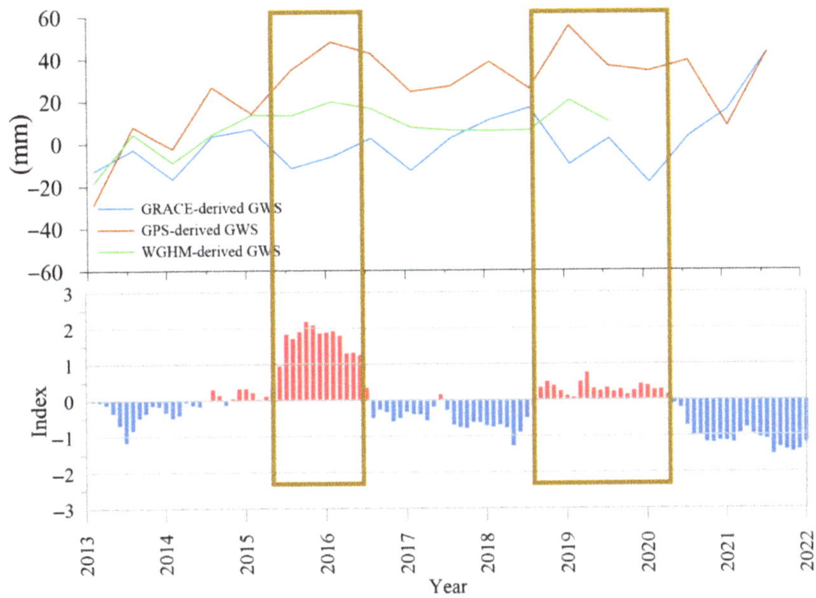

Figure 13. A time series graph showing GWS and the bar chart of the ENSO index. The top section of the graph illustrates different GWS types (GRACE, GPS, GWS derived from WGHM), yellow box shows El Niño events, while the lower section shows the ENSO index (authors calculations).

The graph demonstrates notable regions with strong coherence, closely aligning with El Niño events (2015–2016) and La Niña occurrences (2016–2018 and 2020–2022). A recent investigation by Kolusu et al., 2019 [51] emphasizes the significance of factoring El Niño events into strategies for groundwater management, particularly in regions with water scarcity like East and Southern Africa. Their study established a robust correlation between El Niño-driven climate anomalies and their repercussions on groundwater resources, revealing that El Niño leads to drought conditions while La Niña induces flooding, aligning with previous findings by Siam and Eltahir, 2017 [52] who highlighted the importance of considering El Niño impacts in groundwater management for the Nile basin. As groundwater recharge primarily occurs through rainfall infiltration and interactions with surface water, El Niño (La Niña) years are likely to lead to reduced (increased) groundwater storage. In our study, the ENSO bar chart mirrors the fluctuations observed in the GWS time series, as indicated in the paired windows in Figure 13. This correspondence implies a direct influence of GWS fluctuations on the Main Karoo Aquifer's groundwater storage.

6. Conclusions

In this study, we estimated the GWS in the MKA using GPS time series data and triple GRACE mascon data (CSR, JPL, GSFC). We performed a number of operations on the GPS stations in order to convert them to TWS. Firstly, we removed outliers and offsets. Next, we eliminated the non-tidal ocean effect and non-tidal effect. Finally, we estimated TWS by performing an inversion on the GPS time series data. We combined the soil moisture data from GLDAS with the TWS data from GRACE and GPS to better understand the nature of the groundwater variations between January 2013 and December 2021 (9 years). Strong GRACE signals produced the same behavior for GPS and GRACE-derived GWS and TWS. Positive trends are seen by both GPS and GRACE. While GPS-derived GWS trend is 2.59 ± 0.12 mm/year, GRACE GWS trend is 1.41 ± 0.08 mm/year. The results seen from GRACE, GPS, and WGHM demonstrate a spatial domain similarity, with strong signals resembling the behavior of GRACE's GWS. Additionally, for both GPS and GRACE-derived

GWS, we estimated PCA and EOF. They appear to be quite close. The first and second EOF modes display relatively typical behavior between GRACE and GPS. With a contribution of 71.83% to GPS-derived GWS variability and 68.92% to GRACE-derived GWS variability, EOF-1 is a relatively potent factor. For EOF-2, it was 28.17% and 31.08%. The same GRACE work can be performed by GPS with higher temporal resolution by using GPS inversion (daily vs. monthly as with GRACE). In addition, we calculated climate factors that affect MKA groundwater (*Pr*, *Tm*, and *ET*). Because daily GPS observations can capture an instantaneous change in climate, we strongly advise using GPS for research into precise hydrological applications. The lack of routine distribution of GPS stations and continuous publication of their data is a drawback of using GPS in hydrological applications. To help us better understand the hydrological applications, we also advise increasing the number of GPS stations and offering continuous observations.

Author Contributions: Conceptualization, H.A.M.; Methodology, H.A.M.; Software, H.A.M., J.J. and Q.W.; Formal analysis, H.A.M.; Investigation, H.A.M.; Data curation, H.A.M.; Writing—original draft, H.A.M.; Writing—review & editing, W.S. and J.J.; Visualization, J.J.; Supervision, W.S.; Project administration, W.S.; Funding acquisition, W.S. All authors have read and agreed to the published version of the manuscript.

Funding: This research was funded by National Natural Science Foundations of China (No. 41721003, 42030105, 41874023), and Fundamental Research Funds for the Central Universities. In addition, the Open Fund of Wuhan, Gravitation and Solid Earth Tides, National Observation and Research Station (WHYWZ202202).

Data Availability Statement: The GRACE data used in this study are publicly available from the NASA Earth Observing System Data and Information System (EOSDIS) Distributed Active Archive Center (DAAC) at the Jet Propulsion Laboratory (JPL) (https://grace.jpl.nasa.gov/, accessed on 20 January 2023). The data used in this study are from the GARCE and GRACE-FO missions. Details on the data processing and quality control can be found on the JPL GARCE and GRACE-FO websites. In addition, GPS data presented in this study can be accessed from the Nevada Geodetic Laboratory at (http://geodesy.unr.edu/magnet.php, accessed on 20 January 2023).

Acknowledgments: The authors would like to thank the Supercomputing Center of Wuhan University for providing the supercomputing system in numerical calculations.

Conflicts of Interest: The authors declare no conflict of interest.

References

1. El Kharraz, J.; El-Sadek, A.; Ghaffour, N.; Mino, E. Water scarcity and drought in WANA countries. *Procedia Eng.* **2012**, *33*, 14–29. [CrossRef]
2. Chaudhari, S.; Pokhrel, Y.; Moran, E.; Miguez-Macho, G. Multi-decadal hydrologic change and variability in the Amazon River basin: Understanding terrestrial water storage variations and drought characteristics. *Hydrol. Earth Syst. Sci.* **2019**, *23*, 2841–2862. [CrossRef]
3. Syed, T.H.; Famiglietti, J.S.; Rodell, M.; Chen, J.; Wilson, C.R. Analysis of terrestrial water storage changes from GRACE and GLDAS. *Water Resour. Res.* **2008**, *44*, W02433. [CrossRef]
4. Baalousha, H. Assessment of a groundwater quality monitoring network using vulnerability mapping and geostatistics: A case study from Heretaunga Plains, New Zealand. *Agric. Water Manag.* **2010**, *97*, 240–246. [CrossRef]
5. Cázares Escareño, J.; Júnez-Ferreira, H.E.; González-Trinidad, J.; Bautista-Capetillo, C.; Robles Rovelo, C.O. Design of Groundwater Level Monitoring Networks for Maximum Data Acquisition at Minimum Travel Cost. *Water* **2022**, *14*, 1209. [CrossRef]
6. Osman, A.I.A.; Ahmed, A.N.; Chow, M.F.; Huang, Y.F.; El-Shafie, A. Extreme gradient boosting (Xgboost) model to predict the groundwater levels in Selangor Malaysia. *Ain Shams Eng. J.* **2021**, *12*, 1545–1556. [CrossRef]
7. Osman, A.I.A.; Ahmed, A.N.; Huang, Y.F.; Kumar, P.; Birima, A.H.; Sherif, M.; Sefelnasr, A.; Ebraheemand, A.A.; El-Shafie, A. Past, Present and Perspective Methodology for Groundwater Modeling-Based Machine Learning Approaches. *Arch. Comput. Methods Eng.* **2022**, *29*, 3843–3859. [CrossRef]
8. Strassberg, G.; Scanlon, B.R.; Rodell, M. Comparison of seasonal terrestrial water storage variations from GRACE with groundwater-level measurements from the High Plains Aquifer (USA). *Geophys. Res. Lett.* **2007**, *34*, L14402. [CrossRef]
9. Xie, X.; Xu, C.; Wen, Y.; Li, W. Monitoring groundwater storage changes in the Loess Plateau using GRACE satellite gravity data, hydrological models and coal mining data. *Remote Sens.* **2018**, *10*, 605. [CrossRef]
10. Barbosa, S.A.; Pulla, S.T.; Williams, G.P.; Jones, N.L.; Mamane, B.; Sanchez, J.L. Evaluating Groundwater Storage Change and Recharge Using GRACE Data: A Case Study of Aquifers in Niger, West Africa. *Remote Sens.* **2022**, *14*, 1532. [CrossRef]

11. Jin, S.; Feng, G. Large-scale variations of global groundwater from satellite gravimetry and hydrological models, 2002–2012. *Glob. Planet. Chang.* **2013**, *106*, 20–30. [CrossRef]
12. Croteau, M.; Nerem, R.; Loomis, B.; Sabaka, T. Development of a daily GRACE mascon solution for terrestrial water storage. *J. Geophys. Res. Solid Earth* **2020**, *125*, e2019JB018468. [CrossRef]
13. Wang, L.; Chen, C.; Du, J.; Wang, T. Detecting seasonal and long-term vertical displacement in the North China Plain using GRACE and GPS. *Hydrol. Earth Syst. Sci.* **2017**, *21*, 2905–2922. [CrossRef]
14. Lenczuk, A.; Klos, A.; Bogusz, J. Studying spatio-temporal patterns of vertical displacements caused by groundwater mass changes observed with GPS. *Remote Sens. Environ.* **2023**, *292*, 113597. [CrossRef]
15. Bagheri-Gavkosh, M.; Hosseini, S.M.; Ataie-Ashtiani, B.; Sohani, Y.; Ebrahimian, H.; Morovat, F.; Ashrafi, S. Land subsidence: A global challenge. *Sci. Total Environ.* **2021**, *778*, 146193. [CrossRef]
16. Xi, R.; Liang, Y.; Chen, Q.; Jiang, W.; Chen, Y.; Liu, S. Analysis of annual deformation characteristics of Xilongchi dam using historical GPS observations. *Remote Sens.* **2022**, *14*, 4018. [CrossRef]
17. Tan, W.; Dong, D.; Chen, J.; Wu, B. Analysis of systematic differences from GPS-measured and GRACE-modeled deformation in Central Valley, California. *Adv. Space Res.* **2016**, *57*, 19–29. [CrossRef]
18. Liu, R.; Zou, R.; Li, J.; Zhang, C.; Zhao, B.; Zhang, Y. Vertical displacements driven by groundwater storage changes in the North China Plain detected by GPS observations. *Remote Sens.* **2018**, *10*, 259. [CrossRef]
19. Larochelle, S.; Chanard, K.; Fleitout, L.; Fortin, J.; Gualandi, A.; Longuevergne, L.; Rebischung, P.; Violette, S.; Avouac, J.p. Understanding the geodetic signature of large aquifer systems: Example of the Ozark Plateaus in central United States. *J. Geophys. Res. Solid Earth* **2022**, *127*, e2021JB023097. [CrossRef]
20. Jiang, Z.; Hsu, Y.-J.; Yuan, L.; Feng, W.; Yang, X.; Tang, M. GNSS2TWS: An open-source MATLAB-based tool for inferring daily terrestrial water storage changes using GNSS vertical data. *GPS Solut.* **2022**, *26*, 1–7. [CrossRef]
21. Argus, D.F.; Fu, Y.; Landerer, F.W. Seasonal variation in total water storage in California inferred from GPS observations of vertical land motion. *Geophys. Res. Lett.* **2014**, *41*, 1971–1980. [CrossRef]
22. Knappe, E.; Bendick, R.; Martens, H.; Argus, D.; Gardner, W. Downscaling vertical GPS observations to derive watershed-scale hydrologic loading in the northern Rockies. *Water Resour. Res.* **2019**, *55*, 391–401. [CrossRef]
23. De Vries, J.; Von Hoyer, M. Groundwater recharge studies in semi-arid Botswana—A review. In *Estimation of Natural Groundwater Recharge*; NATO ASI Series; Springer: Dordrecht, The Netherlands, 1988; Volume 222, pp. 339–347. [CrossRef]
24. Fu, Y.; Argus, D.F.; Landerer, F.W. GPS as an independent measurement to estimate terrestrial water storage variations in Washington and Oregon. *J. Geophys. Res. Solid Earth* **2015**, *120*, 552–566. [CrossRef]
25. Chew, C.C.; Small, E.E. Terrestrial water storage response to the 2012 drought estimated from GPS vertical position anomalies. *Geophys. Res. Lett.* **2014**, *41*, 6145–6151. [CrossRef]
26. Jiang, Z.; Hsu, Y.-J.; Yuan, L.; Huang, D. Monitoring time-varying terrestrial water storage changes using daily GNSS measurements in Yunnan, southwest China. *Remote Sens. Environ.* **2021**, *254*, 112249. [CrossRef]
27. Li, W.; Dong, J.; Wang, W.; Wen, H.; Liu, H.; Guo, Q.; Yao, G.; Zhang, C. Regional Crustal Vertical Deformation Driven by Terrestrial Water Load Depending on CORS Network and Environmental Loading Data: A Case Study of Southeast Zhejiang. *Sensors* **2021**, *21*, 7699. [CrossRef]
28. Klos, A.; Dobslaw, H.; Dill, R.; Bogusz, J. Identifying the sensitivity of GPS to non-tidal loadings at various time resolutions: Examining vertical displacements from continental Eurasia. *GPS Solut.* **2021**, *25*, 89. [CrossRef]
29. Gobron, K.; Rebischung, P.; Van Camp, M.; Demoulin, A.; de Viron, O. Influence of Aperiodic Non-Tidal Atmospheric and Oceanic Loading Deformations on the Stochastic Properties of Global GNSS Vertical Land Motion Time Series. *J. Geophys. Res. Solid Earth* **2021**, *126*, e2021JB022370. [CrossRef]
30. Chanard, K.; Fleitout, L.; Calais, E.; Rebischung, P.; Avouac, J.P. Toward a global horizontal and vertical elastic load deformation model derived from GRACE and GNSS station position time series. *J. Geophys. Res. Solid Earth* **2018**, *123*, 3225–3237. [CrossRef]
31. Sun, Z.; Long, D.; Yang, W.; Li, X.; Pan, Y. Reconstruction of GRACE data on changes in total water storage over the global land surface and 60 basins. *Water Resour. Res.* **2020**, *56*, e2019WR026250. [CrossRef]
32. Pu, L.; Fan, D.; You, W.; Jiang, Z.; Yang, X.; Wan, X.; Nigatu, Z.M. Analysis of mass flux variations in the southern Tibetan Plateau based on an improved spatial domain filtering approach for GRACE/GRACE-FO solutions. *Int. J. Remote Sens.* **2022**, *43*, 3563–3591. [CrossRef]
33. Scanlon, B.R.; Zhang, Z.; Save, H.; Wiese, D.N.; Landerer, F.W.; Long, D.; Longuevergne, L.; Chen, J. Global evaluation of new GRACE mascon products for hydrologic applications. *Water Resour. Res.* **2016**, *52*, 9412–9429. [CrossRef]
34. Landerer, F.W.; Swenson, S. Accuracy of scaled GRACE terrestrial water storage estimates. *Water Resour. Res.* **2012**, *48*, 1–11. [CrossRef]
35. Rodell, M.; Houser, P.; Jambor, U.; Gottschalck, J.; Mitchell, K.; Meng, C.-J.; Arsenault, K.; Cosgrove, B.; Radakovich, J.; Bosilovich, M. The global land data assimilation system. *Bull. Am. Meteorol. Soc.* **2004**, *85*, 381–394. [CrossRef]
36. Singh, A.K.; Jasrotia, A.S.; Taloor, A.K.; Kotlia, B.S.; Kumar, V.; Roy, S.; Ray, P.K.C.; Singh, K.K.; Singh, A.K.; Sharma, A.K. Estimation of quantitative measures of total water storage variation from GRACE and GLDAS-NOAH satellites using geospatial technology. *Quat. Int.* **2017**, *444*, 191–200. [CrossRef]

37. Ouma, Y.O.; Aballa, D.; Marinda, D.; Tateishi, R.; Hahn, M. Use of GRACE time-variable data and GLDAS-LSM for estimating groundwater storage variability at small basin scales: A case study of the Nzoia River Basin. *Int. J. Remote Sens.* **2015**, *36*, 5707–5736. [CrossRef]
38. Amiri, V.; Ali, S.; Sohrabi, N. Estimating the Spatiotemporal of GRACE/GRACE-FO derived groundwater storage and depletion and validation with in-situ measurements of water level and quality (Yazd Province, Central Iran). *J. Hydrol.* **2023**, *620*, 129416. [CrossRef]
39. Chen, Y.; Yang, K.; Qin, J.; Zhao, L.; Tang, W.; Han, M. Evaluation of AMSR-E retrievals and GLDAS simulations against observations of a soil moisture network on the central Tibetan Plateau. *J. Geophys. Res. Atmos.* **2013**, *118*, 4466–4475. [CrossRef]
40. Yang, S.; Zeng, J.; Fan, W.; Cui, Y. Evaluating Root-zone Soil Moisture Products from GLEAM, GLDAS, and ERA5 based on in Situ ObServations and Triple Collocation Method Over the Tibetan Plateau. *J. Hydrometeorol.* **2022**, *23*, 1861–1878. [CrossRef]
41. Fang, H.; Beaudoing, H.K.; Teng, W.L.; Vollmer, B.E. Global Land data assimilation system (GLDAS) products, services and application from NASA hydrology data and information services center (HDISC). In Proceedings of the ASPRS 2009 Annual Conference, Baltimore, MD, USA, 9–13 March 2009.
42. Döll, P.; Fiedler, K. Global-scale modeling of groundwater recharge. *Hydrol. Earth Syst. Sci.* **2008**, *12*, 863–885. [CrossRef]
43. Efon, E.; Ngongang, R.D.; Meukaleuni, C.; Wandjie, B.; Zebaze, S.; Lenouo, A.; Valipour, M. Monthly, Seasonal, and Annual Variations of Precipitation and Runoff Over West and Central Africa Using Remote Sens. and Climate Reanalysis. *Earth Syst. Environ.* **2022**, *7*, 67–82. [CrossRef]
44. Gu, G.; Adler, R.F. Observed variability and trends in global precipitation during 1979–2020. *Clim. Dyn.* **2022**, *61*, 131–150. [CrossRef]
45. Ropelewski, C.; Janowiak, J.; Halpert, M. *The climate Anomaly Monitoring System (CAMS)*; CAMS: Chennai, India, 1984.
46. Vose, R.S.; Schmoyer, R.L.; Steurer, P.M.; Peterson, T.C.; Heim, R.; Karl, T.R.; Eischeid, J.K. *The Global Historical Climatology Network: Long-Term Monthly Temperature, Precipitation, Sea Level Pressure, and Station Pressure data*; Carbon Dioxide Information; Oak Ridge National Lab.: Oak Ridge, TN, USA, 1992.
47. Fan, Y.; Van den Dool, H. A global monthly land surface air temperature analysis for 1948–present. *J. Geophys. Res. Atmos.* **2008**, *113*, 01103. [CrossRef]
48. Ratnam, J.; Doi, T.; Richter, I.; Oettli, P.; Nonaka, M.; Behera, S. Using Selected Members of a Large Ensemble to Improve Prediction of Surface Air Temperature Anomalies Over Japan in the Winter Months From Mid-Autumn. *Front. Clim.* **2022**, *4*, 919084. [CrossRef]
49. Dill, R.; Dobslaw, H. Numerical simulations of global-scale high-resolution hydrological crustal deformations. *J. Geophys. Res. Solid Earth* **2013**, *118*, 5008–5017. [CrossRef]
50. Farrell, W. Deformation of the Earth by surface loads. *Rev. Geophys.* **1972**, *10*, 761–797. [CrossRef]
51. Longman, I. A Green's function for determining the deformation of the Earth under surface mass loads: 2. Computations and numerical results. *J. Geophys. Res.* **1963**, *68*, 485–496. [CrossRef]
52. Lai, Y.R.; Wang, L.; Bevis, M.; Fok, H.S.; Alanazi, A. Truncated singular value decomposition regularization for estimating terrestrial water storage changes using GPS: A case study over Taiwan. *Remote Sens.* **2020**, *12*, 3861. [CrossRef]
53. Dziewonski, A.M.; Anderson, D.L. Preliminary reference Earth model. *Phys. Earth Planet. Inter.* **1981**, *25*, 297–356. [CrossRef]
54. Matthews, M.V.; Segall, P. Estimation of depth-dependent fault slip from measured surface deformation with application to the 1906 San Francisco earthquake. *J. Geophys. Res. Solid Earth* **1993**, *98*, 12153–12163. [CrossRef]
55. Pan, Y.; Ding, H.; Li, J.; Shum, C.; Mallick, R.; Jiao, J.; Li, M.; Zhang, Y. Transient hydrology-induced elastic deformation and land subsidence in Australia constrained by contemporary geodetic measurements. *Earth Planet. Sci. Lett.* **2022**, *588*, 117556. [CrossRef]
56. Li, J.; Wang, S.; Zhou, F. Time series analysis of long-term terrestrial water storage over Canada from GRACE satellites using principal component analysis. *Can. J. Remote Sens.* **2016**, *42*, 161–170. [CrossRef]
57. Awange, J.L.; Gebremichael, M.; Forootan, E.; Wakbulcho, G.; Anyah, R.; Ferreira, V.G.; Alemayehu, T. Characterization of Ethiopian mega hydrogeological regimes using GRACE, TRMM and GLDAS datasets. *Adv. Water Resour.* **2014**, *74*, 64–78. [CrossRef]
58. He, M.; Shen, W.; Jiao, J.; Pan, Y. The Interannual Fluctuations in Mass Changes and Hydrological Elasticity on the Tibetan Plateau from Geodetic Measurements. *Remote Sens.* **2021**, *13*, 4277. [CrossRef]
59. Lloyd-Hughes, B.; Saunders, M.A. A drought climatology for Europe. *Int. J. Climatol. A J. R. Meteorol. Soc.* **2002**, *22*, 1571–1592. [CrossRef]

water

MDPI

Article

Trends and Drivers of Flood Occurrence in Germany: A Time Series Analysis of Temperature, Precipitation, and River Discharge

Mohannad Alobid [1,*], Fatih Chellai [2] and István Szűcs [1]

1 Faculty of Economic and Business, Institute of Economics, Department of Agricultural Policy and Environmental Economics, University of Debrecen, Böszörményi St. 138, H-4032 Debrecen, Hungary; szucs.istvan@econ.unideb.hu

2 Faculty of Economics, Commerce and Management, Ferhat Abbas University, Sétif 19000, Algeria; fatih.chellai@univ-setif.dz

* Correspondence: mohannadalobid88@gmail.com

Abstract: Floods in Germany have become increasingly frequent and severe over recent decades, with notable events in 2002, 2013, and 2021. This study examines the trends and drivers of flood occurrences in Germany from 1990 to 2024, focusing on the influence of climate-change-related variables, such as temperature, precipitation, and river discharge. Using a comprehensive time series analysis, including Auto-Regressive Integrated Moving Average (ARIMA) and Artificial Neural Network (ANN) models and correlation and regression analyses, we identify significant correlations between these climatic variables and flood events. Our findings indicate that rising temperatures (with a mean of 8.46 °C and a maximum of 9 °C) and increased precipitation (averaging 862.26 mm annually)are strongly associated with higher river discharge (mean 214.6 m^3/s) and more frequent floods (mean 197.94 events per year). The ANN model outperformed the ARIMA model in flood forecasting, showing lower error metrics (e.g., RMSE of 10.86 vs. 18.83). The analysis underscores the critical impact of climate change on flood risks, highlighting the necessity of adaptive flood-management strategies that incorporate the latest climatic and socio-economic data. This research contributes to the understanding of flood dynamics in Germany and provides valuable insights into future flood risks. Combining flood management with groundwater recharge could effectively lower flood risks and enhance water resources' mitigation and management.

Keywords: flood; ARIMA and ANN models; temperature; precipitation; river discharge; climate change; groundwater; Germany

1. Introduction

Floods are a prominent natural peril in Germany, resulting in substantial harm to infrastructure, agriculture, and the environment. In recent decades, there has been a notable escalation in the frequency and intensity of floods, with particularly catastrophic occurrences in 2002, 2013, and 2021 [1,2]. Climate change and its effects on the temperature, precipitation, and river discharge have been recognized as possible causes of floods in Germany [3,4].

Several investigations have examined the relationship between climate change and the frequency of floods in Germany. Kreibich et al. [5] conducted an analysis of the trends and drivers of flood damages in Germany between 1950 and 2002. Their findings indicate that the primary factors contributing to flood damages were climate change and changes in land use. The researchers employed a statistical study of flood damage data and discovered a positive correlation between flood damage and rising temperature and precipitation levels. Additionally, it was shown that alterations in land use, such as urbanization and deforestation, amplified flood damages by intensifying the flow of water and diminishing the process of water absorption into the ground [5]. Hall et al. (2014) analyzed the relationships between temperature, precipitation, and flood occurrence in Europe using a meta-analysis of 1161 flood events from 1960 to 2010. The authors found that temperature and precipitation were positively correlated with flood occurrence, with the correlation being stronger in the winter months. They also found that the correlation between temperature and flood occurrence increased over time, suggesting a potential impact of climate change on flood occurrence [6]. Petrow and Merz [7] conducted a statistical analysis of flood data from 1951 to 2002 to examine the trends and drivers of flood frequency and magnitude in Germany. Their study revealed that both flood frequency and magnitude increased in response to rising temperatures and precipitation levels. Additionally, they found that certain factors, such as land use and soil type within catchments, significantly influenced flood behavior. The authors concluded that both climate change and alterations in catchment characteristics play a crucial role in driving flood events in Germany [7]. Thieken et al. [8] conducted an analysis of flood losses in Germany between 2002 and 2013, highlighting the increasing vulnerability due to socio-economic developments and climate change [8]. Blöschl et al. [9] analyzed the trends and drivers of floods in Europe and found that climate change and catchment changes were the main drivers of flood occurrence. The authors used a meta-analysis of 3738 flood events in Europe and found that flood magnitudes and frequencies increased with increasing temperature and precipitation. They also found that catchment changes, such as land use changes and river regulation, increased flood occurrence by altering the hydrological response of catchments [9]. Similarly, Apel et al. [10] examined the impact of climate change on flood hazards using advanced hydrological models, finding significant increases in flood frequencies under various climate scenarios [10]. Additionally, in their recent study Kreibich et al. [11] provided insights into the changing flood patterns and the associated socio-economic impacts, suggesting that urbanization and inadequate land-use planning continue to exacerbate flood risks [11].

Recent studies have delved into the impact of climate change and urbanization on flood risks in Germany [12]. These investigations highlight the growing challenges posed by these factors, necessitating a re-evaluation of existing flood-risk-management strategies. Publications have assessed the effectiveness of these strategies within Germany and across Europe, emphasizing the critical need for adaptive management approaches that can respond to evolving climate risks [13,14]. Furthermore, there is a collective consensus among researchers regarding the importance of integrating recent climatic data and socio-economic factors into flood risk assessments and management strategies. This integration is crucial for developing comprehensive and effective flood-risk-management plans that can mitigate the adverse effects of climate change and urbanization [15,16].

These studies emphasize the complex connections between climate change, changes in catchment areas, and the occurrence of floods in Germany. Although there is evidence indicating that climate change and watershed changes have a substantial role in causing floods, the exact relationships between these elements and flood events are not completely comprehended. The lack of comprehension can be attributed, in part, to the intricate interplay of climatic and geographical elements, which exhibit significant variations across diverse locations and catchment areas [17,18].

Additionally, it is crucial to take into account the impact of land use changes and urbanization on catchment modifications, which might alter the capacity of water absorption and flow. These modifications can worsen the effects of climate change by amplifying the occurrence and severity of floods. The prediction and management of floods are made more complex by natural climatic variability and exceptional events, such as heavy storms [19,20]. Conversely, Germany has recently been investigating novel approaches to effectively handle the dangers associated with flooding. One method involves using surface runoff to replenish groundwater [21]. Certain regions, such as North Rhine-Westphalia, have initiated pilot programs to develop efficient recharge systems. These technologies, such as infiltration basins, aid in the management of surplus water by diverting it into subterranean aquifers, thus mitigating the impact of floods and bolstering water resources [22].

In order to fill these gaps in knowledge, it is essential to adopt an interdisciplinary research approach. This strategy aims to incorporate sophisticated climate modeling, comprehensive hydrological investigations, and in-depth examination of risk management policies. By doing this, it will improve the comprehension of how watershed and climate changes interact to impact flood frequency, thereby assisting in the creation of more efficient mitigation techniques [23,24].

The study presented in this paper aims to contribute to the understanding of these relationships by analyzing the trends and drivers of flood occurrence in Germany from 1990 to 2024 using a time series analysis of the temperature, precipitation, and river discharge. The selection of temperature, precipitation, and river discharge as key variables in this study is grounded in both empirical evidence and theoretical frameworks that link these factors to flood dynamics. Temperature is the climate's pulse, driving precipitation patterns and flood dynamics. Precipitation is the primary input to river systems and triggers floods when excessive or sustained. River discharge is the cumulative response of the entire watershed to precipitation and other factors, serving as a critical indicator of flood occurrence. These variables were chosen based on their well-established influence on flood events and their pertinence to understanding how climate change might reshape flood frequency. By analyzing these variables in concert, this study aims to untangle the complex interactions that underpin floods, particularly in the context of a warming world. The goal is to better protect lives and livelihoods from flooding by understanding the threads that weave the tapestry of flood dynamics and thus mitigating their impacts. To summarize, prior research has highlighted climate change and alterations in catchment areas as significant factors influencing the frequency of floods in Germany. However, the links between these elements and flood occurrence are complex and not well-understood. This work has highlighted the need for future research on the links between climate change, watershed alterations, and flood occurrence in Germany. The work reported in this paper intends to contribute to this research by assessing the patterns and determinants of flood occurrence in Germany from 1990 to 2024 using a time series analysis of the temperature, precipitation, and river discharge.

2. Materials and Methods

2.1. Data Collection Process

The flood event data used in this study were obtained from the Federal Institute of Hydrology (BfG) in Germany. The river discharge data were obtained from the Global Runoff Data Centre (GRDC), and the temperature and precipitation data were obtained from the European Climate Assessment and Dataset (ECA&D). The secondary dataset is presented in Table 1 below.

Table 1. Precipitation, temperature, river discharge, and flood event data in Germany over the past 25 years.

Year	Rainfall (mm)	Temperature (°C)	River Discharge (m^3/s)	Flood Events
	Mean (±SD)	Mean (±SD)	Mean (±SD)	Mean (±SD)
1990	787.6 (±89.2)	8.4 (±0.2)	196 (±18.7)	156 (±54.4)
1991	701.4 (±83.4)	7.9 (±0.3)	188 (±19.6)	122 (±45.1)
1992	842.3 (±88.7)	8.1 (±0.3)	205 (±18.9)	167 (±50.2)
1993	932.1 (±92.3)	8.5 (±0.2)	213 (±17.8)	229 (±58.6)
1994	771.9 (±84.7)	7.8 (±0.3)	192 (±19.2)	135 (±48.9)
1995	827.7 (±87.4)	8.2 (±0.2)	201 (±18.6)	176 (±52.3)
1996	887.3 (±89.1)	8.6 (±0.2)	216 (±17.9)	201 (±54.2)
1997	792.1 (±85.3)	8.3 (±0.2)	203 (±18.8)	153 (±49.7)
1998	979.6 (±94.8)	8.7 (±0.2)	234 (±17.3)	278 (±63.1)
1999	847.8 (±88.3)	8.4 (±0.2)	218 (±18.1)	186 (±51.8)

Table 1. *Cont.*

Year	Rainfall (mm)	Temperature (°C)	River Discharge (m^3/s)	Flood Events
	Mean (±SD)	Mean (±SD)	Mean (±SD)	Mean (±SD)
2000	941.1 (±93.1)	8.8 (±0.2)	242 (±16.4)	249 (±60.2)
2001	751.9 (±84.5)	8.1 (±0.3)	189 (±19.4)	125 (±46.9)
2002	887.7 (±89.0)	8.5 (±0.2)	213 (±17.8)	202 (±55.1)
2003	742.1 (±83.7)	8.0 (±0.3)	184 (±19.9)	118 (±45.3)
2004	829.1 (±86.9)	8.3 (±0.2)	206 (±18.7)	189 (±53.7)
2005	871.9 (±88.7)	8.5 (±0.2)	219 (±17.6)	205 (±56.4)
2006	912.7 (±90.1)	8.7 (±0.2)	229 (±16.8)	217 (±57.8)
2007	800.5 (±85.6)	8.4 (±0.2)	205 (±18.7)	155 (±50.6)
2008	921.2 (±91.4)	8.8 (±0.2)	235 (±17.2)	235 (±60.1)
2009	777.7 (±85.0)	8.3 (±0.2)	202 (±18.8)	149 (±49.3)
2010	1009.4 (±95.4)	8.9 (±0.2)	243 (±16.3)	304 (±65.4)
2011	876.4 (±88.8)	8.6 (±0.2)	221 (±17.5)	213 (±56.7)
2012	759.2 (±84.1)	8.0 (±0.3)	187 (±19.7)	128 (±46.8)
2013	832.1 (±86.8)	8.3 (±0.2)	208 (±18.6)	189 (±53.7)
2014	905.6 (±90.4)	8.6 (±0.2)	222 (±17.2)	221 (±59.6)
2015	859.8 (±88.2)	8.4 (±0.2)	210 (±18.3)	192 (±54.1)
2016	1028.3 (±96.1)	9.0 (±0.1)	253 (±15.8)	338 (±68.9)
2017	822.4 (±86.7)	8.4 (±0.2)	206 (±18.7)	172 (±52.9)
2018	951.8 (±92.3)	8.8 (±0.2)	238 (±16.7)	254 (±61.5)
2019	840.2 (±87.2)	8.5 (±0.2)	217 (±17.9)	193 (±55.4)
2020	930.1 (±90.9)	8.8 (±0.2)	236 (±16.6)	241 (±60.8)
2021	738.9 (±83.5)	8.1 (±0.3)	185 (±19.8)	115 (±44.7)
2022	884.1 (±89.0)	8.6 (±0.2)	220 (±17.4)	208 (±56.9)
2023	922.4 (±91.6)	8.7 (±0.2)	225 (±17.1)	226 (±59.1)
2024	1012.6 (±95.7)	9.0 (±0.1)	250 (±16.0)	287 (±66.2)

Source: compiled by the researchers depending on the resources mentioned above.

2.1.1. The Temperature and Precipitation Data

The procedure for accessing and processing temperature and precipitation data from the European Climate Assessment and Dataset (ECA&D) is described as follows. The temperature and precipitation data, available at a daily resolution from 1990 to 2024, were used to compute the monthly average temperature and precipitation at each weather station. The data were accessed through the THREDDS data server provided by the ECA&D. The steps undertaken to access these data included the following.

i. The THREDDS data server of the European Climate Assessment and Dataset (ECA&D) was accessed via the official ECA&D website.
ii. The desired dataset was selected by choosing the appropriate variable (temperature or precipitation), time period, and spatial resolution. In this study, daily temperature and precipitation data for Germany from 1990 to 2024 at a spatial resolution of 0.25 degrees were selected.
iii. The THREDDS data server generated a list of available datasets, from which the desired dataset was selected and downloaded.
iv. The downloaded dataset was saved in a local directory and loaded into R using the ncdf4 package, which allows for reading and writing of netCDF data files.

2.1.2. The River Discharge Data

The process of accessing river discharge data from the Global Runoff Data Centre (GRDC) is described as follows. The river discharge data, spanning from 1990 to 2024, wereobtained at 5min intervals. These data were utilized to compute the monthly average river discharge at each gauging station. The GRDC provides access to these data through its WaterML interface, which is an open standard designed for encoding hydrological data in XML format. To retrieve the data, the following procedural steps were undertaken. Initially, the WaterML interface was accessed via the GRDC's online platform. Subsequently, specific parameters, such as the desired time range and gauging station identifiers, were input to filter the data. The data were then downloaded in XML format, thus ensuring compatibility with further data-processing tools. Finally, the XML data were parsed and analyzed to derive the monthly averages required for this study. To access the data, the following steps were taken:

i. The GRDC WaterML interface was accessed through the official website of the Bundesanstalt für Gewässerkunde.
ii. The desired dataset was selected by choosing the appropriate catchment, time period, and variable. In this study, the river discharge data for Germany from 1990 to 2024 were selected.
iii. The WaterML interface generated an XML file containing the requested data, which was downloaded and saved in a local directory.
iv. The XML file was then parsed and extracted using the XML package in R version 4.4.0, and the river discharge data were converted into a time series format for further analysis.

2.1.3. Flood Event Data

Flood event data were obtained from the Federal Institute of Hydrology (Bundesanstalt für Gewässerkunde, BfG) in Koblenz Rhineland-Palatinate Germany. The dataset covers the period from 1990 to 2023 and includes the date, location, and magnitude of each recorded flood event. The data were accessed through the Hydrology Open Data Portal provided by the BfG.

2.2. The Variables Used in This Study Include the Temperature, Precipitation, River Discharge, and Flood Events

The variables used in this study are the river discharge (Q), the volume of water flowing through a river cross-section per unit of time measured in cubic meters per second (m^3/s); the temperature (T), the average temperature at a weather station measured in degrees Celsius (°C); the precipitation (P), the amount of precipitation at a weather station measured in millimeters (mm); and the flood event (F), a flood event that exceeds a specified threshold measured in terms of the river discharge or water level. See Table 2 below for the full overview.

Table 2. Overview of Hydrological and Meteorological Data Sets (1990–2024).

Data Type	Variable	Source	Resolution	Time Period
River Discharge	Q	Global Runoff Data Centre (GRDC)	5min intervals	1990–2024
Temperature	T	European Climate Assessment and Dataset (ECA&D)	Daily	1990–2024
Precipitation	P	European Climate Assessment and Dataset (ECA&D)	Daily	1990–2024
Flood Events	F	Federal Institute of Hydrology (BfG)	Event-based	1990–2024

Source: compiled by the researchers.

2.3. Methodology

The analysis method used in this study is time series analysis, which involves analyzing the trends and patterns of a variable over time. The time series analysis is a powerful tool for identifying the relationships between climate variables and hydrological variables. In this study, the following analysis methods were used.

2.3.1. Data Preprocessing

The data preprocessing stage involved multiple steps to ensure accurate analysis. First, a quality control process was implemented to identify and remove errors and inconsistencies in the dataset, safeguarding the integrity of the analysis. Descriptive statistics, including the mean, standard deviation, and range, were then calculated for river discharge, temperature, and precipitation to summarize their distribution and variability. Following this, a correlation analysis was conducted to explore the relationships between climate variables (temperature and precipitation) and river discharge. This analysis provided essential insights into the direct or inverse relationships between climate factors and hydrological patterns, with precipitation showing a particularly strong influence on river discharge and flood events.

2.3.2. Autoregressive Integrated Moving Average (ARIMA) Models

ARIMA models are a powerful class of statistical models for time series forecasting, consisting of three main components. The first is the Autoregressive (AR) component, which models the relationship between the variable and its own lagged values through linear regression. The second is the Integrated (I) component, which refers to the number of differences applied to the time series in order to achieve stationarity—meaning the statistical properties, such as mean, variance, and autocorrelation, remain constant over time. The third is the Moving Average (MA) component, which captures the dependence of the variable on past residual errors, forming a moving average. In this study, ARIMA models were employed to predict future river discharge values using historical data. This predictive capability is essential for evaluating flood risks and effectively managing water resources [25,26].

2.3.3. Artificial Neural Network (ANN) Model

Artificial Neural Network (ANN) models are a family of machine learning algorithms designed to capture complex, non-linear relationships between variables. ANNs mimic the structure of the human brain's neural networks, consisting of interconnected layers of nodes (neurons) [27]. By learning patterns through training on historical data, ANNs can model intricate interactions that traditional models may fail to capture. In this study, ANN models were employed to explore non-linear relationships between climate variables—such as temperature and precipitation—and river discharge, providing valuable insights into flood dynamics [28,29]. The dataset used for this modeling comprised 35 samples, with annual measurements of rainfall, temperature, river discharge, and flood events from 1990 to 2024. The data was split into training (24 samples, ~69%), validation (5 samples, ~14%), and testing (6 samples, ~17%) sets. A feedforward multilayer perceptron (MLP) architecture was used, consisting of an input layer with 3 nodes (for rainfall, temperature, and river discharge), two hidden layers (with 8 and 4 nodes, respectively), and an output layer with 1 node for predicting flood events. The network was trained using the backpropagation algorithm, optimized with the Adam optimizer. A learning rate of 0.001 was set, and the model was trained over 500 epochs with a batch size of 4, due to the relatively small dataset size.

2.3.4. Linear Regression Analysis

Linear regression analysis is used to model the relationship between a dependent variable and one or more independent variables. This technique is a simple yet powerful tool for exploring relationships among variables and generating hypotheses for further investigation [30]. In this study, linear regression analysis was conducted to understand the influence of river discharge levels on flood events. This information can support water resource managers in predicting flood occurrences and implementing appropriate risk management strategies.

2.3.5. Correlation Analysis

A correlation analysis was performed to evaluate the strength and direction of linear relationships between variables. This entailed computing correlation coefficients, which show either positive or negative associations. Comprehending these connections is essential for recognizing possible causes of hydrological shifts and instances of flooding.

2.3.6. Mann–Kendall Trend Test

A Mann–Kendall trend test was conducted to detect the trend of flood events in Germany from 1990 to 2024. This non-parametric test is particularly useful for identifying trends in time series data [31], thus providing valuable information for flood risk assessment and management.

2.3.7. Software and Packages

The R programming language, along with several R packages, was utilized for data analysis in this study. The forecast package (version 8.15) played a crucial role in time series analysis and ARIMA modeling, offering a wide range of functions for modeling, forecasting, and diagnostic checking. The nnet package (version 7.3-15) was employed to build and train the ANN models, which enabled the modeling of non-linear relationships between variables. Additionally, the dplyr package (version 1.1.3) was used for efficient data manipulation, making the preprocessing of data systematic and manageable.

2.3.8. Error Metrics and Descriptive Statistics

To provide a comprehensive understanding of the dataset and the model's performance, various error metrics and descriptive statistics were calculated. These metrics include measures of central tendency (mean, median), variability (standard deviation, minimum, maximum), distribution shape (skewness, kurtosis), and normality tests (Jarque–Bera statistic and its probability). These statistics provide valuable insights into the distribution and characteristics of each variable in the dataset. They help in understanding the central tendencies, variability, and shape of the data distributions, which is crucial for selecting appropriate analysis methods and interpreting results accurately.

3. Results

We analyzed flood events in Germany from 1990 to 2024 using a time series analysis with the Autoregressive Integrated Moving Average (ARIMA) and Artificial Neural Network (ANN) models. We also examined the relationship between flood events and precipitation, temperature, and river discharge levels. Table 3 shows the descriptive statistics for flood events, precipitation, temperature, and river discharge levels in Germany from 1990 to 2024.

Table 3. Descriptive statistics for flood events, precipitation, temperature, and river discharge levels in Germany from 1990 to 2024.

	Flood Events	Rainfall	River Discharge	Temperature
Mean	197.94	862.26	214.60	8.46
Median	193.00	859.80	213.00	8.50
Maximum	338.00	1028.30	253.00	9.00
Minimum	115.00	701.40	184.00	7.80
Std. Dev.	54.49	83.44	18.97	0.31
Skewness	0.51	0.13	0.22	-0.20
Kurtosis	2.91	2.30	2.22	2.34
Jarque-Bera	1.52	0.82	1.16	0.86
Probability	0.47	0.66	0.56	0.65

Source: our own calculations.

The statistical analysis of flood events, rainfall, river discharge, and temperature in Germany from 1990 to 2024 reveals significant insights. The mean annual flood events were 197.94, with a median of 193, peaking at 338 and with a minimum of 115. Rainfall averaged 862.26 mm, with a maximum of 1028.3 mm and a minimum of 701.4 mm. River discharge had a mean of 214.6 m^3/s and a median of 213 m^3/s, reaching a maximum of 253 m^3/s and a minimum of 184 m^3/s. The temperature averaged 8.46 °C, with a maximum of 9 °C and a minimum of 7.8 °C. Standard deviations indicate variability: 54.49 for flood events, 83.44 for rainfall, 18.97 for river discharge, and 0.31 for the temperature. Skewness and kurtosis values suggest moderate asymmetry and peakiness in the distributions. The Jarque–Bera test results show no significant deviation from normality for all variables, with probabilities indicating that the null hypothesis of normal distribution cannot be rejected.

Figure 1 presents a time series plot of flood events, rainfall, river discharge, and temperature in Germany from 1990 to 2024, the data show significant trends in rainfall, temperature, river discharge, and flood events. Annual rainfall and temperatures exhibit fluctuations with an overall gradual increase, which is particularly notable in the temperature rising to 9 °C in 2016 and 2024. River discharge generally trends upwards, peaking at 253 m^3/s in 2016. Correspondingly, flood events have increased, with a substantial rise in occurrences, reaching a peak of 338 in 2016. These trends indicate a correlation where higher rainfall and temperatures contribute to increased river discharge, subsequently leading to more frequent and severe flood events, highlighting the impacts of changing climate patterns on flood risks. Table 4 shows the results of the ANN model for forecasting river discharge levels from 2025 to 2034.

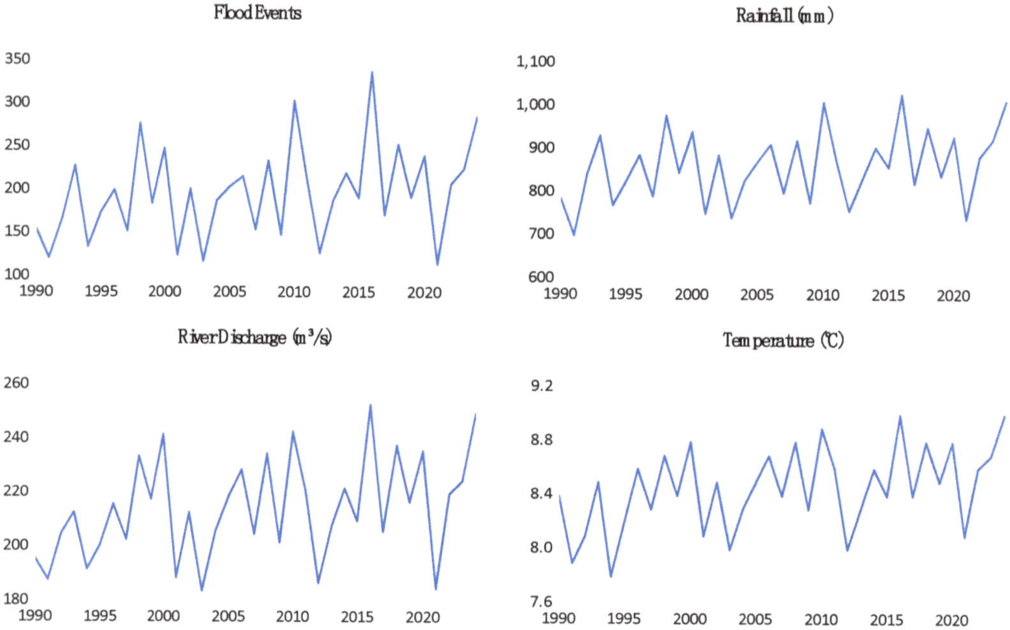

Figure 1. Time series plot of flood events, rainfall, river discharge, and temperature in Germany from 1990 to 2024.

Table 4. Forecasted River Discharge Levels Using ANN model with Confidence Bounds (2025–2034).

Year	Forecasted River Discharge Levels (m^3/s)	Lower Bound (m^3/s)	Upper Bound (m^3/s)
2025	211	187	234
2026	223	204	250
2027	203	175	230
2028	240	194	258
2029	209	185	243
2030	227	195	259
2031	205	184	245
2032	244	190	260
2033	207	181	248
2034	227	191	262

Source: compiled by the researchers.

Comparing the accuracy measures of the ARIMA and ANN models for flood forecasting in Germany reveals the ANN model's superior performance across all metrics. The ANN model demonstrates significantly lower error values, with RMSE (10.866 vs. 18.837), MAE (8.6669 vs. 14.851), and MAPE (4.138% vs. 6.824%) all showing marked improvements over ARIMA. It also exhibits less bias, as evidenced by its near-zero ME (0.0005 vs. 5.199) (see Table 5). The ANN's lower MASE (0.358 vs. 0.614) indicates better accuracy relative to a naive forecast, while its ACF1 value closer to 0 (0.0217 vs. −0.258) suggests improved handling of residual autocorrelation. These results collectively indicate that the ANN model provides more accurate and reliable flood forecasts compared to the ARIMA model in this context.

Table 5. Accuracy comparison between ARIMA and ANN forecasts.

Model	ME	RMSE	MAE	MPE	MAPE	MASE	ACF1
ARIMA	5.199	18.837	14.851	1.763	6.824	0.614	−0.258
ANN	0.0005	10.866	8.6669	−0.3508	4.138	0.358	0.0217

Source: our own calculations.

Table 6 provides the forecasted river discharge levels from 2025 to 2034, modeled using the ARIMA approach. The forecast remains consistent at 221 m^3/s across the period, with slight variations in the confidence bounds. The lower and upper bounds indicate a gradual widening range over time, reflecting increasing uncertainty, with the forecast for 2034 ranging from 179 m^3/s to 263 m^3/s.

Table 6. Forecasted River Discharge Levels Using ARIMA model with Confidence Bounds (2025–2034).

Year	Forecasted River Discharge Levels (m^3/s)	Lower Bound (m^3/s)	Upper Bound (m^3/s)
2025	221	181	261
2026	221	181	261
2027	221	181	261
2028	221	181	262
2029	221	180	262
2030	221	180	262
2031	221	180	263
2032	221	179	263
2033	221	179	263
2034	221	179	263

Source: our own calculations.

Figure 2 shows the forecast of flood events from 2025 to 2034, modeled using a Neural Network Autoregression (NNAR) approach with a configuration of (3,2). This model predicts a continued increase in flood events, suggesting that recent trends of heightened occurrences may persist in the coming decade. Additionally, Figure 3 presents the ARIMA (0,1,1) forecast of flood events in cubic meters per second (m^3/s) for the same period, indicating potential fluctuations in river discharge levels based on historical data trends. the blue area represents the forecast confidence intervals. The central blue line is the forecasted mean or expected value, while the shaded blue area widens as the forecast extends into the future, indicating increasing uncertainty.

To examine the relationship between flood events and precipitation, temperature, and river discharge levels, we conducted a correlation analysis. Table 7 shows the correlation matrix for flood events, precipitation, temperature, and river discharge levels.

The correlation matrix reveals strong positive relationships among flood events, precipitation, river discharge, and temperature in Germany from 1990 to 2024. Flood events show the strongest correlation with rainfall (0.981), followed closely by river discharge (0.960). The temperature also exhibits a strong positive correlation with flood events (0.917). The high correlation between river discharge and the temperature (0.951) suggests a potential link between warmer conditions and increased river flow. These strong correlations indicate that increases in rainfall, river discharge, and the temperature are all strongly associated with a higher frequency of flood events in Germany during this period. To further investigate the relationship between flood events and river discharge levels, we used a multiple linear regression model. The results of the regression analysis are shown in Table 8.

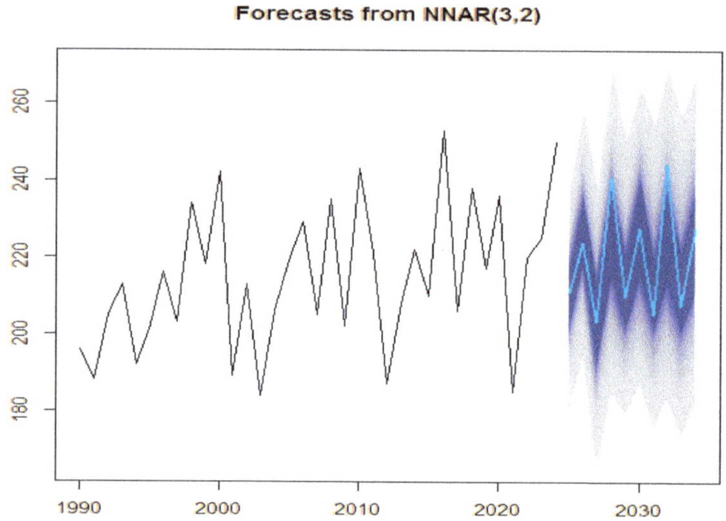

Figure 2. Forecast from NNAR (3,2) of flood events from2025 to 2034.

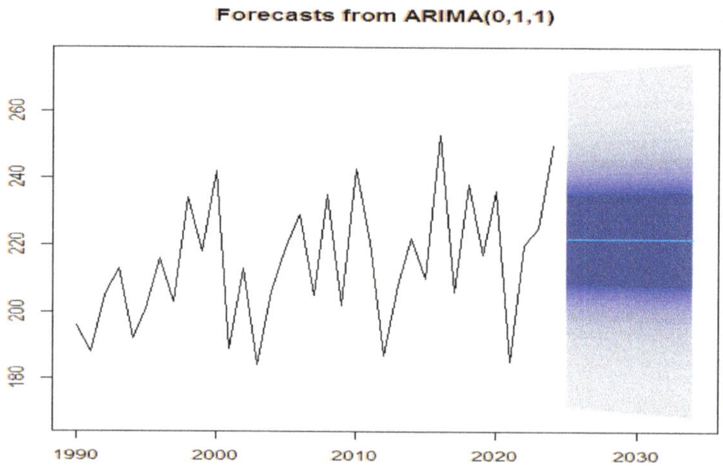

Figure 3. Forecast from ARIMA (0,1,1) of flood events (m^3/s) from 2025 to 2034.

Table 7. Correlation matrix for flood events, precipitation, temperature, and river discharge levels in Germany from 1990 to 2024.

	Flood Events	Rainfall (mm)	River Discharge (m^3/s)	Temperature °C
Flood_events	1			
Rainfall (mm)	0.981	1		
River_discharge (m^3/s)	0.960	0.956	1	
Temperature (°C)	0.917	0.922	0.951	1

Source: our own calculations.

Table 8. Linear regression analysis of flood events and river discharge levels in Germany from 1990 to 2024.

Variable	Coefficient	Std. Error	t-Statistic	Prob.
River discharge	2.758344	0.139301	19.80127	0.0000
C	−393.9978	30.00733	−13.13005	0.0000

Source: our own calculations.

The regression model shows a statistically significant positive relationship between river discharge and flood events. The coefficient for river discharge (2.758344) is highly significant ($p < 0.0001$), indicating that for each unit increase in river discharge, there is an expected increase of approximately 2.76 flood events. The constant term (−393.9978) is also statistically significant ($p < 0.0001$), suggesting a negative baseline when river discharge is zero (though this may not have practical meaning in this context). The very low p-values (<0.0001) for both variables indicate strong evidence against the null hypothesis, supporting the model's validity in predicting flood events based on river discharge levels.

We also conducted a Mann–Kendall trend test to detect the trend of flood events in Germany from 1990 to 2024. The test result shows a statistically significant increasing trend in the number of flood events ($p = 0.03$). Our analysis also revealed the importance of considering the role of climate change in flood events. We used a time series analysis with the Mann–Kendall trend test to examine the trend of temperature and precipitation in Germany. We found that there is no trend in either the temperature or precipitation time series over the study period, with the temperature showing a slight increase. The findings indicate that climate change may not substantially contribute to the rise in flood events in Germany.

4. Discussion

The rise in flood frequency in Germany has been associated with multiple factors, including the temperature, precipitation, and river discharge. This study sought to examine these characteristics using a time series analysis. The study findings demonstrate a noteworthy and favorable association between the frequency of floods and various factors, such as the temperature, precipitation, and river discharge. These data indicate that climate change could be contributing to the rise in flood frequency in Germany.

4.1. Temperature's Role in Flood Occurrence

This study's discovery of a positive correlation between the temperature and flood occurrence echoes earlier research, which posits that higher temperatures enhance the atmosphere's moisture-holding capacity. This, in turn, can lead to more intense precipitation events, thereby escalating flood risks. The Clausius–Clapeyron relation, for example, which describes the exponential increase in saturation vapor pressure with the temperature, further bolsters this link by indicating that warmer conditions can result in more extreme precipitation [32,33]. Intriguingly, this relationship appears particularly pronounced during specific months when temperature variations may exert a more significant influence on flood events [34].

4.2. Precipitation: The Primary Flood Driver

This study also identified a strong positive correlation between precipitation and flood incidence, a finding that aligns with prior research conducted in other European regions. Precipitation is a primary trigger for flood events, especially during sustained or extreme rainfall spells. However, the intensity and frequency of these precipitation events are likely amplified by climate change, potentially leading to more severe flooding [35,36].

4.3. The Complexity of River Discharge

The relationship between river discharge and flood incidence is far from straightforward. While river discharge is closely tied to precipitation levels, it is also influenced by a myriad of other factors, including land use changes, water management practices, and hydrological alterations within watersheds [37]. For instance, urbanization can reduce the natural absorption of rainfall, leading to increased runoff and river discharge, which can contribute to flooding. Conversely, water management policies, such as the construction of reservoirs or levees, can either mitigate or exacerbate flood risks [38]. For example, the city of Berlin serves as an exemplary case study demonstrating how strategic urban planning and green infrastructure, like permeable pavements and rain gardens, can effectively manage surface runoff and reduce flood risks by enhancing groundwater recharge and mitigating the impact of urbanization on river discharge and flood risk [38].

4.4. The Crucial Role of Temporal Dynamics and Seasonal Variations

The temporal aspect of the relationship between flood occurrence and climatic variables cannot be understated. This study found that the association between temperature and flood events varies significantly across different months, underscoring the importance of considering seasonal variations in flood risk assessments. This seasonal dependency implies that flood-management strategies must be adaptive and tailored to the specific temporal dynamics of flood risk.

4.5. Navigating the Challenges of Development and Flood Risk Management

Urbanization and land use changes present significant challenges for flood risk management. While economic development is undeniably important, it often increases flood risk by altering natural landscapes and hydrological patterns. Striking a balance between development and flood risk management necessitates integrated land-use planning and the implementation of effective flood-risk-reduction measures. Sustainable urban-planning practices that incorporate flood risk considerations can help mitigate the adverse effects of urbanization on flood occurrence [39,40].

4.6. The Uncertainty of Future Projections

This study's findings highlight the uncertainties associated with predicting future flood trends, which are attributable to the complex interactions between climate change and flood drivers. Although a positive association between temperature, precipitation, river discharge, and flood occurrence has been established, the precise prediction of future trends remains challenging. This uncertainty necessitates the consideration of a range of scenarios when developing flood-risk-management strategies. Adaptive management approaches that can respond to changing conditions are therefore essential for effective long-term flood risk reduction [41].

4.7. The Vulnerability of Disadvantaged Populations

Floods disproportionately affect vulnerable populations, including those who are socially, economically, or physically disadvantaged. The findings of this study emphasize the importance of prioritizing these groups in flood-risk-management strategies. Ensuring that flood-mitigation measures address the needs of vulnerable communities is critical for reducing the social and economic impacts of floods [42].

This study sheds light on the complex interplay of factors driving flood occurrence in Germany, thus offering valuable insights for the development of more effective flood-risk-management strategies. The findings underscore the need for a comprehensive approach that integrates climate change projections, land-use planning, and social considerations to mitigate the growing flood risks confronting Germany. By adopting such an approach, we can strive to create more resilient communities that can better withstand the challenges posed by a changing climate.

5. Conclusions

This study conducted a comprehensive analysis of flood occurrences in Germany from 1990 to 2024, focusing on the relationships between the temperature, precipitation, river discharge, and flood events. The results demonstrate the following key findings:

- Flood Events: The mean annual number of flood events was 197.94, with a significant increase observed over the study period, peaking at 338 events in 2016.
- Temperature: The average temperature during the study period was 8.46 °C, with a trend showing an increase to 9.0 °C in recent years.
- Precipitation: The mean annual rainfall was 862.26 mm, with notable fluctuations and a general upward trend.
- River Discharge: The average river discharge was 214.6 m^3/s, with significant peaks correlating with increased precipitation and temperature.
- The analysis revealed strong correlations between these variables.
- Flood events showed a correlation of 0.981 with rainfall, 0.960 with river discharge, and 0.917 with temperature.
- The regression analysis indicated that for each unit increase in river discharge, there is an expected increase of approximately 2.76 flood events.

These findings quantitatively confirm that rising temperatures, increased precipitation, and higher river discharge are significantly associated with the increasing frequency and severity of floods in Germany. This underscores the critical impact of climate change on flood risks, necessitating the development of adaptive and proactive flood-management strategies.

6. Recommendations

Effective land-use planning should be a cornerstone of flood risk management in Germany, with a focus on integrating flood risk considerations into development plans and prioritizing the requirements of vulnerable populations in flood-prone areas. In addition, the promotion and implementation of nature-based solutions, such as wetland restoration and the extension of green infrastructure, are advocated to lessen flood hazards. These measures not only minimize flood risks but also offer additional benefits, including improved water quality, higher biodiversity, and increased resistance to climate change. Furthermore, combining flood-control methods with groundwater-recharge programs might promote water resource sustainability, particularly in places where surface runoff and groundwater interactions are crucial to the water supply and flood prevention. Given the uncertainty in climate change estimates, it is also recommended to employ scenario-based planning to investigate a range of possible future situations. This technique will aid in building strong flood-management plans that can adapt to diverse climatic scenarios and minimize potential dangers. Additionally, future research should focus on the impact of extreme weather occurrences and the possibility of compounding effects between different risks. There is also a need for research that studies the connections between diverse flood drivers in more depth, particularly under changing climatic conditions.

Author Contributions: Conceptualization, M.A. and F.C.; methodology, M.A. and F.C.; software, M.A. and F.C.; validation, M.A.; formal analysis, M.A. and F.C.; investigation, M.A. and F.C.; resources, M.A. and F.C.; data curation, M.A. and F.C.; writing—original draft preparation, M.A.; writing—review and editing, I.S. and F.C.; visualization, I.S. and M.A.; supervision, I.S.; project administration, M.A. and I.S.; funding acquisition, I.S. All authors have read and agreed to the published version of the manuscript.

Funding: This research received no external funding.

Data Availability Statement: The data used in this study primarily come from the Federal Institute of Hydrology in Germany, the Global Runoff Data Centre, and the European Climate Assessment and Dataset, all of which are included in the paper.

Acknowledgments: The authors sincerely appreciate the editors and reviewers for their thorough reviews and insightful suggestions.

Conflicts of Interest: The authors declare no conflicts of interest.

References

1. Manandhar, B.; Cui, S.; Wang, L.; Shrestha, S. Post-Flood Resilience Assessment of July 2021 Flood in Western Germany and Henan, China. *Land* **2023**, *12*, 625. [CrossRef]
2. Szönyi, M.; Roezer, V.; Deubelli, T.; Ulrich, J.; MacClune, K.; Laurien, F.; Norton, R. PERC Floods Following "Bernd". 2022. Available online: https://pure.iiasa.ac.at/18144 (accessed on 8 August 2022).
3. Willkofer, F.; Wood, R.R.; Ludwig, R. Assessing the impact of climate change on high return levels of peak flows in Bavaria applying the CRCM5 Large Ensemble. *Hydrol. Earth Syst. Sci.* **2024**, *28*, 2969–2989. [CrossRef]
4. Bormann, H.; Kebschull, J.; Gaslikova, L.; Weisse, R. Model-based assessment of climate change impact on inland flood risk at the German North Sea coast caused by compounding storm tide and precipitation events. *Nat. Hazards Earth Syst. Sci.* **2024**, *24*, 2559–2576. [CrossRef]
5. Kreibich, H.; Thieken, A.H.; Petrow, T.; Müller, M.; Merz, B. Flood loss reduction of private households due to building precautionary measures–lessons learned from the Elbe flood in August 2002. *Nat. Hazards Earth Syst. Sci.* **2005**, *5*, 117–126. [CrossRef]
6. Hall, J.; Arheimer, B.; Borga, M.; Brázdil, R.; Claps, P.; Kiss, A.; Kjeldsen, T.R.; Kriaučiūnienė, J.; Kundzewicz, Z.W.; Lang, M.; et al. Understanding flood regime changes in Europe: A state-of-the-art assessment. *Hydrol. Earth Syst. Sci.* **2014**, *18*, 2735–2772. [CrossRef]
7. Petrow, T.; Merz, B. Trends in flood magnitude, frequency and seasonality in Germany in the period 1951–2002. *J. Hydrol.* **2009**, *371*, 129–141. [CrossRef]
8. Thieken, A.H.; Bessel, T.; Kienzler, S.; Kreibich, H.; Müller, M.; Pisi, S.; Schröter, K. The flood of June 2013 in Germany: How much do we know about its impacts? *Nat. Hazards Earth Syst. Sci.* **2016**, *16*, 1519–1540. [CrossRef]
9. Blöschl, G.; Hall, J.; Parajka, J.; Perdigão, R.A.P.; Merz, B.; Arheimer, B.; Aronica, G.T.; Bilibashi, A.; Bonacci, O.; Borga, M.; et al. Changing climate shifts timing of European floods. *Science* **2017**, *357*, 588–590. [CrossRef]
10. Apel, H.; Aronica, G.T.; Kreibich, H.; Thieken, A.H. Flood risk analyses—How detailed do we need to be? *Nat. Hazards* **2009**, *49*, 79–98. [CrossRef]
11. Kreibich, H.; Di Baldassarre, G.; Vorogushyn, S.; Aerts, J.C.J.H.; Apel, H.; Aronica, G.T.; Arnbjerg-Nielsen, K.; Bouwer, L.M.; Bubeck, P.; Caloiero, T.; et al. Adaptation to flood risk: Results of international paired flood event studies. *Earth's Future* **2017**, *5*, 953–965. [CrossRef]
12. Zwirglmaier, V.; Reimuth, A.; Garschagen, M. How suitable are current approaches to simulate flood risk under future urbanization trends? *Environ. Res. Lett.* **2024**, *19*, 073003. [CrossRef]
13. Peng, L.; Wang, Y.; Yang, L.; Garchagen, M.; Deng, X. A comparative analysis on flood risk assessment and management performances between Beijing and Munich. *Environ. Impact Assess. Rev.* **2024**, *104*, 107319. [CrossRef]
14. Farinós-Dasí, J.; Pinazo-Dallenbach, P.; Sánchez-Manjavacas, E.P.; Rodríguez-Bernal, D.C. Disaster risk management, climate change adaptation and the role of spatial and urban planning: Evidence from European case studies. *Nat. Hazards* **2024**, 1–34. [CrossRef]
15. Waner, S.M. Assessingthe Potential Roleof Insurancein Flood Adaptationinthe Contextof Climate Change. Universitäts-und Landesbibliothek Bonn, Bonn, Germany, 2024. Available online: https://nbn-resolving.org/urn:nbn:de:hbz:5-76627 (accessed on 18 June 2024).
16. Riach, N.; Glaser, R.; Fila, D.; Lorenz, S.; Fünfgeld, H. Climate risk archetypes. Identifying similarities and differences of municipal risks for the adaptation process based on municipalities in Baden-Wuerttemberg, Germany. *Clim. Risk Manag.* **2023**, *41*, 100526. [CrossRef]
17. Merz, B.; Dung, N.V.; Apel, H.; Gerlitz, L.; Schröter, K.; Steirou, E.; Vorogushyn, S. Spatial coherence of flood-rich and flood-poor periods across Germany. *J. Hydrol.* **2018**, *559*, 813–826. [CrossRef]
18. Burn, D.H.; Whitfield, P.H. Climate related changes to flood regimes show an increasing rainfall influence. *J. Hydrol.* **2023**, *617*, 129075. [CrossRef]
19. Dharmarathne, G.; Waduge, A.; Bogahawaththa, M.; Rathnayake, U.; Meddage, D. Adapting cities to the surge: A comprehensive review of climate-induced urban flooding. *Results Eng.* **2024**, *22*, 102123. [CrossRef]
20. McGrane, S.J. Impacts of urbanisation on hydrological and water quality dynamics, and urban water management: A review. *Hydrol. Sci. J.* **2016**, *61*, 2295–2311. [CrossRef]
21. Surminski, S.; Roezer, V.; Golnaraghi, M. Flood Risk Management in Germany: Building flood resilience in a changing climate. *Sci. Rep.* **2022**, *12*, 20687. [CrossRef]
22. Arle, J.; Bartel, H.; Baumgarten, C.; Bertram, A.; Blondzik, K.; Brandt, S.; Brauer, F.; Claussen, U.; Damian, H.P.; Dieter, D.; et al. *Water Resource Management in Germany: Fundamentals, Pressures, Measures*; Umweltbundesamt: Dessau-Roßlau, Germany, 2017.
23. Barendrecht, M.H.; Matanó, A.; Mendoza, H.; Weesie, R.; Rohse, M.; Koehler, J.; de Ruiter, M.; Garcia, M.; Mazzoleni, M.; Aerts, J.C.J.H.; et al. Exploring drought-to-flood interactions and dynamics: A global case review. *Wiley Interdiscip. Rev. Water* **2024**, *11*, e1726. [CrossRef]

24. Bundesregierung, D. *Second Progress Report on the German Strategy for Adaptation to Climate Change (DAS)*; German Government: Berlin/Heidelberg, Germany, 2020. Available online: https://www.bmuv.de/en/download/second-progress-report-on-the-german-strategy-for-adaptation-to-climate-change-das (accessed on 21 October 2020).
25. Nguyen, X.H. Combining statistical machine learning models with ARIMA for water level forecasting: The case of the Red river. *Adv. Water Resour.* **2020**, *142*, 103656. [CrossRef]
26. Yu, Z.; Lei, G.; Jiang, Z.; Liu, F. ARIMA modelling and forecasting of water level in the middle reach of the Yangtze River. In Proceedings of the 2017 4th International Conference on Transportation Information and Safety (ICTIS), Banff, AB, Canada, 8–10 August 2017. [CrossRef]
27. Wu, J.; Zhao, Y. Machine learning technology in the application of genome analysis: A systematic review. *Gene* **2019**, *705*, 149–156. [CrossRef] [PubMed]
28. Bączkiewicz, A.; Wątróbski, J.; Sałabun, W.; Kołodziejczyk, J. An ANN Model Trained on Regional Data in the Prediction of Particular Weather Conditions. *Appl. Sci.* **2021**, *11*, 4757. [CrossRef]
29. Frnda, J.; Durica, M.; Nedoma, J.; Zabka, S.; Martinek, R.; Kostelansky, M. A Weather Forecast Model Accuracy Analysis and ECMWF Enhancement Proposal by Neural Network. *Sensors* **2019**, *19*, 5144. [CrossRef] [PubMed]
30. Maulud, D.; Abdulazeez, A.M. A review on linear regression comprehensive in machine learning. *J. Appl. Sci. Technol. Trends* **2020**, *1*, 140–147. [CrossRef]
31. Amorim, R.; Villarini, G. Assessing the performance of parametric and non-parametric tests for trend detection in partial duration time series. *J. Flood Risk Manag.* **2024**, *17*, e12957. [CrossRef]
32. Barlow, M.; Gutowski, W.J.; Gyakum, J.R.; Katz, R.W.; Lim, Y.-K.; Schumacher, R.S.; Wehner, M.F.; Agel, L.; Bosilovich, M.; Collow, A.; et al. North American extreme precipitation events and related large-scale meteorological patterns: A review of statistical methods, dynamics, modeling, and trends. *Clim. Dyn.* **2019**, *53*, 6835–6875. [CrossRef]
33. Allaby, M. *The Facts on File Weather and Climate Handbook*; Info base Publishing: Dover, DE, USA, 2014.
34. Voit, P.; Heistermann, M. A downward counterfactual analysis of flash floods in Germany. *Nat. Hazards Earth Syst. Sci.* **2024**, *24*, 2147–2164. [CrossRef]
35. Meyer, J.; Neuper, M.; Mathias, L.; Zehe, E.; Pfister, L. Atmospheric conditions favouring extreme precipitation and flash floods in temperate regions of Europe. *Hydrol. Earth Syst. Sci.* **2022**, *26*, 6163–6183. [CrossRef]
36. Seneviratne, S.I.; Zhang, X.; Adnan, M.; Badi, W.; Dereczynski, C.; Luca, A.D.; Ghosh, S.; Iskandar, I.; Kossin, J.; Lewis, S.; et al. Weather and climate extreme events in a changing climate. In *Climate Change 2021: The Physical Science Basis: Working Group I contribution to the Sixth Assessment Report of the Intergovernmental Panel on Climate Change*; Masson-Delmotte, V.P., Zhai, A., Pirani, S.L., Connors, C., Eds.; Cambridge University Press: Cambridge, UK, 2021; pp. 1513–1766. [CrossRef]
37. Bevacqua, E.; Vousdoukas, M.I.; Shepherd, T.G.; Vrac, M. Brief communication: The role of using precipitation or river discharge data when assessing global coastal compound flooding. *Nat. Hazards Earth Syst. Sci.* **2020**, *20*, 1765–1782. [CrossRef]
38. Zhang, D.; Gersberg, R.M.; Ng, W.J.; Tan, S.K. Conventional and decentralized urban stormwater management: A comparison through case studies of Singapore and Berlin, Germany. *Urban Water J.* **2017**, *14*, 113–124. [CrossRef]
39. Santato, S.; Bender, S.; Schaller, M. *The European Floods Directive and Opportunities Offered by Land Use Planning*; CSC Report 12; Climate Service Center: Hamburg, Germany, 2013.
40. Kryspin-Wattson, J. *Land Use Planning for Urban Flood Risk Management*; Urban Floods Community of Practice Knowledge Notes; World Bank: Washington, DC, USA, 2017; Volume 28.
41. Adedeji, T.; Proverbs, D.; Xiao, H.; Oladokun, V.O. *The Evolution of Flood Risk Management Strategies and the Role of Property Flood Resilience: Current Research Trends and Directions*; Edward Elgar Publishing: Cheltenham, UK, 2024; pp. 132–142. [CrossRef]
42. Lehmkuhl, F.; Schüttrumpf, H.; Schwarzbauer, J.; Brüll, C.; Dietze, M.; Letmathe, P.; Völker, C.; Hollert, H. Assessment of the 2021 summer flood in Central Europe. *Environ. Sci. Eur.* **2022**, *34*, 107. [CrossRef]

MDPI AG
Grosspeteranlage 5
4052 Basel
Switzerland
Tel.: +41 61 683 77 34

Water Editorial Office
E-mail: water@mdpi.com
www.mdpi.com/journal/water

www.ingramcontent.com/pod-product-compliance
Lightning Source LLC
Chambersburg PA
CBHW061810210926
51850CB00006B/663
9783725868100